THE Science MATRIX

Level 1 NCEA Science, Physics, Chemistry and Biology

George Hook

Illustrations by Tony Mander

Australia • Brazil • Japan • Korea • Mexico • Singapore • Spain • United Kingdom • United States

The Science Matrix
1st Edition
George Hook

Design: Cheryl Smith, Macarn Design
Book packager: George Hook
Illustrations: Tony Mander
Editor: Eva Chan
Production controller: Siew Han Ong

Acknowledgements
Shutterstock for the images on the cover and pages 1, 4, 5, 6, 7, 8, 9, 10, 11, 12, 13, 14, 15, 16, 20, 22, 23, 24, 26, 27, 28, 29, 31, 32, 36, 37, 38, 39, 40, 41, 42, 43, 44, 46, 50, 53, 54, 56, 57, 58, 60, 64, 66, 69, 70, 71, 72, 74, 75, 76, 77, 79, 82, 83, 84, 86, 87, 88, 89, 90, 91, 93, 94, 95, 97, 98, 99, 102, 103, 104, 105, 106, 107, 108, 109, 110, 111, 114, 116, 117, 118, 119, 120, 122, 123, 124, 125, 126, 128, 129, 130, 131, 132, 134, 135, 136, 137, 138, 139, 141, 142, 144, 145, 146, 147, 148, 152, 153, 155, 156, 159 and 160.
Wikipedia Commons for the images on pages 2, 6, 7, 10, 17, 19, 48, 80, 85, 92, 102, 105, 110, 111, 112, 113, 114, 116, 117, 118, 119, 120, 121, 123, 126, 129, 130, 134, 145, 147 and 153.
NASA for the images on page 9 and 150.
PhotoDisc for the image on page 36.
AliExpress for the image on page 43.
Lawrence Berkeley National Laboratory for the image on page 56.
New Zealand Aluminium Smelters Ltd for the image on page 68.
New Zealand Steel Limited for the images on page 70.
Leslie Garland Pictures for the image on page 75.
Northland Regional Council for the image on page 94.
Shell Todd Oil Services Ltd for the image on page 94.
Methanex New Zealand for the image on page 95.
University of Utah for the image on page 103.
Plant and Food Research for the image on page 109.
NIWA for the image on page 123.
Grahame NZ Photos for the image on page 123.
Other images on pages 17, 20, 21, 22, 23, 24, 25, 26, 27, 28, 29, 30, 31, 35, 39, 43, 51, 57, 64, 66, 67, 76, 77, 78, 79, 84, 86, 87 are the author's.
Every effort has been made to obtain permission for all copyright material. Where the attempt has been unsuccessful, the publisher welcomes information that would redress the situation.

For product information and technology assistance,
in Australia call **1300 790 853**;
in New Zealand call **0800 449 725**

For permission to use material from this text or product, please email
aust.permissions@cengage.com

National Library of New Zealand Cataloguing-in-Publication Data
Hook, George, 1953-
The science matrix : NCEA science, biology, chemistry and physics / George Hook.
Includes bibliographical references and index.
ISBN 978-0-17-026231-6
1. Science—Textbooks. I. Title.
500—dc 23

Cengage Learning Australia
Level 7, 80 Dorcas Street
South Melbourne, Victoria Australia 3205

Cengage Learning New Zealand
Unit 4B Rosedale Office Park
331 Rosedale Road, Albany, North Shore 0632, NZ

For learning solutions, visit **cengage.com.au**

Contents

THE PHYSICAL WORLD

THE MATERIAL WORLD

THE LIVING WORLD

PLANET EARTH AND BEYOND

THE NATURE OF SCIENCE

Coverage of Level 1 NCEA Science, Physics, Chemistry and Biology Standards

	Std	Title of Standard	Relevant Science Matrix Units							
Science	1.1	Demonstrate understanding of aspects of mechanics	1	2	3	4				
Science	1.2	Investigate implications of electricity and magnetism for everyday life	1	5	6	7	8	9	39	40
Science	1.3	Investigate implications of wave behaviour for everyday life	1	12	13	39	40			
Science	1.4	Investigate implications of heat for everyday life	1	10	11	39	40			
Science	1.5	Demonstrate understanding of aspects of acids and bases	14	15	16	20	22			
Science	1.6	Investigate implications of using carbon compounds as fuels	14	23	24	39	40			
Science	1.7	Investigate the implications of the properties of metals for their use in society	14	17	18	39	40			
Science	1.8	Investigate selected chemical reactions	14	21	39	40				
Science	1.9	Demonstrate understanding of biological ideas relating to genetic variation	25	26	27	28				
Science	1.10	Investigate life processes and environmental factors affecting them	25	32	33	34	35	39	40	
Science	1.11	Investigate biological ideas relating to interactions between humans and micro-organisms	25	29	30	39	40			
Science	1.12	Investigate the impact of an event on an NZ ecosystem	25	31	40					
Science	1.13	Demonstrate understanding of the formation of surface features in New Zealand	38							
Science	1.14	Demonstrate understanding of carbon cycling	37	23	24					
Science	1.15	Demonstrate understanding of the effects of astronomical cycles on planet Earth	36							
Science	1.16	Investigate an astronomical or Earth science event.	40							
Physics	1.1	Carry out a practical physics investigation that leads to a linear mathematical relationship, with direction	1	39						
Physics	1.2	Demonstrate understanding of the physics of an application	1	40						
Physics	1.3	Demonstrate understanding of aspects of electricity and magnetism	1	5	6	7	8	9		
Physics	1.4	Demonstrate understanding of aspects of wave behaviour	1	12	13					
Physics	1.5	Demonstrate understanding of aspects of heat	1	10	11					
Chemistry	1.1	Carry out a practical chemistry investigation, with direction	14	39						
Chemistry	1.2	Demonstrate understanding of the chemistry in a technological application	14	40						
Chemistry	1.3	Demonstrate understanding of aspects of carbon chemistry	14	23	24	37				
Chemistry	1.4	Demonstrate understanding of aspects of selected elements	14	15	16	17	18	19		
Chemistry	1.5	Demonstrate understanding of aspects of chemical reactions	14	21						
Biology	1.1	Carry out a practical investigation in a biological context, with direction	25	39						
Biology	1.2	Report on a biological issue	25	40						
Biology	1.3	Demonstrate understanding of biological ideas relating to micro-organisms	25	29	30					
Biology	1.4	Demonstrate understanding of biological ideas relating to the life cycle of flowering plants.	25	32	33					
Biology	1.5	Demonstrate understanding of biological ideas relating to a mammal as a consumer	25	34	35					

THE Physical
WORLD

Basic Physics Concepts

Learning Outcomes - On completing this unit you should be able to explain:
- mechanics concepts such as force, motion, energy and work
- electricity concepts such as charge, current, voltage, resistance and power
- heat concepts such as thermal energy, heat transfer, and modes of heat transfer
- wave properties, the wave equation, wave types, and the fate of light waves
- *summarise and link basic physics concepts on a concept map.*

Mechanics

Motion

- Five quantities define an object's motion: distance travelled d (metres, m); time taken t (seconds, s); speed v (metres per second, ms^{-1}); acceleration a (metres per second per second, ms^{-2}); and direction.
- **Average speed** is given by the formula: $v_{av} = \Delta d/\Delta t$, and **acceleration** by: $a = \Delta v/\Delta t$ (Δ means 'change in').
- The slope of a distance-time graph gives an object's speed, and that of a speed-time graph, its acceleration.

Mass and Weight

- Physical objects have mass m, measured in kilograms (kg) because they are made of atoms. An object's **weight** is the force of gravity acting on its mass.

Force

- Force F is measured in units called newtons (N). A force has both magnitude (size) and direction.
- Often two forces act on an object, in the same or opposite directions. The **net force** F_{net} can be found by adding or subtracting one force to/from the other. Balanced forces are equal but opposite in direction.
- The acceleration of an object, which depends on the size of the force and its mass, is given by the formula: $a = F/m$. (The more common form is $F = m \times a$.)
- A **contact force** requires the object causing the force to be touching the object the force acts on. **Non-contact forces**, such as gravity, act over a distance.

Friction, Gravity and Pressure

- **Friction** is a force that automatically arises when one surface rubs against another. It always slows objects.
- **Gravity** is the force with which Earth attracts objects. As the acceleration g due to Earth's gravity is 10 ms^{-2}, the force of gravity is given by $F_g = m \times 10$.
- Pressure p is the amount of force applied per unit area. The formula is $p = F/a$ and the unit is Nm^{-2}.

Newton's Three Laws

- The resistance of an object to any change in its motion is called **inertia**. A force is needed to overcome it.
- If forces are balanced, then *no change of motion occurs.*
- If forces are unbalanced, then the object *accelerates in the direction of the net force.*
- If a force is applied to a object, then the object will apply an equal but opposite force back (action-reaction).

Energy, Work and Power

- Energy E is defined as the capacity to do work.
- Work W is the amount of energy in joules (J) used to make an object move. The formula is: $W = F \times d$.
- Power P is the rate at which work is done. It is found using the formula $P = W/t$. The units are watts (W).

Kinetic and Potential Energy

- All moving objects have kinetic energy E_k, which can be calculated using the formula: $E_k = \frac{1}{2} m v^2$.
- Objects lifted to a height h gain gravitational potential energy E_P, given by the formula: $E_P = m g h$.

The Law of Energy Conservation

- Energy can be neither created nor destroyed; it can only be transformed from one type to another.
- Useful energy is lost as heat in energy transformations.

Electricity

Electrical Charge

- All objects are made of extremely small particles called **atoms**. Each consists of a central positively charged **nucleus** and negatively charged **electrons** flying around the nucleus at high speed.
- Normally atoms are **electrically neutral**, having equal numbers of negative and positive charges.
- As some objects attract electrons more strongly than others, it is possible to transfer charge (**charge separation**) by rubbing different objects together.
- The object gaining electrons becomes negatively charged and the one losing them, positively charged.
- Charge separation results in an **electrical force field** that acts over a distance, attracting or repelling objects.

Conductors and Insulators

- The atoms of **conductors** have **free electrons** that can transfer charge along the object. The atoms of **insulators** have no free electrons, so any charge stays put.
- **Static electricity** effects occur when an insulator becomes charged and the charge cannot escape.

Electrical Currents

- An **electrical current** occurs when electrons flow along a conducting pathway from a negatively charged location to a positively charged one.
- An **electrical circuit** is a conducting loop with a charge separator (eg battery) and energy users (eg lamps).

ISBN: 9780170262316

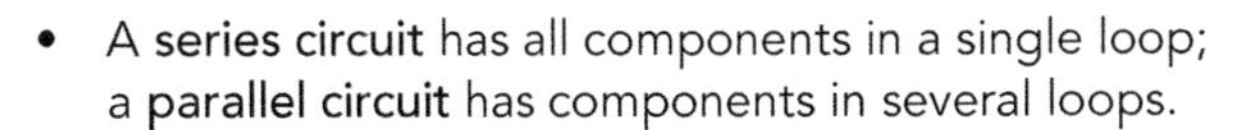

- A **series circuit** has all components in a single loop; a **parallel circuit** has components in several loops.

Current and Voltage

- The flow of electrons in a current transfers charge.
- Current I is measured in units called amperes (A), using an ammeter placed in series in a circuit.
- As current flows through components, the amount of electrical energy gained or lost is known as **voltage**.
- Voltage gain or loss V is measured in units called volts (V), using a voltmeter connected in parallel.

Series and Parallel Circuits

- Series components receive the same current but share the voltage supplied by the battery or mains.
- Parallel components receive the same voltage as the battery or mains supplies but share the current.

Resistance and Power

- Conductors that hold their free electrons more tightly are called resistors, as they resist the current.
- Resistance R is measured in units called ohms (Ω). It can be found using the formula: $R = V/I$.
- The amount of energy a component supplies or gains per second is called its power.
- Power P is measured in units called watts (W). It can be found using the formula: $P = V \times I$.

Heat

Thermal Energy, Heat and Temperature

- **Thermal energy** has to do with vibrations of the particles an object is made of rather than the motion of the object itself. It is the total kinetic energy of the particles.
- Heat Q is the amount of thermal energy, measured in joules (J), transferred from one region to another.
- Heat always flows from hotter to cooler regions.
- **Temperature** T is related to the average kinetic energy of the particles. The unit is the degree Celsius (°C).

Heating Effects

- When a substance is heated, its temperature rises unless its state is changing, eg melting or boiling.
- As an object is heated, its particles vibrate faster and push each other further apart causing expansion.

Heat Capacity and Latent Heat

- Different substances heat up at different rates.
- The **heat capacity** of a substance is the amount of heat needed to raise the temperature of 1 kg by 1 °C.
- The **latent heat** of a substance is the amount of heat needed to change the state of 1 kg of the substance.

Heat Transfer Modes

- **Conduction** is the transfer of heat along an object, caused by energetic particles jostling their neighbours.
- **Convection** is the bulk flow of a liquid or gas caused by hotter, less dense material rising and cooler less dense material sinking due to gravity.
- **Radiation** is the transfer of heat due to hotter objects radiating more infrared waves than cooler ones. These electromagnetic waves travel at the speed of light.

Waves

Waves and Energy

- **Waves** transfer energy from one location to another without matter being transferred. Waves cause regular disturbances (crests and troughs) of the **medium**.
- The wavelength λ is the distance in metres between the successive crests. The period T is the time in seconds between successive crests. The frequency f is the number of waves passing per second. The velocity v is how fast it travels in metres per second.
- The relationship between speed, wavelength and frequency is given by the formula: $v = \lambda \times f$.

Sound Waves

- A **sound wave** repeatedly compresses then lets the air expand as it passes. They are **longitudinal waves** as the disturbance is parallel to the direction of the wave. The waves travel at about 344 $m\,s^{-1}$ at 20 °C.

Electromagnetic Waves

- An **electromagnetic wave** consists of interacting electrical and magnetic disturbances travelling at 300,000 $km\,s^{-1}$ in space but slower in air or water.
- They are **transverse waves,** as the disturbance is at right angles to the direction of the wave.
- Higher frequency waves (eg UV, X-rays) transfer more energy than lower frequency ones (eg light, infrared).
- Atoms that absorb electromagnetic waves become more energetic, heating the substance.

Light Waves

- **Light** consists of electromagnetic waves with frequencies that our eyes can detect. The **spectrum** of visible light is: red-orange-yellow-green-blue-indigo-violet.
- When light waves encounter a new medium (eg glass), the surface may absorb, reflect or refract the waves.
- Absorbed light is re-radiated as infrared waves.
- Reflected light waves bounce off a surface at an equal-sized angle to that at which they struck it.
- Refracted light waves change direction abruptly when they enter a new medium. They bend toward the normal if it is denser and away if it is less dense.

SCIENCE SKILL: Concept Mapping

A useful way of understanding the connections between scientific concepts is to construct a concept map.

Steps:

1. Obtain an A3 sheet and some colour felt-tip pens.
2. Identify the theme of the unit: Physics concepts, and the major topics: the blue headings Mechanics, Electricity, Heat and Waves.
3. Draw a bubble with the theme, with the major topics in bubbles radiating off it.
4. Identify the sub-topics of each major topic: the green sub-headings, eg Motion, Force, etc; and attach them as bubbles to the major topic bubbles.
5. Identify and link key concepts to the sub-topic bubbles, expressing them as briefly as you can, using symbols wherever possible.
6. Use arrows or colour to highlight related concepts.

ISBN: 9780170262316

Speed and Acceleration

Learning Outcomes - On completing this unit you should be able to:
- define speed and acceleration
- provide the appropriate units for speed and acceleration
- use formulas to calculate average speed and average acceleration
- plot and interpret distance-time and speed-time graphs
- *develop a problem-solving method and calculate the slope of a graph.*

Describing Motion

- An object is in **motion** if its location is changing. For NCEA level 1 Science, you will study objects moving in a straight line.
- A **quantity** is something that can be measured in units.
- The units used in this book are the standard metric units. You will need to be familiar with metric prefixes, such as: kilo (x 1000), mega (x 1 000 000), centi (1/100), and milli (1/1000).
- Motion can be described using four different **quantities**.
- **Distance**, symbol ***d***, is used to describe *how far an object has travelled from a starting point.* The units are metres (m) or kilometres (km). 1 km = 1000 m.
- **Time**, symbol **t**, is used to describe *how long a journey has taken*. The units are seconds (s), minutes (min) and hours (h).
- **Speed** (or velocity), symbol ***v***, is used to describe *how fast an object is travelling.* The units for speed are metres per second (m/s or ms^{-1}) or kilometres per hour (km/h or kmh^{-1}).
- **Acceleration**, symbol ***a***, is used to describe *how an object's speed is changing over time*. The units are metres per second per second (m/s/s or m/s^2 or ms^{-2}).
- An object that is speeding up has positive acceleration. An object that is slowing down has negative acceleration or **deceleration**. Negative acceleration is indicated by a minus sign (eg $-10\ ms^{-2}$).

Finding the Speed of an Object

- You can find either the speed of an object at one instant of time, or the average speed of the object over a journey (or part of a journey).
- The **instantaneous speed** of an object can be found using a speedometer, speed camera or speed gun. The instrument will give the units.
- During a journey an object's speed will vary. The **average speed *v*** of an object over a journey (or part of a journey) can be found using a formula or from a distance-time graph.
- To find the average speed over part of a journey, divide the distance travelled by the time taken:

$$\text{Average speed} = \frac{\text{change in distance}}{\text{change in time}} \qquad v = \frac{\Delta d}{\Delta t}$$

(speed in metres per second, distance in metres and time in seconds; or other consistent units). Note: the symbol Δ means 'change in'.
- To change between metres per second and kilometres per hour, use these conversions: $ms^{-1} = 0.28 \times kmh^{-1}$ and $kmh^{-1} = 3.6 \times ms^{-1}$.
- The triangle above gives the formula for finding distance or time if the other two quantities are known. Cover the unknown quantity with your finger and read off the appropriate formula: $\Delta d = v \times \Delta t$ or $\Delta t = \Delta d / v$.
- The units used must be consistent for all three quantities.

Interpreting a Distance-Time Graph

- On a **distance-time graph** the distance an object has travelled from the starting point is plotted against the time elapsed since the start of the journey. The graph line shows how an object's distance changes over time.
- The slope of a straight graph line tells you about the object's speed:

a flat line shows the object is stationary (zero speed)	a straight sloping line shows a constant speed	a steeply/gently sloping line shows a fast/slow speed	a downward sloping line shows going back to the start.

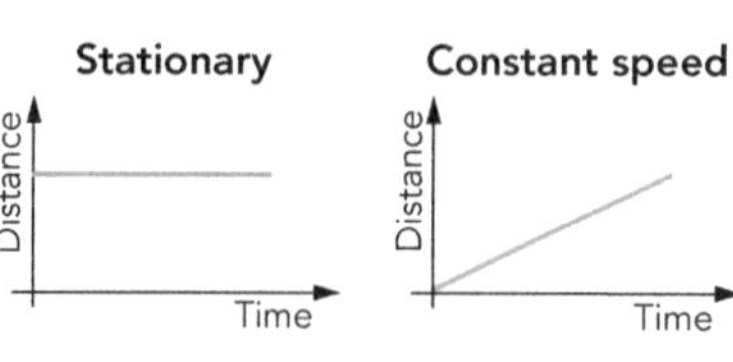

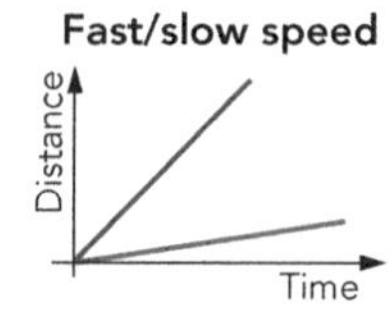

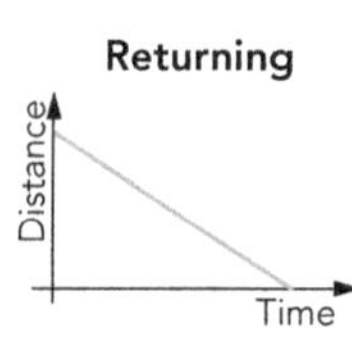

- The slope or **gradient** of a straight graph line can be found by dividing the 'rise' by the 'run':

$$\text{Gradient} = \frac{\text{rise}}{\text{run}} = \frac{\text{change in distance}}{\text{change in time}} = \frac{\Delta d}{\Delta t}$$

- As this is the same formula as that used to calculate average speed ***v***, *the slope of a distance-time graph gives the average speed.*

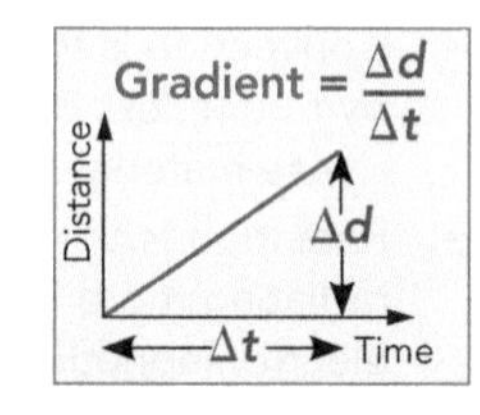

ISBN: 9780170262316

Finding the Acceleration of an Object

- It is possible to measure the **instantaneous acceleration** of an object using an **accelerometer**. Some smart phones have accelerometers built in.
- More typically you will study the **average acceleration** of an object over part of a journey. The **acceleration *a*** of an object can be found using a formula or from a speed-time graph.
- For NCEA level 1 you will only study objects travelling with a constant positive or negative acceleration (deceleration).
- To find the average acceleration over part of a journey, you divide the overall change in speed by the time taken:

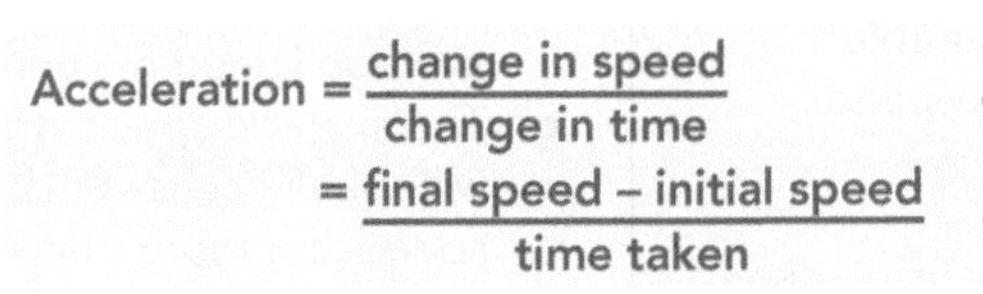

$$\text{Acceleration} = \frac{\text{change in speed}}{\text{change in time}} \qquad a = \frac{\Delta v}{\Delta t}$$

$$= \frac{\text{final speed} - \text{initial speed}}{\text{time taken}} \qquad a = \frac{v_f - v_i}{\Delta t}$$

(acceleration in metres per second per second, speed in metres per second and time in seconds; or other consistent units). Note: v_f stands for the final speed and v_i for the initial speed.

- If the speed is in metres per second (ms^{-1}) and time in seconds (s), then the acceleration will be in metres per second squared (ms^{-2}).
- The triangle above gives the formula for finding the change in speed or time if the other two quantities are known. Cover the unknown quantity with your finger and read off the appropriate formula: $\Delta v = a \times \Delta t$ or $\Delta t = \Delta v / a$.

Interpreting a Speed-Time Graph

- On a **speed-time graph** the instantaneous speed of an object is plotted against the time elapsed since the start of the journey. The graph line shows how an object's speed changes over time.
- The slope of the line tells you about the object's acceleration:

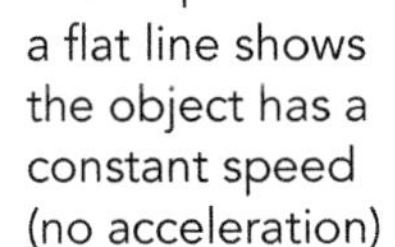

a flat line shows the object has a constant speed (no acceleration)	an upward sloping line shows speeding up	a downward sloping line shows slowing down	a steeply/gently sloping line shows high/low acceleration.

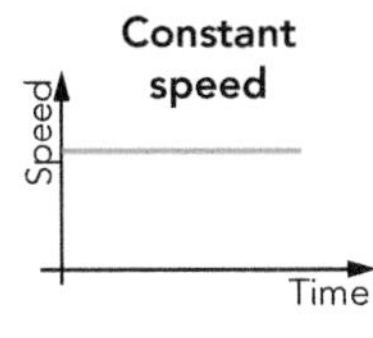

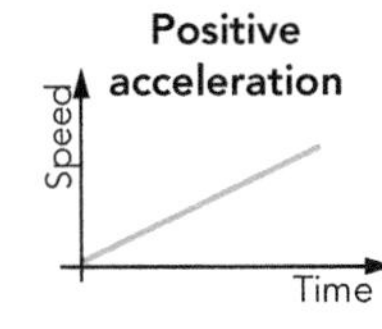

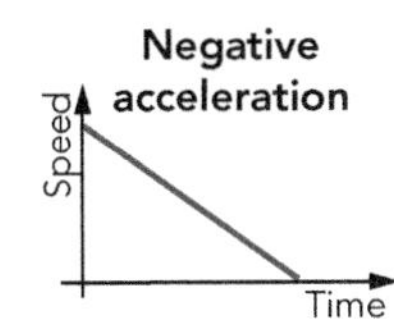

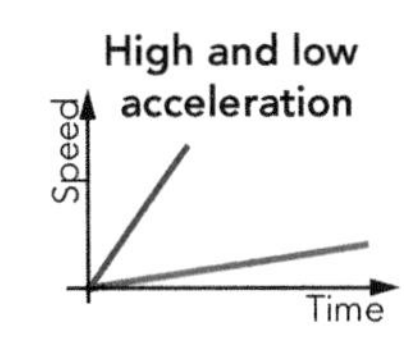

Note: the straight lines mean that the acceleration is constant.

- The **gradient** of a straight line graph can be found by dividing the 'rise' by the 'run' of the whole line:

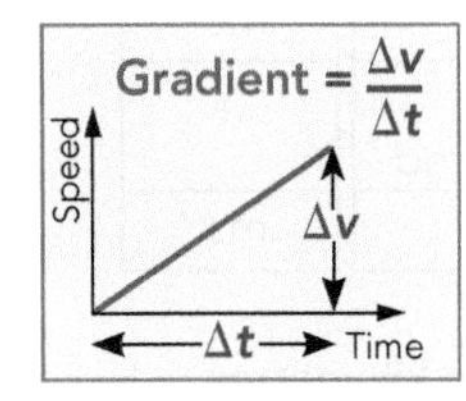

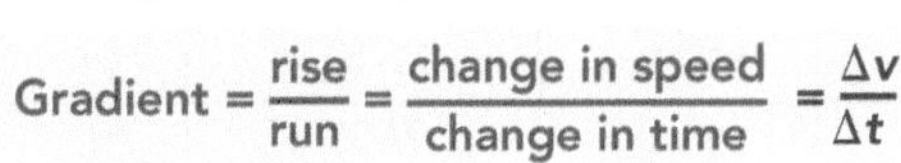

$$\text{Gradient} = \frac{\text{rise}}{\text{run}} = \frac{\text{change in speed}}{\text{change in time}} = \frac{\Delta v}{\Delta t}$$

- For the graph below:

$$\text{Gradient} = \frac{\Delta v}{\Delta t} = \frac{20\ ms^{-1}}{10\ s} = 2\ ms^{-2}$$

- As this is the same formula as that used to calculate acceleration, *the slope of a speed-time graph gives the object's acceleration.*

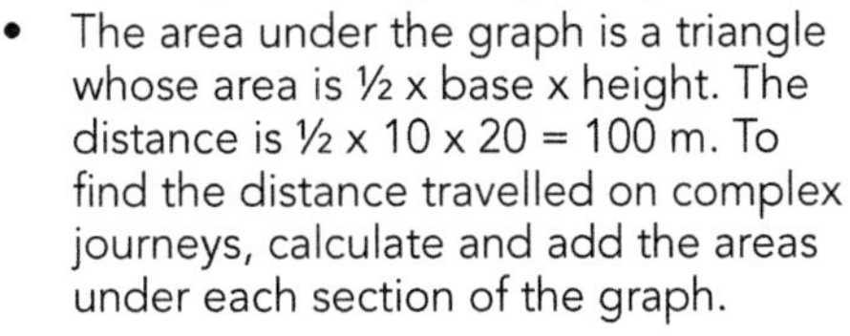

- The area under a speed-time graph represents the distance travelled by the object during the journey.
- The area under the graph is a triangle whose area is ½ x base x height. The distance is ½ x 10 x 20 = 100 m. To find the distance travelled on complex journeys, calculate and add the areas under each section of the graph.

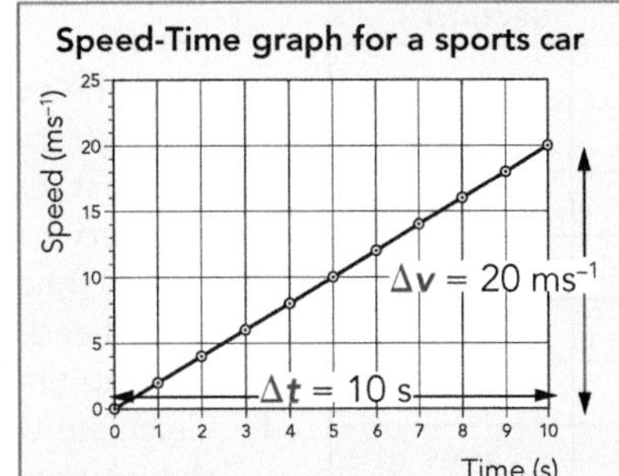

SCIENCE SKILL: Solving a Problem

You will be asked to solve motion problems by applying a formula. It is important to develop a consistent method.

Problem: By 2012 the men's world record for swimming 1.5 km freestyle was 14 min 31 s.

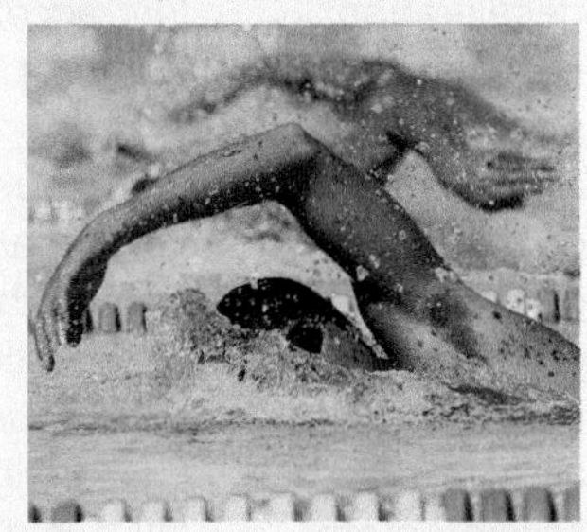

What would the average speed in ms^{-1} have been? Round your answer to 2 decimal places (the number of significant figures in the least accurate measurement).

Applying a method:

1. Write down what you have to find: speed in ms^{-1} to 2 dp
2. Write down **quantities** given: $d = 1.5$ km $t = 14$ min 31 s
3. Write down the **formula** you will need to use: $v = \frac{\Delta d}{\Delta t}$
4. Change the quantities into the appropriate **units**:
   1.5 km = 1500 m
   14 min 31 s = (14 x 60) + 31 = 871 s
5. Substitute these quantities into the formula: $v = \frac{\Delta d}{\Delta t} = \frac{1500\ m}{871\ s}$
6. Use your calculator to find the correct answer: 1500 ÷ 871 = 1.72215844
7. Next **round off** your answer to two decimal places: 1.72215844 = 1.72 (2 dp)
8. Finally record your answer with the correct units: $v = 1.72\ ms^{-1}$

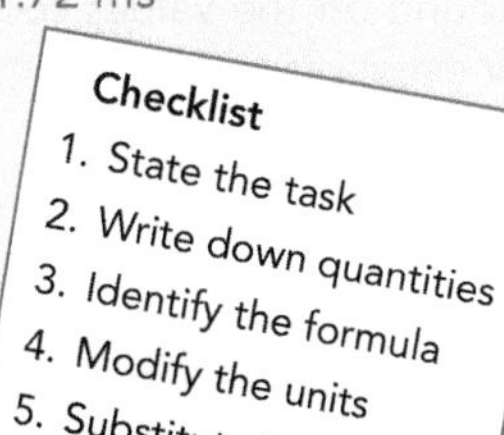

Checklist

1. State the task
2. Write down quantities
3. Identify the formula
4. Modify the units
5. Substitute in formula
6. Complete calculation
7. Round off the answer
8. Give answer and units

ISBN: 9780170262316

Revision Activities

1 Match up the descriptions with the terms.

a motion	A speed of an object at one instant of time
b quantity	B describes how an object's speed is changing
c distance	C graph on which distance travelled is plotted against time elapsed
d time	D used to measure instantaneous acceleration
e speed	E changing location
f acceleration	F acceleration of an object at one instant of time
g deceleration	G property of an object that is measurable
h instantaneous speed	H describes how far an object has travelled
i average speed	I slope of a straight graph line
j distance-time graph	J mean speed over a journey
k gradient	K mathematical relationship between quantities
l instantaneous acceleration	L describes how long a journey has taken
m accelerometer	M reducing a number to a certain number of decimal places
n average acceleration	N what physical quantities are measured in
o speed-time graph	O describes how fast an object is travelling
p unit	P graph on which instantaneous speed is plotted against time elapsed
q formula	Q mean acceleration over a journey
r rounding off	R when the speed of an object is decreasing (also called negative acceleration)

2 Explain the difference between:

a a speedometer and an accelerometer
b acceleration and deceleration
c instantaneous speed and average speed
d negative acceleration and positive acceleration.

3 For each quantity, match up its symbol, the unit it is measured in, and the symbol for that unit.

Quantity	Symbol	Unit	Symbol
distance	*v*	metres per second	m
time	*d*	metre	ms^{-1}
speed	*a*	metres per second squared	s
acceleration	*t*	second	ms^{-2}

4 Using the conversions below, change kilometres per hour into metres per second and vice versa. Round off the values to 1 decimal place.

ms^{-1} = 0.28 x kmh^{-1}
kmh^{-1} = 3.6 x ms^{-1}

Speed in kilometres per hour	Speed in metres per second
10 kmh^{-1}	
50 kmh^{-1}	
80 kmh^{-1}	
100 kmh^{-1}	
	45 ms^{-1}
	60 ms^{-1}
	75 ms^{-1}

5 Decide whether the following statements are true or false. Rewrite the false ones to make them correct.

a A stationary object is not moving.
b 1 m is 1/1000 of 1 km.
c The symbol **v** can represent speed or velocity.
d Negative acceleration means that an object is reversing.
e Deceleration is another word for negative acceleration.
f A speed camera measures the average speed of cars.
g The average speed on a journey will usually be less than the fastest speed and more than the slowest speed.
h The slope of a distance-time graph gives the object's speed.
i To calculate acceleration you need to know initial and final speeds as well as the time taken.
j The gradient of a speed-time graph indicates the object's speed.
k Constant speed means zero acceleration.

6 A triathlon involves swimming, cycling and running.

The distances for each event in a triathlon and the fastest time for that event are recorded in the chart below. Calculate the top competitor's average speed in each event in metres per second. Round your answers off to 1 decimal place.

Event	Distance	Fastest Time	Speed (ms^{-1})
swimming	4000 m	2610 s	
cycling	120 km	9605 s	
running	32 km	9913 s	

a Why is it important that each speed is calculated in the same units?
b What alternative units for speed could have been used?
c Rank the events from the fastest to the slowest average speeds.
d Suggest reasons why such different speeds are reached.
e Use an internet search to find the total distance swum, cycled and ran in the 2012 Olympics triathlon and the winning time.
f Calculate the average speed for the overall journey of the winner of the 2012 Olympics triathlon.

ISBN: 9780170262316

7 A sprinter won a 100 m race is 10.0 s. The distance travelled in each second was electronically recorded and plotted on the graph.

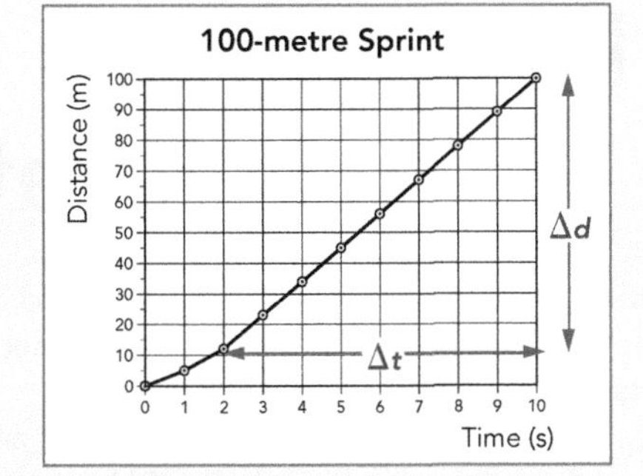

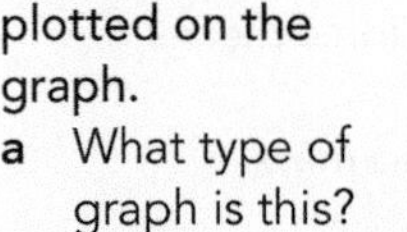

a What type of graph is this?

b What does the straight section of the graph from the 2nd to the 10th second show?

c What does the initial upward-curving part of the graph indicate?

d At t = 2 s the distance travelled was 12 m. Find the slope of the graph during the period from the 2nd to the 10th second (to 1 dp).

e What was the average speed of the sprinter between the 2nd and 10th seconds (to 1 dp)?

f What was the average speed of the sprinter over the whole journey (to 1 dp)?

g Why are results from **e** and **f** above slightly different?

8 The speed of two cars taking off from a standing start was recorded over 10 seconds.

a Plot the speed of the cars against the time elapsed on the same graph.

b Compare the acceleration of each car.

c Calculate the slope of each line after two seconds.

d What is the acceleration of each car after two seconds?

e Why might the acceleration during the first 2 seconds be slower?

Time (s)	Car A Speed (kmh^{-1})	Car B Speed (kmh^{-1})
0	0	0
1	10	5
2	25	15
3	45	30
4	65	45
5	85	60
6	105	75
7	125	90
8	145	105
9	165	120
10	185	135

9 Read the article opposite, then answer the questions below.

a What is the fastest average speed reached over a 100 m sprint?

b Why would the average speed in a race be less than the maximum speed reached?

c What distance did Bolt cover before reaching that maximum speed?

d What force is involved in pushing a sprinter forward?

e What causes air friction or drag?

f When thrust is greater than drag, what happens to the athlete's speed?

g When drag is equal to thrust, what happens to the athlete's speed?

10 The speed of a remotely controlled model racing car was recorded every second during a 10-second period. The results were plotted on the graph.

Model Racing Car Journey

Speed (ms^{-1}) against Time (s)

a What type of graph is this?

b What information does the slope of this type of graph provide?

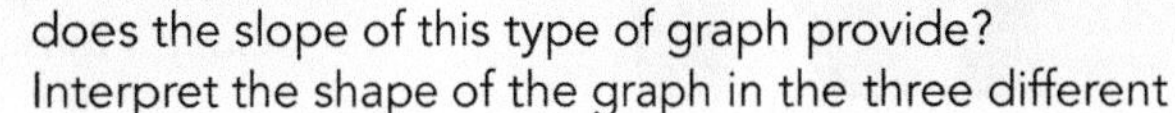

c Interpret the shape of the graph in the three different stages.

d Calculate the acceleration of the car during the first 5 seconds.

e What is the acceleration of the car during the 5th and 6th seconds?

f Calculate the acceleration of the car during the last 3 seconds.

g Calculate the total distance travelled by the model racing car.

11 Use the steps outlined in the science skill section to find the solutions to these problems.

a A farmer took 12 min to plough a furrow 3250 m long. Find his average speed to 1 decimal place (dp).

b A jet plane accelerates in a straight line into the sky when it leaves the runway. When its tyres left the ground the plane was travelling at 62 ms^{-1}; 11 seconds later it was travelling at 140 ms^{-1}. Calculate the plane's acceleration to 2 dp.

c A train decelerates from 100 kmh^{-1} to a standstill in 25 s. Calculate its acceleration in ms^{-2} to 2 dp.

Human Speed Limits

As track athletes have become more highly trained and fitter, the times for field events have continued to fall. The record for the men's 100 m sprint is 9.58 s, held by Usain Bolt of Jamaica.

If times for events are dropping, that means the speeds the athletes are reaching must be increasing. The maximum speed Bolt reached in that race was 12.27 ms^{-1} or 44.17 kmh^{-1}. This occurred between 60 and 70 metres from the start.

These increases in speed are usually very small and it may be that an upper speed limit will eventually be reached.

What is it that prevents athletes from going faster and faster? As an athlete races along a track they apply a thrust force to the track through their running shoes. This thrust force produces an equal and opposite reaction force from the track which propels the athlete forward.

If these were the only forces involved, then the athlete would get faster and faster along the track. But in most short track events top speed is reached within four to five seconds and the athlete continues at that speed for the rest of the race. What force is acting to prevent the athlete from accelerating further?

As the athlete takes off, the air in front is compressed then pushed away on either side. As the air is compressed it pushes back against the athlete. This new force is called air friction or drag and it opposes the thrust force thus reducing acceleration.

As the athlete's speed increases, more air is compressed in front which increases the force of air friction. This force increases rapidly with speed until it equals the thrust force of the athlete. At this point no further acceleration is possible. The athlete can only carry on at a constant speed as the two forces (drag and thrust) are now in balance.

Force, Mass and Acceleration

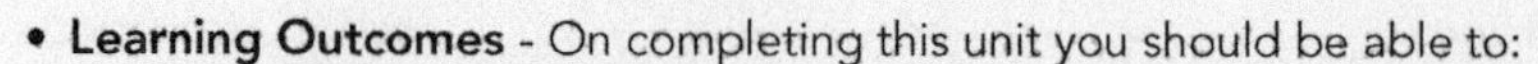

- **Learning Outcomes** - On completing this unit you should be able to:
- describe the effects of force on motion • explain what net force means
- compare the effects of balanced and unbalanced forces
- use the formula $F_{net} = ma$ • distinguish between weight and mass
- explain what pressure is and use the formula $P = F/A$
- *modify a formula into the form that you require.*

Forces

- Moving objects tend to keep moving in the same direction and at the same speed because of **inertia**. To change an object's speed or direction of motion, a push or a pull must be applied to the object. Pushes and pulls are called **forces** in science.
- Forces can squash or stretch fixed objects but for NCEA level 1 you will look at the effects of forces on objects that are free to move.
- A force is not necessarily needed to keep an object moving – the ball shown above keeps moving after it has been hit. But most moving objects slow down because of friction. *In the absence of friction, a moving object will keep on moving.*
- When a force acts on a stationary object and makes it move, the object gains **kinetic energy**. A *force transfers kinetic energy to an object if its motion changes.*
- If whatever is causing a force must touch the object to make it move, then the force is a **contact force**. Rugby involves contact forces.

- There are also **non-contact forces** that act over a distance without touching the object. The force of gravity, magnetic forces and electrostatic forces are three non-contact forces that are encountered every day.

Measuring Forces

- **Force**, symbol ***F***, is a quantity that has both magnitude and direction. The magnitude (size) of a force is measured in **newtons** (N). Larger forces are measured in kilonewtons. 1 kN = 1000 N.
- The mass of an object is the amount of matter in the object. **Mass**, symbol ***m***, is a quantity measured in units called grams (g). As a gram is a small mass, we use **kilograms** (kg) instead. 1000 g = 1 kg.
- A 1 newton force is defined as the force needed to make a free 1 kg object accelerate at 1 ms^{-2} in the absence of any opposing force such as friction.
- On force diagrams, forces are drawn as arrows from the object's centre to indicate the *direction* in which the force acts. The length of the arrow indicates the *magnitude* (size) of the force.
- The size of a force is measured using a spring that extends (or compresses) a certain distance depending on the magnitude of the force. A force meter contains a spring with a marker attached. The stretch of the spring is proportional to the strength of the force, and the marker position indicates the magnitude of the force on the scale.

object F

Forces and Motion

- Usually more than one force acts on an object. When there is more than one force, the **net force** F_{net} can be found by combining the forces.
- If two forces act in the same direction, you add them. If two forces act in opposite directions, you subtract the smaller force from the larger one.
- If two equal-sized forces act in opposite directions, then the net force is zero ($F_{net} = 0$). **Balanced forces** will not change the motion of an object.
- **Newton's 1st law** states that if the forces acting on a free object are balanced, then it will remain stationary or moving at a constant speed.
- If two unequal sized forces act in opposite directions, then a net force ($F_{net} > 0$) acts to change the object's motion. **Unbalanced forces** will change the motion of an object.
- **Newton's 2nd law** states that if an unbalanced force acts on a free object, then it will accelerate in the direction of that force.

Balanced Forces

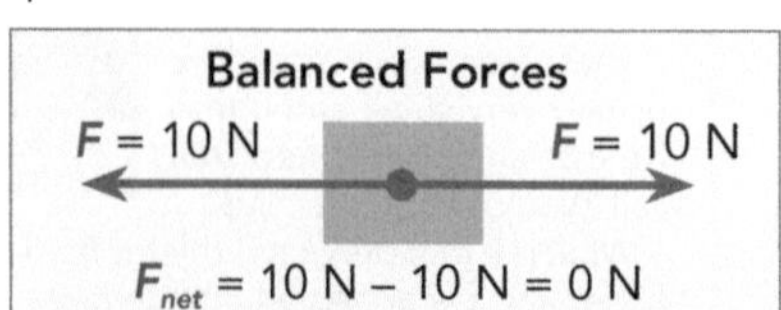

F_{net} = 10 N – 10 N = 0 N

If two equal-sized forces act in opposite directions on an object, then there is no net force. ***Balanced forces** will not change the object's motion.*

Unbalanced Forces

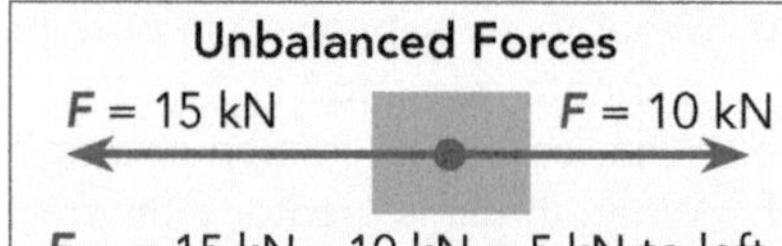

F_{net} = 15 kN – 10 kN = 5 kN to left

If a pair of opposing forces are unequal, then a net force occurs. *The **net force** will cause the object to accelerate in the direction the net force acts.*

ISBN: 9780170262316

Force, Mass and Acceleration

- As the net force on an object increases, its acceleration increases. Rotating the throttle on a scooter increases its thrust force, so its acceleration is greater. *Acceleration is directly proportional to the net force.* (Directly proportional means one quantity increases as the other does.)
- As the mass of an object increases, its acceleration decreases. A scooter with a pillion passenger will not accelerate as fast as a scooter with none. *Acceleration is inversely proportional to mass.* (Inversely proportional means one quantity decreases as the other increases.)
- The **acceleration *a*** resulting from a net force applied to a free object is:

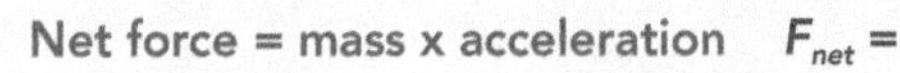

$$\text{Acceleration} = \frac{\text{net force}}{\text{mass of object}}$$

- In NCEA exam papers, the above formula is expressed as:

Net force = mass x acceleration $F_{net} = ma$

(force in newtons, acceleration in metres per second squared, mass in kg).

- The triangle above gives the formula for finding mass or acceleration if the other two quantities are known. Cover the unknown quantity with your finger and read off the appropriate formula: $m = F/a$ or $a = F/m$.

Mass, Weight and Gravity

- The weight of an object is different from its mass. **Weight** is the force exerted by gravity as it pulls an object downward, while mass is the amount of matter in the object. Weight is measured in newtons and mass in kilograms.
- The mass of an object is constant but its weight can change depending on the force of gravity. For example, the force of gravity on the Moon is only one sixth as strong as on Earth.
- Weight, symbol F_g, can be found by modifying the above formula for force:

Weight = mass x acceleration due to gravity $F_g = mg$

- On Earth, gravity causes objects to accelerate downward at 10 ms^{-2} so g = 10 ms^{-2}. On the Moon, g = 1.7 ms^{-2}.
- If your mass is 60 kg, then on Earth your weight is:
  $F_g = mg$ = 60 kg x 10 ms^{-2} = 600 N
  but on the Moon your weight would be:
  $F_g = mg$ = 60 kg x 1.7 ms^{-2} = 102 N.

Types of Forces

- The force that accelerates an object forward is called **thrust**. The force that pulls an object down is called **weight**. The force that holds an object up is called support. The force that lifts less dense objects in a liquid or gas is **buoyancy**. The force that opposes an object's motion as surfaces rub is **friction** or drag (when a solid object moves through a liquid or gas).
- Friction is a contact force that automatically arises whenever a moving object rubs against another surface, which might be a solid, liquid or gas. Friction may equal the force causing the motion but it is never greater.
- As *friction always opposes the motion of an object*, it slows down the object and converts kinetic energy into heat energy.

Action-reaction

- If a force is applied to an object, then the object will apply an equal but opposite force on the object that caused the original force. When you head a ball, you feel the reaction force that the ball applies to your head.
- **Newton's 3rd law** states that for every action force there is an equal but opposite reaction force.

Pressure

- It's more painful to have your foot stood on by a stiletto heel than a flat-heeled shoe, because the woman's weight is concentrated in a very small area.
- The strength of the force exerted per unit of area is called pressure.
- **Pressure**, symbol P, is a quantity defined as the force per unit area experienced by the object that the force is pressing on.
- The unit for pressure is the **pascal** (Pa). One pascal is the pressure caused by one newton of force pressing on one square metre of surface area. As 1 Pa is a very small pressure, the kilopascal (kPa) is used instead. 1 kPa = 1000 Pa.
- Pressure is found by the formula:

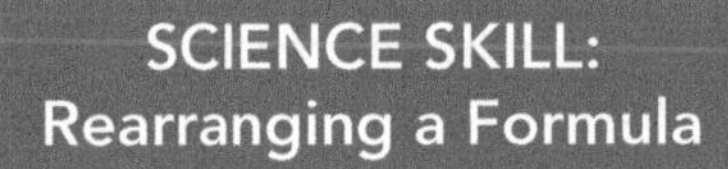

Pressure = force/area
$P = F/A$

(pressure in pascals, force in newtons, area in square metres).

- The triangle gives the formula for force or surface area. Cover the unknown quantity and read off the formula: $F = P \times A$ or $A = F/P$.

SCIENCE SKILL: Rearranging a Formula

Sometimes a formula may not be in the form you require.
Problem: You are given an object's mass and the force applied and you have to find its acceleration. The formula provided is $F = ma$.
How do you rearrange this formula so that a is by itself?
The *strategy* is to take the 'opposite action' and the *rule* is to do the same to both sides.

1. As a is multiplied by m, to get rid of m do the opposite, ie divide by m:
   $$\frac{m \times a}{m}$$
   But you must do the same to both sides:
   $$\frac{F}{m} = \frac{m \times a}{m}$$
2. Cancel to get rid of m on the right:
   $$\frac{F}{m} = \frac{\cancel{m} \times a}{\cancel{m}} = a$$
   which is the same as: $a = F/m$

Revision Activities

1 Match up the descriptions with the terms.

a inertia	A force that is applied to an object to make it move
b force	B unit used for measuring force
c kinetic energy	C something that can change the motion of an object
d contact force	D occurs when the net force on an object is not equal to zero
e non-contact force	E overall effect of combining the forces acting on an object
f newton	F every action force creates an equal but opposite reaction force
g mass	G type of energy possessed by moving objects
h kilogram	H non-zero net force will change an object's motion
i net force	I amount of matter in an object
j balanced forces	J object applying the force must touch the other object
k Newton's 1st law	K caused by unbalanced forces acting on a free object
l unbalanced forces	L force of gravity acting on an object
m Newton's 2nd law	M object applying force doesn't need to touch the other object
n acceleration	N unit used for measuring mass
o thrust	O force caused by two surfaces rubbing against each other
p weight	P unit used for measuring everyday pressures
q buoyancy	Q tendency of less dense matter to rise in a liquid or gas
r friction	R occurs when two forces on an object are equal but act in opposite directions
s Newton's 3rd law	S force exerted on a surface per unit of area
t pressure	T object's motion remains constant unless a net force acts
u kilopascal	U reluctance of a moving object to change its motion

2 Explain the difference between:

a a contact and a non-contact force
b the magnitude and the direction of a force
c mass and weight
d balanced and unbalanced forces.
e pressure and force.

3 Describe the relation between the quantities below by using statements such as: "If the ... increases, then its ... will ..."

a net force on an object and its acceleration
b mass of an object and its acceleration
c mass of an object and its weight
d area on which a force acts and pressure.

4 For each of the diagrams below, work out the net force acting on the object. State the magnitude and direction of the net force.

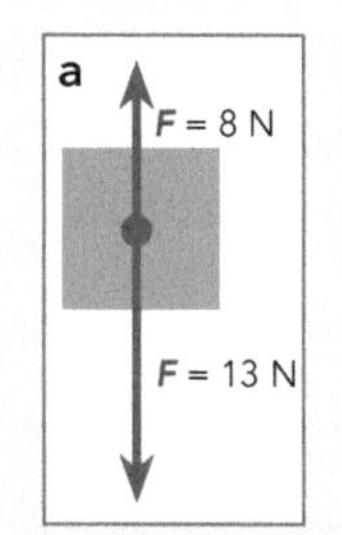

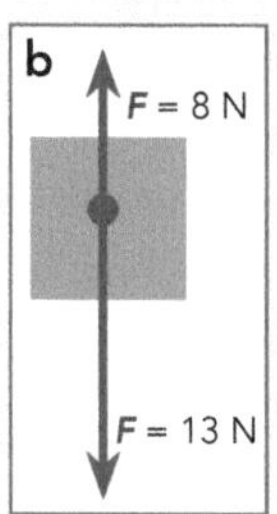

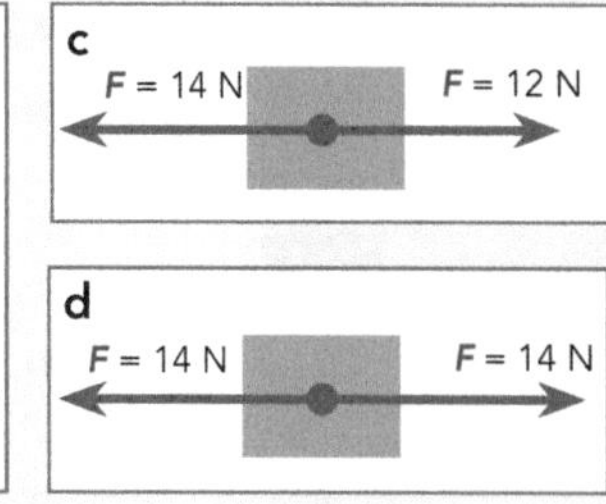

5 Decide whether the following statements are true or false. Rewrite the false ones to make them correct.

a All pushes and pulls are forces.
b A force may change an object's speed or direction of motion.
c A force is always required to keep an object in motion.
d Most moving objects slow down because of friction.
e Forces transfer chemical energy to objects.
f Gravity is an example of a contact force.
g Every force has magnitude and direction.
h Weight and mass are different quantities.
i Weight and support forces act in opposite directions.
j Unbalanced forces will cause a free object to accelerate only.
k Acceleration increases as the net force increases, and decreases as the mass increases.
l The scientific unit for weight is the kilogram.
m Friction opposes the motion of an object.
n When the force acts on a smaller area, the pressure is greater.

6 The box is being pushed sideways. Identify which force is:

a thrust
b friction
c weight
d support.

$F_2 = 10$ N, $F_1 = 15$ N, $F_3 = 8$ N, $F_4 = 10$ N

Answer these questions.

e How did you distinguish between the thrust and the friction forces?
f What is the net force acting on the object in the vertical plane?
g What change in motion will occur in the vertical plane?
h What is the net force acting in the horizontal plane?
i What change in motion will occur in the horizontal plane?
j In which plane are forces balanced? Unbalanced?
k What effect do balanced forces have on a free object's motion?
l What effect do unbalanced forces have on a free object's motion?

7 A snowboarder's total mass is 65 kg and the surface area of the board in contact with the snow is 0.2 m^2.

a Why doesn't the snowboarder sink into soft snow?
b What is the weight force acting on the snowboarder?
c What pressure is the skateboarder applying to the snow?

ISBN: 9780170262316

8 A group of students measured the acceleration of an empty trolley (mass 10 kg) as the net force applied to the trolley was increased in steps. The results are shown opposite. Next, they applied a constant net force to the trolley and measured the acceleration of the trolley as they increased the mass in the trolley in steps. These results are shown opposite.

Force and Acceleration

Acceleration of trolley (ms⁻²)

Net force applied to an empty trolley (N)

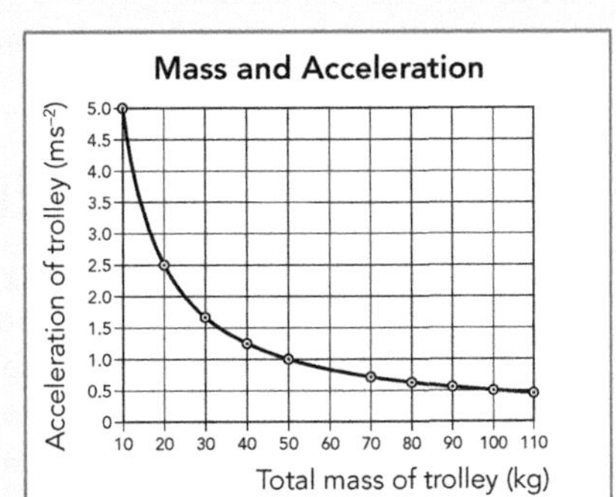

a Describe the slope of the top graph line.
b What is the relationship between net force applied and the acceleration of the empty trolley?
c If the net force doubles, what happens to the trolley's acceleration?
d Is acceleration directly or inversely proportional to net force?
e Describe the slope of the bottom graph line.
f What is the trolley's acceleration when its mass is 60 kg?
g What is the relationship between the total mass of the trolley and its acceleration when a constant force is applied?
h As the trolley's mass is doubled from 10 kg to 20 kg, what happens to the trolley's acceleration?
i Is acceleration directly or inversely proportional to total mass?

9 Read the article opposite, then answer the questions below.
a What type of energy does the jumper have while standing on the bridge?
b When she leaps off, what is the unbalanced force acting on her?
c The weight force acting on her transforms her gravitational energy into what form of energy?
d Why does her acceleration downwards become less than 10 ms^{-2}?
e What force begins to oppose the weight force acting on her?
f How is the magnitude of the force in the rubber rope related to its extension?
g How do you know that the forces on the jumper are balanced when she finally stops moving?
h What is the pattern of energy conversion?

Bungee Jumping

Bungee jumping involves leaping off a high place, such as a bridge, with a latex rope attached around your ankles. The rope stretches then contracts, eventually leaving the jumper dangling above the ground.

Bungee jumping originated in Pentecost Island, off the northeast coast of Australia, as an initiation rite. The islanders built very high platforms, attached vines to their ankles and leapt off. Commercial bungee jumping was initially developed in New Zealand.

Standing on a high spot, the jumper has gravitational potential energy. Gravity is the cause of the weight force acting on the jumper. When the jumper leaps out, gravity makes her accelerate downward at close to 10 ms^{-2}. As she accelerates downward, she gains kinetic energy.

As she falls, the latex rope extends to its normal length and is then stretched further by her weight. When latex is stretched, an 'elastic force' occurs in the rope. This force increases rapidly as the rope stretches.

When the elastic force becomes greater than the weight force, it slows the jumper down.

When the rope is at its maximum stretch, the jumper changes direction. The elastic force accelerates her upward.

As she rebounds upward, the rope shortens and the elastic force weakens. When the elastic force becomes weaker than the weight force, the jumper slows down and eventually stops rising and begins to fall again.

10 The cyclist opposite was pedalling at a constant speed along a level road.
a If the mass of the cyclist and his bike was 80 kg, what would the weight force be?
b What would be the support force acting on the cycle and rider? What direction would it act in?
c How do you know that forces acting on the bike are balanced?
d If the force of friction acting on the bike is 60 N, what thrust force must the cyclist be applying to the pedals?

The cyclist increases the thrust on the pedals, so that the net force acting on the bike is 160 N pushing him forward.
e Are balanced or unbalanced forces acting now? How do you know?
f Calculate the acceleration of the bike and rider.

As he approaches some traffic lights, he brakes and the bike decelerates at 1.5 ms^{-2}.
g What is the size and direction of the net force acting on the bike?
h When the bike is stationary, what would be the magnitude and direction of the weight, support, thrust and friction forces?

11 Study each photograph and discuss the desirable and undesirable effects of friction in each situation. Consider the surfaces rubbing against each other.

a

b

c

Energy, Work and Power

Learning Outcomes - On completing this unit you should be able to:
- define work and relate it to force and energy
- use the formula $W = Fd$ to calculate the work done in moving an object
- calculate the kinetic or potential energy gained when work is done
- use the formula $P = W/t$ to find the rate at which work is being done
- *identify the type of relationship that exists between two variables.*

Energy and Work

- **Energy**, symbol E, is measured in **joules** (J). As 1 joule is a small amount of energy, we also use kilojoules (kJ). 1 kJ = 1000 J.
- Energy is defined as the *capacity to do work* – an object has energy if it is capable of doing work.
- *Work is done when a force moves an object.* When you lift a shopping bag, you do work. When you push a shopping trolley, you do work.
- If you try to push a car and it does not move, then no work has been done as the car has not gained any energy.
- When work is done to an object, the object gains energy. This energy may be **gravitational potential energy** if the object is lifted, or **kinetic energy** if the speed of the object increases.
- The object that has been moved will gain most of the energy that the object causing the force lost. Some energy will be lost as heat due to friction.
- When energy is transferred, the total amount of energy is the same afterwards as before. This is the **Law of Energy Conservation**.
- **Work**, symbol W, involves a transfer of energy so it is measured in joules, the unit for energy. 1 joule is the amount of work done when a 1 newton force moves an object a distance of 1 metre in the absence of friction.

Calculating Work Done

- If you apply a large force to push a trolley, you do more work than when you use a small force. *The greater the force applied, the more work is done.*
- If you push a shopping trolley around all the aisles, you do more work than if you had just pushed it along one aisle. *The further an object is pushed, the more work is done.*
- Therefore the amount of work done is *proportional* to both the size of the force used and the distance over which the force moves the object. (Proportional means one quantity increases as the other does.)

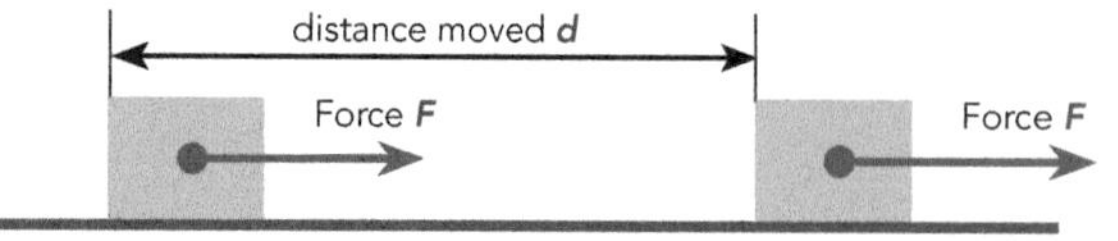

- Work is found by multiplying force applied by distance moved:

Work done = force applied x distance moved

$$W = Fd$$

W
F d

(work in joules, force in newtons, distance in metres).

- The triangle above gives the formula for finding force or distance if the other two quantities are known. Cover the unknown quantity with your finger and read off the appropriate formula: $F = W/d$ or $d = W/F$.

Kinetic Energy

- If a stationary object is made to move horizontally by a force, then work has been done to the object and it gains energy.
- The type of energy the object gains is called **kinetic energy**, symbol E_K.
- As the object is being moved by the force, some of the kinetic energy will be transformed into **heat energy** due to friction.
- The gain in kinetic energy and heat energy is equal to the work done by the force, so:

Gain in kinetic energy + gain in heat energy = work done by the force

- If there is no friction, then:

Gain in kinetic energy = work done by force = force x distance

$$\Delta E_K = Fd$$

- More work must be done to move a full trolley than an empty one. If more work is done to the loaded trolley, it must gain more kinetic energy. *Kinetic energy is proportional to the object's mass.*
- To increase the speed of a trolley, more work has to be done to it. If more work is done to the trolley, it must gain more kinetic energy. *Kinetic energy is proportional to the object's speed squared.*
- The kinetic energy of a moving object can be found by using this formula:

Kinetic energy = 0.5 x mass x speed^2

$$E_K = \tfrac{1}{2}mv^2$$

(kinetic energy in joules, mass in kilograms and speed in metres per second).

ISBN: 9780170262316

Potential Energy

- When you lift a box you do work – you apply a force to move the box over a distance. As you have done work to the box, it will have gained energy.
- The type of energy that a lifted object gains is called **gravitational potential energy**, symbol E_P. The gain in potential energy, symbol ΔE_P, is equal to the amount of work done lifting. (Some of the work done will be changed into heat energy but we will ignore this.)

  Gain in potential energy = work done $\Delta E_P = W$
- The amount of work done in lifting is given by the formula:

  Work done = force x distance $W = Fd$
- The force needed to lift an object at a constant speed is the same size as the **weight force** acting on the object. So the lifting force is given by:

  Force = weight force $F = F_g = mg$

  where m is the object's **mass** and g is the **acceleration due to gravity**.
- Putting these formulas together gives:

  Gain in potential energy

  = work done $\Delta E_P = W$

  = force x distance $= Fd$

  = weight force x distance $= F_g d$

  = mass x g x distance $= mgd$
- Distance is given the symbol h for **height lifted**, and the formula becomes:

Gain in potential energy = mass x g x height lifted

$$\Delta E_P = mgh$$

(potential energy in joules, mass in kilograms, height in metres, g = 10 ms^{-2}).

Power and Work

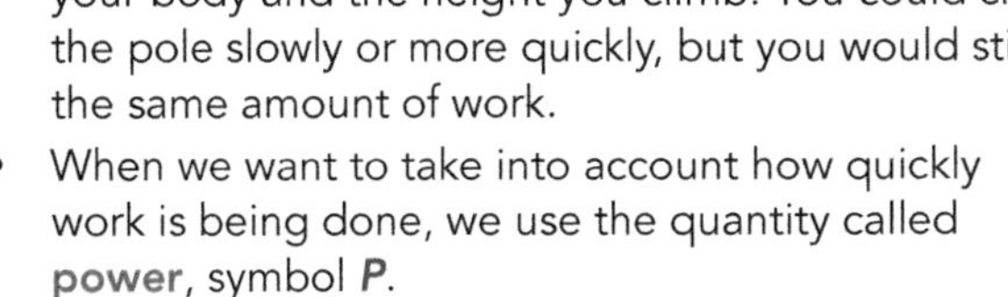

- If you climb a pole, you have done work. The amount of work done depends on the force you apply to lift your body and the height you climb. You could climb the pole slowly or more quickly, but you would still do the same amount of work.
- When we want to take into account how quickly work is being done, we use the quantity called **power**, symbol P.
- Power is the **rate** at which work is done. It tells you how quickly energy is being transferred. *The faster energy is being transferred, the greater the power developed.*
- As work is measured in joules and time in seconds, the units for power is joule per second or **watt** (W). One watt is one joule of energy transferred each second. Larger amounts are measured in kilowatts (kW). 1 kW = 1000 W.
- If you climb a pole in less time, your power will be greater. *Power is inversely proportional to the time taken.* This means power increases as time taken decreases.
- If you had to do more work by carrying a load up the pole at the same time, your power would be greater. *Power is proportional to work done.* This means power increases as the work done increases.
- Power is found by dividing work done by time taken:

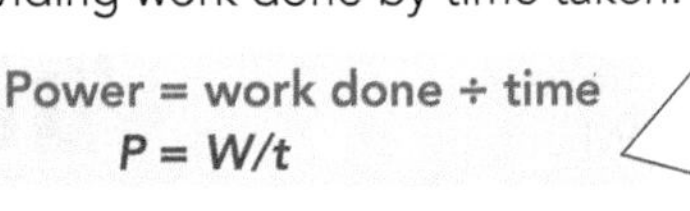

Power = work done ÷ time

$$P = W/t$$

(power in watts, work done joules and time in seconds).
- The triangle above gives the formula for finding work done or time taken if the other two quantities are known. Cover the unknown quantity with your finger and read off the relevant formula: $W = P \times t$ or $t = W/P$.

SCIENCE SKILL: Classifying Relationships

Two **quantities** or **variables** are related if changing one causes the other to change.

The variable you change is called the **independent variable** and the variable that changes as a result is called the **dependent variable**.

If the changes in the dependent variable are measured, you can identify the type of relationship from a graph.

As the independent variable *increases*, the dependent variable may:

a) *rise evenly* – the dependent variable is *directly proportional* to the independent variable, for example:

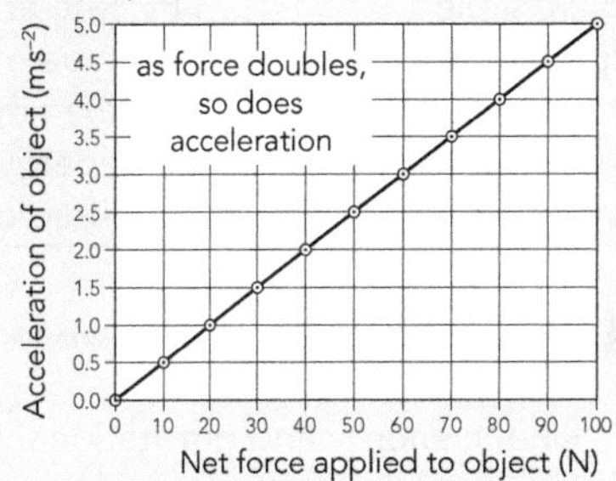

b) *fall at a decreasing rate* – the dependent variable is *inversely proportional* to the independent one, for example:

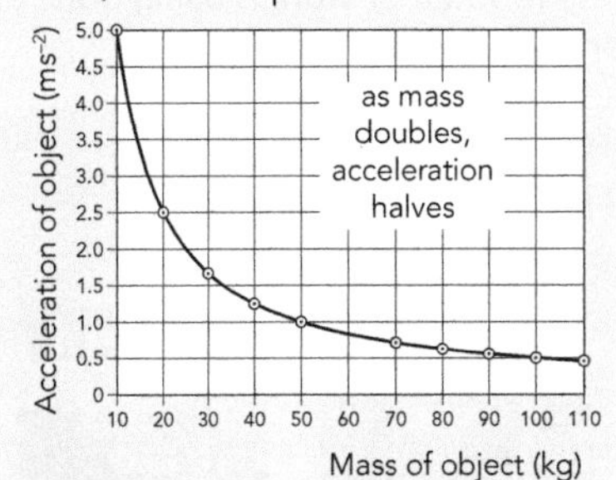

c) *rise at an increasing rate* – the dependent variable is *proportional to the square of* the independent variable, for example:

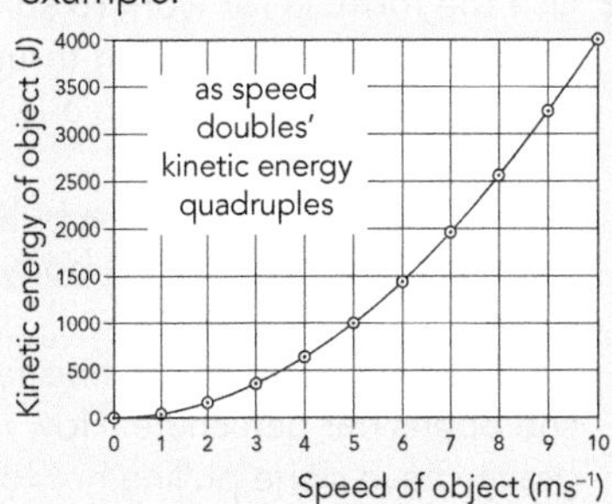

There can be other relationships between variables.

Revision Activities

1 Match up the descriptions with the terms.

a energy	A done when a force is moving an object
b joule	B approximately 10 ms^{-2} on Earth
c force	C how fast something is done
d potential energy	D something that varies in magnitude (size)
e kinetic energy	E total amount of energy is the same after as before
f law of energy conservation	F energy possessed by a moving object
g work	G property of objects that can be measured
h heat energy	H unit that power is measured in
i gravitational potential energy	I having the capacity to do work
j weight force	J variable that causes a change in another variable
k mass	K force of gravity acting on an object
l acceleration due to gravity	L energy associated with a hot object
m power	M needed to change the motion of an object
n rate	N unit that energy is measured in
o watt	O variable that alters because of a change in another
p variable	P rate at which work is done
q quantity	Q type of energy that is stored in an object
r independent variable	R the amount of matter in an object
s dependent variable	S energy possessed by an object that has been lifted up

5 Decide whether the following statements are true or false. Rewrite the false ones to make them correct.

a To be able to do work you must possess energy.
b If you push on a car but cannot move it, you are still doing work.
c When work is done to an object it gains energy.
d The amount of work done depends only on the size of the force used.
e If an object is made to move horizontally it will gain kinetic energy.
f The speed of a moving object affects its kinetic energy more than its mass does.
g The force needed to lift an object at a constant speed is the same size as the weight force acting on it.
h The amount of work done in climbing a ladder does depend on how long you take.
i If your power is high you will be doing work rapidly.
j The dependent variable is the one you alter.
k Rate is how fast something is done.

2 Explain the difference between:

a energy and work
b kinetic energy and gravitational potential energy
c work and power
d an independent variable and a dependent variable.

3 Study each of the photos and the captions below and decide whether work is being done or not. Give a reason for your decision.

A cyclist cruises along the road

A tug is unable to move the ship

A linesman is climbing the pole

4 Use the formula for work to solve these problems.

a A cyclist applies a constant thrust force of 80 N to the pedals as she travels along a level road for 2 km. How much work does she do?
b A linesman climbs 9 m up a pole at a steady speed. If the force he uses to lift his body is 960 N, how much work does he do?
c An 80 kg parachutist is in free-fall for 2400 m before she opens her parachute. How much work will Earth's gravity have done pulling her toward the ground during her free-fall period?
d 240 J of work is done to a trolley as it is pushed 12 m. What was the size of the force applied to the trolley?

6 Describe the relation between quantities using statements in the form: "If the ... is increased, then the ... will ...". For the last gap, choose from: increase/increase rapidly/decrease/be unchanged.

a 'force used' and 'work done' (assume object moves a set distance)
b 'distance moved' and 'work done' (assume a constant force is used)
c 'mass of object' and 'kinetic energy' (assume speed is constant)
d 'speed of object' and 'kinetic energy' (assume mass is constant)
e 'height lifted' and 'potential energy' (assume mass is constant)
f 'time taken to do work' and 'power' (assume work done is constant).

7 The questions below relate to the photographs.

a Is work being done to the bag? How do you know?
b How much work will be done lifting the bag?
c What type of energy will the bag gain?
d How much energy will the bag gain?
e How do you know that work is being done to the trolley?
f How much work is done to the trolley?
g What type of energy will the trolley gain?
h How much energy will the trolley gain while being pushed?

A 5 kg shopping bag is lifted 1 m

A trolley is pushed 15 m by a force of 12 N

ISBN: 9780170262316

8 In the first graph the energy gained by a box was plotted against height lifted.

Height lifted and potential energy

Gain in potential energy (J)

Height box lifted (m)

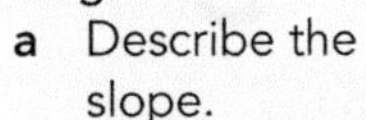

a Describe the slope.

b What is the relationship between height lifted and potential energy?

c In what way is gain in energy proportional to height?

In the second graph the power of a fire-fighter was plotted against time taken to climb a ladder.

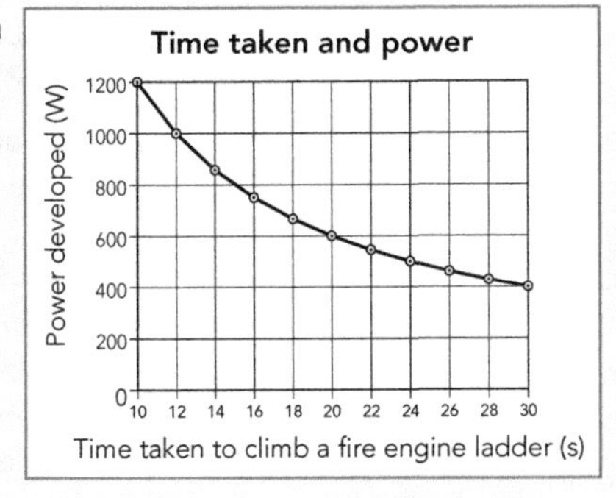

d Describe the slope.

e What is the relationship between power and time?

f In what way is power proportional to climbing time?

In this graph the kinetic energy of a cyclist was plotted against speed.

Speed and a cyclist's kinetic energy

Kinetic energy of cyclist (J)

Speed of a cyclist (ms^{-1})

g Describe the slope.

h What is the relationship between speed and kinetic energy?

i In what way is kinetic energy proportional to speed?

10 The distance a 1020 kg car travelled after the brakes were applied was measured for different speeds.

Speed (kmh^{-1})	Braking Distance (m)	Kinetic Energy (kJ)
0	0	0
10	0.5	4
20	2	16
30	5	36
40	8	64
50	14	100
60	20	144
70	28	196
80	32	256
90	44	324
100	56	400

a Plot a graph of braking distance versus speed. Make sure speed is on the horizontal axis.

b Identify which is the independent variable and which is the dependent variable on your graph.

c Describe the slope of your graph.

d As the speed of the car increases, what happens to the braking distance?

e In what way is braking distance proportional to speed?

The kinetic energy of the car was calculated for the different speeds and recorded in the third column.

f Plot another graph showing kinetic energy versus speed. Make sure speed is on the horizontal axis.

g Why should speed be placed on the horizontal axis?

h Describe the slope of this graph.

i As the speed of the car increases, what happens to the car's kinetic energy?

j In what way is kinetic energy proportional to speed?

k How do the shapes of the two graphs compare?

l Make a statement summarising the effect of speed on kinetic energy and braking distance.

9 Read the article opposite, then answer the questions below.

a What is total stopping distance made up of?

b What is meant by reaction distance?

c What two factors will your reaction distance depend on?

d At 100 kmh^{-1} what is the average reaction distance?

e What must happen for your car to lose kinetic energy?

f What happens to your car's kinetic energy when the brakes are applied?

g What other factors affect braking distance?

h Why will braking distance be quadrupled if you double your speed?

i If your braking distance at 50 kmh^{-1} is 15 m, what will it be at 100 kmh^{-1}?

Stopping a Speeding Car

Imagine you are driving a car along a country road and a sheep starts to cross in front of you. Will you be able to stop before you hit the animal? The total distance you travel before you halt depends on your reaction distance and the car's braking distance. The total distance is the sum of the two.

The distance you travel before the brakes start to engage depends on the time it takes for you to react to the danger and the speed you are travelling at. The average reaction time is about two seconds. The faster your speed, the greater the reaction distance travelled before the car starts slowing. The average reaction distance in metres can be found by multiplying speed in kilometres per hour by 0.56.

A speeding car has kinetic energy. For your car to stop, it must lose its kinetic energy through the action of a force in the direction opposite to your motion. This force is supplied by the brakes. As the brakes are applied, the car's kinetic energy is changed into heat energy due to friction. The brakes and tyres heat up.

As the car slows, more and more kinetic energy is transformed into heat until the car finally loses all of its kinetic energy and you stop.

The amount of kinetic energy your car has to start with is proportional to the square of your speed. If you double your speed the car's kinetic energy will increase by four times. As a consequence, if you double your speed your braking distance will quadruple.

Science 1.2; Physics 1.3

Static Electricity and Charging

Learning Outcomes - On completing this unit you should be able to:
- relate electrical charge to the properties of subatomic particles
- explain how the properties of conductors and insulators affect charging
- distinguish between different methods of charging
- describe the interactions between charged objects
- *investigate the relationship between electrostatic force and distance indirectly.*

Electrical Charge

- Electricity involves the interaction between charged objects, which give rise to forces that either attract or repel such as when your brush your hair.
- There are two types of **charge** – *positive charge and negative charge*. To understand how objects become charged, you need to be familiar with the structure of atoms.
- All **matter** is made of extremely small particles called **atoms** (see page 56). All atoms have the same basic internal structure.
- Atoms are made of sub-atomic particles called **protons**, neutrons and **electrons**. Protons and neutrons have the same mass and are found in the **nucleus** (centre) of the atom. Electrons are much lighter and move around the nucleus at high speed.
- Each electron has a negative electrical charge and each proton has an *equal-sized positive charge*. Neutrons have no electrical charge.
- Atoms normally have *equal numbers of protons and electrons*, for example, the atom shown above has three electrons and three protons. So it has equal numbers of positive and negative charges and is therefore **electrically neutral** overall. It is called an uncharged or neutral atom.
- Electrons are found in distinct spaces around the nucleus, called electron shells. Electrons in the outer shell of certain kinds of atoms can be transferred to outer shells of other kinds of atoms resulting in *charged atoms*.
- Those atoms that lose electrons become positively charged overall as they have more protons than electrons. Those atoms that gain electrons become negatively charged overall as they have more electrons than protons.
- Charged atoms are called **ions**. The diagram below shows a negative ion on the left and a positive ion on the right.

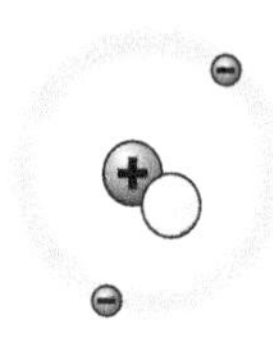

This atom has gained one electron, so it has one extra negative charge – it has become a negative ion.

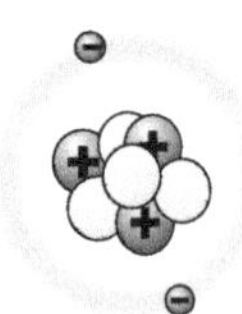

This atom has lost one electron, so it has one extra positive charge – it has become a positive ion.

Conductors and Insulators

- Usually one outer electron of each metal atom is able to *roam freely from metal atom to atom*. This enables metals to act as **conductors**.
- Electrons of nonmetals, such as plastic, rubber, wool and glass, are not free to roam from atom to atom. *Their electrons cannot flow.* For this reason these substances act as **insulators**.
- Although the electrons of an insulator cannot roam, some electrons of atoms on the surface of the insulator can be *transferred* to other objects. This property of insulators allow them to become charged.

Charge Separation

- **Charge separation** occurs when electrons are *transferred from one material to another*, such as when you pull off your jersey.
- When a neutral object is rubbed with an insulator, friction transfers electrons from the surface of one to the surface of the other. This is called **friction charging**.
- If you stroke a glass rod with wool, the rod loses electrons to become positively charged. The wool gains electrons to become negatively charged.
- Charging occurs because different substances differ in their *ability to attract electrons.*
- Charging results in more electrons on the surface of one object and less on the other. This build-up of stationary charge on the surface is called **static electricity**.
- A charged object generates an **electrical force field** that *attracts or repels other charged objects*, such as when a rubbed pen attracts small pieces of paper.
- If two similarly charged objects are brought near each other, the objects repel each other.

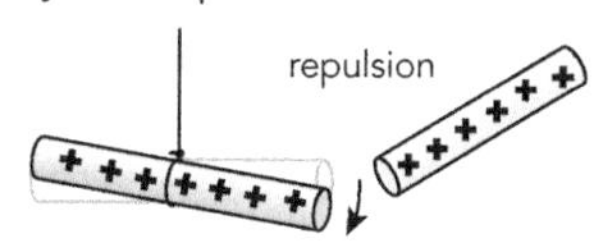

- If two oppositely charged objects are brought near each other, the objects attract each other.

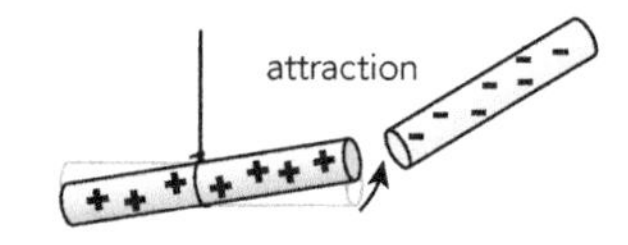

- *Like charges repel, unlike charges attract* due to **electrostatic forces**.

ISBN: 9780170262316

Keeping and Losing Charge

- Both conductors and insulators can be charged by friction through rubbing with an insulator.
- If you hold a plastic rod and rub it with wool, electrons are transferred and both objects become charged. The charge stays on the rod because an insulator does not allow charge to flow off it onto your hand.

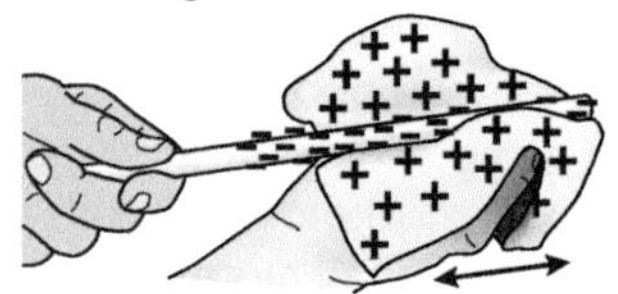

- If you hold a steel rod and rub it with wool, electrons are transferred and both objects become charged. The charge on the rod will flow onto your hand because steel is a conductor. The steel rod discharges.

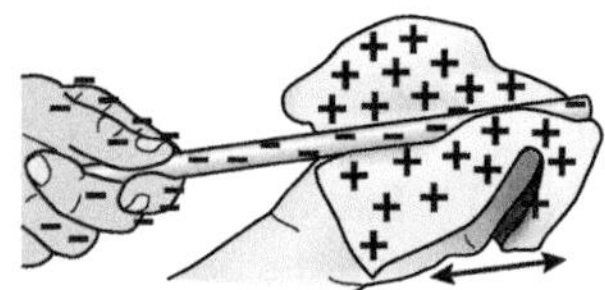

- A charge can be kept on a conductor if it is suspended or supported by an insulator. This stops the charge from flowing off the **insulated conductor**.
- Conductors are sometimes connected to the ground by a conducting wire so that any charge will flow off the object. This is called **earthing**.
- Some buildings have lightning rods that conduct any charge from a lightning strike directly to the ground, thus preventing damage.
- Electrical appliances with a metal case have an earth wire that connects the case directly to the ground. If a fault occurs and the metal frame becomes live, the charge flows to the ground via an earthing rod rather than giving you an electrical shock.

Conduction Charging

- Neutral objects are charged by contact with a charged object. This is **conduction charging** but it is called contact charging in NCEA exams.
- When charge is transferred by contact to an insulated conductor, electrons spread over the conductor's surface to *minimise repulsion between like charges*. This gives a uniform **charge distribution**.
- When charge is transferred by contact to an insulator, the charge remains at the point of contact as the electrons of an insulator are not free to roam. This gives a non-uniform charge distribution.

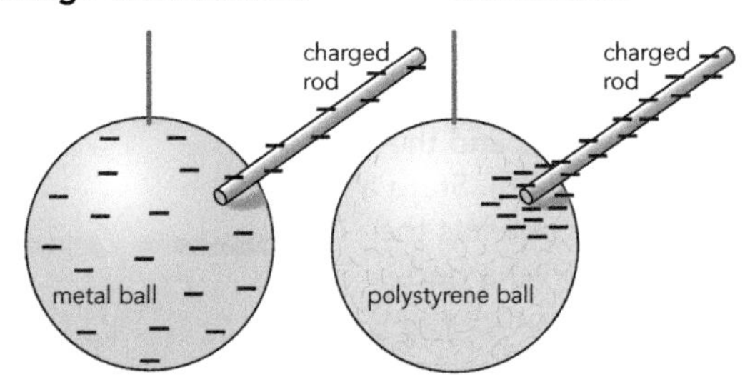

Induction Charging

- A *temporary charge* can be induced on a neutral object by bringing a charged insulator close to it.
- A negatively charged insulator will repel electrons on the near surface of the object as it is brought close by.
- If the object is an insulated conductor, free electrons on the object's near side are repelled toward the far side.
- If the object is an insulator, the electrons are not free to travel but they do move slightly away.

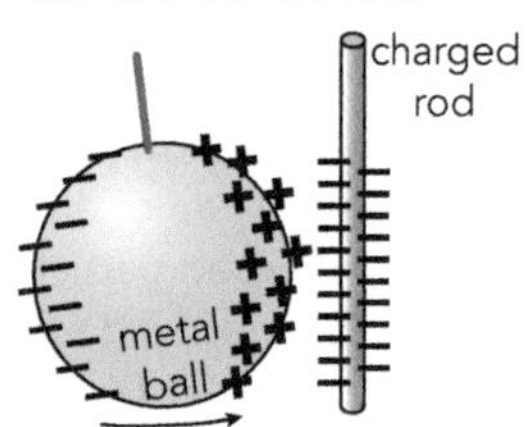

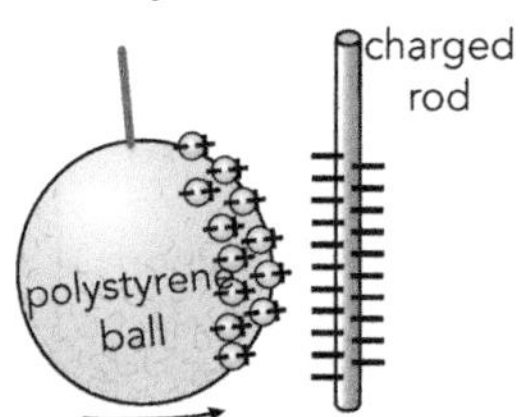

- In either case, the result is a region on the object's near surface that is charged. The positively charged surface is attracted toward the negatively charged rod.
- The effect of **induction charging** on insulators is much smaller than on conductors.
- If the conductor touches the charged rod, then some excess charge will flow onto it. As both rod and conductor are negatively charged they now repel each other.
- If the insulator touches the charged rod, it will stick to it. Electrons cannot flow between insulators, so the charge separation remains in force.

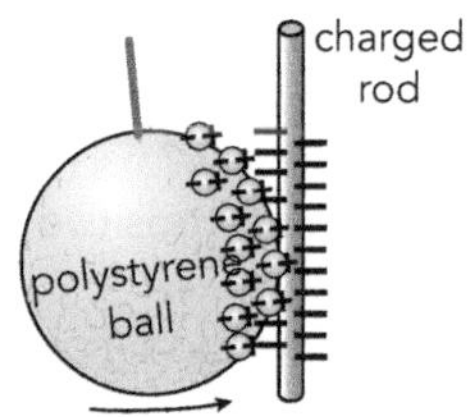

Discharging through Sparks

- Charged objects can be discharged by bringing them close to an object such as a metal tap without actually touching. The discharge occurs as a spark in the air. These **discharges** often occur on dry days, such as when you get a shock stepping off a trampoline.

SCIENCE SKILL: Measuring Effects Indirectly

The strength of electrostatic forces depend on how far apart the charged objects are. The effect of distance on force can be measured indirectly in this activity.

Steps:

1. Comb dry hair with a plastic comb for one minute.
2. Adjust a tap to get a thin, smooth stream of water.
3. Observe what happens when the charged comb is brought close to the stream without touching it.
4. Consider how you could indirectly measure the electrostatic force the stream experiences as you vary the distance away.

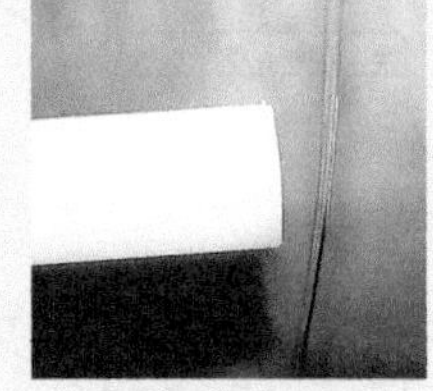

Revision Activities

1 Match up the descriptions with the terms.

a charge	A area in which electrical forces are experienced
b matter	B connecting an object to the ground using a wire
c atoms	C charged atom that has gained or lost electrons
d protons	D temporarily charging a neutral object by bringing a charged insulator near it
e electrons	E atomic property that is either positive or negative
f nucleus	F what all substances and objects are made of
g electrically neutral	G occurs when electrons are transferred
h ion	H substance that allows some electrons to roam
i conductor	I dense central area of an atom
j insulator	J positively charged particles in the atom's nucleus
k charge separation	K attractive or repulsive force in an electrical field
l friction charging	L charging an object by rubbing with an insulator
m static electricity	M how charge is spread over a charged object
n electrical force field	N neither positively nor negatively charged overall
o electrostatic force	O negatively charged particles flying about nucleus
p insulated conductor	P extremely small particles that all matter is made of
q earthing	Q substance that does not allow electrons to roam
r conduction charging	R charging a neutral object by contact with a charged object
s charge distribution	S build-up of charge on an object's surface
t induction charging	T loss of an object's charge by electron transfer
u electrical discharge	U conductor that is insulated from its surroundings

2 Study the structure of the atom shown opposite and identify the parts labelled a to d.

- **e** How many negative charges does it have?
- **f** How many positive charges does it have?
- **g** Does the atom have more or less electrons than it normally would?
- **h** Which electron would it have acquired? Why?
- **i** What will be the overall charge on the atom?
- **j** How could this atom have become charged?
- **k** What is the term for an atom that is no longer electrically neutral overall?

3 Explain the differences between the terms in each of the following lists.

- **a** proton, electron and neutron
- **b** atom and ion
- **c** conductor and insulator
- **d** uniform and non-uniform charge distribution
- **e** friction, conduction and induction charging

4 Describe what will happen in each of the following situations.

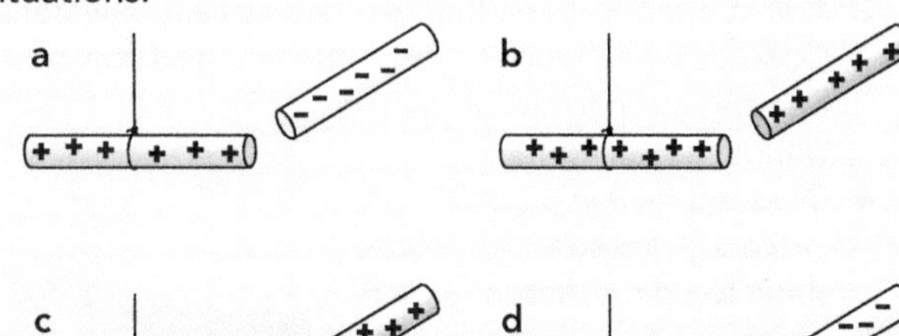

5 Decide whether the following statements are true or false. Rewrite the false ones to make them correct.

- **a** Charged objects always repel each other.
- **b** The nucleus of an atom is positively charged.
- **c** Neutrons are charged particles.
- **d** Electrons and protons have opposite but equal-sized charges.
- **e** A neutral atom has equal numbers of protons and electrons.
- **f** If an atom gains an electron, it becomes positively charged.
- **g** Some outer electrons of metal atoms are able to roam freely.
- **h** Although insulator electrons cannot roam, they can be transferred.
- **i** Friction charging occurs when an object is rubbed with a conductor.
- **j** Static electricity is due to separated stationary electrical charges.
- **k** Like charges attract each other.
- **l** An insulated conductor is supported by an insulator.
- **m** A charged spherical conductor has a uniform charge distribution.
- **n** Touching will discharge objects.

6 Objects can be charged in three different ways. Study each diagram in turn and answer the questions beside it.

- **a** What type of charging is shown in the diagram opposite?
- **b** Should the rubbing material be a conductor or an insulator?
- **c** How are the charges separated by this method?
- **d** Describe an everyday example of this type of charging.

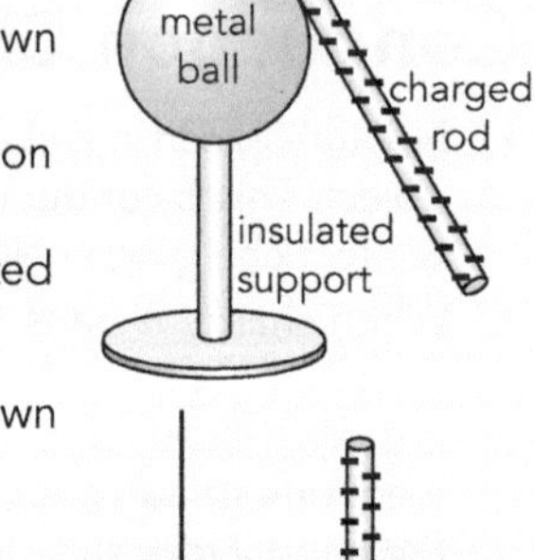

- **e** What type of charging is shown in the diagram opposite?
- **f** Why is the conductor sitting on an insulated base?
- **g** How are the charges separated by this method?

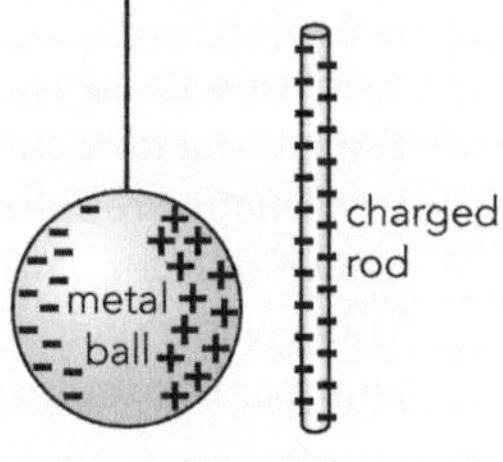

- **h** What type of charging is shown in the diagram opposite?
- **i** Will the charge stay on the ball when the charged rod is withdrawn?
- **j** How are the charges separated by this method?

7 The law of magnetism (see page 32) states the connection between the polarity of magnets and the forces between them. State a similar law that connects the type of charge on charged objects with the forces between them.

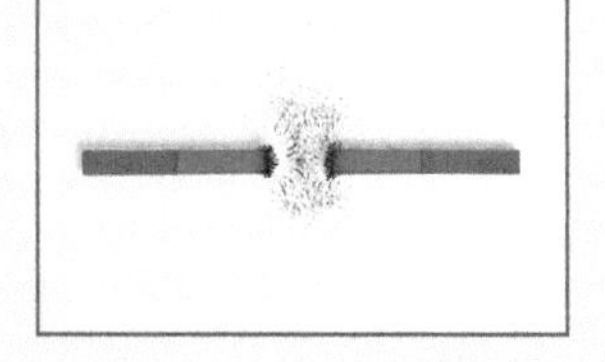

ISBN: 9780170262316

8 The girl shown in the photo charged her skin as she rubbed against the plastic slide.

a How do you know that the girl's hair is charged?

b Is skin a conductor or an insulator? How do you know?

c How do you know that all of her hairs have the same charge?

d What will happen when she touches the other girl?

e What would have happened if the day had been quite humid? Why?

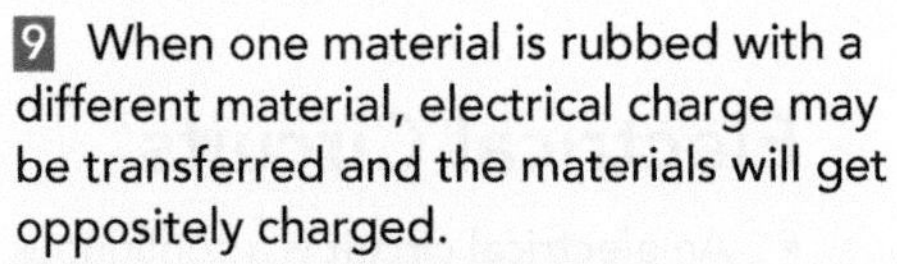

9 When one material is rubbed with a different material, electrical charge may be transferred and the materials will get oppositely charged.

Which gets which charge depends on the series shown in the table. If a material near the top of the table is rubbed by a material near the bottom of the table, then the top material gets a positive charge and the bottom one a negative charge.

+ve
hair
nylon
glass
leather
fur
wool
steel
polysytrene
copper
rubber
polyester
cling film
–ve

a What would happen if a glass rod was rubbed with a polyester cloth?

b What would happen if cling film was wiped with a piece of fur?

c What would happen if leather was stroked with a piece of fur?

d Which material exerts the strongest attraction for electrons? How do you know?

e Which material exerts the weakest attraction for electrons? How do you know?

10 Explain the following situations using static electricity concepts.

a After you have rubbed your pen on your jersey, it picks up a small piece of paper.

b After you have walked across the carpet on a dry day, you get a shock when your hand comes close to a tap.

c When you step out of the car, you get a shock when your finger comes close to the metal roof of the car.

11 Read the article opposite, then answer the questions below.

a Why is the machine called a generator?

b Does it generate static charge or an electrical current?

c Describe the charge on the sphere shown in the diagram.

d Where did the excess electrons come from originally? (Look closely.)

e What two types of charging occur in the machine?

f What would happen if an insulated pad was not used?

g Why would the demonstration not be so dramatic on a humid day?

h If the person moved their other hand toward a tap, what would happen and why?

12 A student rubs a plastic rod and a copper rod with a woollen cloth as shown in the diagrams opposite.

a What charge will the glass rod acquire? (See question 9.)

b Why does the charge stay on the glass rod?

c Why will the copper rod not become charged?

d Where will any charges on the copper rod go to?

e How could you get a charge to stay on the copper rod?

f What do the above activities tell you about the properties of insulators and conductors?

13 A negatively charged glass rod is touched against one side of an insulated metal sphere. The rod is then recharged and touched against the side of a polystyrene sphere.

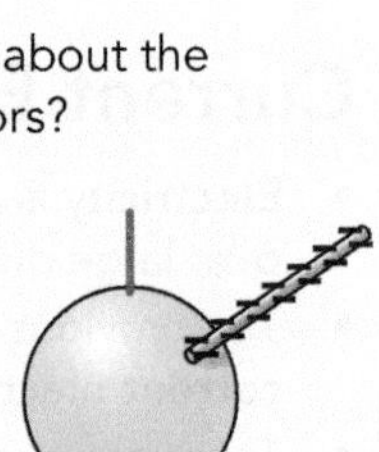

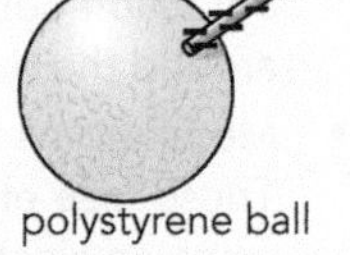

a What charge will the two spheres gain?

b How do you know this?

c Copy each diagram and show the charge distribution over each sphere.

d Which sphere will have a uniform charge distribution? Why does this occur?

e Which sphere will have a non-uniform charge distribution? Why does this occur?

14 Tall buildings are protected from damage by a lightning conductor, which consists of a metal rod mounted at the top of a building connected by a copper wire to a metal stake in the ground.

a Why would the metal rod need to be a conductor?

b What might happen if the connecting wire was accidentally cut?

Van de Graaff Generator

You may have seen a demonstration with a person's hair standing on end because they were touching the metal sphere of a Van de Graaff generator as it was charging.

The machine is able to generate a very large positive or negative charge on the sphere, depending on the design.

Inside the cylinder to which the sphere is attached, a belt is driven very rapidly around two rollers by a motor. The belt and rollers are made of different materials.

As the belt whizzes over the bottom roller it becomes charged. This charge is carried to the top roller, where it is transferred to a metal comb that is connected to the hollow sphere. As the metal sphere is a conductor, the large charge spreads uniformly over the entire surface.

When a person touches the sphere while standing on an insulating pad, some of the excess charge is transferred to their hand.

As the skin is a good conductor, the charge spreads uniformly over its surface because of the repulsion between like charges.

As hairs are light, the repulsion between them forces them apart making them stand up. This effect is more noticeable on low-humidity days.

Direct Current Electricity

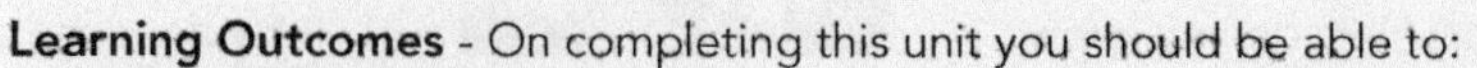

Learning Outcomes - On completing this unit you should be able to:
- describe electrical current and list the requirements of a circuit
- distinguish between conductors and insulators
- identify series and parallel circuits and describe some of their properties
- recognise different electrical components and state their functions
- *draw and interpret simple circuit diagrams.*

Current Electricity

- **Electricity** is a convenient form of energy as it can be transferred rapidly over large distances and readily transformed into other forms of energy.
- The previous unit looked at static electricity but the focus here is on **current electricity**. *An electrical current is a flow of charged particles.*
- There are two types of **electrical charge** – positive and negative. *Like charges repel but unlike charges attract each other.*
- Electrons are the negatively charged particles that normally move at high speed around the positively charged nucleus of an atom because of the attraction between them (see page 56). The electrons are found in 'shells' around the nucleus, with the more energetic electrons found further out.
- Metal atoms have an outer electron that is weakly attracted to the nucleus. These **free electrons** move randomly from atom to atom. A **current** occurs when an electric field makes them drift slowly in one direction.

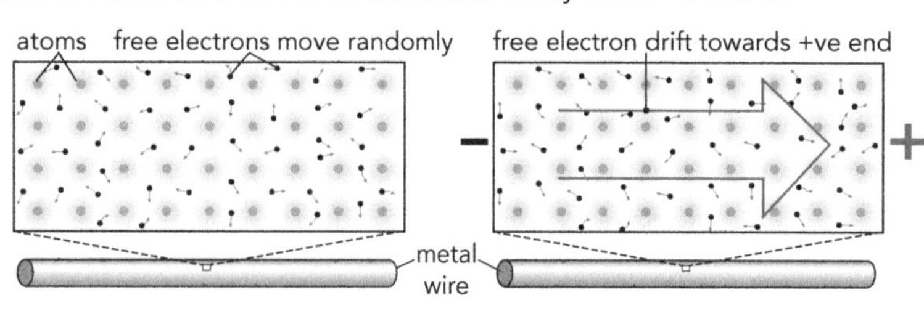

- A substance such as steel, which has free electrons, is called an **electrical conductor**. A conductor allows a current to occur.
- A substance such as plastic, which does not have free electrons, is an **electrical insulator**. An insulator does not allow a current to occur.
- As free electrons move through a conductor they collide with atoms and some of their electrical energy is transformed into light and heat.

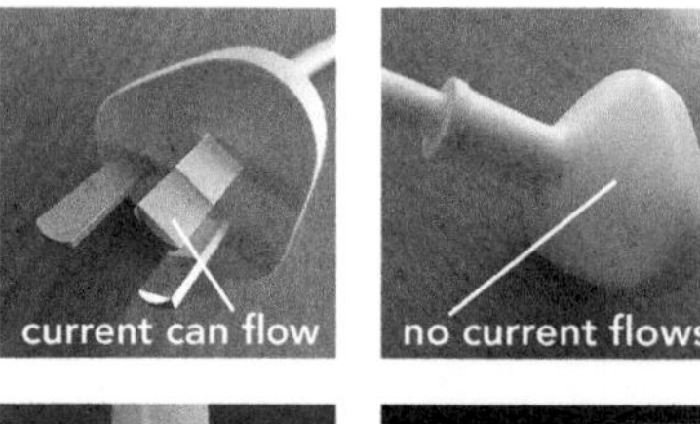

Direct Current

- At NCEA level 1, the focus is on electron drift in one direction only – **direct current** (DC for short). Batteries and power packs supply direct currents.
- Batteries and power packs have a positive (+) **terminal** and a negative (–) terminal.
- A force is needed to make electrons drift. A battery or power pack creates an **electrical force field**. The negatively charged free electrons in the wires of a circuit are repelled from the negative terminal and attracted to the positive terminal.

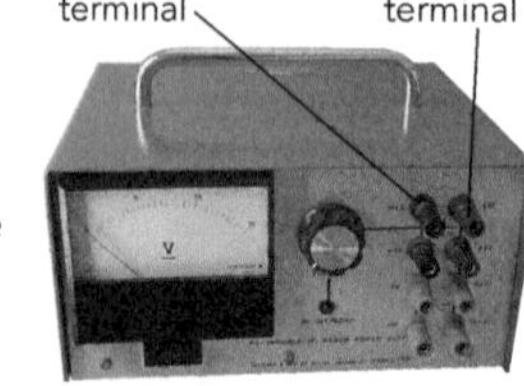

Electrical Circuits

- An **electrical circuit** is a continuous conducting pathway.

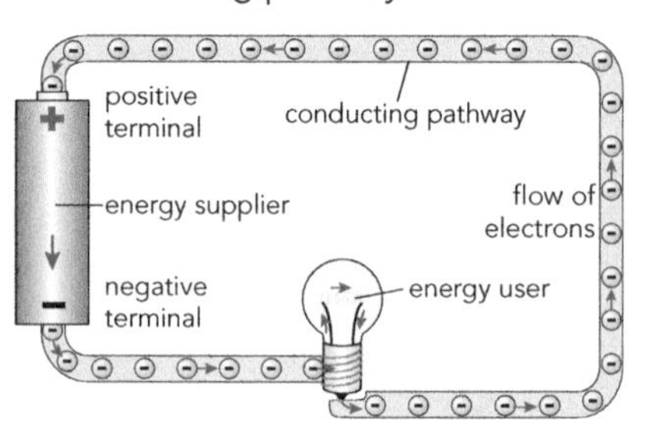

- A circuit must have at least three types of components:
  - a power supply that produces the electrical force field to drive the current (electron drift)
  - conductors that connect the other components into a continuous pathway
  - a device that transforms electrical energy, such as a lamp or heating element.
- When a DC circuit is switched on, all the free electrons around the circuit instantaneously start to drift toward the positive terminal of the power supply.
- As the electrons are being attracted to the positive terminal, the power supply also forces electrons out of the negative terminal and around the circuit because of repulsion.
- As the current is pushed through different devices, electrical energy is converted into heat, light, sound, magnetic energy or kinetic energy.
- The size of current flowing back into the power supply is the same as the size of current leaving the power supply.
- *Electrons may lose energy as they travel around a circuit but they are not used up or lost.*

ISBN: 9780170262316

Series and Parallel Circuits

- The components of a simple circuit can be connected in two different ways – in series or in parallel.
- If the components in a circuit are connected in a single pathway so that the current flows through each component in turn, then the circuit is called a **series circuit**.

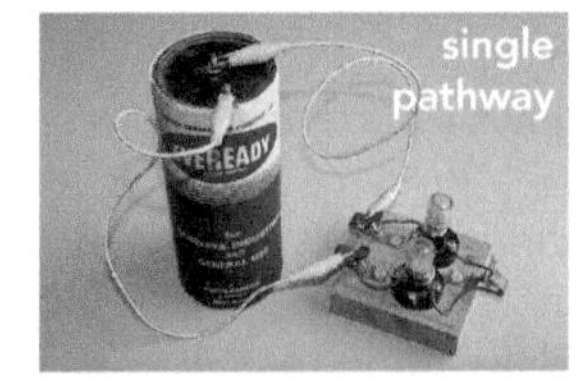

- *In a series circuit there is only one pathway for the current, so the current is the same everywhere around the circuit.*
- If any one component in a series circuit fails, then no current will flow.
- If the components in a circuit are connected in multiple pathways so that the current is shared by several components, then the circuit is called a **parallel circuit**.

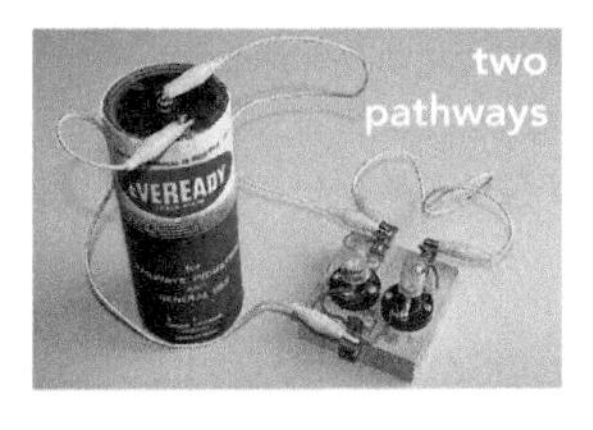

- *In a parallel circuit there are several pathways for the current, so the current is shared out among the pathways. The size of the current in the different pathways may vary.*
- If one component in a parallel circuit fails, then other components in different pathways can still work.
- In more complex circuits, some components will be connected in series (on the same pathway), and other components will be connected in parallel (on different pathways).
- You will learn more about how electricity behaves in series and parallel circuits in the next two units.

Circuit Components

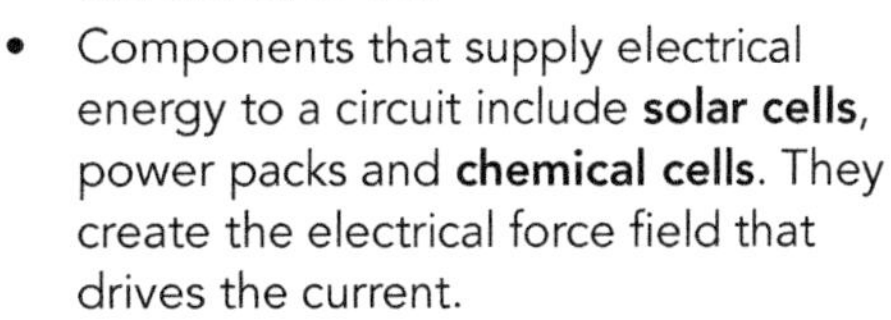

- Wires connect components and allow the current to flow around a circuit.
- Switches are used to turn the current on or off.
- Components that supply electrical energy to a circuit include **solar cells**, power packs and **chemical cells**. They create the electrical force field that drives the current.
- The everyday term for a chemical cell is a battery. A torch battery consists of a single cell; a car battery consists of multiple cells joined in series.
- Lamps are components that convert electrical energy into light and heat.
- **Resistors** are used to limit the amount of current that flows through a circuit or part of a circuit.
- **Variable resistors** are resistors whose resistance can be altered. The greater the resistance, the less current that flows in a pathway.
- A **diode** allows the current to flow in one direction only, the conventional current direction (see page 25).
- Diodes that light up when the current flows in the right direction are called **light-emitting diodes** (LED).
- An **ammeter** is used to measure the size of a current.
- A **voltmeter** is used to measure the amount of electrical energy gained or lost by the current as it passes through a component.
- Direct current (DC) meters usually have one or more red positive terminals and a black negative one.

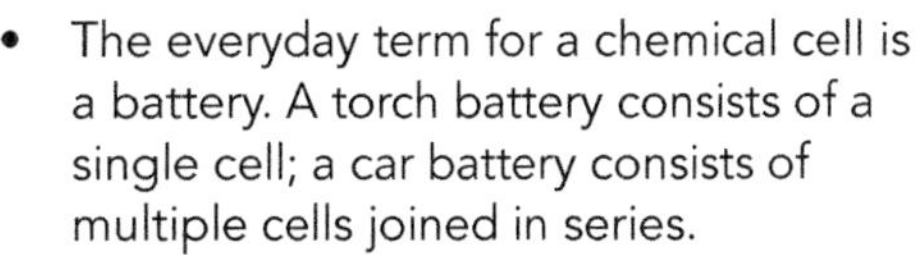

two cells

switch

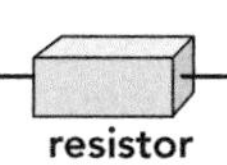

resistor

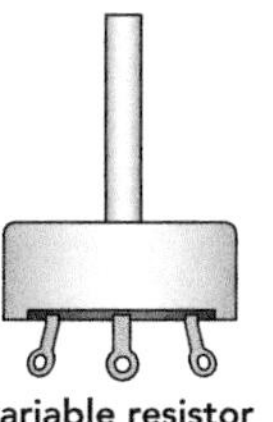

variable resistor

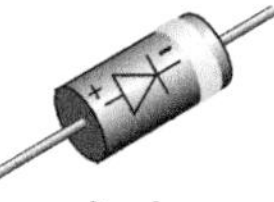

diode

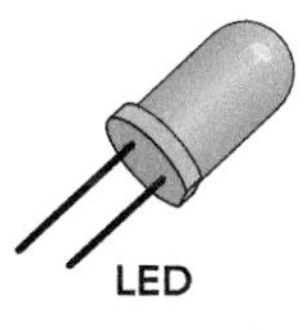

LED

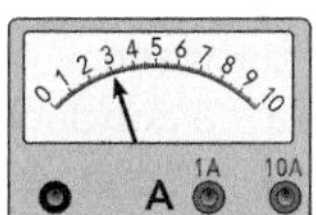

ammeter

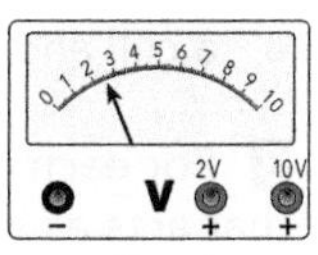

voltmeter

SCIENCE SKILL: Drawing Circuit Diagrams

A circuit diagram represents a circuit. Each type of component has a special symbol. A single pathway is drawn as a rectangle; multiple pathways as connected rectangles.

Drawing a circuit:

1. Decide whether the circuit has components in series or parallel.
2. If all of the components are in series, draw a single rectangle for the circuit.
3. If the circuit has components in parallel, draw connected rectangles, one for each pathway.
4. Identify each component in the circuit and draw its symbol in the correct location on the diagram.
5. Ensure symbols are placed in the correct order.
6. Make sure the lines that represent wires are straight and connect with components – no gaps.
7. Check positive and negative terminals of the power supply and meters are shown the right way round.

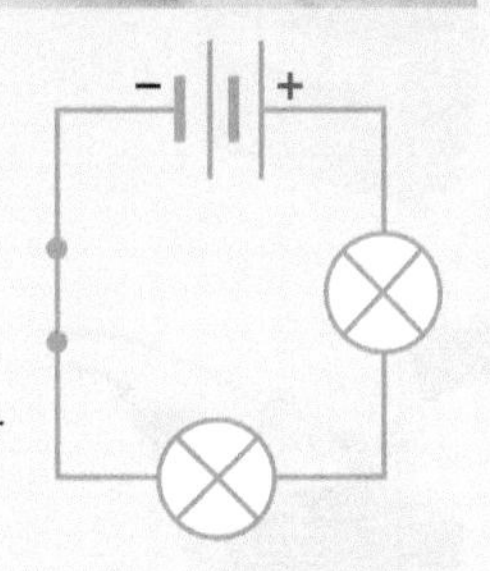

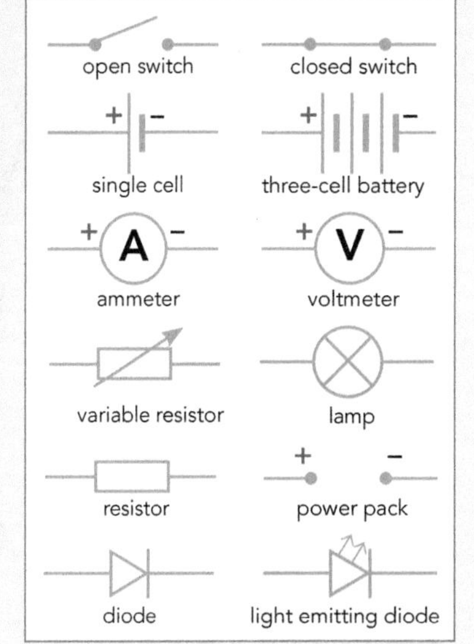

Revision Activities

1 Match up the descriptions with the terms.

	Terms		Descriptions
a	electricity	A	current that flows in one direction only
b	current electricity	B	complete conducting pathway allows a current to flow through it readily
c	electrical charge	C	object whose resistance can be varied
d	free electrons	D	connection point on a component
e	electrical conductor	E	circuit where components are in different loops
f	electrical insulator	F	can be either positive or negative
g	direct current	G	object that transforms chemical energy into electricity
h	terminal	H	form of energy associated with charged particles
i	electrical force field	I	object that transforms light energy into electricity
j	electrical circuit	J	object that opposes the flow of current
k	series circuit	K	instrument that measures the size of a current
l	parallel circuit	L	drift of charged particles along a conductor
m	solar cell	M	measures electrical energy lost or gained by the current
n	chemical cell	N	circuit where all components are in a single loop
o	resistor	O	electrons that are able to move from atom to atom in a metal
p	variable resistor	P	provides the force to drive the current
q	diode	Q	does not allow a current to flow through it
r	light emitting diode (LED)	R	lights up if current flows in the right direction
s	ammeter	S	allows current to flow in one direction only
t	voltmeter	T	object that allows a current to flow through it

2 Explain the difference between:

a a conductor and an insulator
b a solar cell and a chemical cell
c a series and a parallel circuit
d an ammeter and a voltmeter
e a cell and a battery.

3 For each of the following objects, identify the part that acts as a conductor and the part which is designed to act as an insulator.

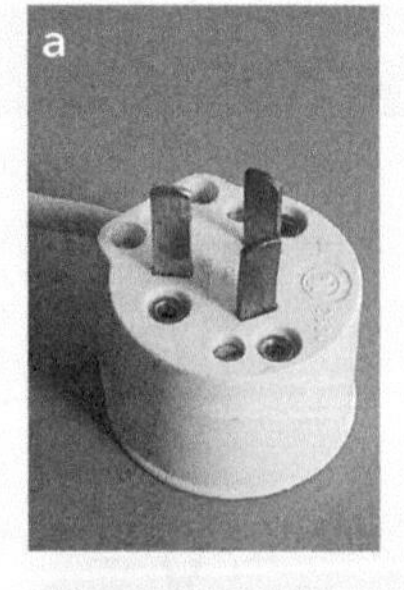

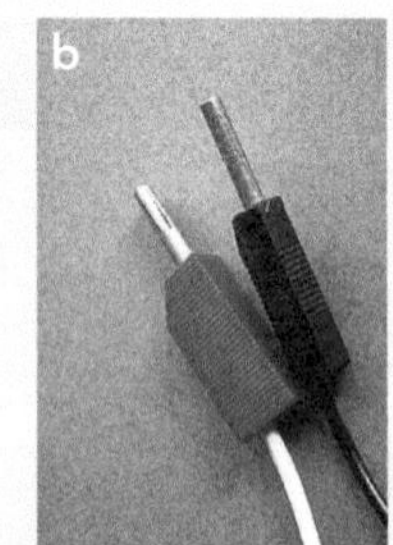

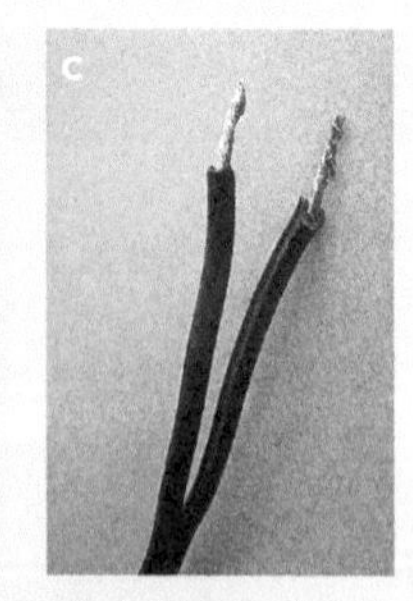

4 Decide whether the following statements are true or false. Rewrite the false ones to make them correct.

a Electricity is a form of energy that enables rapid energy transfer.
b A current involves charged particles drifting in one direction.
c Atoms are the particles that are moving in most electrical currents.
d An electrical insulator will have many free electrons.
e Resistors are components that limit the size of the current.
f Direct current means that the electrons are drifting in one direction only.
g An electrical force field is needed to drive electrons around a circuit.
h Electrons travel from the positive terminal of a battery around to the negative terminal.
i A current needs a complete conducting pathway.
j If two components are in separate branches, they are said to be in series.
k In a series circuit, the current is the same at all points around the circuit.

5 Study the circuit.

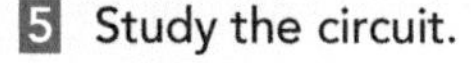

a What types of components are there in this circuit?
b Which component produces electricity?
c Which components conduct electricity?
d Which component transforms electricity?
e Is the circuit a complete conducting pathway? How do you know?

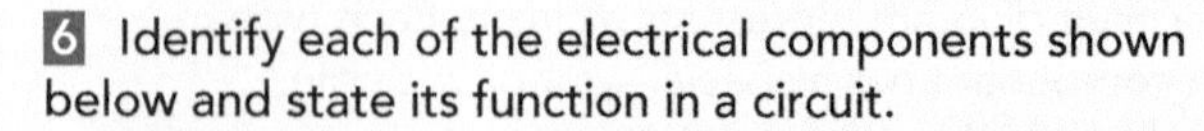

6 Identify each of the electrical components shown below and state its function in a circuit.

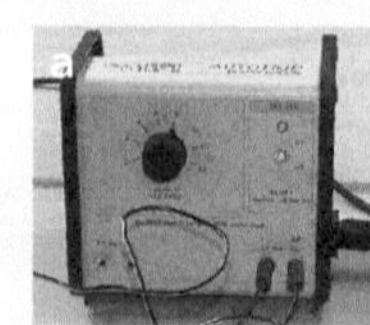

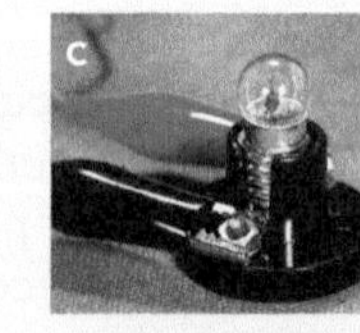

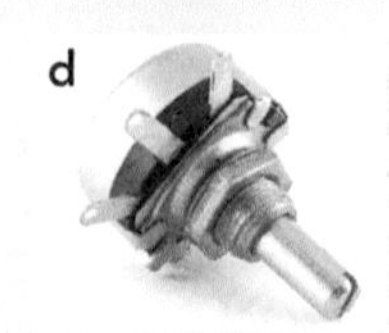

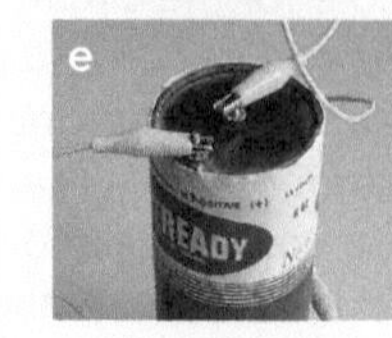

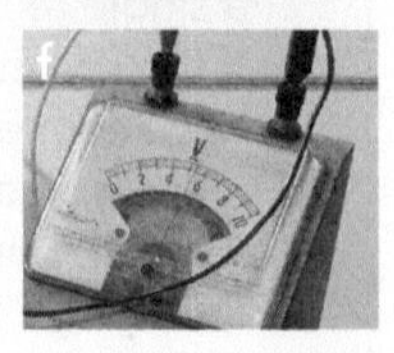

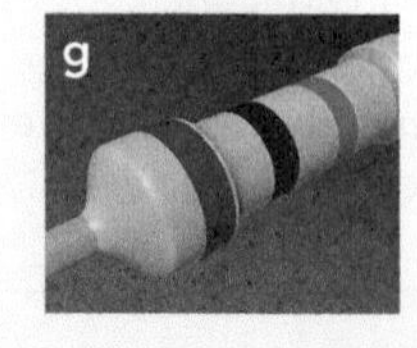

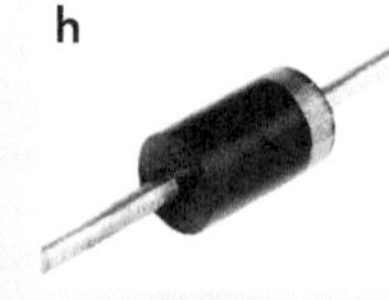

ISBN: 9780170262316

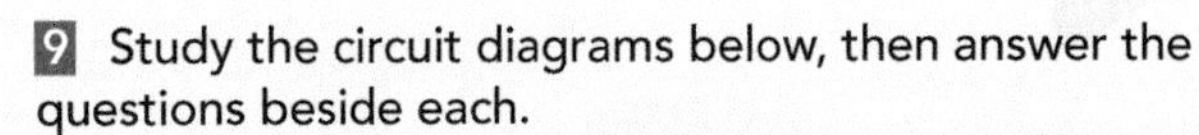

7 Study each circuit in turn, then answer the questions next to that circuit.

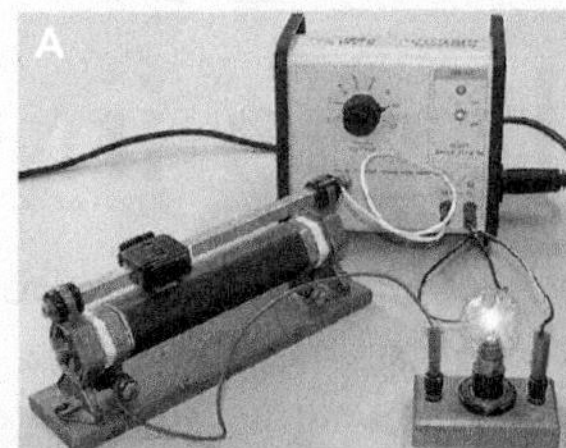

a What is the function of the rheostat shown in circuit A?
b Are the variable resistor and lamp connected in series or parallel?
c How do you know that circuit A is a complete conducting pathway?
d Draw a circuit diagram for circuit A.

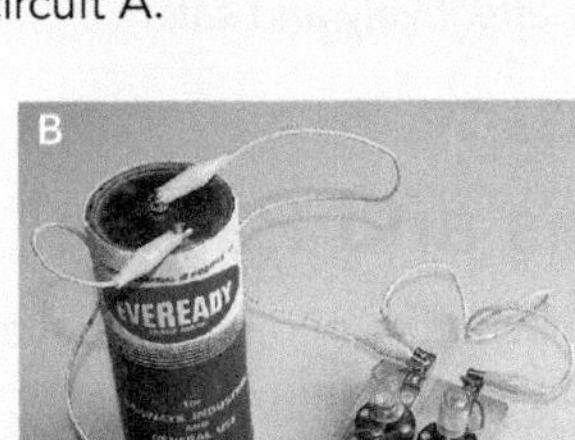

e What is the function of the cell shown in circuit B?
f Are the bulbs connected in series or parallel?
g If the bulbs are equally bright, what can be said about the current going through each?
h Draw a circuit diagram for circuit B.

i What is the function of the switch shown in circuit C?
j Are the two cells in series or parallel? What about the bulbs?
k What would happen if one bulb failed in circuit C?
l Draw a circuit diagram for circuit C.

m How are the two bulbs connected in circuit D?
n Are the cells connected in series or parallel?
o If one bulb failed, what would happen to the other?
p Draw a circuit diagram for circuit D.

9 Study the circuit diagrams below, then answer the questions beside each.

Circuit A

a Is circuit **A** a series or parallel circuit?
b What is the function of the resistor in circuit **A**?
c If the switch in circuit **A** is turned to the off position, what will happen to the lamp?

Circuit B

d How many cells are there in circuit **B**?
e How are the cells connected?
f How are the two lamps connected in circuit **B**?
g What will happen to the two lamps in circuit B if just the top switch is turned to the off position?
h What will happen to the lamps in circuit **B** if just the bottom switch is turned off?

Circuit C

i Which component provides electrical energy in circuit **C**?
j Are the components in series or parallel?
k What will happen to the current flowing in circuit **C** if the resistance of the variable resistor is increased?
l What will happen to the brightness of the lamp if the resistance is decreased?

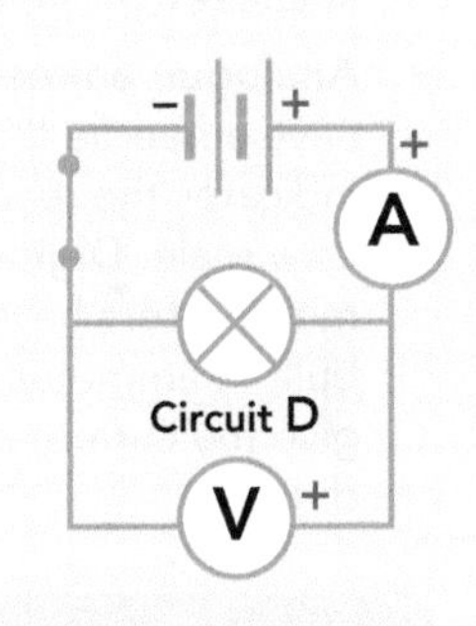

m In circuit **D**, is the ammeter connected in series or parallel with the lamp?
n Is the voltmeter connected in series or parallel with the lamp?
o Which terminal of the battery do the positive terminals of the meters lead to?
p Are the meters connected correctly?

8 Read the article opposite, then answer the questions below.

a Why is lightning said to be a form of current electricity?
b Why is air normally a poor conductor of electricity?
c Where are charges concentrated in a thundercloud?
d What causes the separation of charge in a thundercloud?
e Why is an electrical force field created between the cloud bottoms and the earth?
f What does this powerful electrical force field do to air molecules?
g What makes air conduct?
h Why are tall objects more likely to receive lightning strikes?
i In which direction will the ionised air molecules travel?

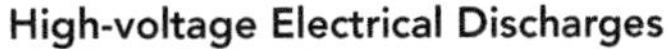

High-voltage Electrical Discharges

Lightning is actually high-voltage electrical currents travelling through the air. Normally, air is a poor conductor because there are no free charged particles to enable a current to flow.

In a thunderstorm, electrical charges in the clouds become separated. The tops of the clouds become positively charged, while the bottoms acquire a huge surplus of negative charge.

Scientists think this separation of charge is caused by convection air currents inside the clouds rubbing ice crystals against water molecules. This is charging by friction.

As the negative charge builds up on the bottoms of clouds, this induces a positive charge on the Earth's surface below the cloud. A huge charge difference develops between the underside of the cloud and the ground below.

This electrical potential difference creates a force field between the clouds and the ground, which is strong enough to pull electrons off air molecules, turning them into ions, which are, of course, charged particles.

Initially, an invisible leader lightning stroke shoots downward in zigzag steps ionising air molecules and releasing free electrons, which together turn the air into a conducting pathway. As the leader stroke nears the ground, another invisible leader leaps up from the highest point on the ground as this is the shortest pathway.

When the two leaders meet, the visible lightning strike occurs. It moves rapidly from the ground along the zigzag ionised pathway back to the clouds. This happens so rapidly that we do not observe the direction in which it travels.

In the lighting strike, large numbers of electrons are transferred to the ground, thus discharging the clouds.

ISBN: 9780170262316

Current and Voltage in Circuits

Learning Outcomes - On completing this unit you should be able to:
- define current and relate it to electrical charge
- describe what happens to the current in series and parallel circuits
- define voltage gain or loss and relate it to electrical energy
- describe what happens to voltage in series and parallel circuits
- *use electrical meters effectively and safely.*

Measuring the Current

- *Current is a flow of electrical charge.* The charge is carried by **electrons** moving along a **conductor**, such as a copper wire.
- **Charge** is measured in units called **coulombs**. A huge number of electrons are needed for a charge of just one coulomb – 6,250,000,000, 000,000,000 in fact!
- **Current** is defined as the number of coulombs of charge passing per second. Current has the symbol *I* and the unit is the **ampere** or **amp** (A).
- One ampere is defined as one coulomb of charge passing by each second.
- To measure the current in a pathway, an **ammeter** is inserted directly into the pathway in series with other components.
- **Analogue ammeters** have a needle that indicates the reading on a scale. **Digital multimeters** have a built-in ammeter and give the current in digits.

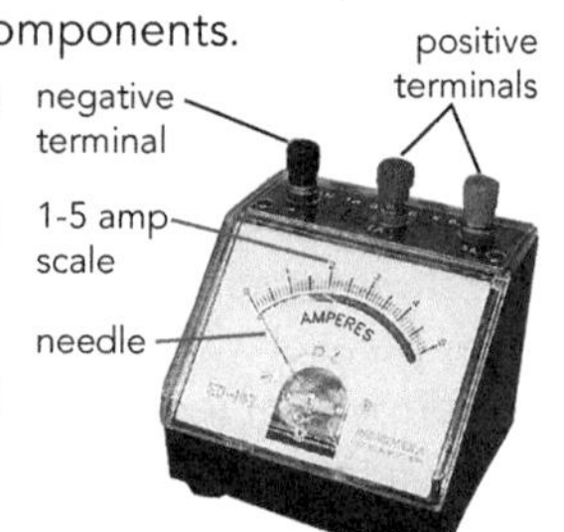

Series and Parallel Circuit Currents

- In a circuit the number of electrons returning to the power supply per second is the same as the number leaving.
- In a simple **series circuit**, all components are connected to form a single pathway. *The current in a series circuit is the same all the way around the circuit.*
- In a circuit with more than one pathway, some components may be connected in series on a particular pathway. The current is the same along that pathway.
- In a **parallel circuit**, when electrons reach a junction some electrons go along one pathway and the rest along the other pathway(s). The current from the power supply is shared amongst the pathways.
- *The currents in the pathways of a parallel circuit add up to equal the current from the power supply, so* $I = I_1 + I_2$.
- The size of the current in each pathway depends on the **resistance** of the pathway. If pathways offer equal resistance, then each receives an equal share of the current. If the resistance of pathways differ, then the pathway with the least resistance gets the biggest share of the current.

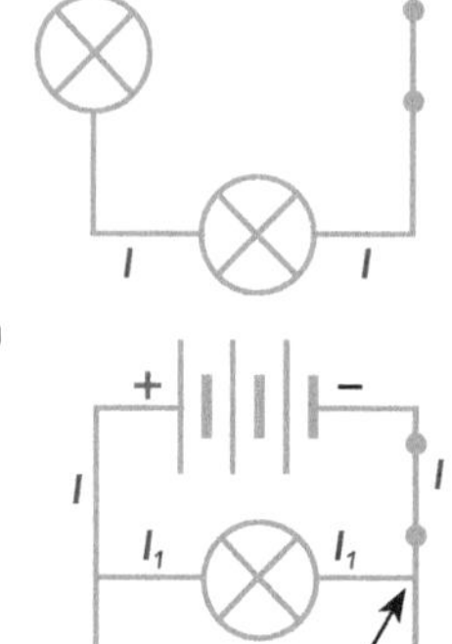

SCIENCE SKILL: Using an Ammeter

An ammeter is connected directly into the circuit at the point of interest. *It is connected in series with other components.*

An ammeter that can measure direct current has a red positive **terminal** and a black negative one. The connection from the red terminal must lead to the red (positive) terminal of the power supply.

Steps:

1. Set up the circuit if it has not been assembled already. Make sure it works, then switch it off.
2. If you are using a digital multimeter, turn the dial to the highest DC amp setting. If the analogue ammeter you are using has several red terminals, use the one with the highest amp setting.
3. Identify where you want to measure the current and disconnect the end of the closest wire.
4. Connect that wire to one of the ammeter terminals and use another wire to connect the other terminal to the rest of the circuit.
5. Check that the connection from the red meter terminal leads toward the red terminal of the power supply. If not, switch the meter wires.
6. Turn on the circuit. If the analogue needle flicks backwards or the digital reading is a negative number, turn off the circuit and swap the meter connections.
7. Read the current off the correct scale on the digital display.
8. If the current is low, switch to a lower setting for a more accurate reading.

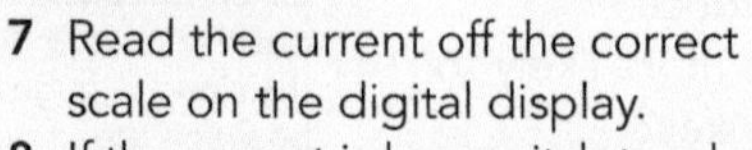

Steps
1 Set up, check and turn off
2 Select setting
3 Find location
4 Insert meter in series
5 Check +ve to +ve
6 Test meter
7 Take reading
8 Adjust setting

ISBN: 9780170262316

Current Direction

- In a circuit, electrons flow from the negative to the positive terminal of the power supply.
- NCEA exams may refer to the **conventional current direction**. This is the direction in which positive charges would travel if they were free to do so, from the positive to the negative terminal as indicated by the → arrows.

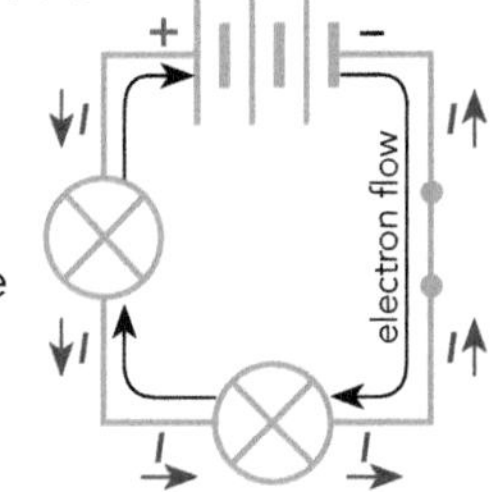

- It is used to find the direction of the magnetic field created by a current (see page 33).

Voltage Gain or Loss

- The power supply in a circuit creates an electrical force field that drives electrons around the circuit. In doing so, the power supply provides these electrons with extra energy. The electrical energy gained by the current as it passes through a power pack or battery is called **voltage gain** (or supply voltage).
- As the current passes through various components (eg LEDs and resistors), its electrical energy is transformed into other forms of energy (eg light and heat). The electrical energy lost by the current as it passes through a component is called **voltage loss** (or voltage drop).
- **Voltage**, *symbol V, is the difference in electrical energy between two points in a circuit.* The unit is the **volt** (symbol V). One volt is defined as one joule of energy gained or lost by each coulomb of charge.
- Voltage is also called **potential difference**.
- A **voltmeter** is used to measure the voltage gain or loss as the current passes through a component, either an analogue voltmeter or a digital multimeter.

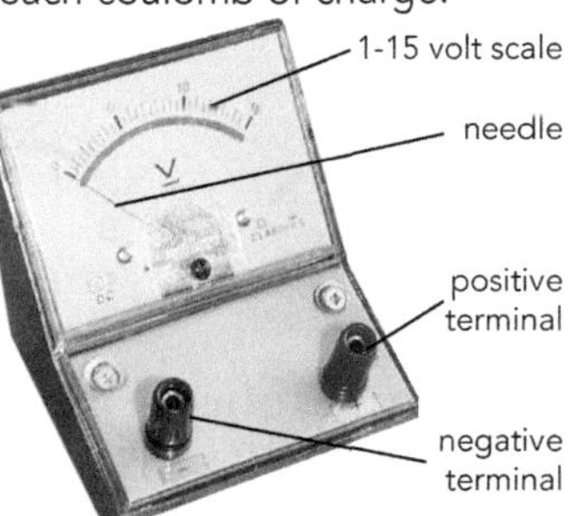

Series and Parallel Voltages

- In a circuit with a single energy user, the voltage loss across that component equals the voltage gain across the power supply.
- If several cells are connected in series, then the total voltage gain of the current is equal to the sum of the voltages supplied by each cell.
- If several components, such as lamps, are connected in series, then each component causes the current to lose a share of the voltage gain from the power supply.
- *The voltage losses as the current passes through each component add up to the voltage gain of the power supply.*
- If components are connected in parallel, then the voltage loss across each equals the voltage supply.
- *The voltage losses of parallel components are identical.*

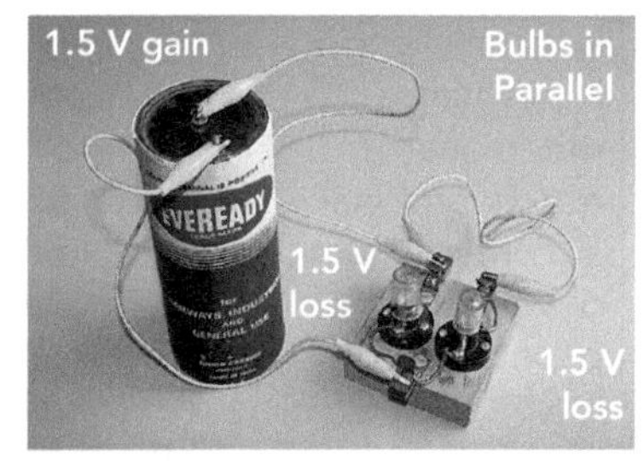

Current and Voltage in Series and Parallel Circuits

In a simple **series circuit**:

- the voltage gain equals the sum of individual voltages
- the current is the same everywhere
- each component loses a share of the voltage gain
- the voltage loss of each component add up to equal the voltage gain of the power supply.

In a simple **parallel circuit**:

- each pathway receives a share of the current from the power supply
- the current in each pathway add up to equal the current produced by the power supply
- the voltage loss of each component is the same as the voltage gain of the power supply.

SCIENCE SKILL: Using a Voltmeter

As a voltmeter is used to measure the difference in electrical energy, it is connected around the component of interest. *It is connected in parallel with the component.* A voltmeter that can measure direct current has a red positive terminal and a black negative one. The connection from the red terminal must lead to the red (positive) terminal of the power supply.

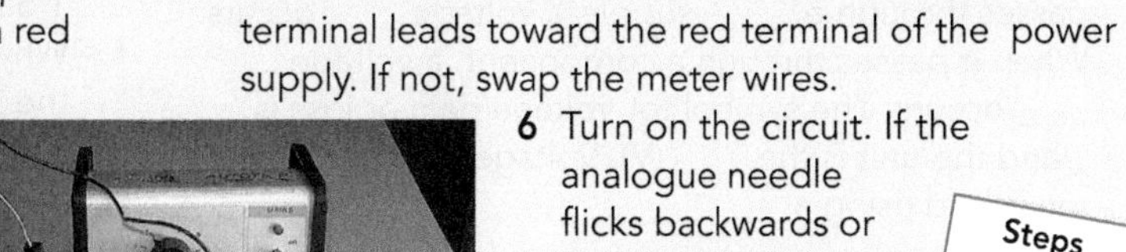

Steps:

1. Set up the circuit if it has not been assembled already. Make sure it works, then switch it off.
2. If you are using a digital multimeter, turn the dial to the highest DC voltage setting. If the analogue voltmeter you are using has several red terminals, use the one with the highest voltage setting.
3. Identify the component whose voltage gain or loss you are going to measure.
4. Use two wires to connect the voltmeter around the component, ie in parallel with the component.
5. Check that the connection from the red meter terminal leads toward the red terminal of the power supply. If not, swap the meter wires.
6. Turn on the circuit. If the analogue needle flicks backwards or the digital reading is negative, turn off the circuit and swap the meter connections.
7. Read the voltage loss or gain off the correct scale or the digital display.
8. If the voltage is low, switch to a lower setting for a more accurate reading.

Steps
1 Set up, check and turn off
2 Select setting
3 Find location
4 Connect meter in parallel
5 Check +ve to +ve
6 Test meter
7 Take reading
8 Adjust setting

ISBN: 9780170262316

Revision Activities

1 **Match up the descriptions with the terms.**

a electrons	A used to measure the current flowing in a circuit
b conductor	B flow of electrical charge
c charge	C meter that gives the reading in digits
d coulomb	D extremely small particles with a negative charge
e current	E electrical energy used by a component
f ampere or amp	F circuit in which components are in several different pathways
g ammeter	G unit for voltage gain or loss
h analogue meter	H substance that allows free electrons to move through it
i digital multimeter	I energy gained or lost by electrons in the current
j series circuit	J opposition to the flow of electrons
k parallel circuit	K property that is either positive or negative
l resistance	L electrical energy provided by a power supply
m terminal	M unit for charge
n conventional current direction	N meter used to measure voltage gain or loss
o voltage gain	O circuit in which all components are in the same pathway
p voltage loss	P meter that has a scale and a needle
q voltage	Q connection point on a component
r volt	R unit for electrical current
s potential difference	S another name for voltage
t voltmeter	T direction in which positive charges would flow

2 **Explain the differences between:**

a electrical charge and current
b an analogue meter and a digital meter
c current and voltage
d voltage gain and loss
e a component and a terminal.

3 **Copy and complete the following statements.**

a Current is a flow of electrical _____. The charge is carried by _________ along __________. The symbol for current is _ and the unit is the ______ or ___ (A). The size of the current is measured using an _______. The current is driven by an electrical _____ _____ created by the _____ supply.

b Voltage is related to the _________ energy gained or lost by _________ in the _______. When the current passes through a _____ supply, a voltage ____ occurs. When it passes through a component, a voltage ____ occurs. The symbol for voltage gain or loss is _ and the unit is the ____ (V). Voltage gain or loss is measured using a _________.

4 **Study the meter shown below, then answer the questions.**

a Is this meter an ammeter or a voltmeter?
b Is it an analogue or a digital meter?
c What unit does it measure in?
d Why does it have two red terminals?
e If the current was expected to be about 4 A, which terminals on the meter would you use?
f What about if the current was expected to be less than 1 amp?

5 **Decide whether the following statements are true or false. Rewrite the false ones to make them correct.**

a Current is a flow of electrical charge.
b 6,250,000,000,000,000,000 electrons give a charge of one coulomb.
c An analogue meter has an electronic display.
d In a simple series circuit, the current is the same everywhere.
e In a parallel circuit, the current is shared amongst the pathways.
f In a parallel circuit, the pathway with the greatest resistance gets the biggest share of the current.
g With a DC meter, the lead from the red terminal must lead back to the black terminal of the power supply.
h An electrical force field drives electrons around a circuit and provides the electrical energy.
i In a series circuit, components use a share of the voltage supply.
j In a simple parallel circuit, the voltage loss across each component equals the voltage gain.

6 **With an analogue meter, the reading is taken to the nearest mark. Study the readings, then answer the questions.**

a What type of meter is being used and what units does it measure in?
b What is the range of the scale on this meter?
c The space between two of the small marks on a scale is called the interval. How large is the interval on this meter?
d How accurately can this meter be read to?
e What is the reading shown in each photo to the nearest 1 decimal place (1 dp)?
f Why is it important to look straight on when reading the scale on a meter?

7 **Look through the instructions for using an ammeter and voltmeter in the Science Skills boxes on the previous two pages, then answer these questions.**

a How is an ammeter connected to measure the current flowing through a component?
b How is a voltmeter connected to measure the voltage loss across a component?
c Where must the wire connected to the red terminal of a direct current meter lead to?
d Why is important to start measuring using the highest scale first?

ISBN: 9780170262316

8 Study this circuit that was set up to measure the current travelling through a lamp and the voltage loss across the lamp. Then answer the questions below.

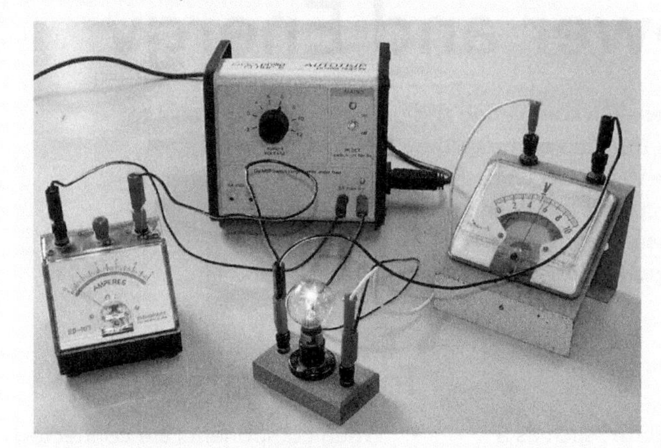

a Trace the wires coming from the ammeter. Is the ammeter connected in series or parallel with the lamp?
b The wire from the red (positive) terminal of the ammeter leads toward which terminal of the power pack?
c How can you tell if the ammeter connections are the right way around?
d What size is the current passing through the lamp?
e Trace the wires coming from the voltmeter. Is the voltmeter connected in series or parallel with the lamp?
f Which terminal of the power pack does the wire from the positive (red) terminal of the voltmeter lead toward?
g What is the voltage drop across the lamp?
h What is the voltage supply from the power pack set to?
i Is the voltage loss across the lamp equal to, more than or less than the voltage gain supplied by the power pack? Suggest a reason to account for any difference.
j Use words to describe the size of the current passing through the lamp and the voltage drop that occurs.
k What is the electrical energy transformed into?

10 In order to answer the questions below, you will need to refer to the summary on page 25.

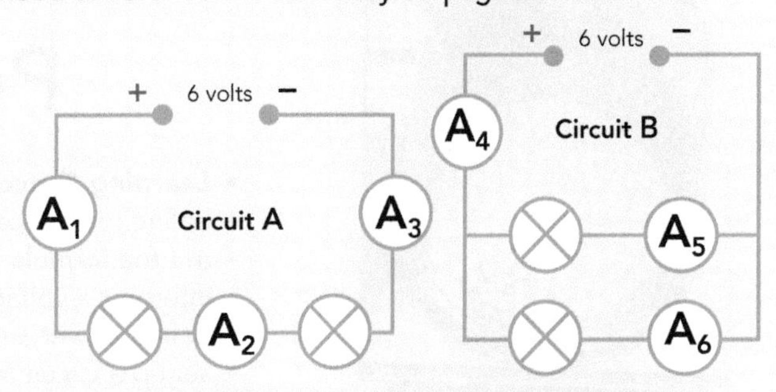

a Are the lamps in circuit **A** in series or parallel?
b If the current passing through ammeter A_1 is 2 amps, what will be the current in ammeters A_2 and A_3?
c Are the lamps in circuit **B** in series or parallel?
d If the current in ammeters A_5 and A_6 is 3 amps, what will be the current in ammeter A_4?
e Write a statement comparing what happens to the current in series and parallel circuits.
f If the two lamps in circuit **A** are identical, what will be the voltage loss across each?
g If the two lamps in circuit **B** are identical, what will be the voltage loss across each?

11 Each cell generates a voltage gain of 1.5 V and the lamps are identical.

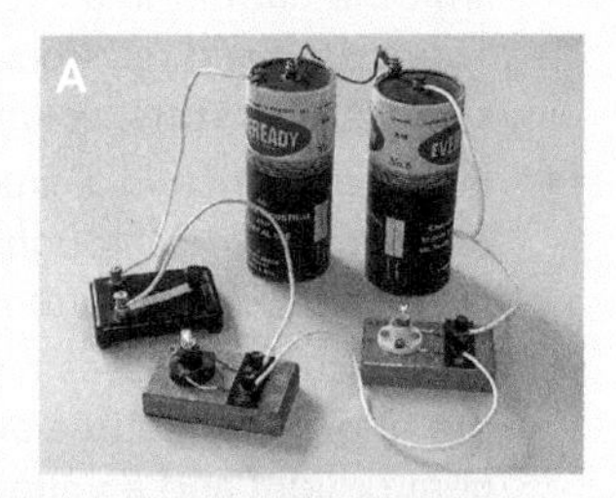

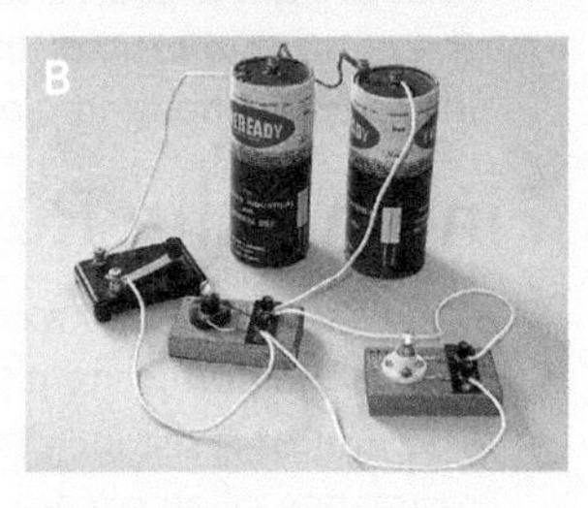

a Are the lamps in circuit **A** in series or parallel? What abut the cells?
b In **A**, what is the voltage gain from the two cells?
c What voltage loss will occur across each lamp in circuit **A**?
d Are the lamps in circuit **B** in series or parallel? What about the cells?
e What voltage loss will occur across each lamp in circuit **B**?

9 Read the article opposite, then answer the questions below.
a What is a cell?
b What is the difference between a chemical cell and a solar cell?
c If 20 joules of light energy fall on a solar cell, how much electrical energy could the cell gain?
d Why are solar cells used to power satellites and remote instruments?
e What are advantages of solar cells?
f What are disadvantages of solar cells?
g What is a semiconductor?
h How is an electrical force field created by a solar cell?
i When the solar cell is connected in a circuit, what two things occur?

Solar Cells

A cell is a device that transforms another type of energy into electricity.

Chemical cells, such as torch batteries, transform chemical energy into electrical energy.

Solar cells transform light into electrical energy. The efficiency of solar cells currently is about 21%.

These cells are used to power satellites, calculators and remote road signs.

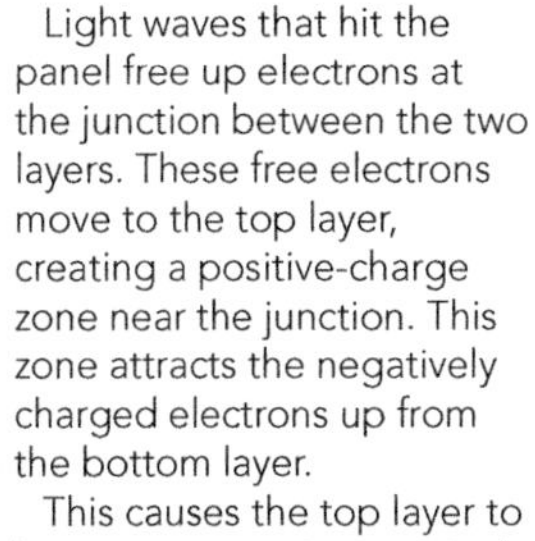

Solar cells are able to use an abundant free energy source – sunlight – but cease to work after dark. A solar cell needs a relatively large surface area to absorb sufficient light to produce the required voltage gain.

Solar cells are flat and consist of two thin layers of different semiconductor materials.

Light waves that hit the panel free up electrons at the junction between the two layers. These free electrons move to the top layer, creating a positive-charge zone near the junction. This zone attracts the negatively charged electrons up from the bottom layer.

This causes the top layer to become a negative terminal, and the bottom layer to become a positive terminal.

When the two terminals are connected in a circuit, they create an electrical force field that drives the current and provides a voltage gain. Some solar cells use batteries to store the energy.

Resistance, Power and Energy

- ***Learning Outcomes*** - On completing this unit you should be able to:
- define resistance as the ratio of the voltage loss to the current
- use the formula $R = V/I$ to calculate the resistance of a component
- interpret a voltage-current graph and relate it to Ohm's Law
- define power and use the formula $P = VI$ to find power
- *set up a circuit to find the resistance and power of a component.*

Resistance and Current

- When an electrical force field is applied to a good **conductor**, such as a copper wire, the free outer **electrons** of the metal atoms are readily pushed through the conductor, creating the current.
- But the atoms of other substances hold on to their 'free' electrons more tightly so it is more difficult to push a current through them. This opposition to the **current** is called **resistance**. *A substance with a high resistance limits the flow of electrons.*
- A circuit component that is specifically designed to limit the current in a circuit is called a **resistor**. There are fixed resistors and **variable resistors**, such as **rheostats**, volume controls and dimmers.

- *If the resistance in a circuit is increased, then the current passing through it will fall. If the resistance is decreased, the current will increase.*
- As the current moves through a resistor, some of the energy of the free electrons is transformed into heat or light and a **voltage loss** occurs. *In opposing the current, a resistor transforms electrical energy into heat.* A heating element is a resistor.

Finding the Resistance

- The **resistance *R*** of a component is defined as the ratio of the voltage loss across the component to the current passing through it.
- The unit of resistance is the **ohm**, symbol Ω. One ohm is defined as the resistance of an object that limits the current to one **ampere** when a voltage drop of one **volt** occurs across the object.
- Resistance is found by dividing the voltage by the current:

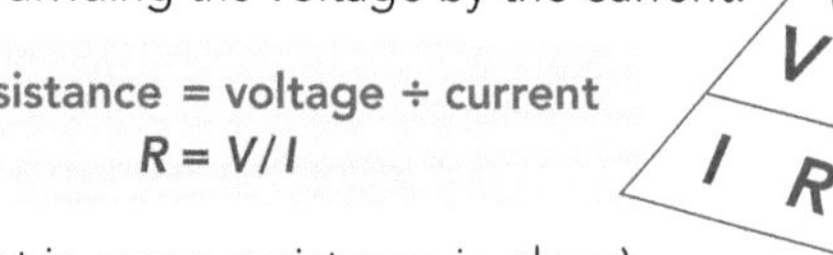

Resistance = voltage ÷ current

$$R = V/I$$

(voltage in volts, current in amps, resistance in ohms).

- The triangle above gives the formula for finding the current or the voltage if the other two quantities are known. Cover the unknown quantity with your finger and read off the appropriate formula: $V = IR$ or $I = V/R$.
- The above relationship is usually given as $V = IR$ in NCEA exam papers.
- The resistance of a component is found by measuring the voltage loss across it and the current passing through it. The voltmeter must be connected in parallel with the component and the ammeter in series with it.
- In the photo opposite, the voltage drop across the lamp is 5.5 V and the current through it is 0.5 A, so the resistance is given by $R = V/I = 5.5 \div 0.5 = 11\ \Omega$.

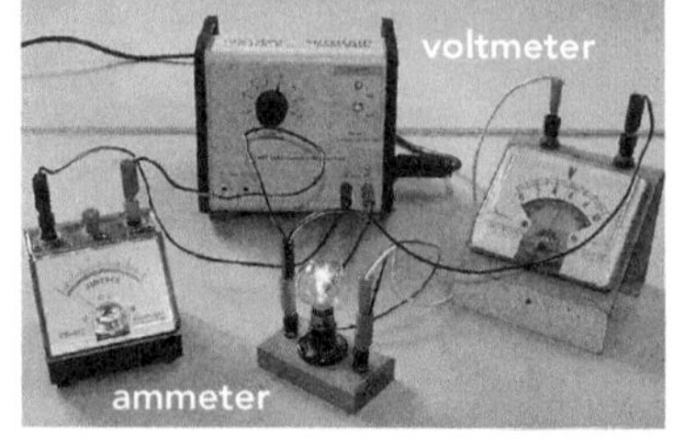

Total Resistance

- When components are connected in series, their **total resistance R_T** is equal to the sum of their individual resistances.

$$R_T = R_1 + R_2 + \ldots$$

Voltage and Current

- The circuit below shows what happens to the voltage loss across a resistor, such as a lamp, when the current through it is increased.

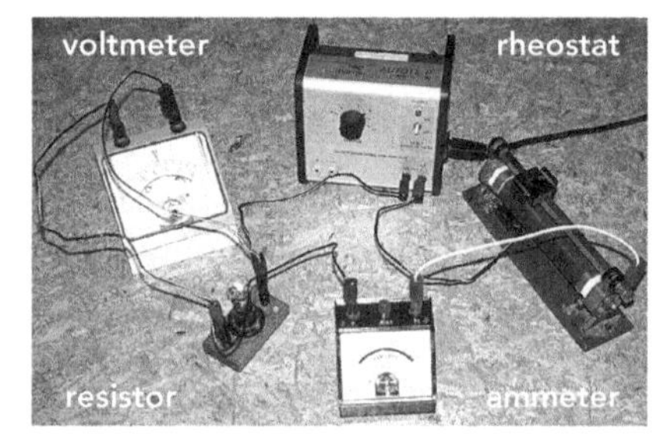

- The rheostat is used to increase the size of the current passing through the resistor (lamp), which is measured by the ammeter. The voltmeter measures the resulting voltage loss across the resistor.
- The voltage loss is plotted against the current below. As you can see, increasing the current passing through the resistor (lamp) causes the voltage loss to rise. *So voltage loss is proportional to the current.*

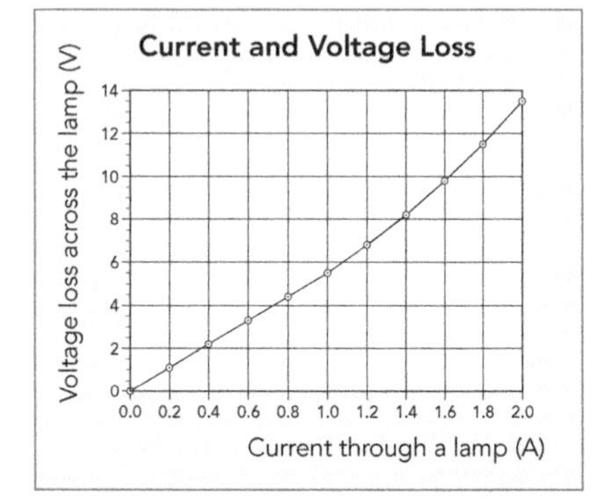

ISBN: 9780170262316

Ohm's Law

- *The slope of a voltage-current graph represents the resistance of the resistor.* If the slope is a straight line, then the resistance of the resistor is constant.
- On the graph at the bottom right of the opposite page, the line is straight initially but then curves upward slightly. This means the resistance is constant at first but then it begins to increase.
- The reason why the resistance increases is because the resistor is heating up. Some of the electrical energy has been transformed into heat energy. *When the temperature of a resistor rises, free electrons find it more difficult to pass through, so its resistance increases.*
- For the resistance to remain constant, the resistor must be kept at a constant temperature.
- If the temperature of the resistor can be kept constant, then **Ohm's law** applies.

Ohm's law
The voltage loss across a resistor is directly proportional to the current passing through it, provided the temperature of the resistor stays constant.

- Directly proportional means that the voltage loss will double when the current doubles.

Power and Energy

light emitting diodes

- **Electrical energy** is required for components (eg light-emitting diodes), appliances (eg toaster) and electronic devices (eg iPod) to work.
- A battery transforms chemical energy into electrical energy. An electrical plug transfers mains electrical energy.
- A torch, toaster, iPod and shaver transform electrical energy into light, heat, sound and kinetic energy, respectively.
- **Energy**, symbol ***E***, is measured in **joules** (J).
- Energy is transferred slowly or rapidly, depending on the **power rating** or **wattage** of the energy supplier or user. If a lot of energy is transferred, then the power rating is high. If little energy is transferred, then the power rating is low.
- **Power**, symbol ***P***, is defined as *the rate at which energy is transferred.* Power is measured in **watts** (W). One watt means one joule of energy is being transferred or transformed per second. For objects with larger power ratings, kilowatts (kW) are used instead. 1 kW = 1000 W.
- As the voltage increases, more energy will be transferred. As the current increases, more energy will be transferred as more electrons are flowing. *Power is therefore proportional to both voltage and current.*
- Power is found by multiplying the voltage by the current:

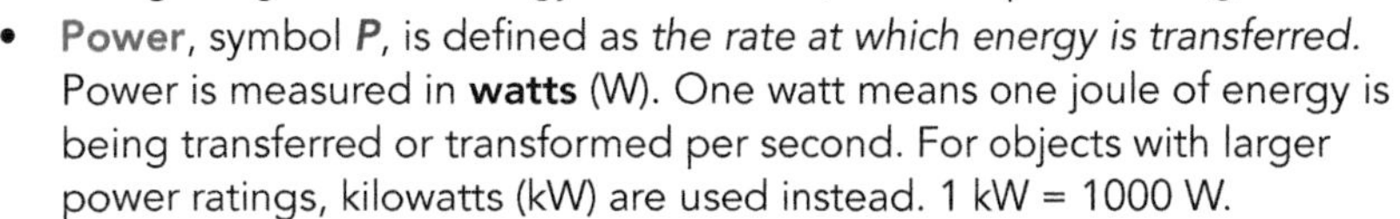

Power = voltage x current $P = VI$

(voltage in volts, current in amps, power in watts).
- The triangle above gives the formulas for finding current or voltage if the other two quantities are known: $V = P/I$ or $I = P/V$.
- The **total energy** transferred or transformed is found by multiplying the power by the time period involved:

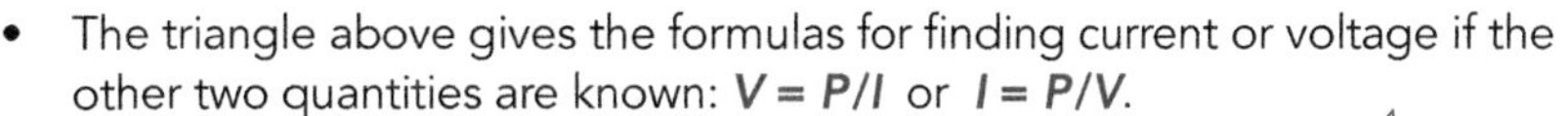

Energy = power x time $E = Pt$

(energy in joules, power in watts, time in seconds).
- The triangle above gives the formulas for finding power or time if the other two quantities are known: $P = E/t$ or $t = E/P$.
- The above relationship is usually given as $P = E/t$ in NCEA exam papers.

SCIENCE SKILL: Finding Resistance and Power

To find the resistance and power rating of a component such as a lamp, the circuit below is used.
The lamp is connected to a suitable power supply. An ammeter is connected in series with the lamp, and a voltmeter is connected in parallel around the lamp.
Notice that the wires from the red terminals of the meters lead to the red terminal of the power supply.

Steps:

1. Take the current reading off the ammeter scale:
   $I = 0.5$ A
2. Record the voltage loss off the voltmeter scale:
   $V = 5.5$ V
3. To calculate the resistance of the lamp, use the formula: $R = V/I$
4. Enter the values into the formula, then complete the calculation mentally or use your calculator:
   $R = 5.5 \div 0.5 = 11$
5. Record the resistance using the appropriate unit:
   $R = 11\ \Omega$
6. To calculate the power of the lamp, identify the correct formula form:
   $P = VI$
7. Enter the values into the formula then complete the calculation using your calculator:
   $P = 5.5 \times 0.5$
   $= 2.75$
8. Round off your answer to the nearest 1 decimal place and record the power used by the lamp using the appropriate unit:
   $P = 2.8$ W

Revision Activities

1 Match up the descriptions with the terms.

a conductor	A unit of current
b electrons	B opposition to the flow of electrons
c current	C unit of energy
d resistance	D substance that allows electrons to flow freely through it
e resistor	E unit of power
f variable resistor	F required for things to work
g rheostat	G unit of resistance
h voltage loss	H rate at which electrical energy is supplied or used
i ampere	I large variable resistor with a slider
j volt	J another term for power rating
k ohm	K electrical energy lost by electrons in the current
l total resistance	L negatively charged particles that flow in a current
m Ohm's law	M energy used by a component over a period of time
n electrical energy	N flow of charged particles such as electrons
o energy	O voltage is proportional to current if the temperature is constant
p joule	P the wattage of a component
q power rating	Q unit of voltage gain or loss
r wattage	R component that is designed to limit current flow
s power	S energy carried by moving charges
t watt	T resistor whose resistance can be altered
u total energy	U overall resistance of several connected resistors

5 Decide whether the following statements are true or false. Rewrite the false ones to make them correct.

- a Electrons can travel through a resistor without losing energy.
- b If the resistance is increased, then more current will flow.
- c A resistor converts electrical energy into heat energy.
- d An ohm is a voltage loss of one volt for each ampere of current.
- e If the current through a component is increased, then the voltage loss across it will also increase.
- f The slope of a voltage-current graph represents the resistance of a component.
- g If the slope of a voltage-current graph is a straight line, then the resistance of the component is constant.
- h The power rating of a component in a circuit is how fast it supplies or uses electrical energy.
- i Power is directly proportional to current and voltage.
- j An object with a high power rating will convert lots of energy each second.

2 Explain the difference between:

- a an insulator and a resistor
- b a fixed resistor and a rheostat
- c resistance and power
- d total energy used and power.

3 Copy and complete the chart below, which summarises different electrical quantities.

Quantity	Symbol	Unit Name	Unit Symbol
Current	*I*	ampere	A
Voltage			
Resistance			
Power			
Energy			

4 Use the formula triangles below to identify the formula to be used in the following situations.

- a resistance given current and voltage loss
- b power given voltage loss and current
- c current given resistance and voltage loss
- d voltage loss given power and current
- e total energy transferred given power and time period.

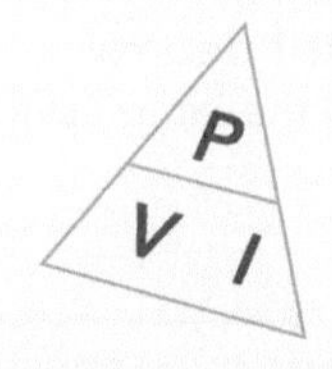

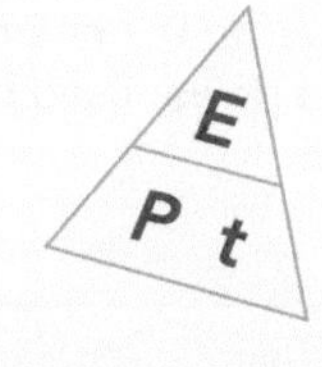

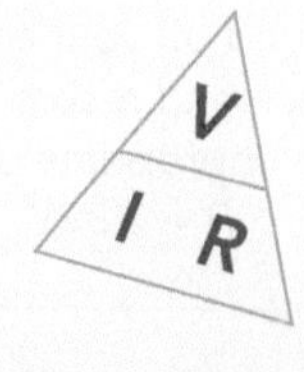

6 The circuit shown opposite was set up to investigate what happens to the current in a circuit when the voltage supply is increased. The voltage was increased by using the voltage selector knob of the power pack. As power packs do not necessarily supply the voltage indicated on the selector scale, a voltmeter was connected across the power pack terminals. The resulting voltages and currents are in the table below.

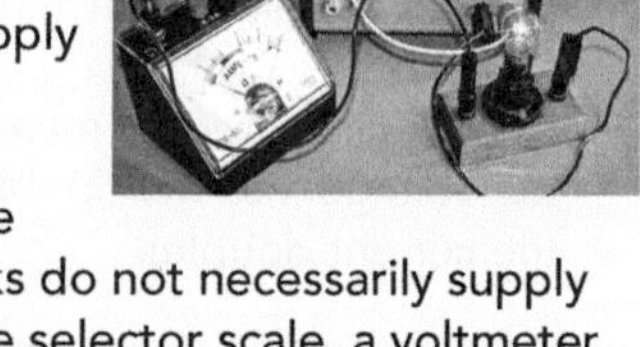

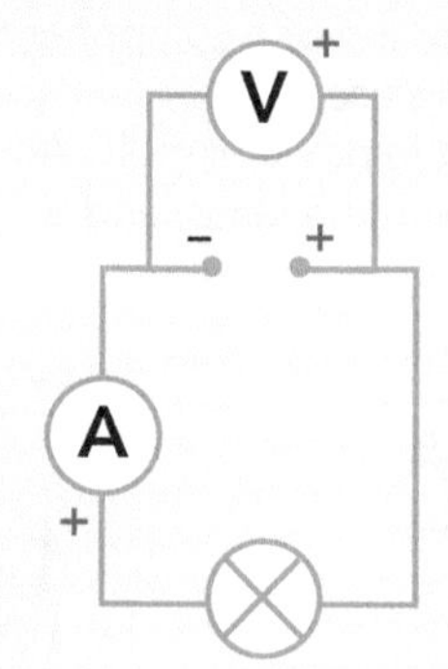

Voltage (V)	Current (A)
0	0
1.8	0.4
3.6	0.8
5.4	1.2
7.2	1.6
9.0	1.9
10.8	2.1

- a From the data, describe what happens to the current as the voltage is increased.
- b Is the current proportional to the voltage gain?
- c Draw a line graph of the data.
- d Describe the shape of the graph.
- e Suggest a reason why the current does not increase as much at higher voltage settings.
- f How has the lamp affected the current in the circuit?

ISBN: 9780170262316

7 In the circuit opposite, the voltage loss across the bulb was 10.2 V and the current in the circuit was 2.5 A.
The circuit was on for 90 seconds.

a What formula would you use to find the bulb's resistance?
b Calculate the resistance of the bulb to the nearest 1 decimal place. Give the unit as well.
c What formula would you use to find the electrical power supplied to the bulb?
d Calculate the electrical power used by the bulb to the nearest 1 decimal place. Give the unit as well.
e What formula would you use to find the total energy used by the bulb?
f Calculate the total energy used by the bulb to the nearest 1 decimal place. Give the unit as well.
g If only 10% of the electrical energy is converted into light, how many joules of light energy will have been emitted?
h How many joules of heat energy will the bulb have produced?

8 Copy and complete the table below by calculating the power used by each appliance given the voltage supply and current.

Appliance	Voltage (V)	Current (A)	Power (W)
Toaster	240	4.0	
Television	240	0.2	
Shaver	6	0.5	
Radio	3	0.5	
Heater	240	10.0	
Fan	240	2.0	
Lamp	240	0.3	
Torch	12	2.4	

9 Read the article opposite, then answer the questions below.

a What is meant by 'mains electricity'?
b What unit is the electrical energy usage of a house measured in?
c How many joules of energy are equal to one kilowatt-hour?
d If a 2 kilowatt heater is left on overnight for 10 hours, how many kilowatt-hours of electricity will it use?
e How many joules of energy will the heater use in that time?
f What will be the cost of the electricity used overnight by the heater?
g What is the advantage and disadvantage of having a hot water cylinder operating on a night-time rate?

10 Copy and complete the chart by finding the total resistance of each appliance given the voltage supply and current.

Appliance	Voltage (V)	Current (A)	Resistance (Ω)
Toaster	240	4.0	
Television	240	0.2	
Shaver	6	0.5	
Radio	3	0.5	
Heater	240	10.0	
Fan	240	2.0	
Lamp	240	3.0	
Torch	12	2.4	

11 Two identical lamps are connected to two cells. In circuit A the lamps are connected in series. In circuit B the same lamps are connected in parallel. The two cells, which are connected in series, provide a total voltage gain of 3 V.

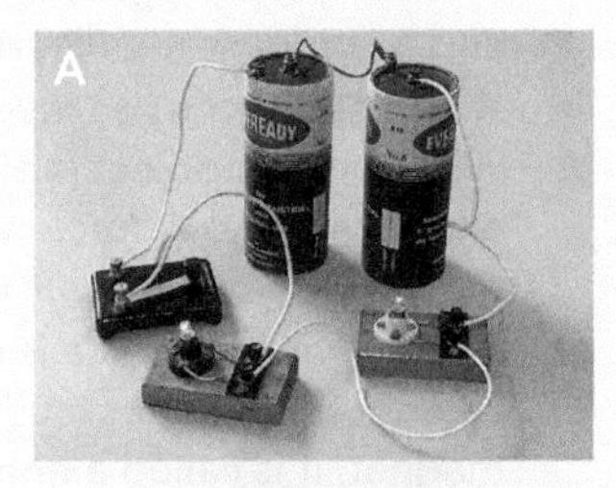

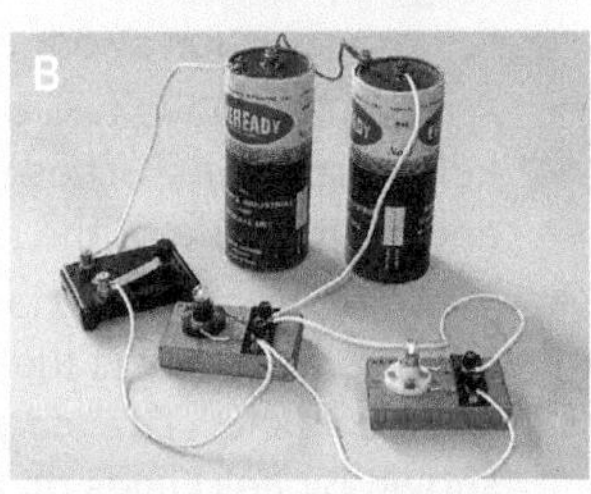

a What would be the voltage loss across each lamp in circuit **A**?
b If the current flowing in circuit **A** was 0.5 A, what would be the resistance of each lamp?
c What would be the total resistance of the two lamps in **A**?
d If in circuit **B** the voltage loss across each lamp was 3 volts, what would be the current flowing through each?
e What would be the total current flowing in circuit **B**? (Remember that the currents in parallel circuits are added to give the total current.)

Power Costs

Appliances that are plugged into the mains use electricity which must be paid for. On the power board of your house there is a meter that measures the amount of electrical energy used.

The meter measures in units called kilowatt-hours. One kilowatt-hour is the amount of energy used by a one kilowatt appliance in one hour.

If a heater was switched on for one hour, then the total energy it would use is given by the formula:

Total energy = power x running time

Power must be in watts and running time in seconds.

1 kilowatt = 1000 watts and
1 hr = 60 min and 1 min = 60 s
so 1 hr = 60 x 60 = 3600 s.

The total energy used by the heater during one hour is given by:

Total energy = 1000 x 3600
= 3600000 J

Therefore a one kilowatt heater uses 3600000 joules every hour and a two kilowatt heater uses 7200000 joules of energy per hour.

The meter on your power board records the running total of kilowatt-hours used by all circuits in the house.

Every month or so a reading is taken from the meter and a calculation is made of the kilowatt-hours used. Your family is then billed for the kilowatt-hours used. The current typical cost per kilowatt-hour is about 14¢. The charging rate can vary; a night store heater only uses power at night and the cost of electricity may be cheaper at that time.

ISBN: 9780170262316

Magnetism and Electromagnetism

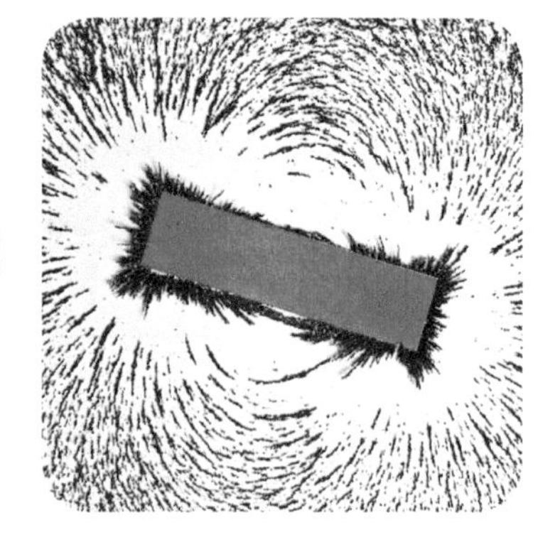

Learning Outcomes - On completing this unit you should be able to:
- interpret the force field diagrams of bar magnets and current-carrying wires
- state the law of magnetism and apply it to the interactions between fields
- calculate the strength of a magnetic field using the formula $B = kI/d$
- explain electromagnetism and how it can be controlled
- *find geographic north using a compass and magnetic declination.*

Magnetic Force Fields

- Some objects will attract things made of iron. We call these objects **magnets**. Examples are a bar magnet and a compass needle, which is free to swing.
- A magnet creates an invisible **magnetic force field**. When an iron object is placed in the field it will experience a pull force. When another magnet is placed in the field it will experience a push or a pull.
- As the magnetic force is experienced over a distance without the magnet touching the iron object or other magnet, it is called a **non-contact force**. Magnetic forces are much stronger than the force of gravity.
- All magnets have two **poles** where the magnetic force field is strongest. These are known as the **north pole** and **south pole** of the magnet.
- The direction of the magnetic field at any point around a magnet is the direction in which a compass needle points. The magnetic field of a magnet is shown by field lines on a diagram.

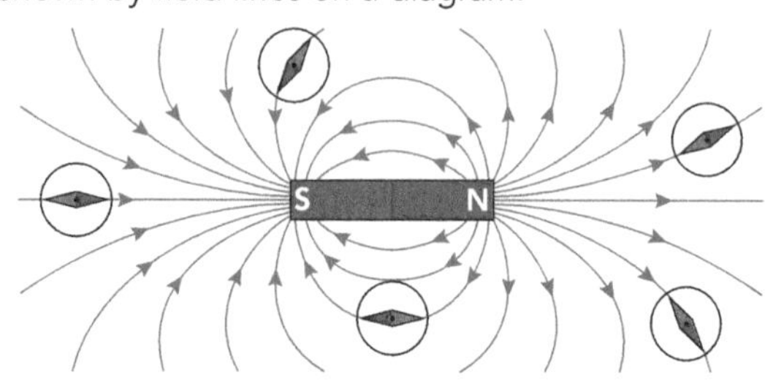

- The arrows on the lines show the **field direction**. These run from the north to the south pole of the magnet.
- The **field strength** is shown by the spacing between the lines. Closer lines indicate a stronger field.

Law of Magnetism

- The diagrams show what happens when one pole of a magnet is moved close to a pole of a freely suspended magnet. In the left diagram, unlike poles are attracted but in the other two diagrams, like poles are repelled.

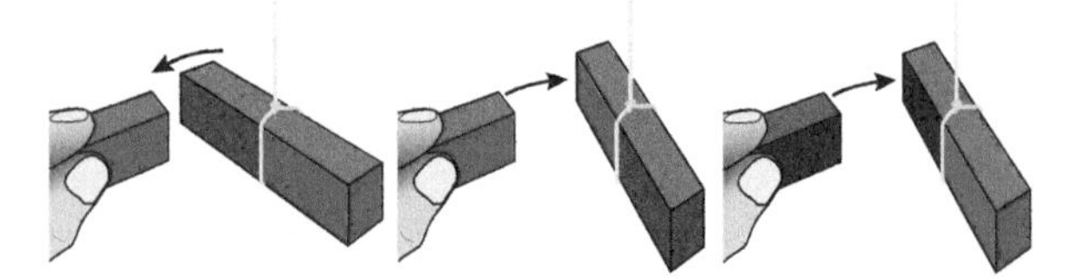

- The **law of magnetism** states that *unlike poles attract but like poles repel each other.*

Interacting Magnetic Fields

- When two magnets are brought close, their magnetic fields interact and a force is exerted on each magnet.

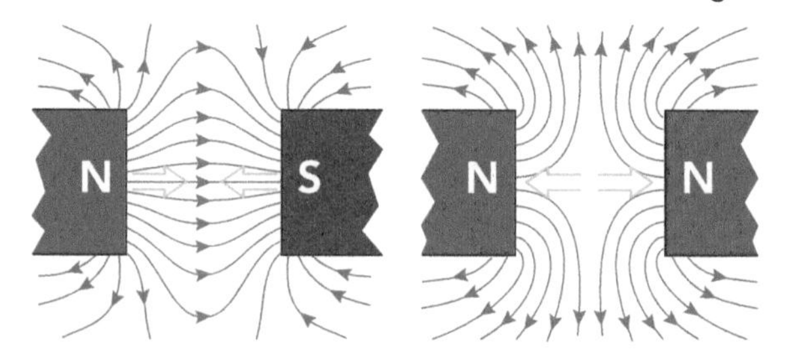

- When different poles are brought together, the fields between the two poles converge causing attraction.
- When similar poles are brought together, the fields between the two poles diverge causing repulsion.

Geomagnetic Field

- The Earth has an enormous magnetic field that is sometimes called the **geomagnetic field**.
- Geomagnetism is caused by huge electrical currents in the molten iron of the **outer core** (see page 150).
- The geomagnetic field can be considered to act like a giant bar magnet. The field is strongest near the poles.

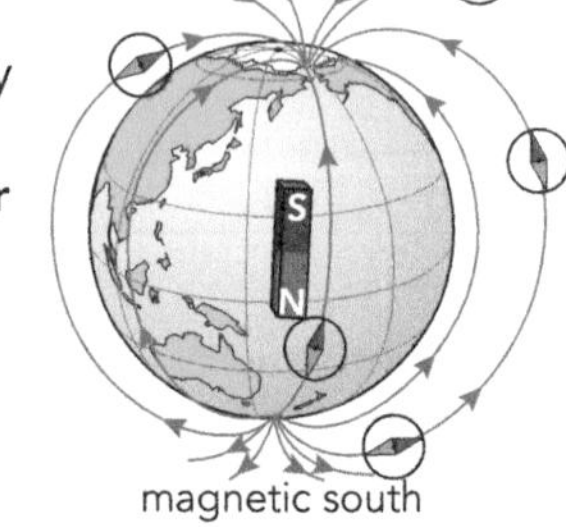

- As the north pole of a compass points northwards, this means that Earth's 'north' magnetic pole is actually a south magnetic pole as shown above.
- Earth's magnetic poles are not directly aligned with the **geographic poles**, which are found along the axis on which Earth spins.
- The geomagnetic field extends well out into space and stops the solar wind (stream of high energy charged particles emitted by the sun) from stripping away the ozone layer, which blocks harmful ultraviolet radiation.
- The geomagnetic field is also responsible for the amazing coloured lights that can seen in the far south of New Zealand, called the **aurora australis** or southern lights.

ISBN: 9780170262316

Electromagnetism

- Electricity and magnetism are closely connected. A current in a wire will create a magnetic field around it, and a moving magnet (eg in a bicycle dynamo) will create an electrical current in a nearby wire.
- If a length of copper wire is connected to a power supply, then a compass needle near the wire will change direction when the power is switched on because of the wire's magnetic field.

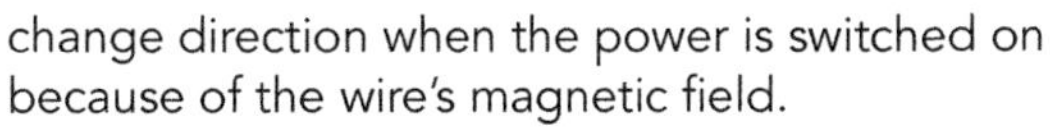

- As charged particles (electrons) move in the current, they create a magnetic force field around the wire. This is called the **electromagnetic effect**.
- The field has a cylindrical shape around the wire and the field lines are drawn as increasingly spaced apart concentric circles.

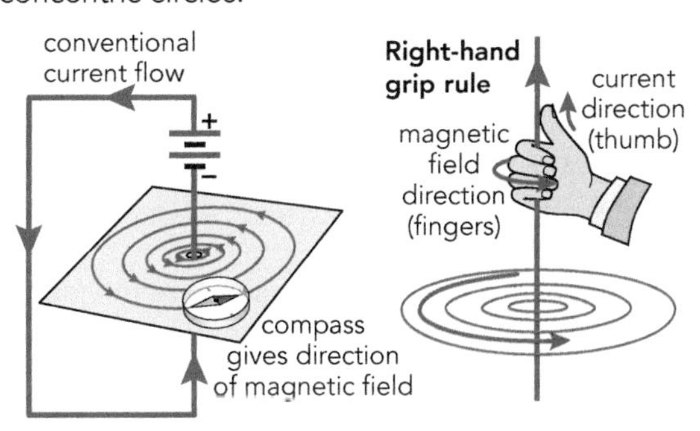

- The field direction is given by the **right-hand grip rule**. If your thumb points in the direction in which **conventional current** would flow in the wire (+ve to –ve), then your curled up fingers show the direction of the magnetic field. (Reversing the terminals will reverse the field direction.)
- **Magnetic field strength *B*** is measured in **teslas** (T).
- The strength of the magnetic field around a current-carrying wire increases as the current increases, and decreases with distance away from the wire.
- The magnetic field strength ***B*** at a distance ***d*** away from a wire with current ***I*** is given by:

Magnetic field strength = k x current ÷ distance

$$B = kI/d \qquad (k = 2 \times 10^{-7})$$

(B in teslas, current in amps, distance in metres).

Interacting Currents

- If two parallel, current-carrying wires are near each other, then their magnetic fields will interact and a force will be applied to each wire.

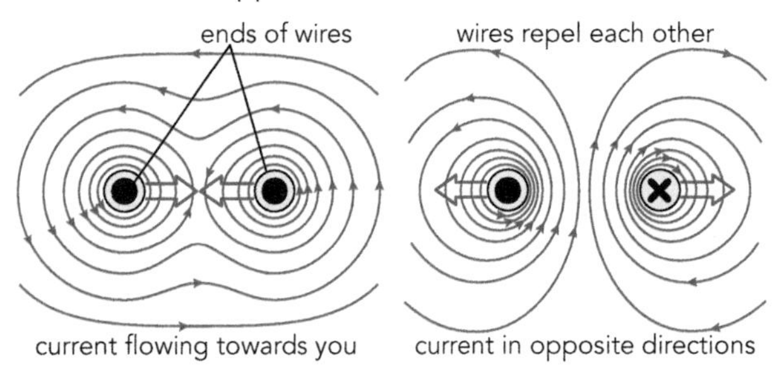

- If the currents are going in the same direction, then the fields around the two wires converge causing attraction.
- If the current are going in opposite directions, then the fields around the two wires diverge causing repulsion.

Solenoids and Electromagnets

- The strength of the magnetic field round a current-carrying wire can be increased by coiling the wire into a **solenoid**. The fields around the individual loops combine to create a much stronger magnetic field.
- The field direction through a solenoid can be found by compasses or by reversing the right-hand grip rule.

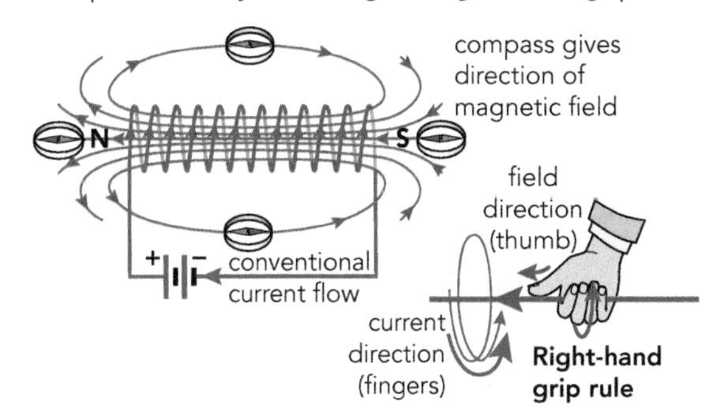

- The field strength can also be increased by inserting a soft iron bar inside the solenoid, which then acts as an **electromagnet**, whose magnetism can be turned on and off. The soft iron bar is temporarily magnetised by the current and its magnetic field combines with the coil's.

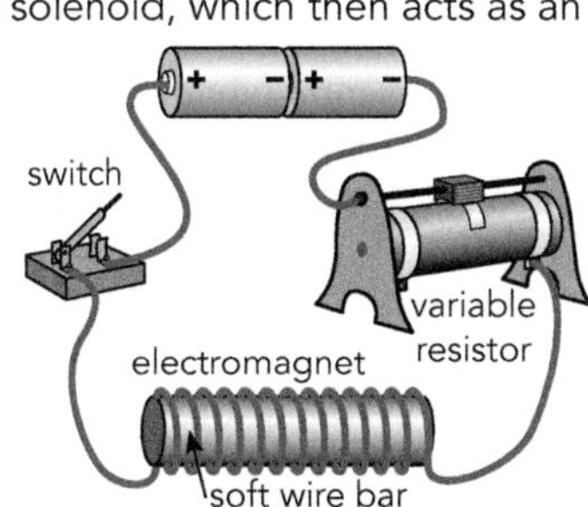

SCIENCE SKILL: Finding Geographic North

As Earth's magnetic poles don't align with its geographic poles, you need to make an adjustment to the direction in which a compass needle points to find geographic north. The angle between magnetic north and geographic north is called **magnetic declination**.

Steps:

1. Go to the website http://magnetic-declination.com and move the map around using the grabber hand until your location is visible, then click on it. The small pop-up window will give you your declination.
2. As the declination in NZ will be 'positive', geographic north is that many degrees west of magnetic north (anticlockwise direction).
3. Take a compass outside and move away from cars and buildings. Let the needle settle, then add the declination to find where geographic north is.

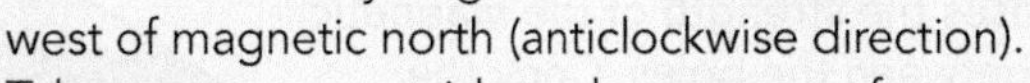

Revision Activities

1 Match up the descriptions with terms.

a magnet	A unlike poles attract but like poles repel
b magnetic force field	B amazing lights seen at night at higher latitudes
c non-contact force	C pole that a compass needle points away from
d poles	D direction in which positive charges would travel if they could
e north pole	E two locations on a magnet where the magnetic force field is strongest
f south pole	F object that attracts things made of iron
g field direction	G molten iron layer found inside Earth
h field strength	H solenoid with a soft iron bar inserted
i law of magnetism	I gives the current and magnetic field directions
j geomagnetic field	J area in which a magnetic force is applied to an iron object or another magnet
k outer core	K a long coil of wire carrying an electrical current
l geographic poles	L moving charged particles create a magnetic field
m aurora borealis	M force that acts over a distance without touching the object
n electromagnetic effect	N unit of magnetic field strength
o right-hand grip rule	O pole that a compass needle points toward
p conventional current direction	P Earth's magnetic field
q tesla	Q direction in which a compass needle points
r solenoid	R found along the axis on which Earth spins
s electromagnet	S how strong the magnetic force will be at a point

2 A bar magnet is hung by a string so that it can rotate freely.

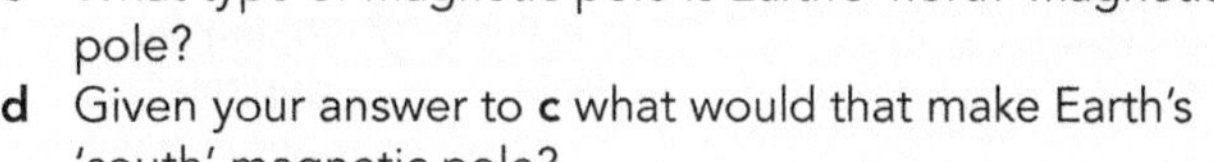

a Which compass direction will the north pole of the magnet point toward?

b Why will it point in that direction?

c What type of magnetic pole is Earth's 'north' magnetic pole?

d Given your answer to **c** what would that make Earth's 'south' magnetic pole?

3 A bar magnet is hung so that it can move freely. Describe what happens when a pole of another bar magnet is brought close by in each of the following situations.

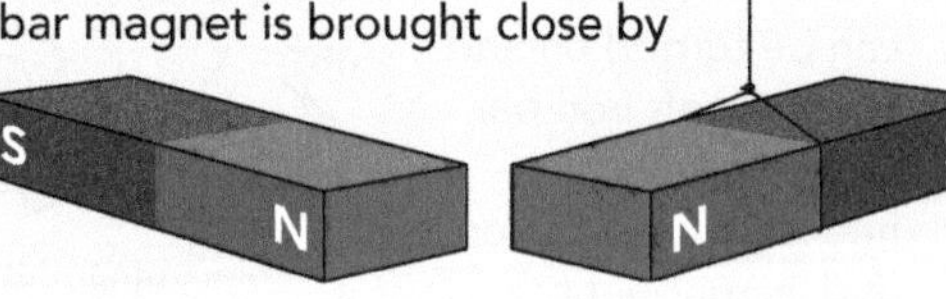

a a north pole is brought close to the north pole of the hanging magnet

b a north pole is brought close to the south pole of the hanging magnet

c a south pole is brought close to the north pole of the hanging magnet

d a south pole is brought close to the south pole of the hanging magnet.

4 Some magnets are U-shaped, which increases the strength of the magnetic field between the poles. Copy the diagram opposite and draw field lines between the poles to show the direction and intensity of the field.

5 Decide whether the following statements are true or false. Rewrite the false ones to make them correct.

a Magnets only attract other magnets.

b A magnetic force field can act over a distance without touching.

c Magnets always have two poles.

d The direction of a magnetic field is shown by a compass needle.

e A strong magnetic field will have well spaced out field lines.

f Like poles attract and unlike poles repel.

g The magnetic fields of two facing north poles will always converge.

h The geomagnetic field of Earth is created by electricity.

i Earth's 'south' magnetic pole is actually a north magnetic pole.

j A moving magnet will create an electrical field.

k In the right-hand grip rule, your thumb gives the field direction.

l Parallel wires with opposing currents will repel each other.

m A soft iron bar increases the strength of an electromagnet.

6 The needle of a compass will align with the magnetic field in which it is placed and point toward the south pole of the field.
Copy the diagram below and draw an arrow in each of the compasses to show the direction of the magnetic field surrounding the bar magnet.

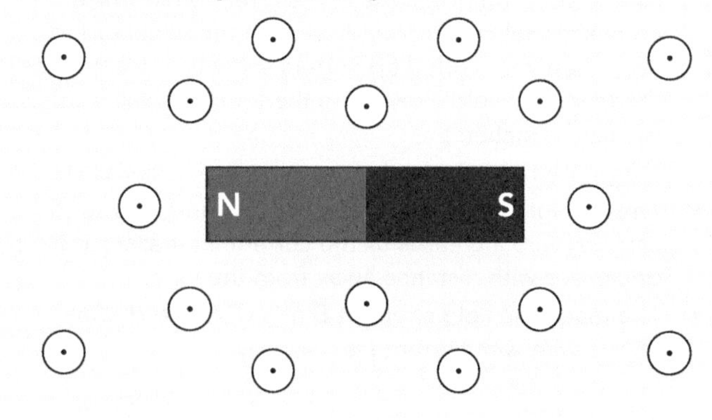

7 The diagram below shows the magnetic field lines around part of a current-carrying wire.

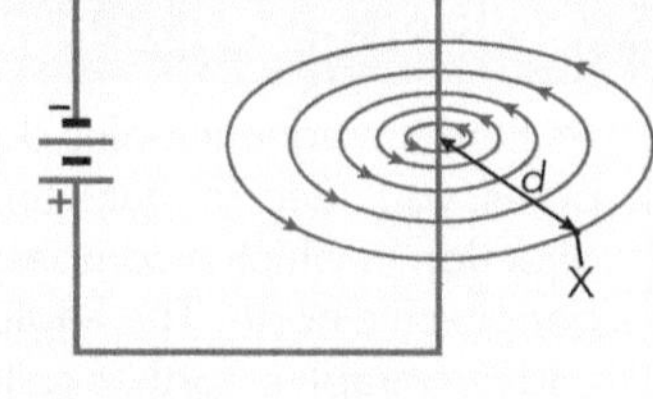

a Which direction would conventional current flow around this circuit?

b Write down the formula you would use to calculate the magnetic field strength around the wire.

c If the current I is 8 A, calculate the magnetic field strength B at point X, which is a distance d 12 cm away from the wire. (Give the unit.)

d How could you increase the field strength at point X?

ISBN: 9780170262316

8 The diagram shows a solenoid with a soft iron core connected in a circuit with a power supply, switch and variable resistor. The solenoid acts as an electromagnet whose strength can be controlled.

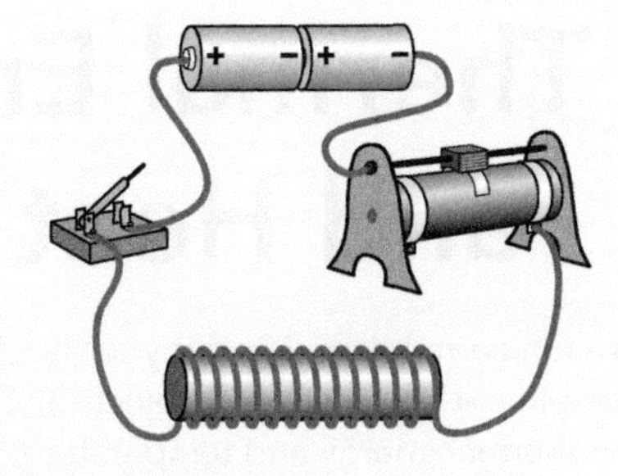

a Draw a circuit diagram of the set up.

solenoid

b How are the components connected?

c Why is the wire coiled into many loops in the solenoid?

d Why does the solenoid have a soft iron core?

e How else could the strength of the magnetic field around the electromagnet be increased?

f How would the action of the variable resistor affect the strength of the electromagnet?

g The electromagnet is used to pick up some iron nails. What would happen to the nails when the switch is turned off?

9 The diagram opposite shows the magnetic field lines around a solenoid.

1 2 3 4

a Which end of the solenoid coil will be the north pole?

b Which end will be the south pole?

c Which direction would the needle be pointing in each of the four compasses (1 – 4)?

d How does the strength of the magnetic field inside the coil compare with that outside?

e How could you reverse the direction of the magnetic field?

The red arrows on the diagram show the direction that positive charges would travel if they could but in reality it is electrons that travel in the current.

f Describe the direction in which the electrons would be flowing.

11 The diagrams below represent two parallel current-carrying wires and the magnetic fields surrounding them.

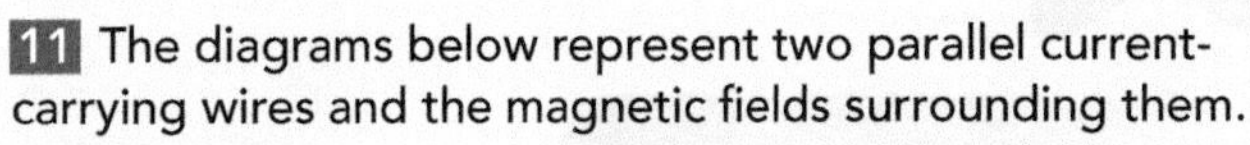

In diagram **A** both currents are going in the same direction.

a Will the two wires attract or repel each other?

b Why does this happen?

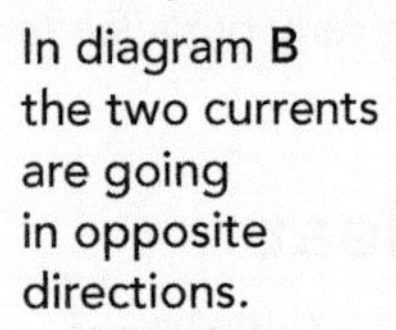

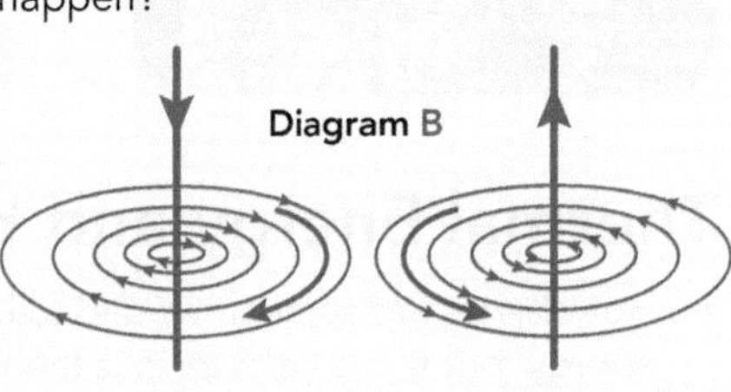

In diagram **B** the two currents are going in opposite directions.

c Will the two wires attract or repel each other?

d Why does this happen?

12 Planet Earth has a huge magnetic field called the geomagnetic field. It acts like a giant bar magnet located in the inner core.

The field lines on the diagram opposite show the direction in which a compass needle would point at various locations about Earth.

a Which direction do compass needles usually point toward?

b What does this make the 'north' pole of Earth?

c Which direction do compass needles usually point away from?

d What does this make the 'south' pole of Earth?

e What is the cause of Earth's geomagnetic field?

f How does the geomagnetic field protect life on planet Earth?

g Why are the geographic north pole and the magnetic 'north' pole found in different locations?

10 Read the article opposite then answer the questions below.

a Why would the globe be called an 'antigravity' toy?

b What is the source of the energy that keeps the globe floating and rotating?

c What energy transformations are occurring in the toy?

d If the bottom of the disc was acting as a south pole, what polarity would the top of the base need to be for repulsion to occur?

e Why would the globe sometimes rise or fall slightly when you release it above the base?

f What two forces must the electromagnets be applying to the disc for the toy to work?

A Floating and Spinning Globe

A recently developed scientific toy involves a globe that levitates (floats) and rotates (spins) in the air.

The toy consists of a globe and metal base plate that is plugged in to the mains.

Inside the globe there is a flat metal disc, consisting of several magnets with different polarities. It sits inside the southern hemisphere. Inside the base plate there are several electromagnets that have different polarities, along with some sophisticated electronics.

The globe needs to be positioned carefully above the plate to ensure that the repulsion caused by the interaction of the magnetic fields of the disc and the electromagnets balance the force of gravity pulling the globe downward.

The globe is made to rotate slowly by switching the direction of the currents through the various electromagnets, which changes the complex magnetic field in the space above the base. This keeps pushing the globe around.

When it is switched off, the globe will drop as the coils in the base no longer act as an electromagnet but it will roll because it has rotational inertia.

Thermal Energy and Heat Flow

Learning Outcomes - On completing this unit you should be able to:
- explain the direction of heat flow between objects at different temperatures
- distinguish between thermal energy and temperature
- describe three different effects of heat flow on substances
- explain why different substances heat up at different rates
- *carry out a fair test to compare the heating or cooling of substances.*

Thermal Energy and Heat

- Hot objects are sometimes said to have lots of 'heat energy' but the correct term is thermal energy.
- Objects and substances are made of extremely small particles that are constantly moving in some way or other. **Thermal energy** is defined as the total kinetic energy of all particles in an object or substance.
- **Heat flow** involves the transfer of thermal energy from hotter to colder regions. *Heat flows spontaneously from hotter to colder regions.* This is sometimes called the **law of heat flow**.
- **Heat**, symbol **Q**, is the amount of thermal energy transferred from a hotter to a colder region. The unit is the **joule** (J) or kilojoule (kJ). 1000 J = 1 kJ.

Kinetic Theory of Matter

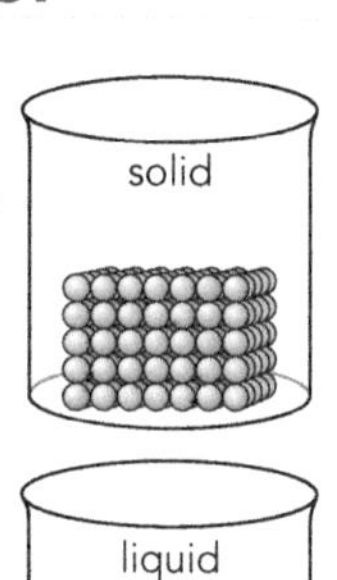

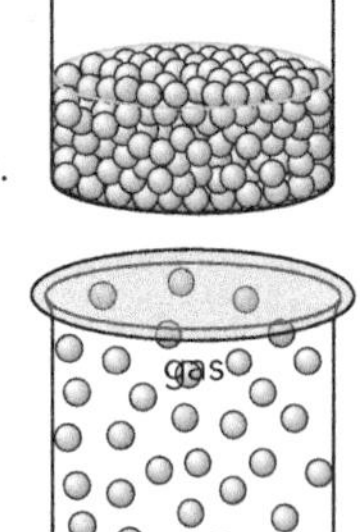

- The kinetic theory of matter explains how matter behaves in terms of the motion of atoms or molecules and the forces of attraction between them.
- Matter is made of particles, either **atoms** or **molecules**. As these particles are in constant motion they have **kinetic energy**. *The particles can vibrate, rotate or move freely.*
- In solids, the attractive forces hold particles close together in a rigid arrangement so they can only vibrate. In liquids, particles are free to move around as well as vibrate and rotate but the attractive forces still hold them close. In gases, particles are far apart and fly about rapidly as well as vibrating and rotating.

Temperature

- We can distinguish between hotter and colder objects. Substances are hotter because their particles are more energetic, ie they have more kinetic energy. Substances are colder because their particles are less energetic.
- **Temperature**, symbol *T*, is *proportional to the average kinetic energy of the substance's particles.* This means that temperature increases as kinetic energy does. The unit used at NCEA level 1 is the **degree Celsius** (°C).
- Notice that temperature of a substance does not tell you how much thermal energy it has.

Effects of Heat Flow

condensation

- Heat flow to or from a substance causes *a temperature, volume or phase change.*

Temperature Change

- When a substance gains thermal energy, the particles become more energetic. The particles have more kinetic energy and the temperature rises to reflect this.
- The particles may vibrate, rotate or speed up, depending on what phase the substance is in.
- When a substance loses thermal energy, the particles become less energetic. The particles have less kinetic energy and the temperature falls to reflect this.

Volume Change

- When a substance gains thermal energy and particle motion increases, the more energetic particles push each other further apart and the substance expands. This is how a thermometer works.
- When a substance loses thermal energy and particle motion decreases, the less energetic particles come closer together and the substance contracts.
- Heat flow causes expansion or contraction in solids, liquids and gases, depending on whether the substance is hotter or colder than its surroundings.
- In general, gases expand and contract more than liquids, which expand and contract more than solids.

Phase (State) Change

- At certain temperatures heat flow will cause a **phase change** rather than a temperature and volume change. Melting, freezing, boiling or condensing (see page 38) may occur depending on the direction of heat flow.
- When a substance is melting (or boiling), *the temperature does not rise* despite the ongoing heat flow to the substance. The absorbed thermal energy is used to overcome the forces that hold the particles together. The kinetic energy of the particles does not change.
- When a substance is freezing (or condensing), *the temperature does not fall* despite the ongoing heat flow to the surroundings. As particles bind together, the heat released flows to the cooler surroundings. The kinetic energy of the particles does not change.
- *Opposite changes occur at the same temperature.* For example, both melting and freezing of a substance occur at the same temperature but during melting heat flows from the surroundings to the substance and during freezing heat flows to the surroundings.

ISBN: 9780170262316

Heating Curves

- If solid ice is placed in a conical flask and the temperature taken regularly over a period of time while heating continues, a **heating curve** such as the one below shows what happens to the temperature of the ice and then the water over time.
- The ice temperature rises until it reaches 0 °C, then it remains constant for several minutes while melting occurs.
- Once the ice has melted, the water temperature rises steadily up to 100 °C. The temperature then remains constant until all of the water has boiled away as the energy is being used to overcome the attraction between particles.

Heating Curve

Temperature of ice/water (°C): -20, 0, 20, 40, 60, 80, 100

Time heated (min): 0 1 2 3 4 5 6 7 8 9 10 11 12 13 14 15

ice · melting · water · boiling

Heat Capacity

- The temperature of a substance rises when heat flows to it from hotter surroundings, and falls when heat flows from it to colder surroundings.
- Different substances need different amounts of thermal energy to change their temperatures by the same amount. For example, 4181 J of thermal energy are needed to warm 1 kg of water by 1 °C but only 466 J will heat 1 kg of steel by the same amount.
- **Heat capacity** is the amount of heat flow required to change a substance's temperature by a set amount.
- The **specific heat capacity** c of a substance is the amount of heat flow needed to change the temperature of a standard mass of a substance by 1 °C. The units used here are J/kg/°C or kJ/kg/°C but other units may be used in NCEA exams.
- *Specific heat capacity is the same whether heat is flowing to or away from a substance.* As much heat is released when cooling as is absorbed when heating.
- How much heat flow is involved depends on the substance's heat capacity and mass, as well as the required temperature change. *The greater the mass, heat capacity or temperature change, the greater the heat flow.*
- The flow of **heat Q** required for **mass m** of a substance with a **specific heat capacity** of c to have a **temperature** change of ΔT is given by:

  Heat = mass x specific heat capacity x temperature change

  $$Q = mc\Delta T$$

  (heat in J, mass in kg, heat capacity in J/kg/°C, temperature change in °C; or other consistent units).
- The formula can be rearranged to find either the mass, specific heat capacity or temperature change if the other three quantities are known as shown below:
  $m = Q/c\Delta T$, $c = Q/m\Delta T$ or $\Delta T = Q/mc$.

Latent Heat

- Heat flow to or from a substance will not necessarily cause its temperature to rise or fall. Sometimes a substance undergoes a phase change instead. *These occur at set temperatures*, such as the melting point (change from solid to liquid) and boiling point (change from a liquid to a gas).
- The amount of thermal energy that is transferred in a phase change is a physical property of the substance.
- The amount of heat absorbed when a substance melts is the same as that released when it freezes. Similarly, the amount of heat absorbed when a substance boils is the same as that released when it condenses.
- The heat involved when 1 kg of a substance either melts or freezes is called its **latent heat of fusion**, symbol L_f. The heat involved when 1 kg of a substance either boils or condenses is called its **latent heat of vaporisation**, symbol L_v. The units used here are J/kg or kJ/kg but other units may be used in NCEA exams.
- *The latent heat of vaporisation is much more than the latent heat of fusion.* For example, it takes 2260000 J to boil 1 kg of water but only 334000 J to melt 1 kg of ice.
- The flow of **heat Q** involved when **mass m** of a substance with a **latent heat** L_f or L_v undergoes a particular phase change is given by:

  Heat = mass x latent heat

  $$Q = mL_f \text{ or } Q = mL_v$$

  (heat in J, mass in kg, latent heat in J/kg; or other consistent units).
- The formula can be rearranged to find either the mass or latent heat if the other two quantities are known:
  $m = Q/L$ or $L = Q/m$.

SCIENCE SKILL: Fair Comparisons

Suppose you have been asked to compare the heat capacities of cooking oil and water. It will be critical to ensure that all conditions are identical and the only thing that varies is the different liquids. *To ensure that an investigation is a fair comparison, all other variables must be controlled.*

Questions to consider:

1. Should you be heating or cooling the two liquids or doesn't it matter?
2. If you are going to heat them, how will you ensure that the heat flow is the same to both? If cooling them, how will you ensure that the heat flow from both is the same?
3. What variables need to be controlled? For example, mass of liquids, size and shape of containers, time heated or cooled, surroundings, etc.
4. How can these variables be made identical? Specify the conditions.
5. How will you measure the temperature change resulting from the heat flows? What instrument will you use and how frequently will you read it?
6. What else needs to be fair?

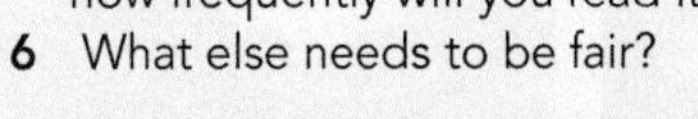

Revision Activities

1 Match up the descriptions with the terms.

a thermal energy	A energy associated with moving objects
b heat flow	B unit of temperature
c law of heat flow	C graph showing the temperature of a substance as it is being heated
d heat	D total kinetic energy of the particles of a substance
e joule	E heat flow needed to melt/freeze 1 kg of substance
f atom	F transfer of thermal energy
g molecule	G proportional to the average kinetic energy of the particles of a substance
h kinetic energy	H heat flow needed to boil/condense 1 kg of substance
i temperature	I heat flows spontaneously from hotter to colder regions
j degree Celsius	J heat required for 1 kg of a substance to have a 1°C change in temperature
k phase change	K amount of thermal energy transferred
l heating curve	L change of state
m heat capacity	M unit of energy
n specific heat capacity	N amount of matter in a substance
o mass	O group of atoms bonded together
p latent heat of fusion	P basic building block of all substances
q latent heat of vaporisation	Q amount of heat flow required to change a substance's temperature by a set amount

2 Explain the difference between the following terms:

a atoms and molecules
b thermal energy and heat
c temperature and thermal energy
d melting point and freezing point
e boiling point and condensing point
f heat capacity and latent heat.

3 Describe the state changes occurring in the lava opposite and relate it to heat flow.

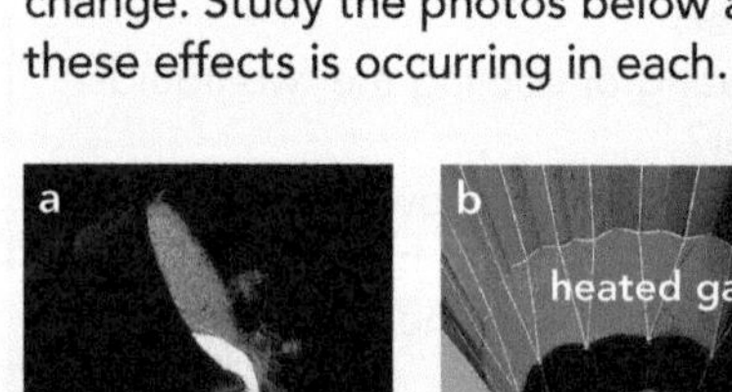

4 Heat flow may cause a temperature, size or phase change. Study the photos below and identify which of these effects is occurring in each.

c
liquid wax

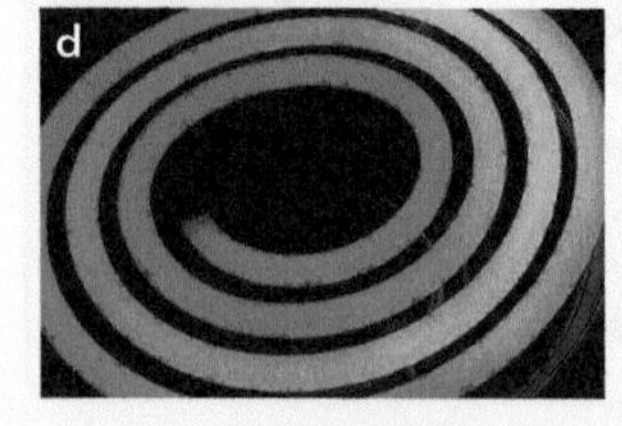

5 Decide whether the following statements are true or false. Rewrite the false ones to make them correct.

a Heat flows from colder to hotter objects.
b The particles that make up matter are in constant motion.
c The thermal energy possessed by an object is the total kinetic energy of its atoms or molecules.
d Temperature relates to the average kinetic energy of particles.
e The particles of a solid can only vibrate.
f Gas particles have the freest movements.
g Melting and freezing of a substance occur at different temperatures.
h Objects expand when heated because their particles push each other further apart.
i A substance with a high heat capacity will heat up quickly.
j When a change of state is occurring, the temperature will keep rising (or falling).

6 Complete this table comparing the arrangement and behaviour of the particles in different states of matter.

Feature	Solid	Liquid	Gas
Particle spacing?		close	
Fixed particles?	fixed		
Kinetic energy?			high
Vibrating particles?		yes	
Rotating particles?	no		
Free motion?		slower	

7 Copy the diagram, label the phase changes a to f and add key colours. Two phase changes not shown in the diagram are evaporation and sublimation. Add labelled arrows to the phase change diagram to show where these would occur.

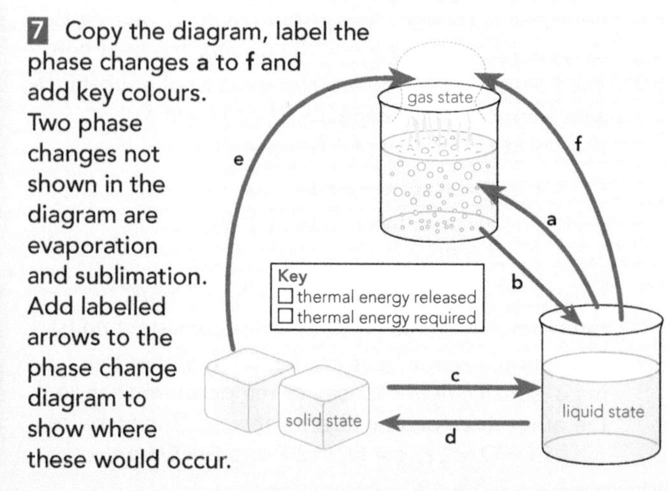

ISBN: 9780170262316

8 A beaker of ice was heated over a 15-minute period and the temperature in the beaker recorded every minute. The results are plotted on the heating curve shown.

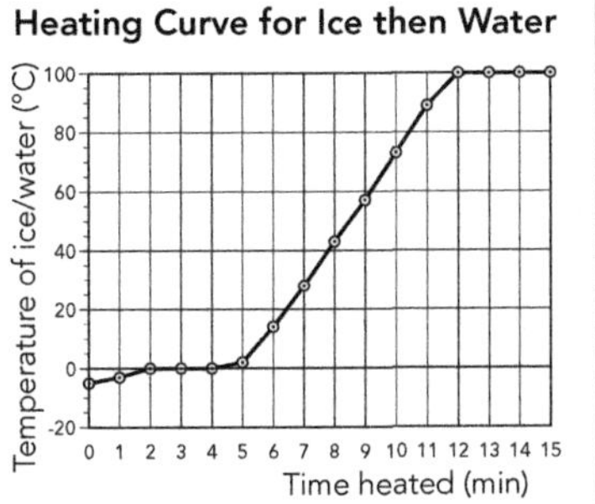

a What process was occurring between the second and fourth minutes?
b What happened to the temperature in this period?
c Was the beaker being heated during this period?
d What happened to the thermal energy supplied during this period?
e What state was water in from the fifth to the twelfth minute?
f What was the effect of continued heating during this period?
g What happened to the temperature of the water between the 12th and 15th minutes?
h What process was occurring in those minutes?
i What happened to the thermal energy supplied between the 12th and 15th minutes?
j Describe the typical pattern shown by a heating curve.

9 Naphthalene was heated till it melted and the temperature reached 130 °C. The temperature was then recorded every minute for 10 minutes as it cooled.

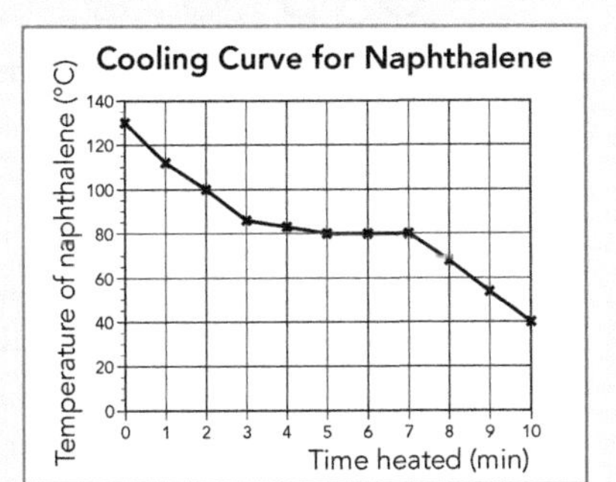

a Describe the pattern shown by the cooling curve and explain what is happening.
b What phase change is occurring during the fifth and sixth minutes? How do you know?
c Why was the temperature constant during that time?
d What would happen to the surroundings in this phase?

10 Read the article opposite, then answer the questions below.
a What states does carbon dioxide occur in?
b Why is solid carbon dioxide called dry ice?
c Why is dry ice used to keep ice cream cold rather than ice made from water?
d What is the specific heat capacity of dry ice?
e What is the latent heat of sublimation of dry ice?
f Will warming or subliming dry ice cool frozen foods more?
g How does dry ice cool frozen goods when it is subliming?
h Why will water make dry ice sublime much more rapidly?

Dry Ice: a Sublime Solid

Dry ice is carbon dioxide in the solid state. It is often used to keep frozen foodstuffs at about –20 °C for a period of time when refrigeration is not available.

Dry ice does not melt into a liquid, instead it turns directly into a gas, hence the name dry ice. The process is called sublimation.

The sublimation point of carbon dioxide is –78 °C. At this temperature dry ice will turn into a gas and carbon dioxide gas will turn into a solid, depending on whether heat is flowing to or from the substance.

The dry ice pellets used to keep frozen foods cold are colder than –78 °C to start with. A kilogram of dry ice will warm by 1 °C if it is supplied with 0.735 kilojoules of heat.

When 1 kg of dry ice has warmed to –78 °C, it takes 571 kJ to make it sublime. This heat comes from it surroundings, usually the frozen goods that the ice is packed amongst, making them colder.

It usually takes a few hours for the dry ice surrounding the frozen foods to sublime. Dry ice can be made to sublime rapidly if the pellets are dropped into water. As shown in the photo, the dry ice rapidly turns into a dense cloudy gas that spreads across the bench top. The water does not freeze but it does become very cold.

11 A candle-maker needed to heat and melt 0.5 kg of a particular kind of paraffin wax so that she could pour it into a mould. The specific heat capacity of the paraffin wax is 2.9 kJ/kg/°C, its latent heat of fusion is 210 kJ/kg, and its melting point is 62 °C.

molten wax

a If the wax was 22 °C to start with, how much heat would the wax need to absorb to reach 62 °C?
b How much heat will the wax need in order to melt once it has reached a temperature of 62 °C?
c How much heat has the wax absorbed in total and why would the candle-maker have supplied more heat than that?

12 The table below gives the melting point (MP), latent heat of fusion (L_f), boiling point (BP) and latent heat of vaporisation (L_v) for each of six substances. Study the data, then answer the questions below.

Substance	MP (°C)	L_f (kJ/kg)	BP (°C)	L_v (kJ/kg)
Alcohol	–114	108	78.3	855
Ammonia	–75	339	–33.34	1369
Lead	327.5	23.0	1750	871
Nitrogen	–210	25.7	–196	200
Oxygen	–219	13.9	–183	213
Water	0	334	100	2260

(Data source: Wikipedia article on Latent Heat)

a Identify what state each substance will be in at room temperature (20 °C)?
b How much heat is needed to melt 5 kg of solid lead?
c How much heat is needed to vaporise 5 kg of liquid lead?
d What mass of alcohol could be boiled if 3420 kJ of heat was transferred to it?
e Compare the amounts of heat needed to boil and melt a substance? Why is there such a difference?
f Which substance would need to lose the least amount of heat to condense it? To freeze it?

ISBN: 9780170262316

Heat Transfer and Insulation

Learning Outcomes - On completing this unit you should be able to:
- explain the differences between thermal energy, heat and temperature
- distinguish between conduction, convection and radiation
- explain why different surfaces heat or cool at different rates
- account for how thermal insulation works in practical examples
- *apply principles of heat to solving a puzzling sensory experience.*

Heat Transfer

- The **thermal energy** of a substance is the total kinetic energy of its particles.
- Thermal energy is transferred from one location to another as **heat**. We use the term **heat transfer** to describe *thermal energy in transit*.
- Heat will flow from a higher to lower temperature location providing there is a **thermal pathway**.
- When heat transfer occurs, the atoms at one location lose kinetic energy while the atoms at the other location gain kinetic energy.
- As **temperature** is related to the average kinetic energy of the particles, the temperature of the substance that loses thermal energy will fall, while the temperature of the substance that gains thermal energy will rise.
- *Eventually the two locations will reach the same temperature.* They are now in **thermal equilibrium**.

Heat Transfer Modes

- Heat transfer modes are conduction, convection and radiation. All three can occur with a bar heater.
- Which mode occurs depends on the state of the substances involved, whether they are in contact or not, and if not, what lies between them.

Conduction

- Thermal energy can be transferred by conduction through a solid or from one solid to another if they are in direct contact, eg if your hand touches a bar heater.
- **Conduction** is the transfer of thermal energy through interactions between neighbouring particles (atoms, molecules and electrons) that are vibrating.
- In a hotter region, energetic atoms vibrate more, jostling neighbours causing them to vibrate more as well. Those atoms transfer kinetic energy by jostling their neighbours, and so on. The overall effect is for heat to flow from a hotter to a cooler region. The atoms stay in the same place while kinetic energy is transferred from atom to atom.
- Substances that conduct heat well are called **thermal conductors**. Substances that conduct heat poorly are called **thermal insulators**.
- Metals are much better conductors than nonmetals, although surprisingly diamond is the best of all. Metal objects often feel cold because they rapidly conduct heat away from your hand.
- Metals are good conductors because energised **free electrons** that roam through the metal cause a *rapid transfer of kinetic energy* as they interact with atoms and other electrons.

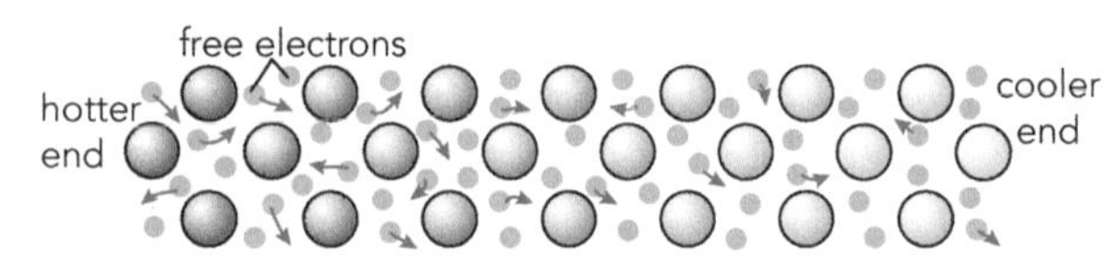

- Conduction occurs best in solids as the atoms are in close contact. Liquids are poor conductors and gases are basically insulators as their atoms are further apart.

Convection

- Thermal energy can also be transferred from a hotter to a colder region by a **convection current**.
- **Convection** is the transfer of thermal energy through the bulk flow of a heated gas or liquid from a hotter to a colder region. *As solids can't flow, convection only occurs in liquids and gases.*
- When air or water is heated at a particular location, it expands in volume as the molecules move faster. It becomes *less dense than the surrounding air or water.* The lighter, hotter air or water rises, transferring thermal energy to a colder location.

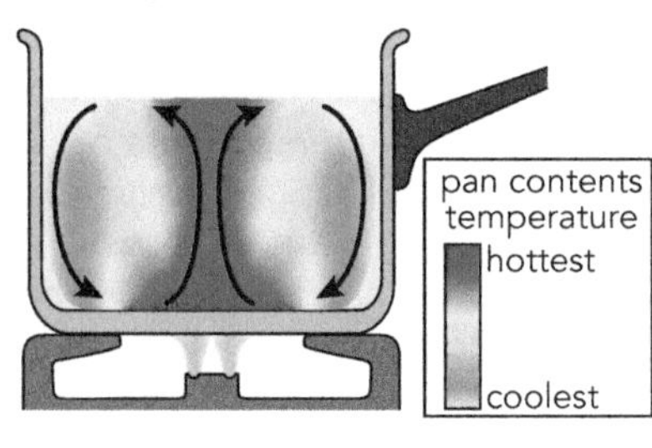

- As the less dense air or water rises, denser cooler air or water moves in to take its place. This bulk movement of air or water can create a circular convection current.
- A bar heater warms the air in front of it, which expands and rises toward the ceiling. Nearby colder, floor-level air moves in to take its place in front of the heater.
- Convection currents can also be used to cool a room.
- *Convection is faster than conduction at transferring heat over a distance.* When a warm current reaches a colder object, the heat still has to be transferred by conduction.

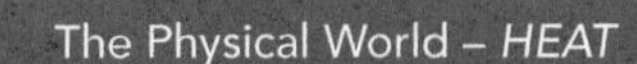

ISBN: 9780170262316

Radiation

- Thermal energy can be transferred from a hotter to a colder region by **radiation** (electromagnetic waves).
- **Thermal radiation** is the transfer of thermal energy by invisible **infrared waves** emitted by molecules and atoms. The image shows a hand emitting infrared radiation.

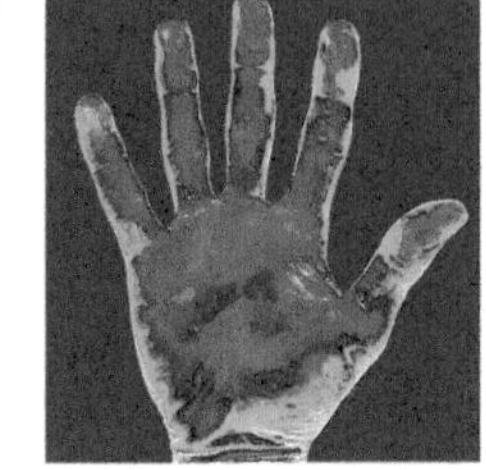

- All substances above absolute zero (the coldest possible temperature) emit infrared waves. If a substance is hotter than its surroundings, it will lose thermal energy to its surroundings by infrared radiation and cool down.
- Infrared waves spread out in all directions and travel in straight lines at the speed of light. These waves can travel through air and also through the vacuum of space, which is how they reach us from the Sun.
- A hotter object can only transfer energy to a colder object by radiation if the waves strike the object.

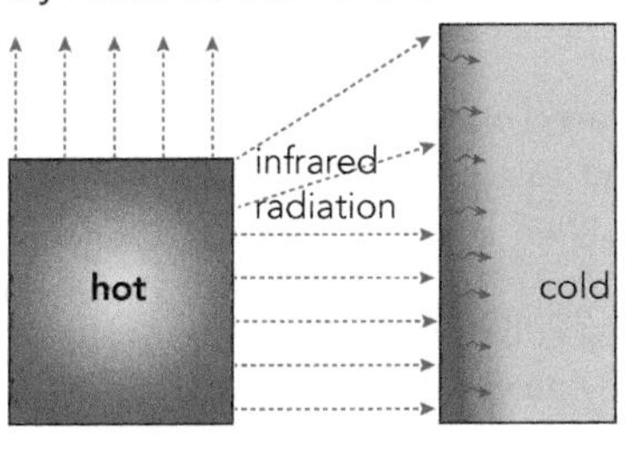

- When infrared waves strike an object's surface they will be **reflected**, **absorbed** or **transmitted**. *Absorbed infrared radiation will cause an object's temperature to rise.*

Keeping Cool

- Whether thermal radiation is absorbed or reflected depends on the *substance's colour and surface texture*.
- Dark and rough surfaces are good absorbers of thermal radiation. White, shiny or smooth surfaces are poor absorbers and reflect radiant heat to the surroundings.

- Hot dark surfaces are good emitters of infrared waves and cool quickly when the surroundings are colder. Hot white objects are poor emitters of infrared waves and cool slowly.

Thermal Insulation

- **Thermal insulation** is used to prevent heat entering or escaping from a container or building. It acts as a *barrier preventing thermal equilibrium from occurring rapidly*.
- As thermal energy can be transferred through conduction, convection and radiation, all three need to be minimised. Usually reducing conduction is the key.
- An insulator is basically a poor conductor of heat, so surrounding a substance with an insulator will reduce conduction. Most insulators rely on trapped air.
- Gases are good insulators, so argon gas or air is used between panes of glass in double-glazed windows to reduce heat loss by conduction.
- A rubber wet suit reduces heat loss by conduction to the water.
- Neither conduction nor convection can occur through a vacuum, which is why a vacuum (thermos) flask works so well.
- You can keep a room warm or cool by closing windows and doors as this stops convection currents transferring heat.

insulated cap and supports
glass vessel
silvered inside
vacuum
outer case

- Heat transfer by radiation can be reduced by using a reflective material. For example, the silvered glass layer of a flask reflects heat back into the liquid.

Rate of Heat Transfer

- How rapidly thermal energy is transferred between substances depends on the substances, the temperature difference, the distance apart, and the transfer mode.
- **Power**, symbol *P*, is a measure of the *rate at which thermal energy is transferred*. The unit is the **watt** (W).
- Power is found by dividing the heat transfer (or energy transfer) by the time period:

Power = heat transfer ÷ time period

$P = Q/t$ (or $P = E/t$ in NCEA exams)

Q
P t

(power in watts, energy in joules, time in seconds).

- The formula can be rearranged to find **Q** or **t** if the other two quantities are known: $Q = Pt$ or $t = Q/P$.

SCIENCE SKILL: Applying a Physics Principle to an Everyday Situation

In an NCEA exam you may be asked how a physics principle applies to an everyday situation. The question may test whether you correctly understand the physics principle. In this activity the heat receptors in your skin will send conflicting messages to your brain. You then need to apply your knowledge of heat transfer to work out what is actually going on.

Steps:

1. Half fill one ice-cream container with cold water and add 10 ice cubes.
2. Half fill a second with cold tap water and a third with warm water (not too hot).
3. Wait 3 minutes then put your left hand in the ice water and your right hand in the warm water for 2 minutes.
4. Immediately put both hands in the cold water. What are the heat receptors on each hand telling you? Can this be true?

Solving the puzzle:

The water in the container can't be both hot and cold, which means that the heat receptors in your skin must be sensing something else. Heat flows from your warmed hand to the cooler water, so that makes your right hand feel cold. Heat flows the cold water to your chilled hand, so that makes your left hand feel warm.

So, heat receptors in your skin sense heat flow rather than the temperature of your surroundings.

Revision Activities

1 Match up the descriptions with the terms.

a thermal energy	A transfer of thermal energy from atom to atom
b heat	B occurs when wave energy heats a substance
c heat transfer	C outer shell electrons that roam freely
d thermal pathway	D transfer of thermal energy
e temperature	E bulk movement of a heated gas or liquid from a hotter to a colder region
f thermal equilibrium	F total kinetic energy of particles of a substance
g conduction	G rate at which thermal energy is transferred
h thermal conductor	H invisible waves that heat objects when absorbed
i thermal insulator	I substance or object that conducts heat well
j free electrons	J unit of power
k convection current	K available mode of transferring thermal energy
l convection	L occurs by conduction, convection or radiation
m radiation	M related to the average kinetic energy of particles
n thermal radiation	N occurs when locations reach same temperature
o infrared waves	O barrier to unwanted heat transfer
p reflection	P transfer of thermal energy by infrared waves
q absorption	Q occurs when waves bounce off the surface
r transmission	R current in a liquid or gas caused by a temperature difference
s thermal insulation	S occurs when waves pass through a substance
t power	T electromagnetic waves emitted by atoms
u watt	U substance or object that conducts heat poorly

2 Explain the differences between the terms in each of the items below.

a thermal energy and temperature
b heat and thermal energy
c a thermal insulator and a thermal conductor
d conduction, convection and radiation
e infrared and light radiation
f power and heat transfer.

3 What is the main mode of heat transfer in each image?

a

b

c

d

e

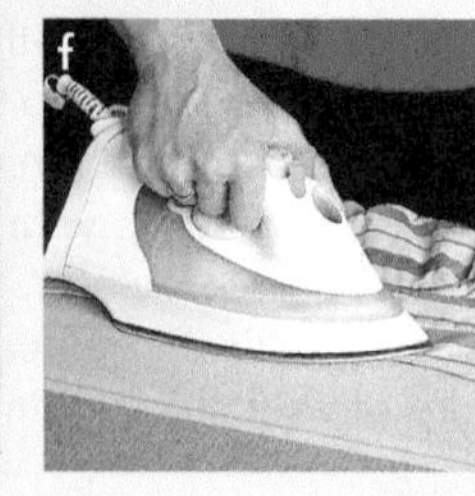
f

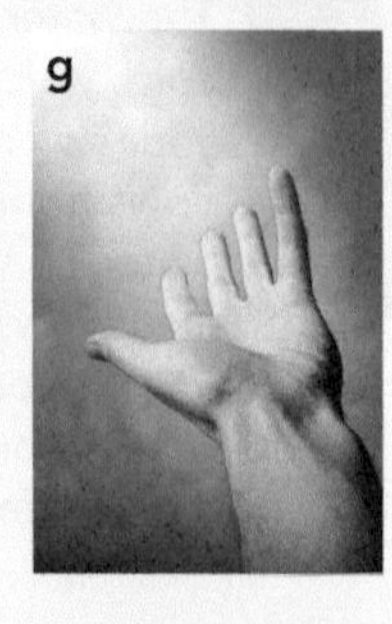
g

4 Decide whether the following statements are true or false. Rewrite the false ones to make them correct.

a Heat is thermal energy in transit.
b An object and its surroundings that are at the same temperature are in thermal equilibrium.
c Conduction can only occur when substances are in direct contact.
d Conduction is more effective in solids than liquids or gases.
e Convection can occur in solids, liquids and gases.
f Sea breezes are an example of convection currents in action.
g Infrared radiation can be seen by the human eye.
h All atoms emit infrared radiation.
i Dark objects are good absorbers of thermal radiation but poor emitters.
j Thermal insulation allows thermal equilibrium to be reached.
k Only radiation can transfer thermal energy through a vacuum.
l Heat transfer can be found by multiplying power by time.

5 Study the diagrams labelled **A**, **B** and **C**, then answer the questions below.

a In which diagram does heat transfer depend upon bulk movement to a cooler location?
b In which diagram does heat transfer depend upon the movement of electromagnetic waves?
c In which diagram does heat transfer depend upon the movement of energised electrons?
d Describe the heat transfers occurring in diagram **D**.

heat flow
A

hottest
coolest
B

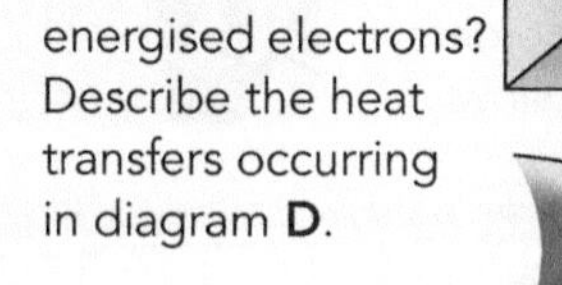

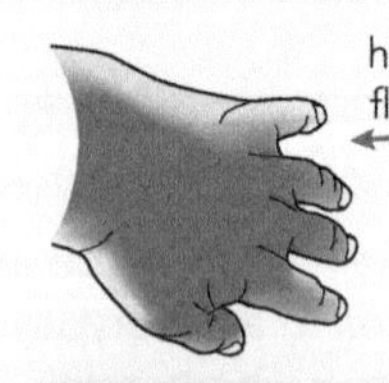

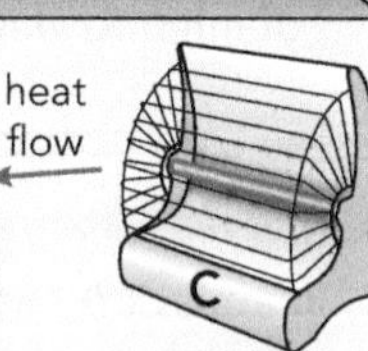

6 One kilogram of water at 22 °C was placed in a kettle. The water absorbed 209 kJ of heat from its surroundings during a 1½-minute period.

a Calculate the heating power of the kettle (ie, the rate at which thermal energy is transferred to the water).
b After 1½ minutes, how hot will the water be? (The specific heat capacity of water is 4.18 kJ/kg/°C.)

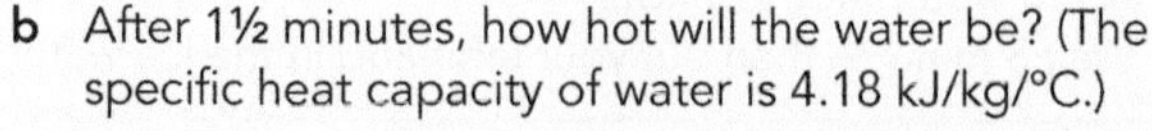

ISBN: 9780170262316

7 In an experiment to compare the ability of different-coloured surfaces to radiate heat, three identical silver- coloured drink cans were obtained.

One can was painted white, another black, and third left untouched. Each can was filled with water at 80 °C through a small opening in the top.
All three were placed at the same location on a bench and the temperature of the water inside each can was taken every two minutes during a 20-minute period.
The results are shown in the table below.

Time (min)	White can temp. (°C)	Black can temp. (°C)	Silver can temp. (°C)
0	80	80	80
2	77	75	78
4	75	71	76
6	73	67	75
8	71	62	73
10	68	58	72
12	66	53	70
14	63	49	69
16	59	45	67
18	56	40	65
20	52	36	64

a Plot the data on a single line graph. Add a key to show which can each line refers to.
b List four variables that were kept identical to ensure the comparison was a 'fair test'.
c What was the only variable that differed between cans?
d Compare the cooling trends shown on your graph.
e Predict which will reach room temperature (20 °C) first.
f Write a conclusion comparing the ability of each surface colour to radiate heat energy.
g Predict what would happen to the temperature of the cans after several hours.

8 Read the article opposite, then answer the questions below.
a How is energy transferred from the sun to the solar shower bag?
b Why is the top of the bag clear plastic?
c What would happen if the top surface was black?
d What effect do infrared waves have on water molecules?
e Why is the inside bottom surface of the bag black?
f Why is the outside bottom surface of the bag silver-coated?
g What would happen to the water if the bag was placed upside down in the sun?
h How could the water be kept warm after the sun disappears?

9 When you exercise, your core body temperature rises and your body will attempt to cool you to prevent overheating. The main way it does this is through sweating.

a When the sweat on your skin disappears into the air, what phase change is occurring?
b Is energy absorbed or released by the sweat when this phase change occurs?
c How would that help lower your body temperature?
d Why would you not cool down so quickly on a humid day?
e Would you cool down quicker on a windy or on a still day? Why?

10 When you breathe on a cold window or mirror, droplets of water form on the surface.

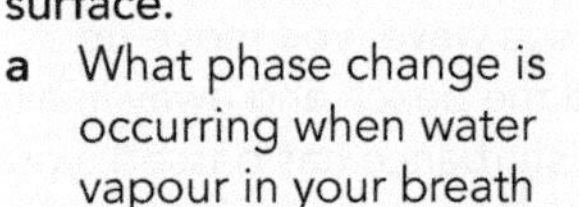

a What phase change is occurring when water vapour in your breath changes into water droplets on the glass?

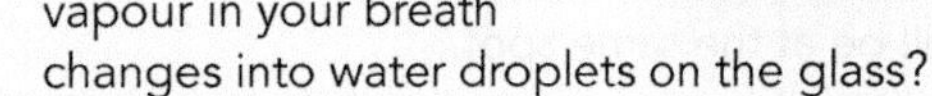

b Does water vapour release or absorb thermal energy when this phase change occurs?
c What happens to the surrounding air when this phase change occurs?
d Why doesn't this phase change occur on a hot day?

11 When you heat milk using a steamer you must be careful to avoid getting a steam burn, which is worse than a boiling water burn.
a Why will the temperature of the steam be 100 °C?
b What process turns steam into water when it touches your hand?
c Where does the extra heat come from that makes a steam burn worse than a boiling water burn?
d What should you do if you do get a steam burn?

Solar Showers

A solar shower is a useful device enabling campers to enjoy a warm or even a hot shower. In the absence of electricity, radiant energy from the sun is used to heat water.

The solar shower consists of a strong, flat rectangular plastic bag that can be filled with about 5 litres of water.

The bag is laid flat on the ground and left to heat up in the sun during the day.

The top surface of the bag is clear plastic, which allows infrared radiation from the sun to pass through into the water. Some of that radiation strikes water molecules and is absorbed. This increases their kinetic energy and therefore the temperature of the water.

The inside of the bottom surface of the bag is black, which absorbs much of the radiant heat that has passed through the water. The black surface heats up rapidly.

As hot, black surfaces are good at radiating heat, this surface radiates heat back into the water, further increasing its temperature.

The outside of the bottom surface of the bag is silver-coated and, as silver surfaces are poor radiators of heat, little heat is lost from the bottom surface to the ground.

By applying heat principles, these cleverly designed bags allow the water temperature to rise to nearly 50°C at times. The preferred shower temperature for adults is about 40 °C.

ISBN: 9780170262316

Properties and Types of Waves

Learning Outcomes - On completing this unit you should be able to:
- classify different kinds of waves
- explain the various properties of waves and how they are related
- use the wave equation $v = f\lambda$ to find either speed, frequency or wavelength
- describe the key differences between water, light and sound waves
- *convert basic metric units into larger or smaller ones.*

Waves

- You will have seen ocean waves rolling in and heard them break on the beach. Seeing depends on receiving light waves and hearing depends on detecting sound waves, so waves are all around you.
- When you encounter a big sea wave, you move up and down as well as toward the beach and away in a circular motion. After the disturbance has passed, you would still be at the same spot.
- Water, light and sound waves all transfer **energy** (see page 12) *from source to destination by disturbances.* Waves transfer vibrational energy but not **matter**.
- A **wave** is a regular disturbance or vibration, either of matter or force fields, that **propagates** (spreads out).

Classification of Waves

- Waves that need a **medium** (eg water) to propagate are called **mechanical waves**. The medium temporarily deforms or moves as the disturbance passes.
- Waves that do not require a medium to propagate are called **electromagnetic waves** (see page 33).
- If the disturbance is *perpendicular to the propagation direction*, as in light waves, the wave is called a **transverse wave**. The wave moves in one direction but the disturbance is in another.

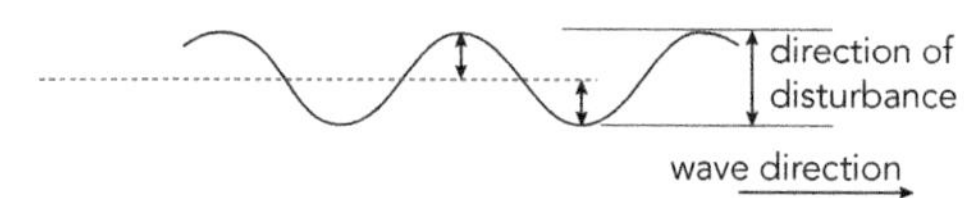

- If the disturbance is *parallel to the propagation direction*, as occurs in sound waves and can occur with a slinky spring, the wave is called a **longitudinal wave**. The disturbance occurs in the same direction as the wave.

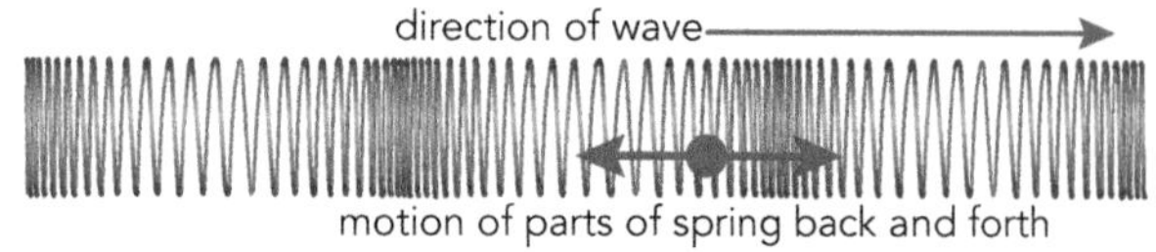

- If the wave front is straight and perpendicular to the propagation direction, it is called a **plane wave front**.
- If the wave fronts propagate as ever-increasing circles (spheres), it is a **circular** (or spherical) **wave front**.
- Waves that travel across the top of a liquid or solid are called **surface waves**. Waves that spread out through a substance rather than across its surface are called **body waves.**

Properties

resting position, wavelength, crest, amplitude, trough

- All waves have points of maximum disturbance called crests and troughs. The midline between them is the position before being disturbed.
- For NCEA level 1, the observer is always stationary.
- The distance between the resting position and a crest is the **amplitude**, symbol A. The basic unit is the metre (m).
- *The greater the amplitude, the more energy transferred.* The amount of energy transferred is proportional to the square of the amplitude.
- The distance between two successive crests is the **wavelength**, symbol λ. The basic unit is the metre (m).
- A **wave cycle** lasts from the arrival of one crest to the arrival of the next. The time for one wave cycle is the **period**, symbol T. The basic unit is the second (s).
- The number of crests passing per second is the wave's **frequency**, symbol f. The basic unit is the **hertz** (Hz). *The higher the frequency, the more energy transferred.*
- The frequency f of a wave of period T is given by:

  Frequency = 1 ÷ period $\quad f = 1/T$

  (frequency in Hz, period in s). Conversely, $T = 1/f$.
- The **speed** v of a wave that travels **distance** d in **time** t is given by:

  Speed = distance ÷ time
  $v = d/t$

  (d / v t)

  (speed in ms^{-1}, distance in m, time in s; or other consistent units).
- The triangle gives formulas for distance and time if the other two quantities are known: $d = vt$ and $t = d/v$.
- If we consider just a single wave cycle, then the distance travelled will be the same as the wavelength, and the time take will be the same as the period.
- The formula becomes $v = \lambda/T$, but $f = 1/T$, so $v = f\lambda$.
- The **speed** v of a wave of **frequency** f and **wavelength** λ is given by the **wave equation**:

  Speed = frequency x wavelength
  $v = f\lambda$

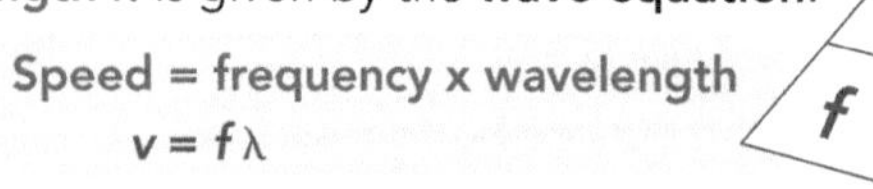

  (speed in ms^{-1}, frequency in Hz, wavelength in m).
- The triangle gives the formulas: $f = v/\lambda$ and $\lambda = v/f$.
- The wave equation shows that *speed is proportional to wavelength given that frequency remains constant.* This means wave speed increases as wavelength does.

ISBN: 9780170262316

Wave Types

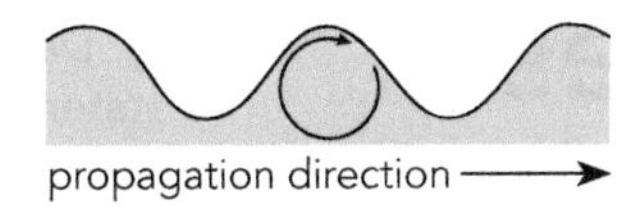

Water Waves

- Water waves are surface waves as the disturbance travels across the surface of the medium.
- The water moves in a circular motion as each wave passes. Sea waves are both transverse and longitudinal, as the motion is up and down and back and forth.
- Ocean waves usually have plane wave fronts. Pond ripples from a falling object have circular wave fronts.
- The energy source for ocean waves is the kinetic energy of the wind blowing on the surface. For pond ripples, it's the kinetic energy of the falling object.
- As a medium is required, they are mechanical waves.
- The amplitudes of ocean waves typically range from a few centimetres up to 10 metres. The *strength of a wave is related to its amplitude.*
- The wavelengths of ocean waves range from 5 to 100 m.
- The period of an ocean wave is typically between 1 and 10 s, so frequencies range from 0.1 to 1 Hz.
- *Waves vary in speed mostly because wind speeds vary.* Ocean waves slow down when they hit shallow water. As frequency can't change, the wavelength must decrease according to the wave equation.

Visible Light Waves

- The energy source for visible light waves is the *kinetic energy of vibrating atoms in a very hot object*, such as the glowing filament of a light bulb or the sun.
- No medium is necessary for propagating light waves, as they are electromagnetic waves.
- An electromagnetic wave consists of *vibrating electrical and magnetic fields* that *propagate at light speed.*

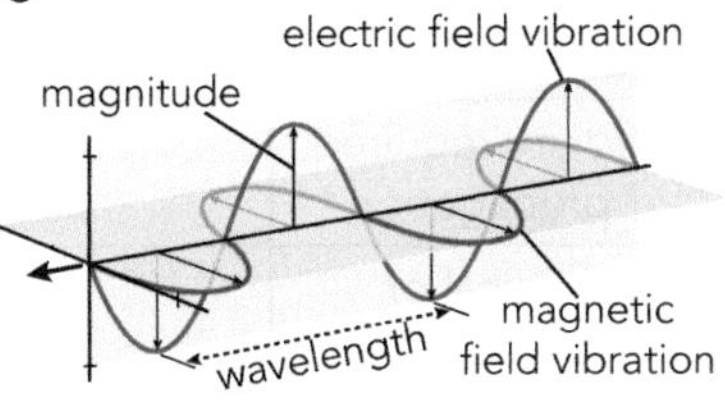

- The wave is a transverse wave, as the vibrations of the two fields are perpendicular to the propagation direction.
- The *greater the amplitude* of the light waves that reach your eye, the *brighter the light.*
- The wavelengths of visible light waves range from 380 nanometres for violet light up to 750 nm for red light. (1 nm = 1/1 000 000 000 m.)
- The *frequency of a light wave affects the colour* your eyes will see. Visible light waves range from 480 terrahertz for red light up to 670 THz for violet light. (1 THz = 1 000 000 000 000 Hz.)
- Light waves travel at a speed of 300 000 kms^{-1} in space but they slow down as they pass through a transparent medium such as air, glass or water.
- As the wave's frequency can't change, the wavelength must decrease as the wave slows, according to the wave equation. Even though the wavelength changes, the colour you see remains the same.

Sound Waves

- Sound travels through air, water and solids objects as body waves radiating in all directions.
- The *energy source is a vibrating object*, such as a plucked guitar string or drum struck by a drumstick.
- Sound waves regularly squash the medium as they pass, which then expands out again.
- A **compression** occurs where the medium is squashed and a **rarefaction** where the medium is expanded.
- Sound waves are longitudinal waves, as the vibrations of the molecules occur in a direction parallel to the propagation direction.

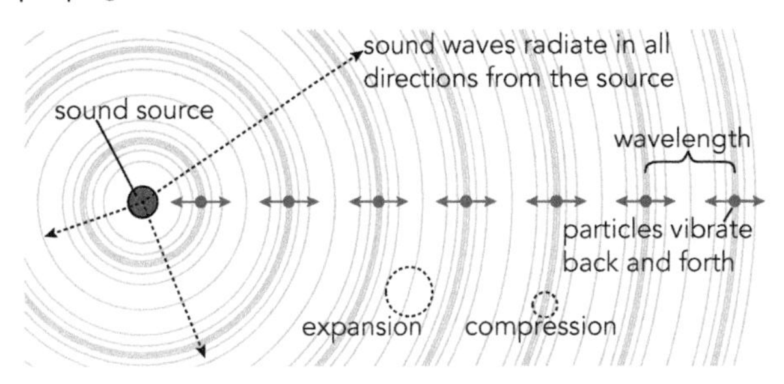

- As a medium is required for propagation, sound waves are classed as mechanical waves.
- The *greater the amplitude* of the sound waves that reach your ear, the *louder the noise*.
- The wavelengths of audible sound waves range from 17 mm to 17 m. (1 mm = 1/1000 m.)
- The periods of audible sound waves range from 0.05 to 50 milliseconds (1 ms = 1/1000 s), so frequencies range from 20 Hz to 20 kHz. (1 kHz = 1000 Hz.)
- The *frequency of a sound wave determines the* ***pitch*** of the sound your ears hear (middle C is 264 Hz).
- Sound waves have different speeds in different media. In dry air at 15°C, they travel at about 340 ms^{-1}.
- Sound waves travel faster through water than air, and faster still through solids as the particles are closer together. As the frequency can't change, the wavelength increases when the wave speeds up in a denser medium, according to the wave equation.

SCIENCE SKILL: Converting Basic Metric Units

The diagram shows the operations needed to convert a basic unit into a smaller or larger unit and the reverse operation.

Step:

1 Identify the operation needed to convert the unit, then use your calculator to arrive at the correct answer.

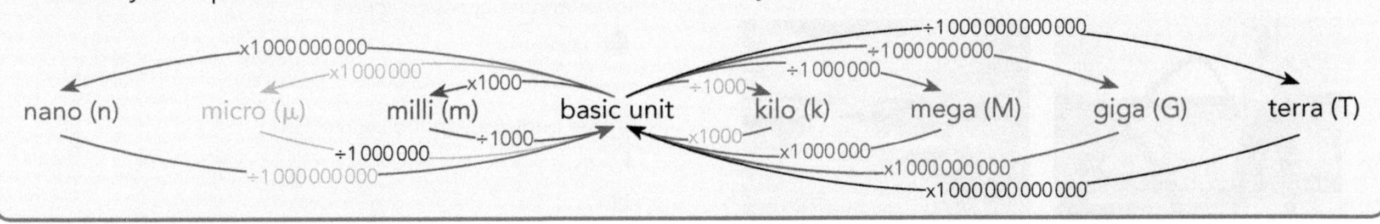

Revision Activities

1 **Match up the descriptions with the terms.**

a energy	A waves that rely on force fields to propagate
b matter	B time for one wave cycle to occur
c wave	C wave front that radiates out as a growing circle
d propagation	D wave that travels across surface of a substance
e medium	E regular disturbance or vibration of a medium or force fields that propagates
f mechanical wave	F region where the medium is expanded
g electromagnetic wave	G disturbance is perpendicular to propagation
h transverse wave	H number of waves passing per second
i longitudinal wave	I matter that some waves require to propagate
j plane wave front	J distance between two successive crests
k circular wave front	K event lasting from passing of one crest to next
l surface wave	L straight wave front perpendicular to propagation
m body wave	M required in order to be able to do things
n amplitude	N highness or lowness of a sound
o wavelength	O disturbance is parallel to propagation
p wave cycle	P wave that travels through a substance
q period	Q stuff that all substances and objects are made of
r frequency	R distance between resting position and a crest
s hertz	S travelling away from the wave source
t compression	T waves that deform a medium as they propagate
u rarefaction	U region where the medium is squashed
v pitch	V basic unit of frequency

2 **Explain the differences between the terms in each of the following items:**

a a mechanical wave and an electromagnetic wave
b a transverse wave and a longitudinal wave
c a plane wave front and a circular wave front
d a surface wave and a body wave
e a crest and a trough
f the amplitude and the wavelength of a wave
g the period and the frequency of a wave
h a compression and a rarefaction.

3 **Classify the type of wave generated by each of the sources shown below. Use the options: mechanical electromagnetic wave; transverse/longitudinal wave; surface/body wave. (Not all options may be relevant.)**

a

b

c

d

e

4 **Decide whether the following statements are true or false. Rewrite the false ones to make them correct.**

a Some waves can move matter from one location to another.
b All waves require a medium in order to propagate.
c The disturbance caused by a transverse wave is perpendicular to the propagation direction.
d A stone dropped into a pond will create plane wave fronts.
e Both crests and troughs are maximum disturbances.
f The greater the amplitude of a wave, the more energy it transfers.
g Wave speed is proportional to wavelength.
h The frequency of a wave does not change when it passes through a new medium.
i A sea wave will push a floating person closer to the shore.
j Light wave are propagated by interacting electrical and magnetic fields.
k Pitch is determined by frequency.

5 **Classify the three wave types by copying and completing the table below. (Not all features may apply.)**

Feature	Water waves	Light waves	Sound waves
Medium required for propagation?			
Mechanical or electromagnetic?			
Transverse or longitudinal wave?			
Plane or circular wave front?			
Surface or body wave?			
Increasing the amplitude affects?			

6 **Study the diagram below showing the sound waves generated by a lower and a higher frequency source.**

a Which source will generate longer wavelength waves and which shorter wavelength waves?
b Which source will generate a higher pitched sound and which a lower pitched sound?

ISBN: 9780170262316

7 A pool wave has an amplitude of 0.54 m, a wavelength of 2.50 m, and a period of 4.0 s.

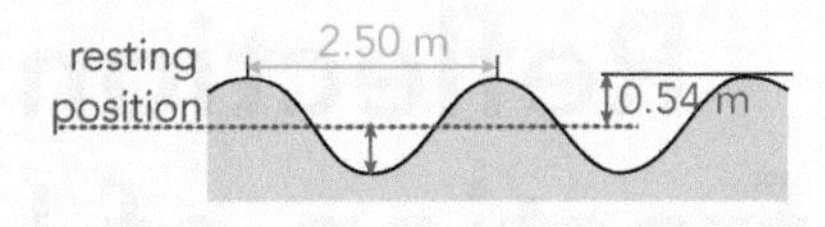

a What is the wave height (difference in height between a crest and a trough)?
b If the amplitude was halved, how would this affect the energy transferred by the wave?
c Is there any relationship between amplitude and wavelength?
d What will be the frequency of the wave?
e Calculate the speed of the wave by modifying the wave equation.

The wave enters shallower water and slows to two thirds of its previous speed.

f What is its new speed?
g Given that the frequency of a wave can't change, what will its wavelength be now?

8 A sound wave has an amplitude of 0.1 m and a frequency of 264 Hz. It travelled through the air at a speed of 340 ms^{-1}.

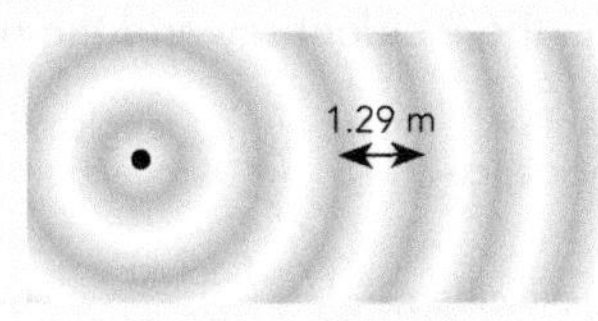

a If the amplitude of the sound waves was doubled, how would this affect the energy it transferred?
b What is the relationship between the amplitude of a sound wave and how loud it will be to our ears?
c What is the period of this sound wave?
d Calculate the wavelength of the wave by modifying the wave equation.
e How does the frequency of a sound wave affect its pitch?

The sound wave entered the water of a swimming pool. (Sound waves travel at 1497 ms^{-1} in water at 20 °C.)

f How many times faster do sound waves travel in water than in air?
g Why would sound waves travel faster in water?
h Given that the frequency of a wave can't change, what will its wavelength be now?
i Would the note sound different to an underwater swimmer as compared to a person sitting by the pool?

9 Read the article opposite, then answer the questions below.

a What are seismic waves?
b What are they caused by?
c What is the energy source of seismic waves?
d Which waves are body waves and which are surface waves?
e Compare the speed of the three types of seismic waves.
f Which waves are transverse and which are longitudinal?
g Describe, in order, the ground movements that a trained observer might notice when a earthquake occurs nearby.
h Explain why L-waves cause the most damage even though they arrive last.

Seismic Waves

Seismic waves are caused by earthquakes. Huge forces build up in the rock beneath Earth's surface because of the movements of tectonic plates (see page 150).

When the strain exceeds the strength of the rock, a fracture occurs as two blocks of the crust grind past each other. The massive jolt is the cause of the earthquake.

The energy released by the fracture is transported away by seismic waves. It is these waves that we experience in a quake. They radiate through the rock or across the surface from the epicentre.

P-waves: compress & stretch ground

S-waves: shake ground up & down

L-waves: move surface side to side

Primary (P) waves are body waves that travel through Earth's crust. They are the fastest waves. P-waves compress and stretch the underground rock as they pass by.

Secondary (S) waves are also body waves that travel through Earth's crust. They arrive after the P-waves and shake the underground rock up and down as they pass.

Love (L) waves radiate across the surface of the ground. Being the slowest waves they arrive last but they do cause the most damage because of their large amplitudes and longer durations.

10 A red light wave with a wavelength of 750 nanometres travels through air at 300 000 kilometres per second.

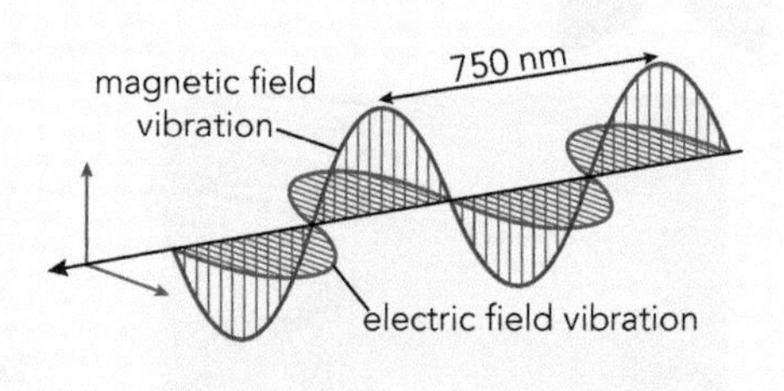

a If nanometres (nm), kilometres per second (kms^{-1}) and Terrahertz (THz) are consistent units, calculate the wave frequency by modifying the wave equation.
b If a violet light wave has double the frequency of the red light, what will its frequency be?
c Calculate the wavelength of the violet light wave given that it will travel at the same speed as the red light wave.
d Given that light of whatever frequency travels at the same speed in the same medium, what can you say about the relationship between wavelength and frequency for different-coloured lights.

Light travels through glass at about 200 000 kms^{-1}.

e Given that the frequencies of particular light waves do not change, calculate the wavelengths of red and violet waves as they travel through glass.

11 The diagram shows the energy transfers associated with an ocean wave, from its source to its destination on the beach.

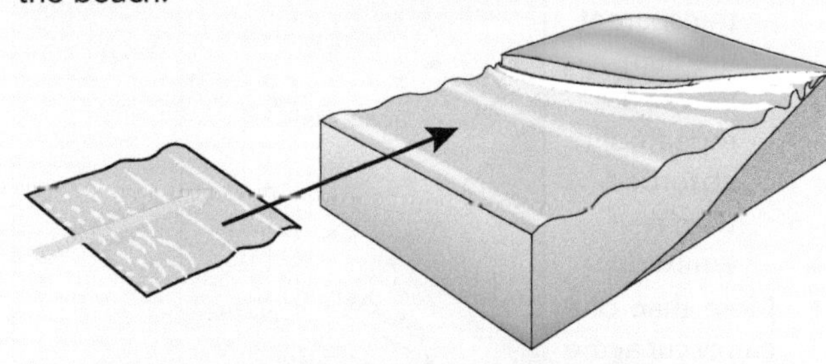

a What is the energy source of the ocean wave?
b What kind of energy did the source have?
c What two kinds of energy are associated with the wave as it travels across the ocean?
d When the wave hits the shore, what forms of energy is the wave's energy converted into?

ISBN: 9780170262316

Reflection and Refraction of Light

Learning Outcomes - On completing this unit you should be able to:
- explain the possible fates of light waves when they encounter a new medium
- distinguish between reflection, refraction and dispersion
- recall the principles of reflection and refraction
- explain why total internal reflection occurs
- *draw ray diagrams to show the location and orientation of an image.*

The Fate of Light Waves

- Light waves travel extremely fast in a straight line through air, but when they encounter a new **medium**, such as water or glass, three events can happen.
  1. Light waves are *absorbed* by surface atoms of an opaque medium, heating up the substance (see page 36). The light waves are not re-emitted.
  2. Light waves are *reflected* by surface atoms of a reflecting medium. When **reflection** occurs, waves are absorbed then re-emitted into the original medium.
  3. Light waves are *transmitted* from atom to atom through a transparent medium. Waves are repeatedly absorbed then re-emitted.

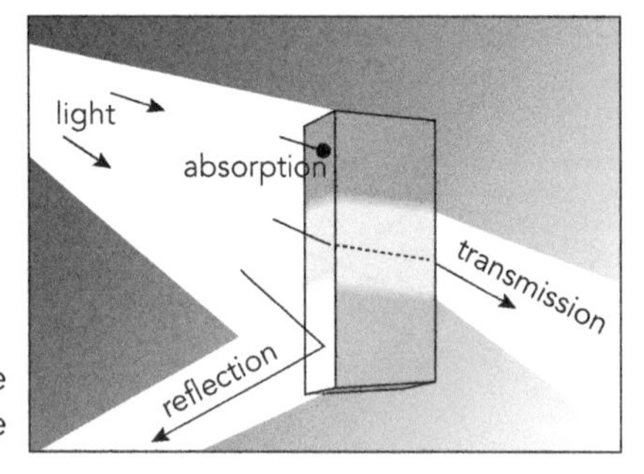

- More than one can occur at the same time.
- To make the study of light behaviour easier, we draw **light rays** (arrows) to show the direction in which wave fronts are propagating and consider only **plane** (flat) **boundaries** (interfaces) between media.

Reflection

- The surfaces of most substances reflect some light waves otherwise we wouldn't see them. Most surfaces reflect light diffusely so we don't see an image.
- Smooth surfaces of glass, shiny metal and water reflect sharp images as the reflected waves are parallel. (Mirrors have a silvered backing that reflects all light waves.)

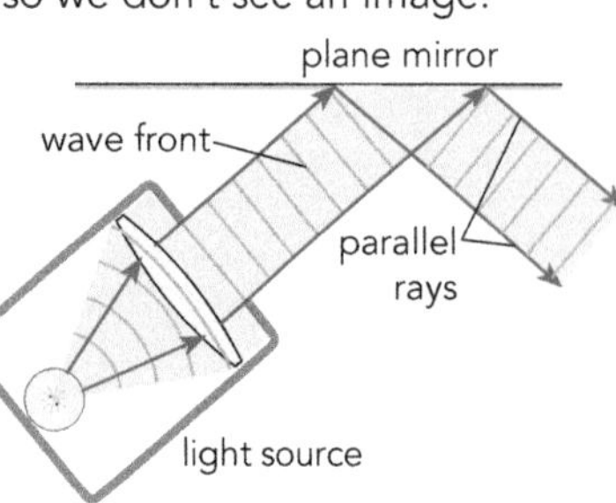

Principles of Reflection

- If the incoming (incident) wave hits a reflecting surface straight on (at 90°) it will reflect straight back.
- If the **incident ray** hits the surface at an angle, the **reflected ray** will propagate at a new angle.
- The **angle of incidence** lies between the incident ray and the **normal** (an imaginary line perpendicular to where the ray hits a surface). The **angle of reflection** lies between the reflected ray and the normal.
- The reflected ray always lies in the plane formed by the incident ray and the normal.

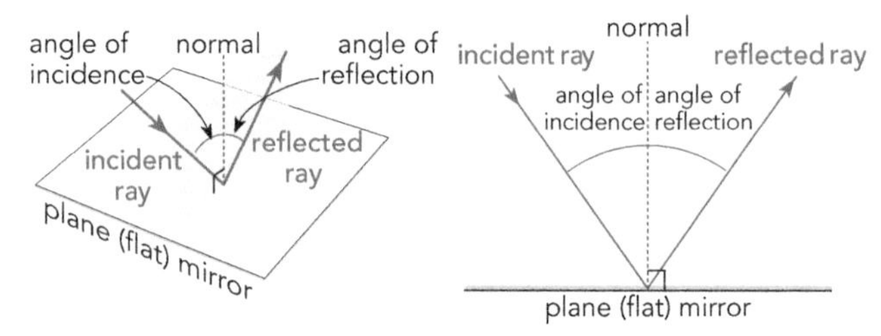

- *The angle of reflection always equals the angle of incidence.*
- When reflection occurs, the wave direction changes but not its speed, as it travels in the same medium.
- As the frequency of a wave does not change, its wavelength will stay the same (see page 44).

Reflected Images

- Mirror-like surfaces form a sharp **image** of the **object** that the light originally came from.
- For a plane mirror surface the image:
  - a appears to be behind the mirror
  - b is the same size as the object
  - c is the same distance from the mirror as the object
  - d is upright but **laterally inverted** (left-right swapped).

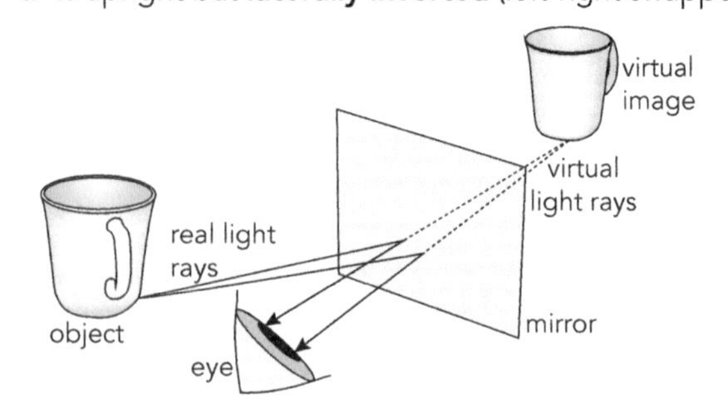

- If the light rays do not actually come from where the image appears to be, then it is called a **virtual image**.

Transmission

- Light waves travel through transparent media at specific speeds related to the medium's **optical density**. In a denser medium the atoms are slower to re-emit light waves, so overall light travels slower.

ISBN: 9780170262316

- Light travels at about 299 000 kms^{-1} through air, 225 000 kms^{-1} through water and 200 000 kms^{-1} through glass.

Principles of Refraction

- When a light wave enters a medium with a different optical density, its *speed and direction change abruptly*.
- The bending of light when a wave crosses the boundary between two media is called **refraction**.
- When a light wave passes from a less into a more optically dense medium it *slows down*. This causes it *to bend toward the normal*.
- As frequency can't change, the *wavelength shortens as the wave slows*.

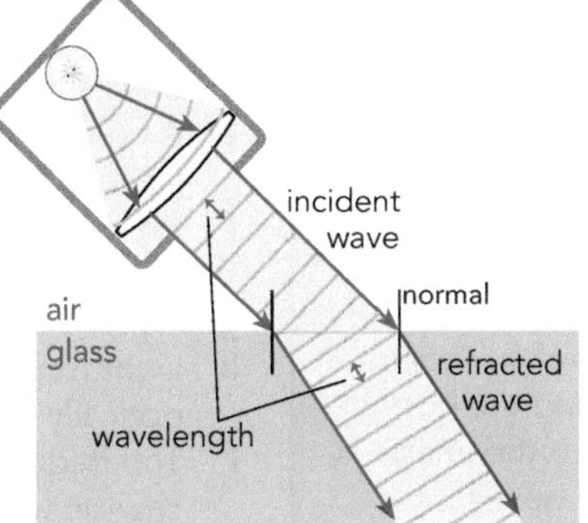

- Conversely, when a light wave passes from a more into a less optically dense medium it *speeds up*. This causes the **refracted ray** *to bend away from the normal* and the *wavelength lengthens as the wave speeds up*.
- When light enters a less dense medium, the **angle of refraction** is greater than the angle of incidence, but when it enters a more dense medium it is less.

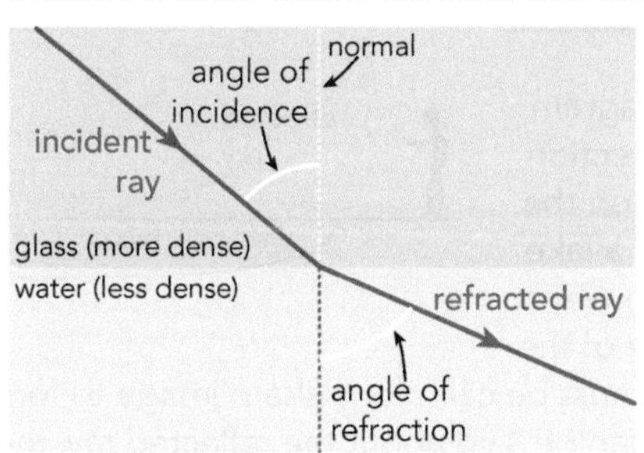

- *The greater the difference between the optical densities of two media, the more the light will refract.*
- If a light wave enters a new medium straight on, then it will carry on in the same direction but either speed up or slow down, depending on the relative optical densities.

Dispersion

- Sunlight is made up of light waves of different colours. These colours make up the **visible spectrum**.
- The colour of different light waves is due to their frequencies. Red light has the lowest frequency and violet the highest (see page 45).

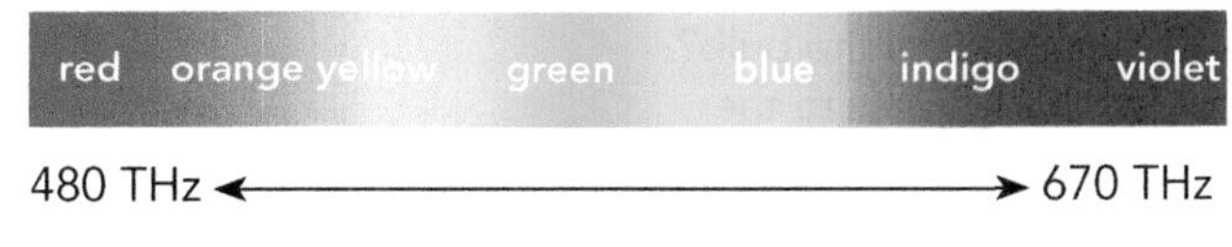

480 THz ←——→ 670 THz

- The extent to which light waves refract depends not only on the relative optical densities of the two media involved but also on the frequency of the light waves themselves.
- Higher frequency light waves refract (bend) more than lower frequency light waves. This property can be used to separate sunlight into a visible spectrum.
- The separation of different frequency (or wavelength) waves is called **dispersion**.

- The effect can be amplified by using a **prism** as the light waves are refracted twice, once when they enter the prism and again when they exit.

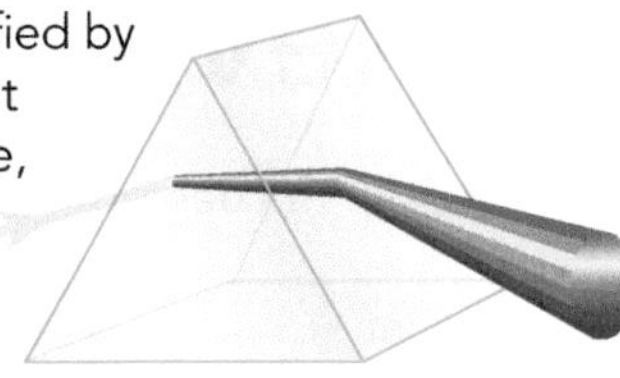

Total Internal Reflection

- When light enters a less dense medium, some waves are reflected and the rest are refracted. The angle of refraction will be greater than the angle of incidence.

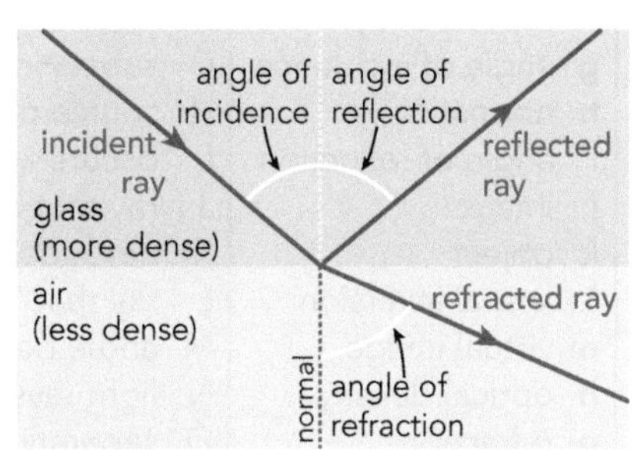

- If the angle of incidence increases to a **critical angle**, then refracted light bends so much that it travels parallel to the surface of the medium.

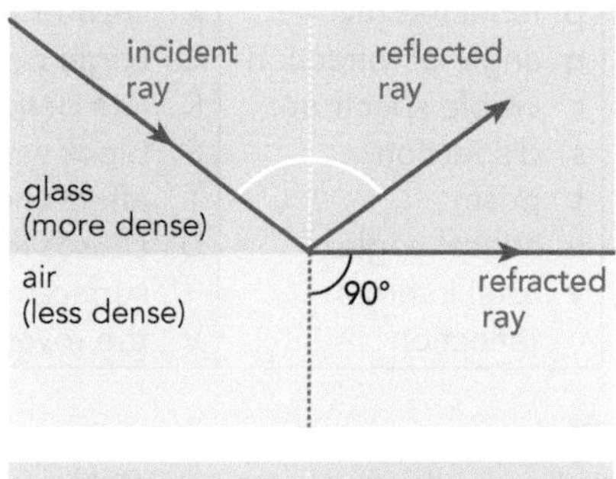

- If the angle of incidence is greater than the critical value, then all light is reflected back into the incident medium. This is **total internal reflection**.

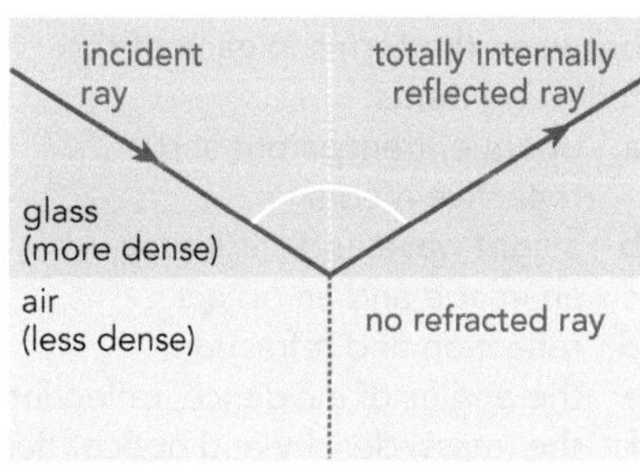

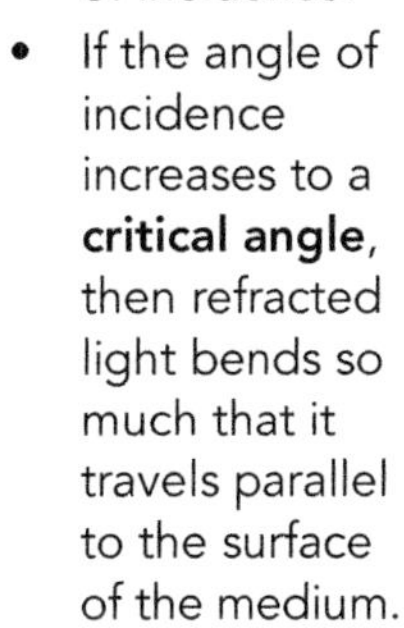

SCIENCE SKILL: Drawing a Ray Diagram

In NCEA exams you have to draw light ray diagrams.

Steps:

1. Rule a straight line to represent the mirror.
2. Draw the object in front of the mirror at an appropriate size and distance.
3. Draw a flipped inline image of the object, the same size and distance behind the mirror
4. Rule a line from one end of the object to the mirror and a dotted line from where it met the mirror to the same end of the image.
5. Repeat step 4 starting from the other end.
6. Place a ruler on each dotted line and draw a line from the mirror to where they meet.
7. Draw an eye at the intersection.
8. Add arrows to show the wave directions.

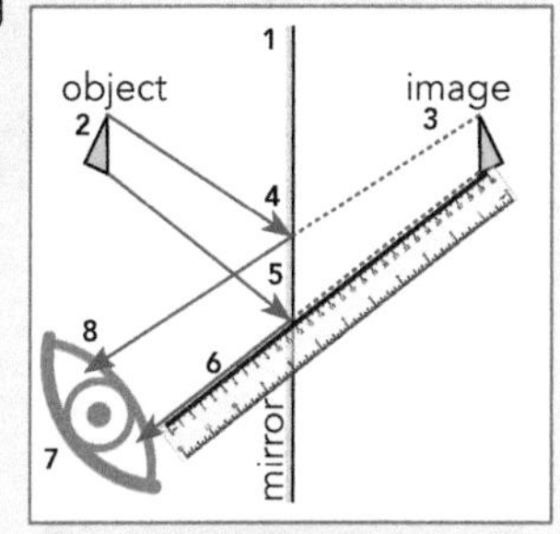

ISBN: 9780170262316

Revision Activities

1 Match up the descriptions with the terms.

a medium	A angle between the reflected ray and the normal
b reflection	B occurs when all light waves are reflected
c light ray	C bending of waves as they enter a new medium
d plane boundary	D arrow used to show propagation direction
e incident ray	E range of colours visible to the human eye
f reflected ray	F incoming ray from an object or light source
g angle of incidence	G substance that light waves can travel through
h normal	H source of incident light waves
i angle of reflection	I occurs when light waves 'bounce' off a surface
j image	J ray that bends abruptly as it enters a medium
k object	K flat surface between two media
l lateral inversion	L ray that 'bounces' off the surface of a medium
m virtual image	M angle between the incident ray and the normal
n optical density	N light rays do not come from where the image is
o refraction	O separation of different frequency waves
p refracted ray	P view of an object other than where it is
q angle of refraction	Q angle between the refracted ray and the normal
r visible spectrum	R line perpendicular to where a wave hits a surface
s dispersion	S block with the same cross-section throughout
t prism	T affects how easily waves travel through a medium
u critical angle	U causes refracted waves to travel parallel to the surface of the new medium
v total internal reflection	V the reversal of an object, as seen in a mirror

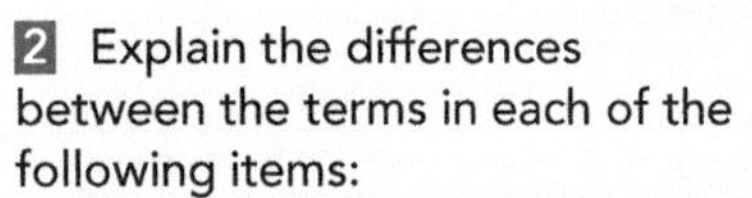

2 Explain the differences between the terms in each of the following items:

- a opaque, transparent and reflective media
- b a light wave and a light ray
- c an image and an object
- d reflection and refraction
- e the angles of incidence, reflection and refraction
- f the (mass) density and optical density of a substance
- g refraction and dispersion.

3 Describe a situation in which reflection, absorption and transmission are all occurring. (Hint: see photo above.)

4 For each image below, decide whether reflection, absorption, transmission, or diffraction (see page 51) is occurring. (More than one may occur for some.)

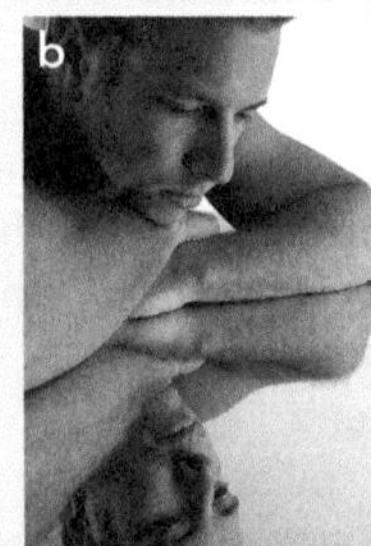

5 Decide whether the following statements are true or false. Rewrite the false ones to make them correct.

- a Emitted light waves always travel in straight lines.
- b Reflection and refraction can never occur together.
- c A plane surface can never be curved.
- d The angle of reflection always equals the angle of incidence.
- e A reflected wave travels at the same speed as the incident wave.
- f An image in a mirror is identical in all ways to the object.
- g A plane mirror image is virtual.
- h Light waves travel faster in a more optically dense medium.
- i The frequency of a wave does not change when it is refracted.
- j A refracted wave travels at a different speed in a new direction.
- k When entering a denser medium, light bends away from the normal.
- l When a prism reflects white light, it is dispersed in the visible spectrum.
- m Total internal reflection means that no light is refracted.

6 The diagram shows reflection occurring on the surface of a lake.

reflected ray
incident ray
object
water surface
image

- a Explain why the surface of the water must be calm for a sharp image to form.
- b What can be said about the reflected ray, the normal and the incident ray?
- c What is the relationship between the angle of reflection and the angle of incidence?
- d Why is a dotted line drawn from the image of the tree to the lake surface?
- e What five things can be said about the image?

7 The diagram shows a light box emitting a light wave that crosses a plane boundary into a less optically dense medium.

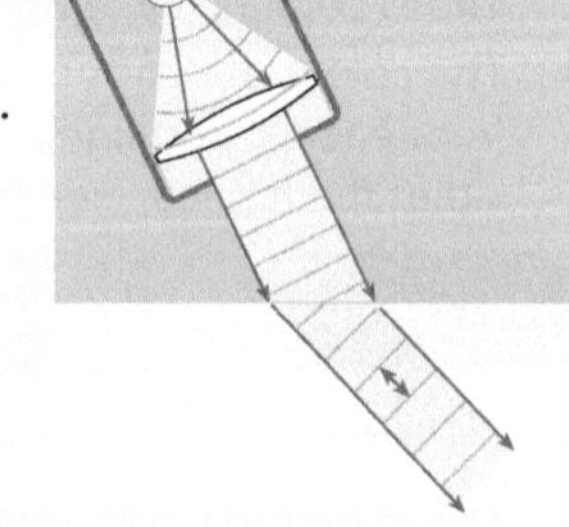

- a What does the lens do to the circular wave fronts coming from the bulb?
- b What happens to the direction of propagation as the waves enter the less dense medium?
- c What is the relationship between the angle of refraction and the relative optical density?
- d What happened to the speed of the wave as it entered the less dense medium?
- e What is the relationship between wave speed and optical density?
- f What happened to the wavelength?
- g Will the frequency of the wave have changed?

ISBN: 9780170262316

8 Copy the diagram and complete the following.

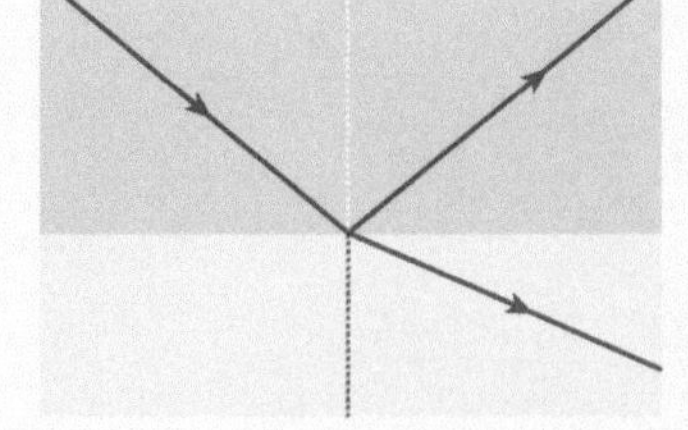

- **a** Label the incident, reflected and refracted rays, and the normal.
- **b** Mark the angles of incidence, reflection and refraction.
- **c** Describe the relationships between the three angles.
- **d** Label the less and more optically dense mediums.
- **e** Compare the speeds of the incident, reflected and refracted light waves, giving reasons for your answers.
- **f** How does the angle of incidence compare with the critical angle for the bottom medium? Explain why.

9 The diagram shows a bird's eye view of a light wave entering a semi-circular glass prism, then reflecting and refracting as it exits the prism back into the air.

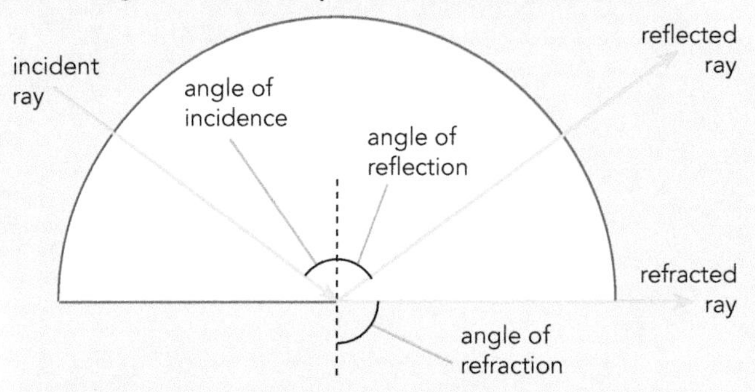

- **a** Why does the incident ray carry straight on when it enters the glass? And the reflected ray when it exits?
- **b** What is the angle of reflection equal to?
- **c** Why does the refracted ray travel parallel to the boundary when it emerges from the glass?
- **d** What is size of the angle of refraction?
- **e** What can be said about the size of the angle of incidence?
- **f** What would happen if the angle of incidence is decreased?
- **g** What would happen if the angle of incidence is increased?
- **h** What is the term used to describe the situation if the angle of incidence was increased?

10 Read the article opposite, then answer the questions below.

- **a** What kinds of situations will cause waves to diffract?
- **b** Do all wave types diffract?
- **c** Draw a bird's eye sketch of the waves diffracting on the beach in the photo.
- **d** What else other than diffraction would help us to hear sounds being emitted in another room?
- **e** What happens if a gap is large compared to the wavelength?
- **f** What happens if the wavelength is small compared to the size of a gap?
- **g** What happens when a longer wavelength wave goes around an obstacle?

Diffraction of Waves

Diffraction is the bending of a plane (straight) wave front as it goes around an obstacle or through a narrow gap. For example, straight wave fronts generate circular wave fronts as they pass through a narrow gap.

Diffraction occurs with water, sound, light and seismic waves.

In the photo, plane wave fronts can be seen to generate circular wave fronts radiating from the gap toward the beach.

Diffraction is part of the reason why we can hear sounds coming through an open doorway even if we are not in a direct line with the source.

Light waves also diffract and this is what is responsible for the different-coloured lights you see when you hold the blank surface of a DVD up to face the sunlight.

Diffraction is greatest when the size of the gap or obstacle is close to the wavelength of the incident waves. In the photo, the gap is about 80 m wide and the wavelength is about 20 m.

A smaller gap would cause the wave fronts to bend even more.

The larger the wavelength, the more the more pronounced the circular wave fronts will be as they go through a gap or round an obstacle.

The speed, frequency and wavelength of diffracted waves will be the same as the incident wave.

11 Copy and complete the table below for red light travelling from a vacuum into a glass block. (Assume the angle of incidence is less than the critical angle.) You will need to use the wave equation (see page 44).

Feature	Incident wave	Reflected wave	Refracted wave
Speed	300 000 kms^{-1}		
Direction			toward normal
Colour	red		
Frequency	500 THz		
Wavelength	600 nm		

12 The box opposite lists different transparent media from the most to the least optically dense. Use this information to answer the questions below.

diamond
glass
perspex
water
ice
air
vacuum

- **a** In which medium is light slowest? Fastest?
- **b** Which way is light bent in a more optically dense medium? A less dense medium?
- **c** What would happen to the direction of a light wave travelling from air into water?
- **d** What would happen to the direction of a light wave travelling from glass into ice?
- **e** What would happen to the direction of a light wave travelling from diamond into air then into water?
- **f** Which media boundary would bend light the most?

13 The diagram shows parallel red and violet light rays passing through a triangular prism.

incident red and violet rays
refracted red and violet rays

- **a** Which ray bends the most? Why?
- **b** What is the relationship between frequency and the angle of refraction?
- **c** What is the diagram an example of?
- **d** What two factors influence how much a particular light wave will bend when it enters a new medium?
- **e** What would happen if white light is used instead?

THE Material WORLD

Basic Chemistry Concepts

Learning Outcomes - On completing this unit you should be able to:
- explain how the structure of atoms accounts for their behaviour
- interpret word/ symbol equations, and explain how factors affect reaction rates
- describe and explain physical and chemical properties of metals and nonmetals
- describe different kinds of reactions, such as neutralisation and oxidation
- *summarise and link basic chemistry concepts on a concept map.*

Atoms, Elements and Compounds

Atoms

- All **matter** consists of extremely small particles called **atoms**, assembled out of **sub-atomic particles**.
- Atoms have a central **nucleus** consisting of positively charged **protons** and uncharged **neutrons**. Negatively charged **electrons** fly around the nucleus.
- Atoms are normally electrically neutral overall, as they have equal numbers of electrons and protons.
- Electrons fill **electron shells** around the nucleus from the innermost, lowest energy shell outwards. The first accommodates up to 2, the second and third, up to 8.
- Those atoms that have full outer shells are stable.
- There are 98 different kinds of atoms, each with a different number of protons (eg chlorine atoms have 17; sodium atoms, 11), known as its **atomic number**.

Elements

- An **element** is a chemical made of only one type of atom. As there are 98 different kinds of atoms, there are 98 naturally occurring elements.
- Each has a unique name (eg sodium) and symbol (Na).
- All atoms of an element have the same number of protons, and normally an equal number of electrons.
- The **Periodic Table** (see page 59) lists elements by their atomic number but arranges them in **periods** (rows) according to which electron shell is being filled.
- All elements in a **group** (column) have atoms with the same number of electrons in the outermost occupied shell and, as a consequence, they react in similar ways.
- Metallic elements are found on the left and middle of the table, and nonmetallic elements on the right.

Ions and Molecules

- As having a full outer shell is much more stable than a having a partially filled one, atoms take, share or donate electrons to fill or empty their outer shell.
- Atoms of metallic elements that have a few electrons in their outer shell (eg Na with 1) readily lose them to other atoms. As they now have more protons than electrons, the atoms have a positive charge overall (eg Na will have a charge of +1).
- Atoms of nonmetallic elements that are a few electrons short of a full outer shell (eg Cl with 7) pull electrons off metal atoms. As they now have more electrons than protons, the atoms have a negative charge overall (eg Cl will have a charge of –1).
- Charged atoms are called **ions**, and the charge is written as a superscript (eg Na^+, Mg^{2+}, Cl^-, O^{2-}).
- Opposite charges attract, so the atoms that gained electrons and those that lost them are held together by **ionic bonding** (eg Na^+ and Cl^- ions form bonds).
- Atoms of nonmetallic elements also share electrons with other nonmetal atoms in order to fill shells and become more stable. The attraction the nuclei have for the shared electrons is called **covalent bonding**.
- The group of atoms sharing electrons is called a **molecule** (eg H_2, O_2). (The subscripts give the number of atoms in the molecule.)

Compounds

- A **compound** is an electrically neutral substance formed when different kinds of atoms bond together.
- In an **ionic compound** (eg salt), oppositely charged ions are held together in a fixed array called a **lattice**.
- The formula of an ionic compound gives the *ratio of ions* in the lattice (eg $MgCl_2$ means 2 Cl^- to every 1 Mg^{2+}).
- In a **molecular compound** (eg carbon dioxide), different kinds of atoms are bonded together in the molecules.
- The formula of a molecular compound gives the *actual number of atoms* (eg H_2O has 2 H atoms and 1 O atom).

Chemical Reactions

Reactions and Equations

- A **chemical reaction** occurs when existing **chemicals** (the **reactants**) react to form new ones (the **products**).
- At the atomic level, a chemical reaction involves the transfer or sharing of electrons when atoms collide.
- A reaction can be written as a **word equation**, eg: Hydrogen gas + oxygen gas → water.
  The '+' means 'and' and the arrow, 'changed into'.
- A reaction can also written as a **symbol equation**, eg: $H_2 + O_2 \rightarrow H_2O$.
- A **balanced equation** shows the ratio of atoms, ions or molecules involved. The number of each type of atom must be the same on both sides of the arrow, eg: $2H_2 + O_2 \rightarrow 2H_2O$ has 4 H's and 2 O's on both sides. Large numbers give the number of molecules or ions.

Exothermic and Endothermic Reactions

- In a chemical reaction, existing bonds between atoms are broken and new ones formed.
- As chemical bonds have **chemical potential energy**, forming bonds absorbs energy, usually in the form of

ISBN: 9780170262316

heat, and breaking bonds releases energy as heat.

- An **exothermic reaction** (eg the combustion of H_2 gas to form water) releases more heat than it absorbs, so products are hotter than reactants.
- An **endothermic reaction** (eg baking soda reacts with vinegar to give CO_2) absorbs more heat than it releases, so products are cooler than reactants.

Reaction Rate

- The **reaction rate** can be found by measuring how fast a reactant disappears or a product forms.
- At the atomic level, chemicals react if their atoms, ions or molecules collide with sufficient force.
- The reaction rate is affected by factors such as reactant **concentration**, temperature and particle size.
- Increasing a reactant's concentration results in more collisions. Raising the temperature makes reactant particles move faster, increasing the force and frequency of collisions. Smaller solid reactants (eg powders) expose more surface atoms to collisions.
- Some reactions start spontaneously, but most require some **activation energy** (heat) to start.
- **Catalysts** are chemicals that speed up reactions by reducing the required activation energy, which means reactions occur faster at lower temperatures.

Metals and Nonmetals

Metals

- Metals, with a few exceptions, are: shiny, silvery-grey solids with high melting and boiling points; **malleable** and **ductile**; good conductors of heat and electricity.
- Metallic elements are found on the left and middle of the Periodic Table, with partially filled outer shells.
- Metal atoms have at least one electron that is free to wander from atom to atom. The attraction that the positively charged atomic nuclei have for the **free electrons** is called **metallic bonding**. It is responsible for all of the above **physical properties**.
- Most metal atoms are reactive, as they readily lose a few outer shell electrons to become more stable ions.
- The **chemical properties** of metals include reacting with: oxygen to form metal oxides; water to form metal hydroxides and H_2 gas; and acid to form a **salt** and H_2 gas.
- The far left group of elements on the Periodic Table are the most reactive; moving across the table the metals become less and less reactive.
- When the reactions of freshly cut metals are compared, a **reactivity series** can be established:\
  K > Na > Li > Ca > Mg > Al > Zn > Fe > Pb > Cu > Ag > Au.

Nonmetals

- Nonmetals are found on the right of the Periodic Table, which means they have nearly full or full shells.
- The physical properties of nonmetal elements are more varied. Nonmetals have low melting and boiling points, so most are gases. In the solid state they are often weak and brittle, and mostly poor conductors.
- Most of these physical properties are related to the fact that the nuclei of nonmetal atoms *attract their outer electrons strongly so there are no free electrons.*
- Some nonmetals exist as **allotropes** (alternative forms).
- The far right group of nonmetals are unreactive gases, existing as monatomic molecules. The others exist as polyatomic molecules (eg N_2, O_2, S_8) or lattices (C).
- Atoms of reactive nonmetals take outer electrons off metal atoms to form negative ions (eg O^{2-}), or share them with other nonmetal atoms to form molecules (eg SO_2), becoming more stable with full outer shells.
- Reactive nonmetal atoms combine with oxygen to form **oxides** (eg $C + O_2 \rightarrow CO_2$). When these molecules dissolve in water, they form acidic solutions.

Acid and Alkali Reactions

- **Acids** are chemicals that release hydrogen ions (H^+) when dissolved in water (eg $HCl \rightarrow H^+ + Cl^-$). The H^+ ions result in an **acidic solution**.
- **Alkalis** are chemicals that release hydroxide ions (OH^-) when dissolved in water (eg $NaOH \rightarrow Na^+ + OH^-$). The OH^- ions result in an **alkaline solution**.
- The strength of an acidic or alkaline solution depends on the extent to which the acid or alkali **ionises** in water.
- The pH scale show how acidic or alkaline a solution is.

0	1	2	3	4	5	6	7	8	9	10	11	12	13	14
strongly acidic						weakly		weakly						strongly alkaline
							neutral							

- A pH of 7 is neither acidic nor alkaline - it is neutral.
- When hydrogen and hydroxide ions collide in a solution they form water molecules ($H^+ + OH^- \rightarrow H_2O$).
- **Neutralisation** occurs when an acid (eg HCl) reacts with an alkali (eg NaOH) to give a salt plus water (eg $H^+ + Cl^- + Na^+ + OH^- \rightarrow NaCl + H_2O$).

Oxidation and Combustion

- Oxygen atoms, which are two electrons short of a full outer shell, strongly attract outer electrons of other atoms, either taking them or gaining a share of them.
- This makes oxygen a very reactive chemical, which combines readily with other elements, ions or compounds in a reaction called **oxidation** (eg $2Mg + O_2 \rightarrow 2MgO$). Oxidation is a highly exothermic reaction.
- The **combustion** of fuels is very rapid oxidation that results in CO_2 and H_2O if sufficient O_2 is present.

SCIENCE SKILL: Concept Mapping

A useful way of understanding the connections between scientific concepts is to construct a concept map.

Steps:

1. Obtain an A3 sheet and some colour felt-tip pens.
2. Identify the theme of the unit: the orange title, and the major topics: the blue headings.
3. Draw a bubble with the theme and the major topics in bubbles radiating off it.
4. Identify the sub-topics of each major topic: the green sub-headings; and attach them as bubbles to the major topics.
5. Identify and link key concepts to the sub-topic bubbles, expressing them as briefly as you can, using symbols wherever possible.
6. Use arrows or colour to highlight related concepts.

ISBN: 9780170262316

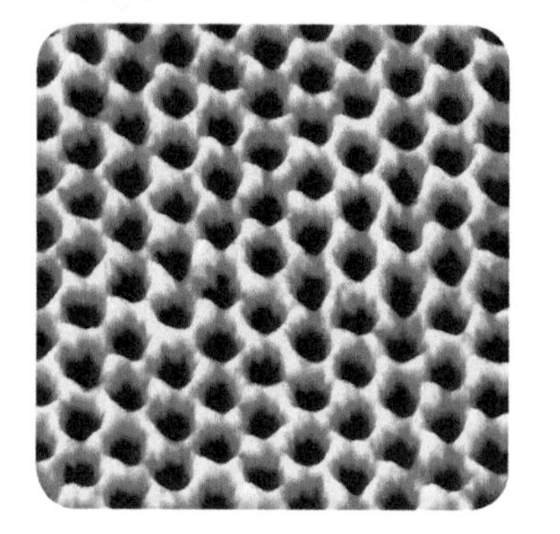

Atoms and Elements

Learning Outcomes - On completing this unit you should be able to:
- draw a typical atom to show the location of the sub-atomic particles
- state the differences between protons, neutrons and electrons
- describe how electrons are arranged in shells around the nucleus
- define an element and explain how they are grouped on the Periodic Table
- *determine the electron configuration of a given atom.*

The Structure of Atoms

- The term **matter** is used to describe all substances found in our world. They may be solids, liquids or gases.
- Scientists have discovered that all matter is made up of extremely small particles called **atoms**. To get an idea of how small atoms are, there would be about 850 000 000 000 000 000 000 000 atoms in a 1 centimetre copper cube.
- The amazing image above shows the location of actual atoms of a substance called graphene, a form of carbon.

Sub-atomic Particles

- Atoms can be considered to be tiny balls but scientists have found that they have an *internal structure.*
- All atoms have a tiny central area called the **nucleus**, where nearly all the mass of the atom is concentrated. Most of an atom's volume is empty space occupied by a few fast-moving objects vibrating around the nucleus.
- Atoms are made of a variety of **sub-atomic particles**, of which there are three you need to know about for NCEA level 1: protons, neutrons and electrons.
- **Protons**, symbol p or ⊕, are relatively heavy particles with a single positive electrical charge. They are closely packed in the nucleus along with the neutrons.
- **Neutrons**, symbol n or ○, have a similar mass as protons, but are **neutral** (uncharged).

electron cloud
electron
neutron
proton
central nucleus

- Protons in the nucleus repel each other (like charges repel, page 16), but a stronger **nuclear force** holds the protons and neutrons together, preventing the nucleus from disintegrating.
- **Electrons**, symbol e or ⊖, have a single negative charge, but are *much lighter than protons or neutrons.*
- Electrons vibrate around the nucleus at high speeds, occupying a space called the **electron cloud**.

Electrically Neutral Atoms

- If an atom has equal numbers of electrons and protons (eg the atom opposite has 5 e and 5 p), then it has *no overall charge* and is said to be a **neutral atom**.
- The negative charge of the electron cloud cancels out the positive charge of the nucleus.

Electron Shells

- Although they move very rapidly, electrons are bound to the nucleus by the attractive force between the positively charged protons and the negatively charged electrons (unlike charges attract).
- Electrons are in constant motion but can have only *discrete amounts of energy.* Those with the lowest possible energy levels are found closest to the nucleus. Electrons with higher energy levels are found further out.
- The region where electrons with the same amount of energy are most likely to be found is called an **electron shell** (energy level).
- Higher energy electron shells enclose lower energy electron shells.
- Electrons do not travel in fixed orbits around the nucleus like planets around the sun, rather they exist as *energy waves confined to a specific region* (energy shell).

Filling the Shells

- *Electrons prefer to have the least amount of energy possible, as this is the most stable state.* The shells closest to the nucleus have the lowest energy levels, but they can only be occupied if an electron has the right amount of energy and if there is a vacancy.
- Electrons with higher energy levels can emit a discrete amount of energy as an electromagnetic wave (see page 44) and then occupy a vacant lower energy shell.
- Up to 2 electrons can coexist in the innermost (first) shell, up to 8 electrons in the second shell, and up to 8 in the third shell.
- These limits are due to **electrostatic forces** between electrons (like charges repel). Electrons repel each other, so only a limited number can fit into each shell.
- The innermost electrons fill first: up to 2 in the 1st, then up to 8 in the 2nd and up to 8 (or 18) in the 3rd shell.
- The atom opposite has 13 electrons: 2 have occupied the innermost shell, 8 the second shell, and the remaining 3 are found in the third shell.

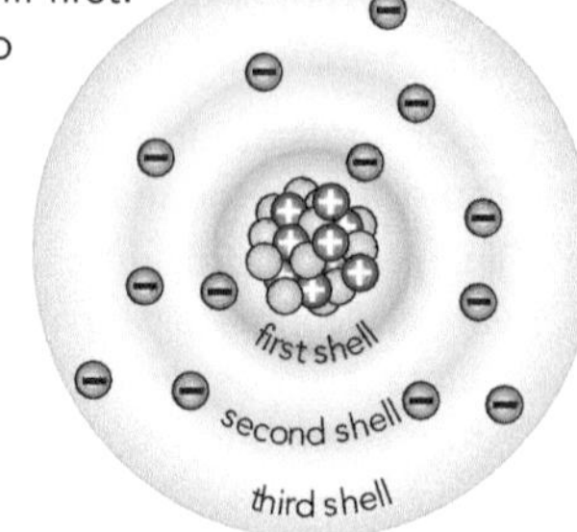

- The electron arrangement of an atom is *most stable* when the outermost occupied shell is filled with the maximum possible number of electrons.

ISBN: 9780170262316

Unique Atoms

- Atoms are not all identical. Chemists have identified 98 different kinds of naturally occurring atoms.
- Each kind of atom has a unique number of protons in its nucleus. For example, helium atoms all have 2 protons, oxygen atoms all have 8, and aluminium atoms, 13.
- The number of protons in the nucleus of an atom is called its **atomic number**, symbol ***Z***. The atomic number of helium atoms is 2, that of oxygen atoms is 8, and for aluminium atoms, it's 13.
- Atom also have neutrons in the nucleus. The number of neutrons in the nucleus of an atom is called its **neutron number**, symbol ***N***.
- The **mass number**, symbol ***A***, of an atom is the number of protons and neutrons in the nucleus.

$$A = Z + N$$

The above formula can be rearranged to find the number of protons or neutrons in an atom: $Z = A - N$ and $N = A - Z$.

- This atom has 17 protons and 20 neutrons, so $A = 17 + 20 = 37$.

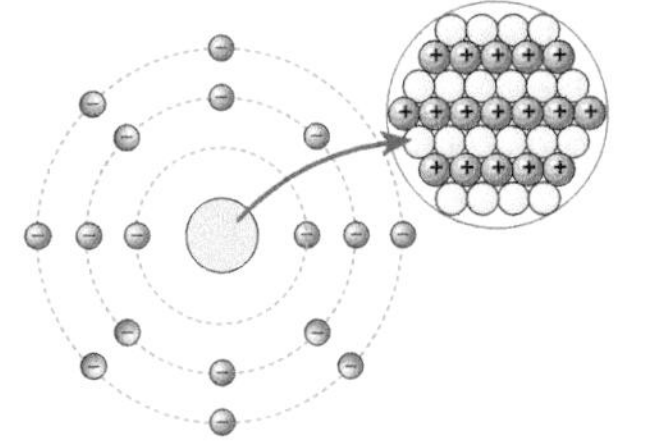

- Different kinds of atoms vary in mass because they have *differing numbers of nuclear particles.*
- If an atom is electrically neutral (equal numbers of electrons and protons), then its *atomic number also gives the number of electrons.* The neutral atom shown above will therefore have 17 electrons.
- Different kinds of atoms vary in size because they have *differing numbers of electrons.* (An atom's size is determined by the size of its electron cloud.)
- Each kind of atom is given a unique name and chemical symbol, eg all atoms with 8 protons are called oxygen atoms, symbol O.
- *Atoms with the same number of protons have the same symbol.*

Elements

- The substances you can see or touch are made up of huge numbers of atoms. Helium gas in a balloon, a drop of mercury and a lump of sulfur are all made of vast numbers of atoms.
- A substance made of one kind of atom only is called an **element**. The helium gas in the balloon is an element, as it is made of helium atoms only; the drop of mercury is an element, as it is made of mercury atoms only; and the lump of sulfur is an element, as it is made of sulfur atoms only.
- As there are over 100 kinds of atoms, *there are over 100 elements.* The same symbol is used for an element and its atoms, eg He for the element helium and for helium atoms.

The Periodic Table

- Chemists arrange the elements on the **Periodic Table**. The elements are placed in order, according to the atomic number ***Z*** of their atoms.
- Hydrogen is the first element, as the atomic number of H atoms is 1. Helium is the second element, as the atomic number of He atoms is 2.
- The elements are placed in rows called **periods** from left to right. When a period is full, the next element is placed on the left of the period below.
- The number of elements that can go across a period relates to the number of electrons that can fill a particular shell. The first period has just 2 elements, the second and third periods have 8 elements, and so on.
- The metal elements are on the left and the nonmetal elements on the right. (Hydrogen is neither as it behaves like a metal and a nonmetal.)
- Each column of the table is called a **group**. All *elements in a group react in similar ways*, eg the elements helium, neon and argon in the far right group are all unreactive.

Periodic Table of the First 20 Elements

The small numbers give the atomic number ***Z***.

	Group 1	Group 2	Group 13	Group 14	Group 15	Group 16	Group 17	Group 18	
Period 1	Hydrogen **H** 1							Helium **He** 2	Period 1
Period 2	Lithium **Li** 3	Beryllium **Be** 4	Boron **B** 5	Carbon **C** 6	Nitrogen **N** 7	Oxygen **O** 8	Fluorine **F** 9	Neon **Ne** 10	Period 2
Period 3	Sodium **Na** 11	Magnesium **Mg** 12	Aluminium **Al** 13	Silicon **Si** 14	Phosphorus **P** 15	Sulfur **S** 16	Chlorine **Cl** 17	Argon **Ar** 18	Period 3
Period 4	Potassium **K** 19	Calcium **Ca** 20							

For NCEA level 1 you need to learn the symbols of the first 20 elements in order.

SCIENCE SKILL: Determining Electron Configurations

For each of the first 20 elements, you need to be able to work out the **electron configuration** of its atoms. This tells you how many electrons are in each shell. *Remember, inner shells fill first.* The configuration is written in the format: x, y, z, ..., where x is the number in the first shell, y the number in the second shell, etc.

Problem:

Work out the electron configuration of sulfur atoms.

Steps:

1. Locate the element sulfur on the table and record its atomic number: ***Z*** = 16 (ie, each atom has 16 protons).
2. As an atom is considered to be neutral, the atomic number also gives the number of electrons: sulfur atoms have 16 electrons.
3. Now proceed to fill each electron shell in order until you run out of electrons.
   - **a** The innermost shell will take up to 2 electrons: S will have 2 in the first shell, which leaves 14 electrons (16 – 2 = 14).
   - **b** The second shell will take up to 8 electrons: S will have 8 in the second shell, which leaves 6 electrons (14 – 8 = 6).
   - **c** The third shell will take up to 8 as well: S has 6 electrons left to go into the third shell.
4. Write the electron configuration in the standard format: 2, 8, 6.

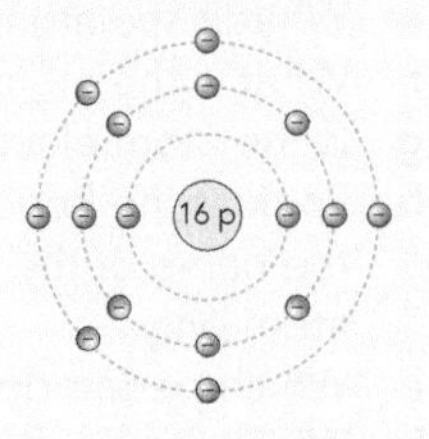

Revision Activities

1 Match up the descriptions with the terms.

Terms	Descriptions
a matter	A number of neutrons in an atom's nucleus
b atom	B atom with no overall charge
c nucleus	C positively charged particle found in the nucleus
d sub-atomic particle	D central area of an atom with protons and neutrons
e proton	E number of protons in the nucleus of an atom
f neutron	F negatively charged particle in motion around the nucleus
g nuclear force	G all the substances in our world
h electron	H row of the Periodic Table
i electron cloud	I substance made of one kind of atom only
j neutral atom	J particles that atoms are made of
k electron shell	K force between electrically charged objects
l electrostatic force	L uncharged particle found in the nucleus
m atomic number	M region around the nucleus that equal-energy electrons occupy
n neutron number	N extremely small particles all matter is made of
o mass number	O number of protons and neutrons in the nucleus
p element	P space where an atom's electrons are likely to be found
q Periodic Table	Q chart of elements arranged according to atomic number
r period	R force binding protons and neutrons in the nucleus
s group	S column of the Periodic Table

2 Explain the difference between the terms in each of the items below:

a a nucleus and an electron cloud
b a proton, a neutron and an electron
c atomic, neutron and mass numbers
d an atom and an element
e a period and a group on the Periodic Table.

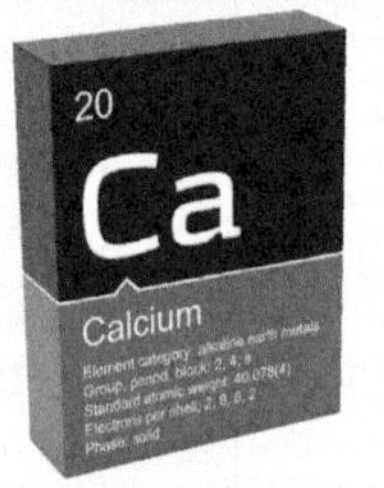

3 Copy and complete the table below summarising the properties of sub-atomic particles.

Particle	Mass	Charge	Location
proton	heavy		
neutron			nucleus
electron		negative	

4 In the diagram opposite, all of the sub-atomic particles of the atom are shown.

a How many protons are there?
b How many neutrons are there?
c How many electrons are there?
d What is the atom's atomic number?
e What is the atom's neutron number?
f What is the atom's mass number?
g Is the atom electrically neutral? How do you know?
h Look at the Periodic Table on page 57 and identify the name of the element that is made of this type of atom only.
i Which period does the element belong to?
j Which group does the element belong to?

5 Decide whether the following statements are true or false. Rewrite the false ones to make them correct.

a All matter is made of atoms.
b An atom's central area is called the nucleus and the surrounding area the electron cloud.
c Protons have a negative charge and electrons a positive charge.
d Both protons and neutrons are much heavier than electrons.
e Like charges attract and unlike charges repel.
f Neutral atoms have equal numbers of protons and neutrons.
g The higher the energy of an electron, the closer it will be to the nucleus.
h Electron shells are filled by equal energy electrons.
i Different types of atoms have different numbers of protons.
j The number of electrons in a neutral atom is equal to the atomic number.
k An element is made of identical atoms.
l Elements in a group react in similar ways.

6 The diagram opposite shows a neutral atom with 10 neutrons.

a How many electrons does it have?
b How many protons will it have?
c What is its atomic number?
d What kind of atom is it? (Look at the Periodic Table on page 57.)
e What is its mass number?
f Is the element a metal or a nonmetal?
g What group does it belong to?
h What other element reacts similarly to this element?

7 Explain why the right diagram is a more accurate depiction of the location of electrons than the left one.

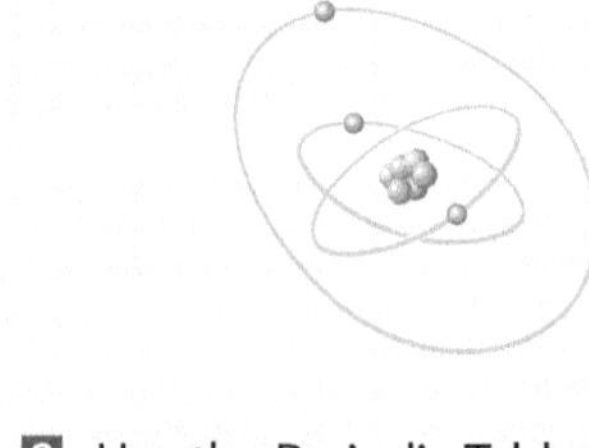

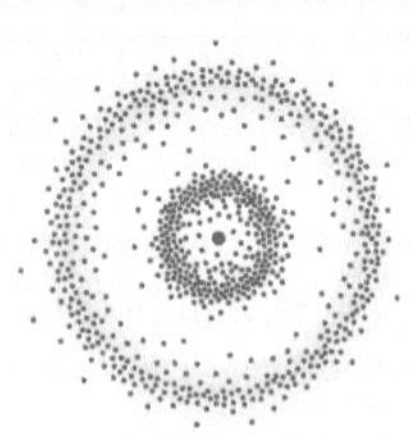

8 Use the Periodic Table on the opposite page to find the symbols for the following elements.

a fluorine
b sodium
c iron
d bromine
e calcium
f lead
g mercury
h silver
i iodine
j zinc
k copper
l nickel
m chlorine
n potassium
o argon
p aluminium
q carbon
r nitrogen
s neon
t lithium
u helium
v magnesium
w zinc
x silicon
y phosphorus
z oxygen

ISBN: 9780170262316

9 Use the Periodic Table below to name these elements.

a Fe	**h** Ar	**o** He	**v** Mg
b Ca	**i** Ag	**p** C	**w** Hg
c K	**j** Si	**q** Br	**x** Cu
d Pb	**k** P	**r** Li	**y** Al
e Zn	**l** S	**s** O	**z** Au
f H	**m** N	**t** F	
g Be	**n** Ne	**u** Na	

10 Use the Periodic Table below to find the names of elements whose atoms have these atomic numbers.

a 1	**e** 10	**i** 16	**m** 29
b 2	**f** 11	**j** 17	**n** 30
c 6	**g** 12	**k** 20	**o** 82
d 7	**h** 13	**l** 26	

11 How many protons do the following atoms have?

a hydrogen	**e** aluminium	**i** lithium
b carbon	**f** calcium	**j** sodium
c magnesium	**g** sulfur	**k** chlorine
d nitrogen	**h** fluorine	**l** zinc

12 How many electrons do these neutral atoms have?

a iron	**e** hydrogen	**i** mercury
b lead	**f** magnesium	**j** helium
c sulfur	**g** carbon	**k** bromine
d oxygen	**h** neon	**l** nitrogen

13 Refer to the Periodic Table below when answering these questions.

a Why is the chart called the Periodic Table?
b There are over 100 elements. How many elements are shown on this reduced version of the Periodic Table?
c How are the elements arranged on the table?
d What does the term 'period' refer to?
e Which elements are found in period 2 of the table?
f What does the term 'group' refer to?
g Which elements are found in group 2 of the table?
h Which elements are found in group 18 of the table?
i Where are the metals found on the table?
j Where are the nonmetals found?
k Which group has atoms with completely filled shells?
l Which group has atoms with one outer shell electron?
m Which group has atoms that are one electron short of a full shell?

14 Determine the atomic, neutron and mass numbers (***Z***, ***N*** and ***A***) of each of the atoms shown opposite.

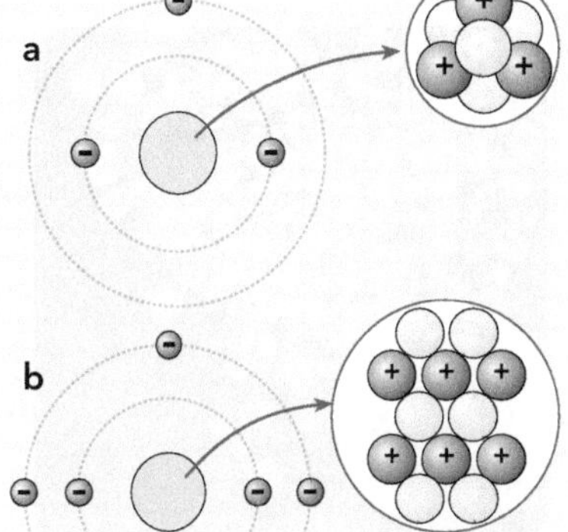

15 Decide whether each of the three atoms shown opposite is electrically neutral or not.

16 Determine the electron configuration of each of the three atoms shown opposite.

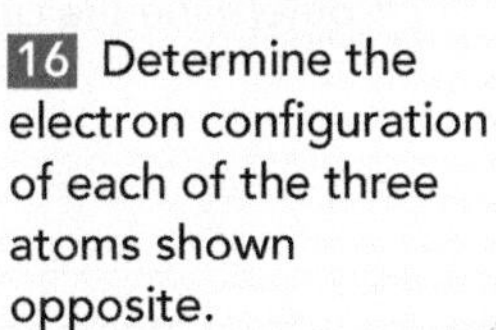

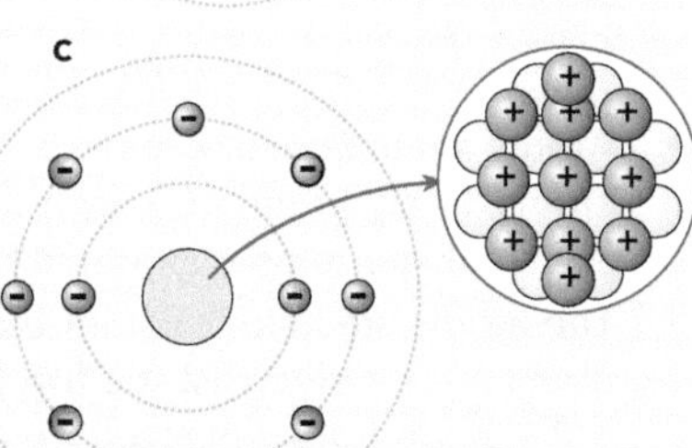

17 Identify the name of each atom and the group to which it belongs on the Periodic Table.

18 Copy and complete the chart below.

Element	Atomic number	Neutron number	Mass number
lithium			7
carbon			12
fluorine			19
aluminium		14	
chlorine		18	

19 The diagram opposite shows an electrically neutral atom that is unreactive because each of the occupied electron shells is full.

a What atom is it?
b What is its electron configuration?
c What group does it belong to?

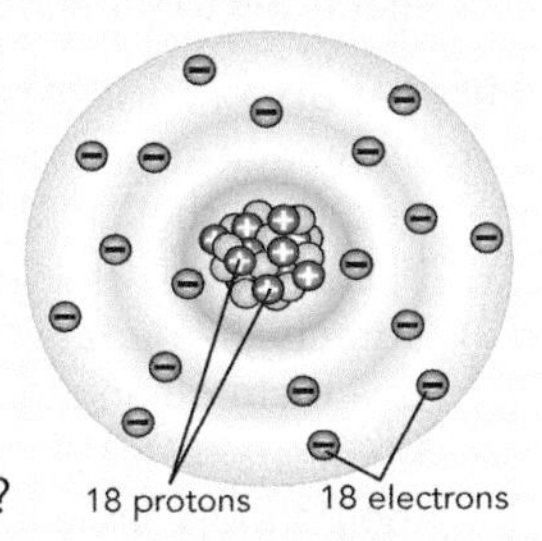

The Periodic Table of the Elements

Key: name / symbol / atomic number — Potassium K 19 **Metals**; Chlorine Cl 17 **Nonmetals**

	Group 1	Group 2	Group 3	Group 4	Group 5	Group 6	Group 7	Group 8	Group 9	Group 10	Group 11	Group 12	Group 13	Group 14	Group 15	Group 16	Group 17	Group 18
Period 1	Hydrogen H 1																	Helium He 2
Period 2	Lithium Li 3	Beryllium Be 4											Boron B 5	Carbon C 6	Nitrogen N 7	Oxygen O 8	Fluorine F 9	Neon Ne 10
Period 3	Sodium Na 11	Magnesium Mg 12											Aluminium Al 13	Silicon Si 14	Phosphorus P 15	Sulfur S 16	Chlorine Cl 17	Argon Ar 18
Period 4	Potassium K 19	Calcium Ca 20	Scandium Sc 21	Titanium Ti 22	Vanadium V 23	Chromium Cr 24	Manganese Mn 25	Iron Fe 26	Cobalt Co 27	Nickel Ni 28	Copper Cu 29	Zinc Zn 30	Gallium Ga 31	Germanium Ge 32	Arsenic As 33	Selenium Se 34	Bromine Br 35	Krypton Kr 36
Period 5	Rubidium Rb 37	Strontium Sr 38	Yttrium Y 39	Zirconium Zr 40	Niobium Nb 41	Molybdenum Mo 42	Technetium Tc 43	Ruthenium Ru 44	Rhodium Rh 45	Palladium Pd 46	Silver Ag 47	Cadmium Cd 48	Indium In 49	Tin Sn 50	Antimony Sb 51	Tellurium Te 52	Iodine I 53	Xenon Xe 54
Period 6	Caesium Cs 55	Barium Ba 56	Lanthanum Lu 71	Hafnium Hf 72	Tantalum Ta 73	Tungsten W 74	Rhenium Re 75	Osmium Os 76	Iridium Ir 77	Platinum Pt 78	Gold Au 79	Mercury Hg 80	Thallium Tl 81	Lead Pb 82	Bismuth Bi 83	Polonium Po 84	Astatine At 85	Radon Rn 86
Period 7	Francium Fr 87	Radium Ra 88																

ISBN: 9780170262316

Ions and Compounds

Learning Outcomes - On completing this unit you should be able to:
- explain how chemical bonds hold atoms together
- define the terms element, compound, molecule, ion and lattice
- explain why atoms react to achieve full outer electron shells
- predict the behaviour of atoms from the number of valence electrons
- *determine the chemical formulas of ions and ionic compounds.*

Atom Behaviour

- Atoms are the building blocks of matter. Most atoms are joined to other atoms, to either the same or a different kind. Atoms may be joined together in a small group or a huge number in a fixed array.
- The atoms are held together by *the attraction between opposite electrical charges*. The **electrostatic force** that pulls and holds atoms together in a definite arrangement results in a **chemical bond**.

The Behaviour of Atoms

- When the outermost occupied **electron shells** of atoms are not full, the atoms are in a higher energy state than when they are full. As a high energy state is unstable, *atoms react to fill shells, thus becoming more stable.*
- The **electron configurations** (arrangement) of group 18 atoms are helium 2, neon 2,8 and argon 2,8,8. As all three kinds of atoms have full outer shells, they are in low energy, stable states and those elements are unreactive.
- Other kinds of atoms with partly filled shells are in high energy states. They will achieve low energy, stable states as those *shells fill or empty.*
- The electrons in the outermost occupied shell of an atom are called **valence electrons**. These are the ones involved in chemical reactions.
- A **chemical reaction** occurs when atoms take electrons from, lose electrons to, or start sharing electrons with other atoms.

Metal and Nonmetal Atom Interactions

- Nonmetal atoms on the right of the Periodic Table have a nearly full outer occupied shell. If they gain sufficient electrons, that shell will be filled.
- The positively charged nucleus of a nonmetal atom (eg chlorine) strongly attracts the negatively charged valence electrons of nearby atoms, and can pull them off an atom with weak nuclear attraction.
- Metal atoms on the left of the Periodic Table have a nearly empty outer occupied shell. If they lose sufficient electrons, they will end up with full inner shells.
- The nucleus of a metal atom (eg sodium) weakly attracts its own valence electrons, and can lose them to a nearby nonmetal atom that strongly attracts valence electrons.
- In a reaction between a metal and nonmetal element, the metal atoms lose their valence electrons to the nonmetal atoms. *Both kinds of atoms will now be in lower energy states and are therefore more stable and less reactive.*

Chlorine atom Cl — incomplete shell (7e) — 17 p — 17e

Chlorine ion Cl^- — 17 p — 18e

lone electron — full outer shell (8e)

Sodium atom Na — 11 p — 11e

Sodium ion Na^+ — 11 p — 10e

Ions

- A neutral atom has the same number of electrons as protons. The negative and positive charges balance and the atom has no charge overall.
- When atoms gain or lose electrons, they become more stable, *but the positive and negative charges no longer balance*. These charged atoms are called (monatomic) **ions**.
- An ion's charge is shown by the superscript (small raised text) in its **chemical formula**, eg Na^+ has 1 positive charge and O^{2-} has 2 negative charges.
- An atom that gains electrons now has more electrons than protons. As it has more negative charge than positive, it becomes a **negative ion**, eg S^{2-}.
- An atom that loses electrons now has more protons than electrons. As it has more positive charge than negative, it becomes a **positive ion**, eg Mg^{2+}.
- Different kinds of atoms gain or lose electrons depending on the most direct step to getting a full outermost occupied shell.
- Metal atoms with nearly empty outer shells need to lose electrons; nonmetal atoms with nearly full outer shells need to gain them.
- You can predict what ion will occur from the atom's **valence number**, which is the number of electrons in its outer shell (see the table top left on the next page).
- Group 1 atoms will lose their 1 valence electron, eg $H - 1e \rightarrow H^+$, $Na - 1e \rightarrow Na^+$, $K - 1e \rightarrow K^+$.
- Group 2 atoms will lose their 2 valence electrons, eg $Mg - 2e \rightarrow Mg^{2+}$, $Ca - 2e \rightarrow Ca^{2+}$.

ISBN: 9780170262316

Valence Numbers of the First 20 Elements

	Group 1	Group 2	Group 13	Group 14	Group 15	Group 16	Group 17	Group 18	
Period 1	Hydrogen **H** 1							Helium **He** 2	Period 1
Period 2	Lithium **Li** 1	Beryllium **Be** 2	Boron **B** 3	Carbon **C** 4	Nitrogen **N** 5	Oxygen **O** 6	Fluorine **F** 7	Neon **Ne** 8	Period 2
Period 3	Sodium **Na** 1	Magnesium **Mg** 2	Aluminium **Al** 3	Silicon **Si** 4	Phosphorus **P** 5	Sulfur **S** 6	Chlorine **Cl** 7	Argon **Ar** 8	Period 3
Period 4	Potassium **K** 1	Calcium **Ca** 2	For NCEA level 1 you need to be able to work out the valence number for each of the first 20 elements.						

- Group 13 metal atoms will lose their 3 valence electrons, eg $Al - 3e \rightarrow Al^{3+}$.
- Group 14 atoms usually don't form ions.
- Group 15 atoms have 5 valence electrons and need to gain three more, eg $N + 3e \rightarrow O^{3-}$, $P + 3e \rightarrow P^{3-}$.
- Group 16 atoms have 6 valence electrons and need to gain two more, eg $O + 2e \rightarrow O^{2-}$, $S + 2e \rightarrow S^{2-}$.
- Group 17 atoms have 7 valence electrons and need to gain one more, eg $F + 1e \rightarrow F^-$, $Cl + 1e \rightarrow Cl^-$.

Compounds

- An **element** is a substance *made of one kind of atom only*. The atoms may exist singly as in helium gas; or be bonded in pairs as in oxygen gas; or be bonded in a lattice as billions of carbon atoms are in a diamond.
- A **compound** is a substance *made of different kinds of atoms or ions bonded together* (chemically combined).
- Water is a compound in which two hydrogen atoms are bonded to each oxygen atom.
- Salt is a compound in which each sodium ion is bonded to six surrounding chlorine ions and vice versa.

Ionic Compounds

- When metal and nonmetal elements react, electrons are transferred. The nonmetal atoms gain electrons to become negative ions and the metal atoms lose electrons to become positive ions.
- The positive and negative ions are attracted to each other and form a **lattice**, which is a fixed array of a large number of atoms. Each negative ion is surrounded by positive ions and vice versa.
- The attraction between oppositely charged ions holds the lattice together. This is called **ionic bonding**, and substances with ionic bonds are **ionic compounds**.
- The positive and negative charges of ionic compounds balance, as *compounds are neutral*.
- Positive ions are named by adding the word ion after the atom's name (eg magnesium ion Mg^{2+}). Negative ions are given the suffix '-ide' (eg oxide ion O^{2-}, sulfide ion S^{2-}, chloride ion Cl^-).
- Ionic compounds are written with the positive ion first, (eg magnesium oxide $Mg^{2+}O^{2-}$) but the charges are usually omitted (eg MgO).
- The subscript numbers (small lower numbers) in an ionic compound's formula *give the ratio of ions*. So $MgCl_2$ means 1 Mg^{2+} ion to every 2 Cl^- ions (1 Mg^{2+} : 2 Cl^-). (Note: no subscript means 1 of that atom, eg $MgCl_2$ is really Mg_1Cl_2.)

- A table of ions is used to find the ratio of ions in a compound by balancing charges. Included are some more complex ions (ammonium, carbonate, sulfate, hydroxide, nitrate and bicarbonate).

+1	+2	+3	−3	−2	−1
H^+	Be^{2+}	Al^{3+}	N^{3-}	O^{2-}	F^-
Li^+	Mg^{2+}	Fe^{3+}	P^{3-}	S^{2-}	Cl^-
Na^+	Ca^{2+}				
K^+	Fe^{2+}			CO_3^{2-}	OH^-
	Cu^{2+}			SO_4^{2-}	NO_3^-
NH_4^+	Zn^{2+}				HCO_3^-
	Pb^{2+}				
	Ba^{2+}				

Molecular Compounds

- Nonmetal atoms attract valence electrons of identical or different nonmetal atoms. The electrons are shared, resulting in full outer shells and more stable atoms.
- When a pair of atoms share valence electrons, those electrons move around the nuclei of both atoms. The attraction the two nuclei have for the shared electrons is the **covalent bond** that holds the atoms together.
- A **molecule** is a neutral group of covalently bonded atoms. Each has a definite number of atoms that share electrons with a neighbouring atom (or atoms). A charged molecule (eg NO_3^-) is called a **polyatomic ion**.
- Simple molecules involve just two identical atoms. Each atom shares one or more valence electrons. This helps to fill the outer shells of both atoms. Examples are H_2, O_2, N_2 and Cl_2 molecules.

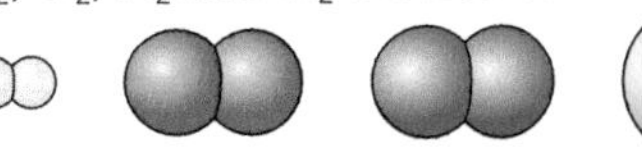

- **Molecular compounds** involve different kinds of atoms sharing electrons. In a water molecule (H_2O), the central O atom shares electrons with two H atoms.

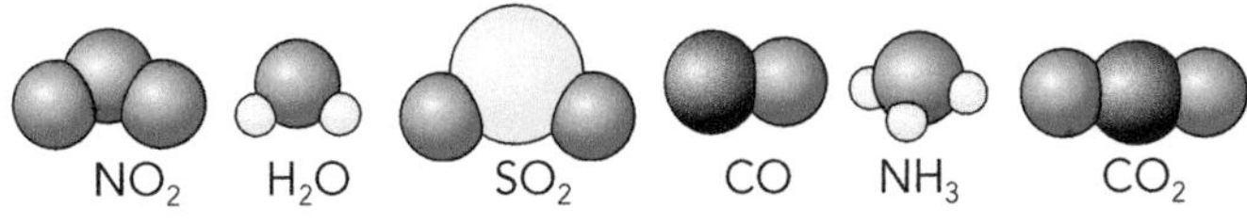

- In formulas, the subscript number gives the number of that type of atom in the molecule (no number means 1).

SCIENCE SKILL: Balancing Formulas

In NCEA exams you will be given a table of ions and asked to write the formula of an ionic compound. The key thing to recall is that compounds are neutral.

Problem: Find the formula of iron (Fe^{3+}) sulphate.

1. Identify the formulas of the positive and negative ions: Fe^{3+} and SO_4^{2-}.
2. Write out the basic formula for the compound, putting the positive ion first: $Fe^{3+}SO_4^{2-}$.
3. If the charges balance, then the ions are in the correct ratio but drop the charges. The two charges (3+ and 2–) don't balance.
4. If the charges don't balance, find the correct ratio by **dropping and swapping** the charge numbers:

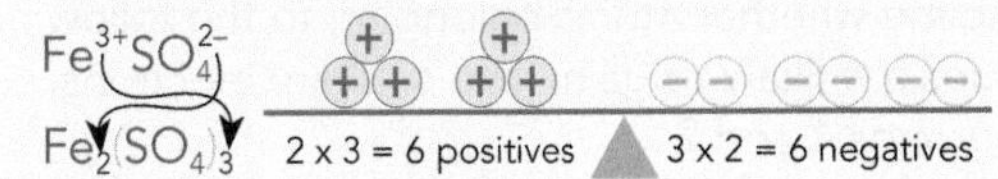

5. Write the formula without charges but with brackets for multiple polyatomic ions: $Fe_2(SO_4)_3$.

Revision Activities

1 Match up the descriptions with the terms.

Terms	Descriptions
a electrostatic	A occurs when electrons are transferred or shared
b chemical bond	B bond between atoms that have transferred electrons
c electron shell	C occurs when electrostatic attraction binds atoms
d electron configuration	D space around the nucleus in which electrons with identical energy levels are found
e valence number	E describes number and type of atoms in a chemical
f chemical reaction	F substance in which different kinds of atoms are bonded together/chemically combined
g ion	G regular array of atoms or ions in a solid
h chemical formula	H atom that has more protons than electrons
i negative ion	I describes how electrons are arranged in shells
j positive ion	J molecule made of different kinds of atoms
k valence number	K pure substance in which different kinds of ions are held together by ionic bonds
l element	L atom that has more electrons than protons
m compound	M electrons in the outermost occupied shell of an atom
n lattice	N electrically charged atom or group of atoms
o ionic bond	O group of atoms that are involved in sharing electrons
p ionic compound	P substance made of one kind of atom only
q polyatomic ion	Q number of electrons in the outer occupied shell
r molecule	R attraction between opposite electrical charges
s molecular compound	S bond between atoms that are sharing electrons
t covalent bond	T molecule with an overall electrical charge

2 Explain the differences between the terms in each of the items below:

a atoms and ions
b a negative and a positive ion
c an element and a compound
d a lattice and a molecule
e an ionic and a molecular compound
f ionic and covalent bonding.

3 Complete the following table for the first 20 elements by referring to the Periodic Table on page 61, then answer the questions below.

Element	Symbol	No. of electrons	Electron config.	Valence no.	Gp no.
Hydrogen	H	1	1	1	1
Helium	He	2	2	2	18
Lithium	Li	3	2,1	1	1
etc					

a Which atoms are already in a low energy state and are therefore stable?
b Which kinds of atoms will become more stable by:
- **i** losing one electron?
- **ii** losing two electrons?
- **iii** losing three electrons?
- **iv** gaining two electrons?
- **v** gaining one electron?

c Describe whether atoms belonging to the following groups are likely to gain, lose or share electrons:
- **i** groups 1 and 2
- **ii** groups 14 and 15
- **iii** groups 16 and 17
- **iv** group 18.

4 Decide whether the following statements are true or false. Rewrite the false ones to make them correct.

a Most atoms are not bonded to other atoms.
b Chemical bonds hold atoms together in definite arrangements.
c Atoms with full electrons shells are unstable.
d Atoms may gain, lose or share electrons to achieve full shells.
e Monatomic ions are charged atoms.
f Some nonmetal atoms are able to pull electrons off metal atoms.
g A compound is made of identical atoms bonded together.
h Most metal atoms tend to lose electrons to achieve filled shells.
i Ions are formed when electrons are transferred between atoms.
j Ionic bonding involves attraction between same-charge ions.
k Metal elements form compounds with nonmetal elements.
l A molecule consists of a fixed number of atoms.
m A lattice consists of a fixed number of atoms in a fixed array.

5 Study the diagram, which shows electron transfer occurring between a magnesium and a sulfur atom, then answer the questions below it.

Sulfur atom S

Sulphide ion S^{2-}

16 p

12 p

Magnesium ion Mg^{2+}

Magnesium atom Mg

a Write down the electron configurations for sulfur and magnesium atoms.
b What are the valence numbers for sulfur and magnesium atoms?
c What do sulfur atoms need to do in order to become more stable?
d What do magnesium atoms need to do in order to become more stable?
e Which atoms exert the strongest pull on valence electrons?
f How many electrons are transferred and from which atoms to which atoms?
g Write down the electron configurations for sulfide and magnesium ions.

ISBN: 9780170262316

6 Study the table opposite, then answer the questions below.

a Which atoms have full shells?
b Describe how atoms with full shells behave.
c How do metal atoms behave?
d How many valence electrons will calcium atoms (Ca) lose in order to acquire full shells?
e Describe how carbon atoms are likely to behave.
f How many electrons do oxygen atoms need to gain to have full shells?
g Will hydrogen atoms form ions or molecules?
h Summarise the behaviour of nonmetal atoms.
i Which nonmetal atoms form ions?

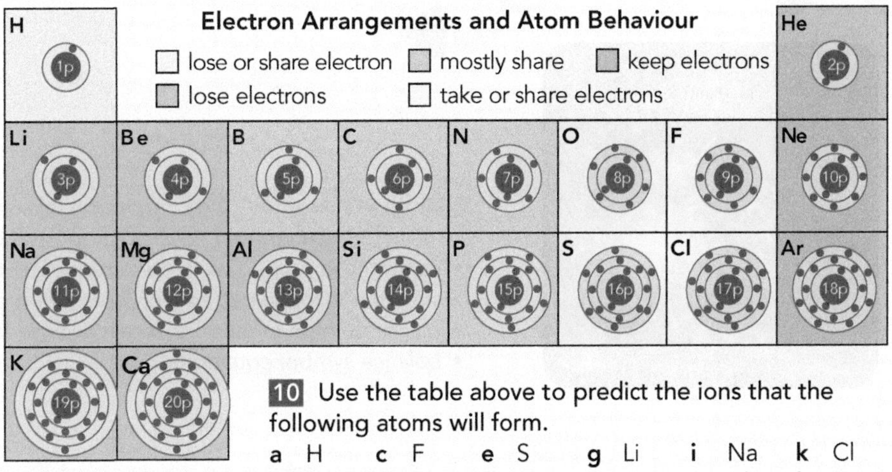

Electron Arrangements and Atom Behaviour

☐ lose or share electron ☐ mostly share ☐ keep electrons
☐ lose electrons ☐ take or share electrons

H (1p)							He (2p)
Li (3p)	Be (4p)	B (5p)	C (6p)	N (7p)	O (8p)	F (9p)	Ne (10p)
Na (11p)	Mg (12p)	Al (13p)	Si (14p)	P (15p)	S (16p)	Cl (17p)	Ar (18p)
K (19p)	Ca (20p)						

10 Use the table above to predict the ions that the following atoms will form.

a H c F e S g Li i Na k Cl
b Be d Mg f K h O j Al l Ca

7 Write the formula of each resulting ion:

a F gains 1e
b Ca loses 2e
c O gains 2e
d Mg loses 2e
e Al loses 3e
f K loses 1e.

Finding the Formula of an Ion

1 If atom A gains x electrons, then it has x more negative charges than positive charges, so its formula will be A^{-x}.
2 If atom B loses y electrons, then it has y more positive charges than negative charges, so its formula will be B^{+y}.

11 Polyatomic ions are molecules that have gained or lost electrons overall. Name these polyatomic ions and state how many electrons have been gained or lost.

a OH^- c CO_3^{2-} e SO_4^{2-}
b NO_3^- d HCO_3^- f NH_4^+

12 Refer to the science skill panel on page 61 and the table of ions below to determine the formula of:

a sodium chloride
b potassium chloride
c calcium oxide
d magnesium oxide
e magnesium chloride
f copper hydroxide
g magnesium carbonate
h calcium hydroxide
i ammonium chloride
j magnesium nitrate
k aluminium hydroxide
l aluminium carbonate
m iron (Fe^{3+}) sulfate
n lithium nitride.

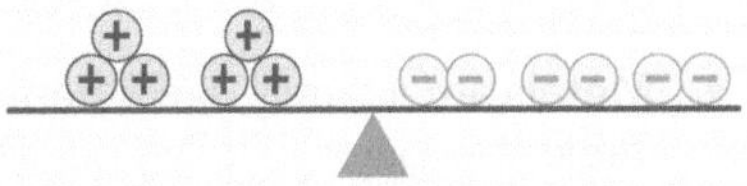

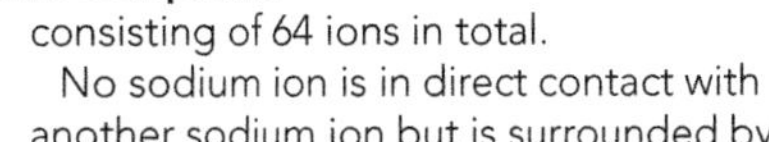

Table of Charges on Ions

+1	+2	+3	−3	−2	−1
H^+	Be^{2+}	Al^{3+}	N^{3-}	O^{2-}	F^-
Li^+	Mg^{2+}	Fe^{3+}	P^{3-}	S^{2-}	Cl^-
Na^+	Ca^{2+}				
K^+	Fe^{2+}			CO_3^{2-}	OH^-
	Cu^{2+}			SO_4^{2-}	NO_3^-
NH_4^+	Zn^{2+}				HCO_3^-
	Pb^{2+}				
	Ba^{2+}				

8 Write interpretations of these ionic formulas:

a $CaCl_2$
b $Mg(OH)_2$
c $Mg(NO_3)_2$
d K_2O
e $CaCO_3$
f NH_4OH
g Na_2SO_4
h $Al(OH)_3$
i $ZnSO_4$.

Interpreting an Ionic Formula

- chemical symbols give the ions
- subscript after an ion gives the number of monatomic ions in ratio
- subscript after brackets gives number of polyatomic ions in ratio
- no subscript means just 1
- charges on ions are not shown

Example: formula $Ca(NO_3)_2$ means 1 Ca^{2+} : 2 NO_3^-, $Ca(NO_3)_2$ has 1 calcium ion to every 2 nitrate ions

9 Read the article opposite, then answer the questions below.

a What kind of element is sodium?
b What kind of element is chlorine?
c Which ion has one more electron and which has one less than their respective atoms?
d What electron transfer occurs when sodium chloride is made by reacting sodium with chlorine?
e Which kind of atom has the strongest attraction for valence electrons?
f Describe the arrangement of sodium and chloride ions in a crystal.
g Why are no chloride ions found immediately next to any other chloride ion?
h What keeps the ions fixed in place?

Table Salt: an Ionic Compound

Sodium chloride consists of sodium ions bonded to chloride ions. Each sodium ion has 1 positive charge (Na^+), and each chloride ion has 1 negative charge (Cl^-).

The formula for table salt is NaCl, which means there is 1 Na^+ ion for every 1 Cl^- ion. The ratio is 1 Na^+ : 1 Cl^-.

The space-filling model opposite shows the relative sizes, the numbers, and the arrangement of the two kinds of ions.

The (green) chloride ions are relatively larger than the (purple) sodium ions, because they have more filled shells.

The top diagram shows that there is an equal number of each ion within a 'crystal' consisting of 64 ions in total.

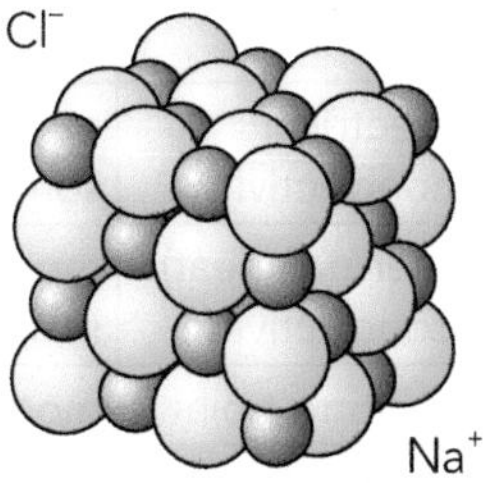

Cl^-

Na^+

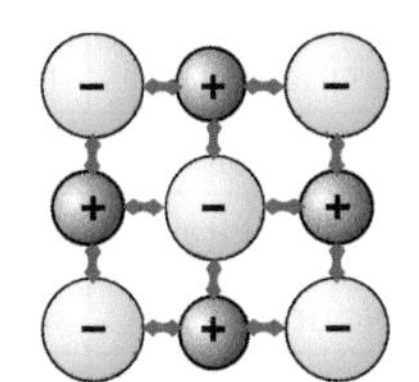

No sodium ion is in direct contact with another sodium ion but is surrounded by 6 chloride ions (1 above, 1 below, 1 in front, 1 behind and one on either side).

Similarly no chloride ion is in direct contact with another chloride ion but is surrounded by 6 sodium ions.

The electrostatic attractions (↔) between oppositely charged ions are the ionic bonds that hold the ions in the fixed array shown.

The sodium and chloride ions can be separated by dissolving salt crystals in water. The water molecules pull the ions apart.

Properties of Metals

Learning Outcomes - On completing this unit you should be able to:
- describe the common physical properties of metals
- relate these properties to the arrangement of, and bonding between, atoms
- state the reaction patterns of metals with oxygen, water and acid
- place common metals into a reactivity series
- *balance symbol equations.*

Physical Properties

- Metals are very important in everyday life: bike frames and car bodies are made of metal; appliances have metal casings; utensils and cooking pots are made of metal; scissors and nails are made of metal.
- Metals are mostly strong, dense, shiny solids that can be worked into different shapes. They are also good conductors of heat and electricity.
- These features are called the **physical properties** of metals, as no chemical reactions are involved. Physical properties are properties that can be measured without changing the composition of a substance.

Atomic Arrangement and Bonding

- **Metals** are a family of elements on the left and middle of the Periodic Table (see page 57). The physical properties of metals can be explained by considering how their *atoms are arranged and bonded together.*
- Metal atoms are regularly packed into layers that make up a **lattice**. The tightly packed atoms result in most metals having a relatively high density.
- Some of the **valence electrons** (see page 60) of metal atoms are weakly attracted to the nuclei and roam freely throughout the metal. This 'sea' of electrons gives metals good **electrical** and **thermal** (heat) **conductivity**.
- Because of the roaming electrons, metal atoms have a positive charge overall. The attraction between the positively charged metal 'ions' and the 'sea' of negatively charged free electrons moving around them holds the atoms together. This strong attraction is called **metallic bonding**.

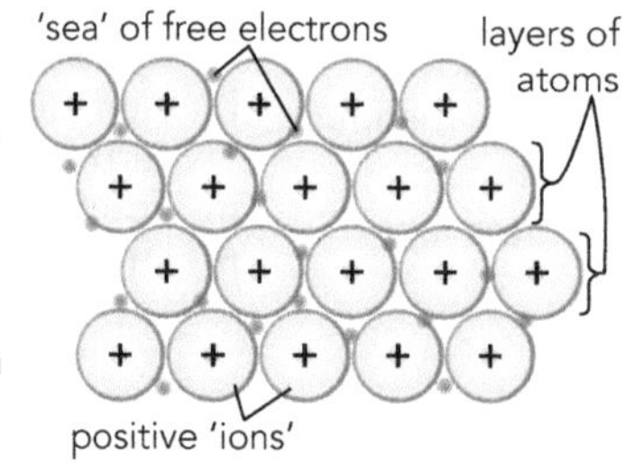

- Metals have high melting points, as much heat energy is needed to overcome the strong bonding. Most metals are in a solid state at room temperature (20 °C).
- The strength of the metallic bonding makes metals **ductile** (stretchable) and moderately hard.
- Metals are **malleable** (deformable), as the layers of atoms slide easily over each other.
- As free electrons reflect light well, most metals are silvery-grey and all have a metallic **lustre** (shine) when cut.

Property	Metals	Exceptions
density	mostly heavy, a few are light	Al, Li, Na, Mg, K, Ca
electrical conductivity	mostly good, some excellent	Hg (mercury)
thermal (heat) conductivity	mostly good, some excellent	Pb (lead), Hg
melting point	mostly high	Hg, K, Na
state (20 °C)	nearly all solid	liquid Hg
hardness	moderately hard	soft Pb
ductility	very ductile	none
malleability	very malleable	none
colour	silvery-grey	Cu, Au (gold)
lustre when cut	metallic shine	none

Elements and Alloys

- Pure metals are **elements** – substances made up of one type of atom only. There are over 80 different kinds of metallic elements (see page 59).
- For NCEA Science you need to know the properties of calcium (Ca), magnesium (Mg), aluminium (Al), zinc (Zn), iron (Fe), lead (Pb), copper (Cu), and silver (Ag).
- For NCEA Chemistry you also need to know about the properties of potassium (K), sodium (Na), and lithium (Li).
- Most of the metallic substances found in everyday life are actually **alloys**. These are made by mixing another element or several elements (see page 69) into the hot, molten metal. Steel, brass and coins are all alloys.

brass weight

- An alloy is neither an element nor a compound – it is a mixture.
- Alloying gives the modified metal better properties, eg increased hardness or strength.

Chemical Properties

- **Chemical properties** relate to how metal elements react with other chemicals.
- You need to know the pattern of reactions that metals have with oxygen, water and acids.
- Most metal atoms have a few loosely held valence electrons that flow throughout the metal. *These electrons are lost altogether when the metal reacts with another element or compound.*
- Most kinds of metal atoms readily lose a few electrons (up to 3 for some) to become positive ions.

ISBN: 9780170262316

Metals and Oxygen

- Most kinds of metal atoms react to form **ionic bonds** with oxygen atoms. The reaction is called **oxidation**.
- **Combustion** is the *rapid oxidation that may occur when a metal is heated.*
- Magnesium ribbon (Mg) combusts with an intense white flame when heated in oxygen (O_2). This results in white smoke and powder.
- When metal atoms react with oxygen, they lose a few electrons and become positive ions (eg $Mg - 2e \rightarrow Mg^{2+}$). The oxygen atoms gain those electrons and become negative oxide ions ($O + 2e \rightarrow O^{2-}$).
- The oppositely charged ions (eg Mg^{2+} and O^{2-}) are attracted to *form an ionic compound called a metal oxide* (eg the white MgO smoke and powder).
- The general word equation for the reaction pattern is:

Metal + oxygen gas → metal oxide

- Word and symbol equations for the above reaction are:
Magnesium + oxygen gas → magnesium oxide
$2Mg + O_2 \rightarrow 2MgO$ (balanced equation).

Metals and Water

- A few metals react with cold water, more react with steam, but most don't react.
- When a calcium (Ca) granule is dropped into water (H_2O), it fizzes and gradually disappears. The gas pops when ignited, proving it is hydrogen. The liquid in the test-tube turns red litmus blue, proving that it is an alkaline solution. If the liquid is heated, water evaporates leaving colourless crystals.
- When metal atoms react with water, the water molecules pull electrons off the metal atoms to form oxide ions (O^{2-}) and hydrogen gas (H_2). The metal atoms *lose electrons to become positive ions* (eg $Ca - 2e \rightarrow Ca^{2+}$).
- The oxide ions react with other water molecules to give hydroxide ions (OH^-), which form an alkaline solution.
- As the water evaporates, the oppositely charged metal and hydroxide ions (eg Ca^{2+} and OH^-) are *attracted to form an ionic compound called a metal hydroxide* (the colourless $Ca(OH)_2$ crystals).
- The general word equation for the reaction pattern is:

Metal + water → metal hydroxide + hydrogen gas

- Word and symbol equations for the above reaction are:
Calcium + water → calcium hydroxide + hydrogen gas
$Ca + 2H_2O \rightarrow Ca(OH)_2 + H_2$ (balanced equation).

Metals and Acids

- Many metal elements react with acids. Some react violently, others slowly, a few not at all.
- **Acids**, such as hydrochloric (HCl) and sulfuric (H_2SO_4), *release hydrogen ions (H^+) when dissolved in water.* These ions are the 'acid particles' (see page 84) that react with metal atoms.
- When zinc (Zn) granules are placed in hydrochloric acid (HCl), a colourless gas is released as the granules slowly disappear. The gas pops when ignited, proving that it is hydrogen. When the excess liquid is evaporated by heating, colourless crystals remain.
- When metals react with acids, the hydrogen ions pull electrons off metal atoms to form hydrogen gas ($2H^+ + 2e \rightarrow H_2$). The metal atoms *lose electrons, becoming positive ions* (eg $Zn - 2e \rightarrow Zn^{2+}$).

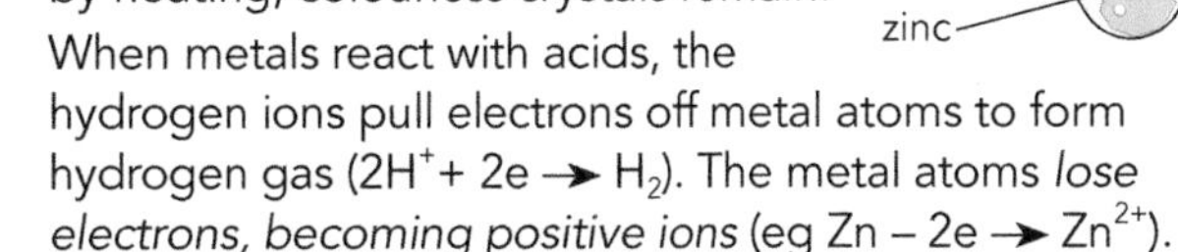

- As the water is evaporated, the positive metal ions and the negative ions that made up the other part of the acid (eg Cl^-) are attracted to form an ionic compound called a **salt** (eg the colourless $ZnCl_2$ crystals).
- Table salt, NaCl, is one such compound, formed when sodium metal reacts with hydrochloric acid.
- The general word equation for the reaction pattern is:

Metal + acid → salt + hydrogen gas

- Word and symbol equations for the above reaction are:
Zinc + hydrochloric acid → zinc chloride + hydrogen gas
$Zn + 2HCl \rightarrow ZnCl_2 + H_2$ (balanced equation).

Reactivity of Metals

- Metal atoms tend to lose a few of their loosely held valence electrons to form positive ions, which have more stable energy levels (see page 60). *This tendency varies among the metal elements.*
- The **reactivity** of a metal element depends on how tightly its positively charged atoms hold on to their free electrons (see page 60). Atoms of unreactive metals hold on very tightly, but the atoms of reactive metals lose free electrons readily or even explosively.
- The reactivity of freshly cut metals with oxygen, water and acid can be compared by experiment. The following **reactivity series** has been established:
K > Na > Li > Ca > Mg > Al > Zn > Fe > Pb > Cu > Ag
The symbol > means more reactive. (The additional elements are included for NCEA Chemistry.)
- Not all metals react as expected as they may have acquired a *resistant coating* (see page 68) that prevents a reaction from occurring until the coating is removed.

SCIENCE SKILL: Balancing Equations

The number of each type of atom found in the reactants and products must balance.
Problem: Balance the following symbol equation.
$Mg + HCl \rightarrow MgCl_2 + H_2$
Steps:
Write down the numbers of each type of atom on both sides of the equation as ratios:

2 Identify the atoms that are not balanced: H and Cl
3 Using trial and error, increase the number of reactants or products (don't alter any subscripts) and record the numbers of atoms on each side: for example $Mg + 2HCl \rightarrow MgCl_2 + H_2$ gives the ratios 1 Mg : 2 H : 2 Cl 1 Mg : 2 H : 2 Cl
4 When the numbers match, the reaction is balanced.

ISBN: 9780170262316

Revision Activities

1 Match up the descriptions with the terms.

Terms	Descriptions
a physical property	A pure substance made of one kind of atom only
b metal	B how strongly a substance reacts with other chemicals
c lattice	C property that does not involve a chemical reaction
d valence electron	D substance that releases hydrogen ions in water
e electrical conductivity	E bond formed by the attraction between oppositely charged ions
f thermal conductivity	F shiny solids that are able to conduct heat and electricity
g metallic bonding	G able to be stretched into a wire
h ductile	H regular arrangement of atoms or ions in a solid
i malleable	I how light is reflected by the surface of a substance
j lustre	J property that describes how a chemical reacts
k element	K reaction of a substance with oxygen gas
l alloy	L capacity of a substance to conduct electricity
m chemical property	M compound formed when metal reacts with an acid
n ionic bond	N outer shell electron that can be removed
o oxidation	O metal with another element mixed into it
p combustion	P ability of a substance to conduct heat
q acid	Q occurs when a substance burns/oxidises rapidly
r salt	R able to be hammered into a different shape
s reactivity	S attraction between free electrons and charged metal atoms

2 Explain the differences between:

a metals and nonmetals
b a physical and a chemical property
c ductile and malleable
d oxidation and combustion
e a pure metal and a metal alloy.

3 Consider the data below as you answer the questions.

a Most metals are silvery-grey. Which are the exceptions?
b Which metal is the hardest? The softest?
c Which metals would be liquids if heated to 100 °C?
d Which metals would float in water? (Water has a density of 1 $g\,cm^{-3}$.)
e Which is the densest metal listed on the table?
f Which metal is the best electrical conductor? The worst?
g Which metal is the best heat conductor? The poorest?
h Which is the strongest metal? The weakest?
i Which metals will remain shiny in the air?

4 Decide whether the following statements are true or false. Rewrite the false ones to make them correct.

a Pure metals are rarely used to make everyday objects.
b All metals are silvery-coloured solids.
c Physical properties also include the reactivity of the metal.
d Most physical properties of metals can be explained by the bonding between metal atoms.
e Metal atoms are held together by metallic bonding.
f Metals can be hammered into new shapes and stretched into thin wires.
g An alloy is a compound.
h Metal atoms lose electrons to become negative ions.
i Combustion is the rapid oxidation of a metal when heated in air.
j All metals react with water to release hydrogen gas.
k A salt is a compound formed when metals react with acids.
l A reactive metal holds on tightly to its free electrons.

5 Study the diagram below and use it to explain the following properties of *most* metal elements:

a metals are dense
b metals are good conductors
c metals are solids
d metals have high MPs
e metals are ductile
f metals are malleable
g metals are shiny
h metals are fairly hard.

'sea' of free electrons
layers of atoms
positive 'ions'

6 Use the data in the table below to decide on the best metal to use in making each item. Justify your answer. Remember that cost comes into it too.

a fishing sinker
b bridge
c aircraft wing
d electrical wire
e thermometer
f metal jewellery
g ladder
h car body
i saucepan
j scissors

Metal	Ag	Al	Au	Ca	Cu	Fe	Hg	K	Li	Mg	Na	Pb	Zn
Colour	silvery	silvery	yellow	grey	orange	grey	silvery	silvery	silvery	silvery	silvery	grey	silvery
Hardness (Mohs scale)	2.5	2.8	2.5	–	3.0	4.0	liquid	–	–	2.5	–	1.5	2.5
Melting point (°C)	962	660	1064	842	1084	1538	−39	63	180	650	98	327	419
Density ($g\,cm^{-3}$)	10.49	2.70	19.30	1.55	8.96	7.87	13.5	0.86	0.53	1.74	0.97	11.34	7.14
Electrical conductivity (S/nm)	63.0	35.0	45.2	28.2	58.0	10.4	1.04	0.014	11.7	21.5	21.0	4.87	16.8
Thermal conductivity ($Wm^{-1}K^{-1}$)	429	242	318	201	401	80.4	8.3	102.5	84.8	154	142	35.3	116
Strength (MPa)	170	40-50	100	–	210	350	N/A	–	–	–	–	12	110-200
Left in the air	tarnishes	resistant	unreactive	tarnishes	tarnishes	rusts	–	tarnishes	tarnishes	resistant	tarnishes	tarnishes	tarnishes

ISBN: 9780170262316

7 Clean 1 cm squares of different metals are placed in dilute hydrochloric acid. The reactions are shown below.

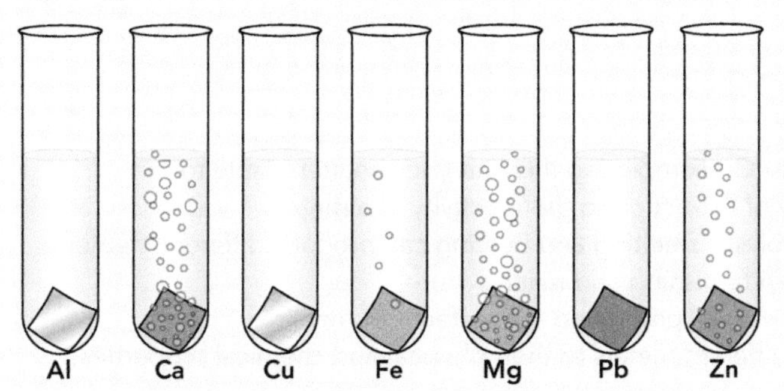

a Which metals did not react with the acid?
b Place the seven metals in order of their reactivity.
c Which metal is not in its expected place?
d Explain why that metal did not react.
e What is the gas being produced in four of the test-tubes? How would you test for it?

In the test-tubes in which reactions occurred, salts were left behind after the liquid evaporated. Give the names and formulas of salts formed in the test-tubes labelled:

f Ca
g Fe
h Mg
i Zn.

When metals react with acids, hydrogen gas and a salt are made. Complete word equations, then write balanced symbol equations for the following:

j Calcium + sulfuric acid
k Iron + hydrochloric acid
l Magnesium + hydrochloric acid
m Zinc + sulfuric acid

magnesium ribbon

9 Refer to the information on page 65 as you answer these questions.

a Explain what happens to the valence electrons of metals that react.
b What do metals atoms become when a reaction occurs?
c When a metal reacts with oxygen, what is the product?
d Write a word equation for the reaction pattern between reactive metals and oxygen gas.
e When a metal reacts with water, what are the final products after excess water has evaporated?
f Write a word equation for the reaction pattern between reactive metals and water.
g When a metal reacts with an acid, what are the final products after excess liquid has evaporated?
h Write a word equation for the reaction pattern between reactive metals and acids.

10 The table below summarises observations of the reactivity of different metal elements.

a Explain why reactions of K, Na, Li and Ca are usually demonstrated by a teacher or not permitted.
b Describe the tendency of different metals to lose some of their valence electrons.
c Explain why reactions of most metals are more vigorous when heated in oxygen than when heated in air.
d Explain why the reactions of most metals are more vigorous when placed in heated steam than when placed in cold water.
e Identify the metal that doesn't react as strongly as it ought to according to the reactivity series. Explain why.

Element	K	Na	Li	Ca	Mg	Al	Zn	Fe	Pb	Cu	Ag	Au
In air at 20°C	rapidly tarnishes	rapidly tarnishes	rapidly tarnishes	no observable reaction								
Heated in oxygen	burns vigorously	burns vigorously	burns strongly	burns quite fast	burns vigorously	goes white	goes yellow	moderate reaction	slow reaction	surface blackens	no reaction	
In water at 20°C	reacts very violently	reacts violently	fast reaction	reacts readily	reacts very slowly	no observable reaction						
In heated steam	explosive	explosive	very violent	violent	burns rapidly	little reaction	moderate reaction	reacts slowly	no observable reaction			
In dilute HCl, and H_2SO_4	explosive reaction	explosive reaction	violent reaction	reactive with HCl	reactive with both	slow reaction	quite reactive	reacts slowly	no observable reaction			

8 Read the article opposite, then answer the questions below.

a Are most metallic substances alloys or metal elements?
b What is an alloy and why are they used?
c How and when is an alloy made?
d What properties of a metal can be changed by alloying?
e What is bronze and why was it important early in human history?
f What is brass and why was it important?
g Why is steel so much more useful than iron?
h What is stainless steel used for?
i What are 'designer alloys' and why are they important in technology?

Metal Alloys

Most of the metallic substances you use in everyday life are likely to be metal alloys rather than pure metal elements.

An alloy is a mixture of pure chemical elements, the main one being a metal. A small amount of another element is mixed with the metal. The added element is usually a metal, although non-metal elements such as carbon may be added.

The other element is added in controlled amounts to the molten metal element when it is being smelted.

An alloy will have typical metallic properties (eg shiny, flexible, good conductivity, etc) but it may be harder, stronger or more resistant to corrosion than the original metal. Some alloys will have a lower melting point than the metal, which will make it easier to work.

One of the earliest alloys used by humans was bronze, which has some tin mixed into copper to make the copper stronger.

Brass, a hard alloy of copper and zinc, was used for many machine parts (see photo) during the Industrial Revolution.

Iron is abundant, but pure iron is not very strong and rapidly rusts. Steel, which is an alloy of iron with small amounts of carbon and manganese, is much stronger and more resistant to corrosion. Stainless steel, which does not rust, contains chromium and nickel.

Chemists can produce 'designer alloys' for very specific purposes, eg light, strong, heat-resistant alloys for spacecraft and aircraft.

ISBN: 9780170262316

The Extraction and Uses of Metals

Learning Outcomes - On completing this unit you should be able to:
- relate the difficulty of extraction to the reactivity of a metal
- explain how electrolysis, smelting and heating can produce different elements
- distinguish between tarnishing and rusting
- describe different ways of protecting the surfaces of metals
- relate the uses of different metals to their physical and chemical properties.

Extracting Metals

- An **element** is a substance made of one type of atom. Few metal elements exist naturally, but metal **compounds** are abundant in Earth's rocky crust and seas.
- **Metal compounds** consist of metal ions bonded to ions, such as oxide ions (O^{2-}).
- Metal **ions** are positively charged as they have lost some electrons. For metal ions to become neutral atoms, they must *gain electrons from another source.* This reaction is called **reduction**.
- The ease with which a metal is extracted is related to its **reactivity**. Ions of reactive metals must be 'forced' to accept electrons, as the resulting metal atoms are less stable (see page 60). *The more reactive the metal, the more energy is needed to reduce its ions.*
- The reactivity series (see page 65) includes:
  a highly reactive metals (K, Na, Li, Ca, Mg and Al)
  b moderately reactive metals (Zn, Fe and Pb)
  c low reactivity metals (Cu, Ag and Au).

Highly Reactive metals

- As it is difficult to reduce ions of highly reactive metals, a *huge electrical current and heat are used.* The current 'forces' electrons onto the metal ions, reducing them to metal atoms in a process called **electrolysis**.
- In NZ, aluminium is produced from alumina (aluminium oxide, Al_2O_3). Alumina is chemically separated from bauxite, an **ore** mined in Weipa, Australia.
- At the Bluff 'smelter', electrolysis is carried out in 'pots' (see diagram below). Alumina is dissolved in molten cryolite at 970 °C to give a conducting solution. The Al^{3+} ions migrate to the negative **electrode**, where they are 'forced' to accept electrons to become Al atoms. The O^{2-} ions are pulled to the positive carbon electrodes, where they react to give CO_2 gas.

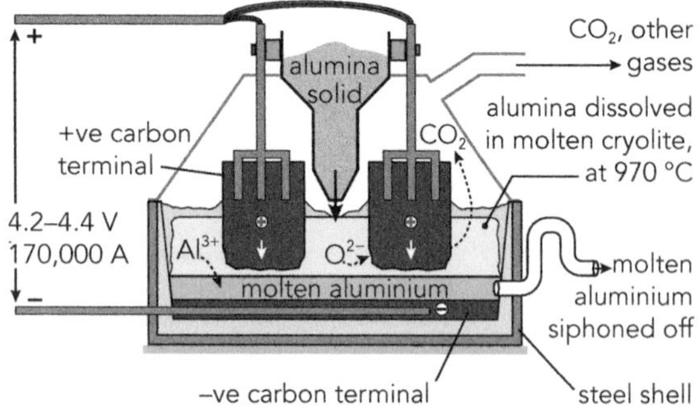

- The overall reaction is: $\mathbf{2Al_2O_3 + 3C \longrightarrow 4Al + 3CO_2}$.

Moderately Reactive Metals

- Moderately reactive metals are extracted from their compounds *by heating them with carbon.* A source of carbon (eg coal) is burnt to produce carbon monoxide gas. The metal ions are then 'forced' to accept electrons from carbon monoxide, which reduces them to metal atoms in a process called **smelting**.
- In NZ, iron is produced by smelting magnetite (iron oxide, Fe_3O_4). Ironsand ore is excavated at the mouth of the Waikato River, and magnetite is magnetically separated from the lighter-coloured quartz sand.
- At the Glenbrook steel mill, magnetite and coal are heated in a furnace, then fed into a rotating kiln where reduction occurs at 950 °C. The Fe^{2+} and Fe^{3+} ions are reduced to 'sponge' iron by carbon monoxide (CO) gas, which is oxidised to carbon dioxide (CO_2).

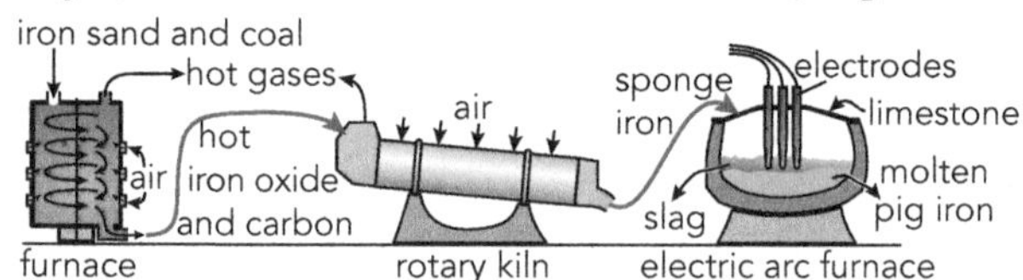

- The overall reaction is: $\mathbf{Fe_3O_4 + 4CO \longrightarrow 3Fe + 4CO_2}$.
- The 'sponge' iron is melted at 1500 °C in an electric arc furnace to give molten iron, which is drained off separately from floating impurities called **slag**.
- Finally, in a steel-making furnace, carbon and other elements are added to the molten iron to form alloys.

Low Reactivity Metals

- Low reactivity metals, such as gold, silver and copper, exist as **native elements** in Earth's crust. Although not abundant, they are mined. They can also be *extracted by heating their ores*, eg $\mathbf{2Cu_2O \longrightarrow 4Cu + O_2}$.

Tarnishing and Rusting

- The surface atoms of most metals react with oxygen, moisture or carbon dioxide in the air, to form a compound. For some metals, that compound acts as an *impervious layer preventing any further reactions.*
- This reaction is called **tarnishing**. Calcium, magnesium, aluminium, zinc, lead and copper all tarnish in the air.
- Freshly cut aluminium tarnishes to give a transparent, impervious oxide layer that stops other reactions.
- **Corrosion** is the gradual destruction of a metal by chemical reaction with its environment. **Rusting** is the type of corrosion that occurs when iron is exposed to air and moisture, forming a flaky orange oxide layer.

ISBN: 9780170262316

Protecting Metals

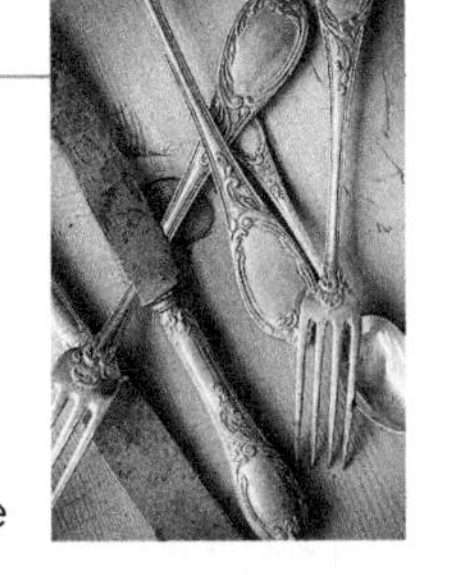

- Different methods prevent reactive metals from corroding and, in particular, iron from rusting.
- Objects made of low reactivity metals, such as copper and silver, may need protection. In Rotorua, silver objects turn grey or black due to hydrogen sulfide gas in the air.
- Aluminium objects usually do not need extra surface protection because of the impervious oxide coating.

Alloying

- **Alloying** is mixing another element into a molten metal. Some alloys prevent corrosion from occurring.
- Magnesium alloys used for mag wheels have manganese for corrosion resistance.
- Carbon and chromium are added to iron to form the strong, corrosion-resistant alloy called stainless steel.

Coating

- Coating makes metals corrosion-resistant by stopping oxygen, water vapour and other reactive gases from reaching the metal atoms. Examples include:
  - **a** painting wrought iron fences
  - **b** covering steel machinery parts in oil or grease
  - **c** enclosing outdoor furniture in a plastic layer
  - **d** chrome-plating bumpers and headlights (chromium is a fairly unreactive metal)
  - **e** galvanising boat trailers (zinc forms an impervious protective coating when it tarnishes)
  - **f** anodising outdoor aluminium surfaces by making the impervious oxide coating thicker
  - **g** enamelling whiteware (eg refrigerators).

Galvanising

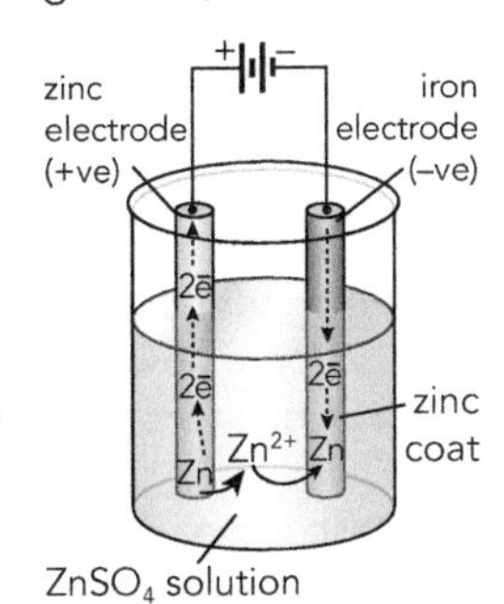

- Iron objects are **galvanised** (zinc-plated) by putting them in a solution of zinc sulfate and setting up a voltage difference. The Zn^{2+} ions from the positive zinc electrode are attracted to the negative iron electrode, where they gain electrons to form a zinc layer.
- Most iron objects are galvanised by dipping them into molten zinc (hot dipping).

Sacrificial Protection

- The rusting of steel can be limited by attaching a more reactive metal (eg Zn). Free electrons flow from the more reactive metal to the steel, which stops the iron atoms from becoming ions, thus preventing rusting.
- As the *reactive metal corrodes instead of the steel*, the process is called **sacrificial protection**.
- Zinc or magnesium blocks are often fixed to the hulls of steel ships to limit the rusting that occurs at sea.

Use of Metals

- The uses of different metals relate to their **physical** and **chemical properties**. Most everyday metallic objects are alloys and have protection from corrosion.

Highly Reactive Metals

- Potassium, sodium, lithium are too reactive to have many practical uses. However, all are used in alloys of other metals. Na is used in sodium vapour street lights and Li in lithium batteries.

Magnesium

Properties	Uses
Low density, yet strong (in alloys containing Al)	Car parts, cell-phones, laptops
Moderately reactive metal	Protects steel hulls
Combusts readily	Marine flares

Aluminium

Properties	Uses
Light, corrosion-resistant and strong when alloyed	Aircraft frames, ladders, foil, cans
Great electrical conductor	Power pylon wires

Moderately Reactive Metals

Zinc

Properties	Uses
Forms a protective coating	Galvanising
Moderately reactive metal	Protecting steel hulls

Iron

Properties	Uses
Hard ('high' carbon alloy)	Tools, scissors
Malleable/ductile ('low' C)	Structural steel, wire
Resistant (chromium alloy)	Pans, cutlery, sinks

Lead

Properties (toxic)	Uses
Malleable and soft	Plumbing (now banned)
Heavy and resists corrosion	Sinkers, shotgun pellets

Low Reactivity Metals

Copper

Properties	Uses
Mostly unreactive	Pipes
Excellent heat conductor	Saucepan bottoms
Great electrical conductor	Electrical wiring

Silver

Properties	Uses
Malleable, ductile, rare	Jewellery
Best electrical conductor	Expensive electronics

Gold

Properties	Uses
Unreactive	Teeth fillings
Malleable, ductile, rare	Jewellery, gold leaf

ISBN: 9780170262316

Revision Activities

1 Match up the descriptions with the terms.

Terms	Descriptions
a element	A slow destruction of a metal by chemical reactions
b compound	B occurs when ions or atoms gain electrons
c metal compound	C adding a protective layer of zinc to a metal
d reduction	D surface corrosion of iron and steel objects
e reactivity	E unwanted material produced during smelting
f electrolysis	F able to be stretched out, eg into a wire
g ore	G pure substance made of one kind of atom only
h electrode	H reducing metal ions to atoms using electricity
i smelting	I able to be hammered or squashed
j slag	J element found in an uncombined state in crust
k native element	K chemical in which different atoms are bonded
l tarnishing	L naturally occurring rock or mineral
m rusting	M pure substance in which positive metal ions are bonded to negative ions
n corrosion	N how reactive an element is with other chemicals
o alloying	O attaching a reactive metal to steel so that the reactive metal corrodes instead
p galvanising	P reducing metal ions to neutral atoms using heat and carbon
q sacrificial protection	Q forming a protective coating on the metal surface
r malleable	R adding other elements to a molten metal
s ductile	S conductor through which current enters or leaves a solution of ions

2 Explain the differences between the terms in each of the items below:

a an element and a compound
b an neutral atom and an ion
c an ore and a native element
d electrolysis and smelting
e tarnishing and rusting
f a pure metal and an alloy
g galvanising and sacrificial protection.

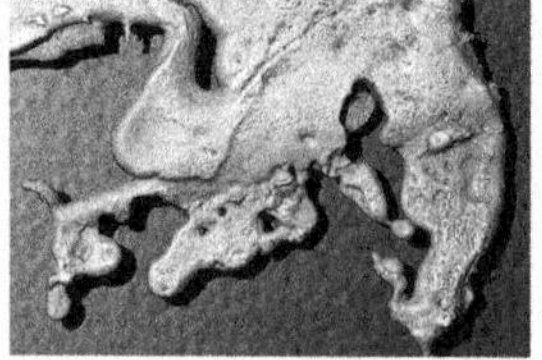

Mined native copper

3 Copy and complete the table to compare the reactivity of metals with methods of extraction.

Reactivity	Elements	Extraction
Highly reactive		
Moderately reactive		
Low reactivity		

a Explain how the ease with which a metal element is obtained is related to its reactivity.

4 Copy the maps and mark these locations after Googling them:

a Weipa bauxite
b Bluff aluminium smelter
c Waikato Heads iron sand
d Glenbrook steel mill.

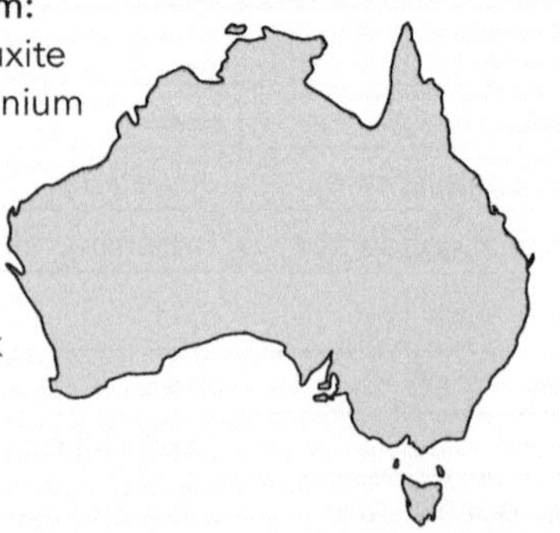

5 Decide whether the following statements are true or false. Rewrite the false ones to make them correct.

a Many metal elements exist naturally.
b Reactive metals atoms are less stable than their ions.
c The less reactive a metal is, the harder it will be to extract it.
d As aluminium is a very reactive metal, it can only be extracted by electrolysis.
e Moderately active metals can be obtained by smelting their ores.
f Magnetite is separated from quartz sand ore by magnets.
g Tarnishing damages the surface of metals.
h The rusting of iron requires both oxygen and water.
i Not all reactive metals corrode.
j Unreactive metals don't require protective coatings.
k Steel hulls are protected by blocks of a less reactive metal.
l Galvanising protects steel objects, as the zinc forms an impervious layer.

6 Metals are extracted from ores mined from Earth's crust. The photos show the ironsand and bauxite ores that grey magnetite and white alumina are extracted from respectively.

a Explain why open-cast mining is used in both cases.
b Describe how each ore would be mined.
c What impurity is mixed in with magnetite in ironsand?
d Describe how is it removed.
e How do you know there are impurities in bauxite?
f Compare how those impurities are removed.

ironsand

7 Describe each of the steps of the steel-making process shown here.

ISBN: 9780170262316

8 **Refer to this diagram and the text on page 68 when answering the questions.**

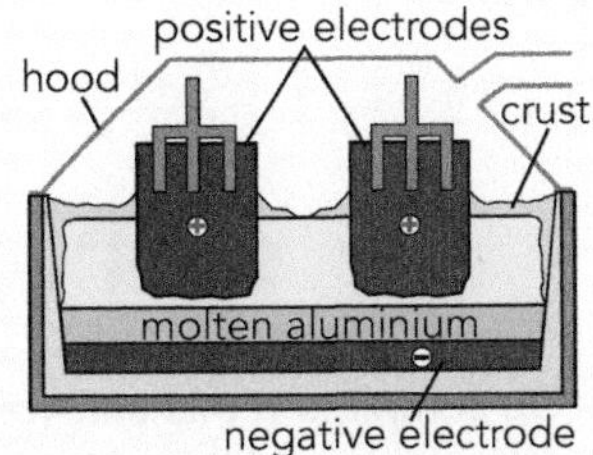

a State the process that reduces Al^{3+} ions to atoms.
b Explain why it must be used.
c State the chemical name and formula of alumina.
d Identify the charge on aluminium ions (see page 61).
e Describe what molten cryolite does.
f Which terminal are aluminium ions attracted to? Why?
g Describe what happens to the Al^{3+} ions at that terminal.
h Write down a word equation for the reaction that occurs at the negative terminal.
i What is this type of reaction called?
j Explain why the negative terminal is at the bottom.
k Describe the purpose of the hood.

9 **Refer to this diagram and the text on page 68 when answering the questions.**

a What is the chemical name and formula of magnetite?
b Describe how is it separated.
c Explain why coal is mixed in with the magnetite.
d What are the charges on the two kinds of iron ions found in magnetite?
e How hot does it get in the rotary kiln?
f What is the heat source?
g What do C atoms in coal supply to the iron ions?
h What process reduces iron ions to iron atoms?
i Explain why is it used instead of electrolysis.
j Write symbol equations for reduction of Fe^{2+} and Fe^{3+}.
k Write an equation for the reaction involving CO.
l What is this type of reaction called?
m Explain how and why slag is removed from the melter.
n Find out why 'sponge iron' and 'pig iron' are so called. (Check on the internet.)

10 **Read the article opposite, then answer the questions below.**

a What is igneous rock?
b Why would the Waihi area have been like Rotorua today?
c Why is the gold- and silver-bearing ore found in seams?
d What are the two main steps involved in getting the gold?
e How many tonnes of ore are needed for 1 kg of Au? 1 kg of Ag?
f Why is hydrogen cyanide added?
g What environmental issues can be caused by hydrogen cyanide?
h What's the role of carbon granules?
i How does electrolysis reverse the action of hydrogen cyanide?
j Why is the bullion transported to Perth?

11 **In an experiment designed to identify which of four conditions (air, moisture, light and warmth) are required for rusting to occur, five test-tubes were set up. The results after 24 hours are shown below.**

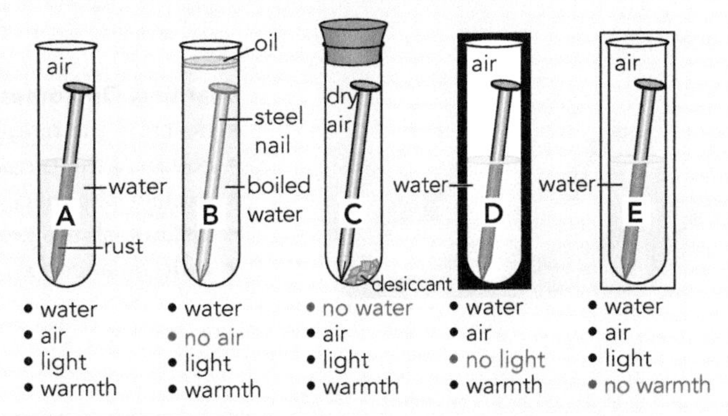

a Spot the test-tube provided with all four conditions.
b Identify which test-tubes are set up as controls.
c Explain the purpose of each of test-tubes B to E.
d Compare what each of test-tubes B to E proved.
e Identify the combination of conditions required for steel nails to rust.

12 **For each item listed below, identify the most suitable metallic element or alloy to use and explain why.**

a hammers
b kitchen foil
c light serving spoons
d saucepan bottoms
e metal fence
f fishing sinkers
g wedding rings
h sensitive circuits
i carving knives
j orange street lamps
k sparklers (firework)
l battery electrodes

13 **For each object listed below, identify the most suitable way of making it corrosion resistant.**

a fridges
b roofing iron
c steel hulls
d clothes racks
e bicycle chains
f iron gates
g aluminium ladders
h spanners
i aluminium window frames
j steel buckets
k car bodies
l NZ coins
m silver rings
n copper roofing
o mag wheels
p scissors
q power lines

Gold and Silver Mining at Waihi

Millions of years ago, lava flows occurred in the Waihi district. Earthquakes vertically fractured the igneous rock and hot geothermal water rose up through the cracks bearing dissolved minerals.

Dissolved native gold and silver rapidly crystallised onto the surface of the fractures when further quakes caused sudden pressure drops.

Gold and silver do not exist as nuggets at Waihi, rather they are found as many small crystals embedded in veins of the rock and must be extracted by chemical means. On average, 1 tonne of gold-bearing ore contains 1 gram of Au and 30 grams of Ag. The main steps involve concentrating then separating.

The Martha Mine is a huge open-cast mine. The ore at the bottom is blasted and excavated, then huge trucks transport it to the crushers.

The rock is ground into fine particles the size of sand grains. Water is added to make a slurry, which is pumped to tanks where added hydrogen cyanide pulls electrons off the silver and gold atoms so they become dissolved ions.

Gold and silver ions stick to the surface of carbon granules, and the rest of the slurry is removed. The ions are then 'washed' off the granules using super-heated steam.

The ion solution then undergoes electrolysis, with the Au^{2+} and Ag^{+} ions being reduced to a metal sludge. The sludge is then heated to 1200 °C, and the molten metallic mixture is separated from the slag as it is poured into moulds to form 20 kg bars that consist of 75-90% Ag and 10-25% Au.

The bars are shipped to the Perth Mint in Australia, where the gold and silver metals are separated before being sold.

ISBN: 9780170262316

Properties and Uses of Nonmetals

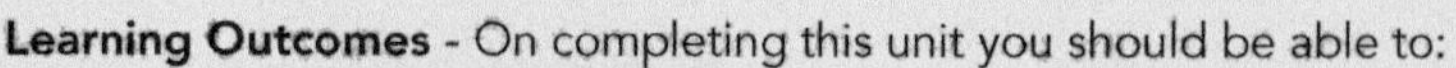

Learning Outcomes - On completing this unit you should be able to:
- describe the range of properties exhibited by nonmetal elements
- compare the properties of nonmetals with those of metals
- explain why reactive nonmetal atoms tend to share or take electrons
- distinguish the properties of different allotropes of nonmetal elements
- *write balanced symbol equations for reactions involving nonmetal elements.*

Physical Properties

- Of the 98 naturally occurring **elements**, 72 are metals, 17 are nonmetals, and 9 are **metalloids** (having both metal and nonmetal properties).
- Some nonmetal elements occur freely in nature, eg sulfur deposits (see photo above), but others occur naturally only in a combined state as part of a **compound**, eg chlorine ions in sodium chloride, NaCl.
- Nonmetals can be produced by reactions (eg H_2 gas, by adding a reactive metal to acid) or by **electrolysis** (eg Cl_2 gas, by passing a current through molten NaCl).
- At room temperature (20°C), 11 of the 17 non-metal elements are gases (H, He, N, O, F, Ne, Cl, Ar, Kr, Xe, Rn), 1 is a liquid (Br), and 5 are solids (C, P, S, Se, I).
- For NCEA Chemistry you need to know about carbon, nitrogen, oxygen, sulfur, and chlorine in particular.
- The physical properties *contrast with those of metals*. Nonmetals:
  - a are less dense in the solid state
  - b are dull in appearance when solid
  - c are mostly poor electrical and thermal conductors
  - d have low melting and boiling points (mostly gases)
  - e are mostly weak and brittle in the solid state
  - f often have different forms called **allotropes**.

Chemical Properties

- All nonmetal elements are located at the *right end of the Periodic Table*, except for hydrogen.

Periodic Table of the Elements

(Key: metals, metalloids, nonmetals)

H																	He
Li	Be											B	C	N	O	F	Ne
Na	Mg											Al	Si	P	S	Cl	Ar
K	Ca	Sc	Ti	V	Cr	Mn	Fe	Co	Ni	Cu	Zn	Ga	Ge	As	Se	Br	Kr
Rb	Sr	Y	Zr	Nb	Mo	Tc	Ru	Rh	Pd	Ag	Cd	In	Sn	Sb	Te	I	Xe
Cs	Ba	Lu	Hf	Ta	W	Re	Os	Ir	Pt	Au	Hg	Tl	Pb	Bi	Po	At	Rn
Fr	Ra																

- The elements in the far right column (He, Ne, Ar, Kr, Xe, Rn) have atoms with full outer electron shells, hence they are unreactive and called inert gases.
- Most of the other nonmetal atoms have nearly full outer **electron shells** and a strong attraction for the **valence electrons** of other atoms (see page 60).
- These nonmetal atoms can *pull the free electrons off metal atoms*, which results in negatively charged nonmetal ions and positively charged metal ions.
- In the solid state, the oppositely charged ions are attracted to form a **lattice** of ions. This attraction is called **ionic bonding**.

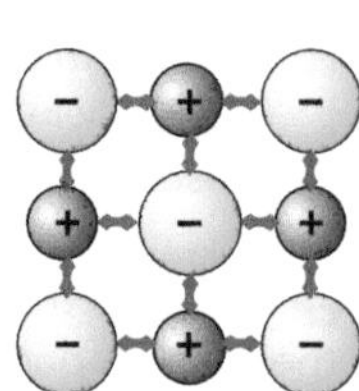

- Nonmetal atoms strongly attract the valence electrons of other nonmetal atoms, *but these electrons are shared*. The attraction of nuclei for the shared electrons is called **covalent bonding**, which binds the atoms in discrete **molecules**.
- Many nonmetal elements exist as molecules (N_2, O_2, Cl_2, F_2, P_4, S_8 and Cl_2). Different nonmetal atoms are also bonded together in molecules (eg CO, CO_2, H_2O, NO, NO_2, NH_3, SO_2, SO_3, H_2SO_4 and HCl).
- Nonmetals atoms take electrons off metal atoms to form **ionic compounds**. They also share electrons with other nonmetal atoms to form **covalent compounds**.

Carbon

- Carbon is a solid element that has many allotropes, eg graphite, diamond, fullerenes and amorphous carbon. Diamond exists as extremely hard, transparent crystals and graphite is the soft, grey substance in pencil 'lead'.
- C atoms have 4 electrons in their outer shells. To become stable, each shares 3 or 4 electrons with other C atoms, giving each a 'full' outer shell.
- In diamond, each C atom is covalently bonded to 4 neighbours in a lattice. This bonding gives diamond its hardness and also makes it an excellent thermal conductor.
- In graphite, each C atom is covalently bonded to 3 others in a sheet. These slide over each other making graphite soft. The bonding makes it an excellent electrical conductor.
- In **amorphous carbon** (eg coal, soot, carbon powder), the C atoms are not arranged in a lattice or a sheet. It combusts readily when ignited in oxygen gas to form oxides. *If abundant oxygen is present*, it burns with a clear blue flame to give carbon dioxide (CO_2) gas:

  Carbon + oxygen ⟶ carbon dioxide

  $C + O_2 \rightarrow CO_2$.

  This reaction is **complete combustion**.
- CO_2 is a colourless, odourless gas used in fizzy drinks. As it is a heavy gas that does not support combustion, it is used in extinguishers to smother flames.

ISBN: 9780170262316

- When CO_2 dissolves in water, it turns blue litmus red, proving it is an **acidic solution** (see page 84).
- In *limited oxygen,* amorphous carbon combusts with a yellow flame to give carbon monoxide (CO) gas:
  Carbon + oxygen → carbon monoxide
  $2C + O_2 \rightarrow 2CO$.
  This reaction is **incomplete combustion**.
- Carbon monoxide is also a colourless, odourless gas but it is highly toxic.
- In fullerenes, C atoms are bonded in hexagons and pentagons to form a hollow tube or sphere ('bucky balls'). The molecules are made by heating graphite with an inert gas in an electric arc furnace. A variety of nanotechnology uses is being developed.

Nitrogen

- Nitrogen is a gaseous element that forms 78.1% of air. It is a colourless, odourless gas.
- Nitrogen atoms have five electrons in their outer shell. In order to become more stable, each shares three electrons with another, thus giving each a 'full' outer shell. The triple-bonded molecules (N_2) are not very unreactive as a consequence.
- In the *extreme heat* caused by lightning or combustion in car and jet engines, N_2 gas reacts with O_2 gas to give colourless nitric oxide (NO) gas:
  Nitrogen + oxygen → nitric oxide
  $N_2 + O_2 \rightarrow 2NO$.
- The NO gas rapidly reacts further with O_2 gas in the atmosphere to give nitrogen dioxide (NO_2) gas:
  Nitric oxide + oxygen → nitrogen dioxide
  $2NO + O_2 \rightarrow 2NO_2$.
  Yellow-brown NO_2 is a toxic, choking gas.
- Nitrogen dioxide gas also dissolves in moist air to form an acidic solution. This contributes to the problem of **acid rain** in industrialised countries.
- N_2 gas is used to produce ammonia (NH_3) gas:
  $N_2 + 3H_2 \rightarrow 2NH_3$.
- Ammonia is a pungent gas that dissolves in water, giving an **alkaline solution** that is used as household bleach. NH_3 is also used to make nitrogen-based fertilisers.
- Nitrogen-fixing bacteria (see page 121) in the roots of some plants combine N and H atoms to give ammonia.

Oxygen

- Oxygen is a gaseous element that exists as two distinct allotropes - oxygen gas and ozone gas. 20.7% of air is colourless, odourless oxygen gas.
- Oxygen atoms have 6 electrons in their outer shell. In order to become more stable, they share electrons with other O atoms to effectively gain 'full' outer shells.
- Each O atom shares 2 electrons with another O atom in an oxygen molecule (O_2). The two covalent bonds give each atom 'full' shells.
- *Oxygen is a reactive gas.* It reacts with metals to give ionic oxides (eg MgO), which form *alkaline solutions in water.* It reacts with nonmetals to give covalent oxides (eg SO_2), which form *acidic solutions in water.*
- Oxygen gas is produced by plants and required by all living things for respiration (see page 137).
- In an O_3 molecule, the central atom shares 2 electrons with 2 other atoms, each of which shares 2 electrons with the central atom. This electron-sharing arrangement makes ozone (O_3) gas less stable, and therefore more reactive than O_2.
- Ozone purifies and disinfects water (see page 75).
- Ozone in the upper atmosphere absorbs dangerous ultraviolet radiation from the sun. It is changed into oxygen in the reaction: $2O_3 \rightarrow 3O_2$.

Sulfur

- Sulfur is a brittle, odourless, yellow solid that occurs in volcanic deposits, from which it can be mined.
- Sulfur atoms have 6 electrons in their outer shell. In order to become more stable, each shares an electron with two other S atoms in an 8-atom ring molecule, S_8. All atoms get 'full' outer shells from the covalent bonds.

- Sulfur is used in the manufacture of sulfuric acid.
- In a chemical plant, sulfur is combusted with oxygen. It burns with a blue flame to give sulfur dioxide (SO_2):
  Sulfur + oxygen → sulfur dioxide
  $S_8 + 8O_2 \rightarrow 8SO_2$.
  SO_2 is a colourless, pungent, choking gas.
- Next, the SO_2 is reacted with more oxygen to form toxic sulfur trioxide (SO_3) gas:
  Sulfur dioxide + oxygen → sulfur trioxide
  $2SO_2 + O_2 \rightarrow 2SO_3$.

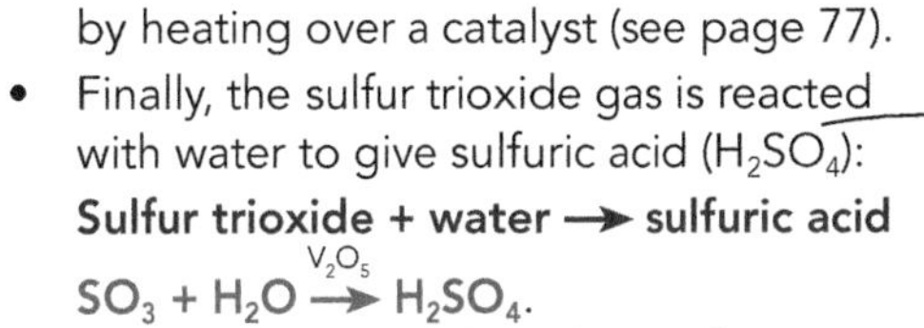

- The rate of this slow reaction is increased by heating over a catalyst (see page 77).
- Finally, the sulfur trioxide gas is reacted with water to give sulfuric acid (H_2SO_4):
  Sulfur trioxide + water → sulfuric acid
  $SO_3 + H_2O \xrightarrow{V_2O_5} H_2SO_4$.
  Vanadium pentoxide is the catalyst.
- H_2SO_4 is a colourless, thick liquid that readily dissolves in water to give a strongly acidic solution (see page 84). It is used to make the fertiliser superphosphate.

Chlorine

- Chlorine is an irritating green gas not found naturally as an element. It is made by the electrolysis of molten salt.
- Chlorine atoms have 7 electrons in their outer shell. In order to become more stable, pairs of Cl atoms share 2 electrons to form chlorine molecules (Cl_2).
- Cl strongly attracts valence electrons of other atoms, so it is a highly unreactive gas that reacts with most elements to give chlorides.
- Sodium burns in chlorine (see page 60) to give an ionic sodium chloride (NaCl):
  Sodium + chlorine → sodium chloride
  $2Na + Cl_2 \rightarrow 2NaCl$.
- Chlorine disinfects water (see page 75) in the reaction:
  Chlorine + water → hydrochloric acid + hypochlorous acid
  $Cl_2 + H_2O \rightarrow HCl + HClO$.

Revision Activities

1 Match up the descriptions with the terms.

- **a** element
- **b** metalloid
- **c** compound
- **d** electrolysis
- **e** allotrope
- **f** amorphous solid
- **g** electron shell
- **h** valence electrons
- **i** lattice
- **j** ionic bonding
- **k** covalent bonding
- **l** molecule
- **m** ionic compound
- **n** molecular compound
- **o** complete combustion
- **p** incomplete combustion
- **q** acidic solution
- **r** alkaline solution

- A compound in which atoms are sharing electrons
- B electrons in the outer occupied electron shell
- C bonds formed by atoms sharing electrons
- D burning a substance in abundant oxygen
- E solid that does not have a crystalline structure
- F solution resulting from dissolving a metal oxide in water
- G extracting an element from a compound using electricity
- H regular array of a large number of atoms or ions
- I alternative forms of an element
- J discrete group of atoms sharing electrons
- K pure substance made of one type of atom only
- L solution resulting from dissolving a nonmetal oxide in water
- M bonds formed between oppositely charged ions
- N space around nucleus that electrons occupy
- O element that has both metal and nonmetal properties
- P compound formed out of ions or polyatomic ions
- Q burning a substance in limited oxygen
- R pure substance in which different kinds of atoms are chemically combined

2 Explain the differences between the following terms:

- **a** metal and nonmetal elements
- **b** an element and a compound
- **c** ionic and covalent bonding
- **d** a crystalline and an amorphous solid
- **e** an ionic and a molecular compound
- **f** incomplete and complete combustion
- **g** an acidic and an alkaline solution.

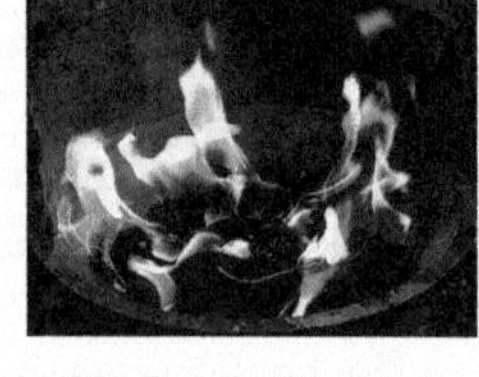

3 Draw up a table as shown below and compare the physical properties of metal and nonmetal elements.

Property	Metal elements	Nonmetal elements
State	Nearly all solids	Mostly gases and solids
etc.		

4 Decide whether the following statements are true or false. Rewrite the false ones to make them correct.

- **a** Most metals are elements.
- **b** Some nonmetals exist naturally in elemental form.
- **c** All nonmetal properties differ from those of metals.
- **d** A particular property of nonmetals is the existence of allotropes.
- **e** All nonmetals are located in the far right of the Periodic Table.
- **f** The outer occupied electron shells of inert gases are full.
- **g** Having a full outer electron shell makes atoms less stable.
- **h** Reactive nonmetals are able to pull electrons off metal atoms.
- **i** Ionic bonding involves the transfer of electrons.
- **j** In molecules, atoms are involved in sharing electrons.
- **k** Carbon's unusual properties relate to it having 4 valence electrons.
- **l** N_2 molecules have triple bonds.
- **m** In O_2, two electrons are shared.
- **n** Sulfur molecules are ring-shaped.
- **o** Chlorine is highly reactive.

5 Consider the data relating to the physical properties of nonmetal elements given in the table below as you answer these questions.

- **a** Which element is in an atypical state for a nonmetal?
- **b** How many of the nonmetals listed have allotropes?
- **c** Which elements exist as monatomic molecules?
- **d** Which element exists as a network of covalently bonded atoms rather than discrete molecules?
- **e** Which element has an atypical lustre for a nonmetal?
- **f** How do the colours of nonmetals differ from metals?
- **g** Which elements would be solids at 100 °C?
- **h** Which element would be a liquid at 100 °C?
- **i** What is unusual about the state change of graphite?
- **j** What other physical property is density related to?
- **k** Which nonmetal is extremely hard? How do you know?
- **l** Which element has very atypical electrical conductivity?
- **m** How good are most nonmetals at conducting heat?
- **n** Why is the conductivity of carbon exceptional?

Element	H	He	C		N	O		F	Ne	P	S	Cl	Ar	Br	I
Allotrope	n/a	n/a	graphite	diamond	n/a	oxygen	ozone	n/a	n/a	'white'	rhombic	n/a	n/a	n/a	n/a
State	gas	gas	solid	solid	gas	gas	gas	gas	gas	solid	solid	gas	gas	liquid	solid
Formula	H_2	$He_{(1)}$	n/a	n/a	N_2	O_2	O_3	F_2	$Ne_{(1)}$	P_4	S_8	Cl_2	$Ar_{(1)}$	Br_2	I_2
Lustre	n/a	n/a	dull	glassy	n/a	n/a	n/a	n/a	n/a	dull	dull	n/a	n/a	n/a	dull
Colour	clear	clear	black	clear	clear	clear	clear	pale	clear	pale	yellow	green	clear	red	grey
MP (°C)	–259.14	–272.20	sublimes at 3642	3550	–210.00	–218.79	–192	–219.62	–248.59	44.2	115.21	–101.5	–189.35	–7.2	113.7
BP (°C)	–252.87	–268.93		4827	–195.79	–182.95	–112	–188.12	–246.08	280.5	444.6	–34.04	–185.85	58.8	184.3
Density ($g\,cm^{-3}$)	0.0001	0.0002	2.267	3.53	0.0012	0.0014	0.0021	0.0016	0.0090	1.823	1.96	0.0032	0.0018	3.1028	4.94
Hardness (Mohs)	n/a	n/a	0.5	10	n/a	n/a	n/a	n/a	n/a	0.5	2.0	n/a	n/a	n/a	0.5
Electrical conductivity (S/m)	n/a	n/a	1×10^{5}	1×10^{-3}	n/a	n/a	n/a	n/a	n/a	1×10^{7}	1×10^{-15}	n/a	n/a	1×10^{-10}	1×10^{-7}
Heat conductivity ($Wm^{-1}K^{-1}$)	0.1805	0.1513	140	900-2300	0.0258	0.0265	–	0.0277	0.0491	0.236	0.205	0.0089	0.0177	0.122	0.449

ISBN: 9780170262316

6 The photos below relate to the combustion of three nonmetals.

© Leslie Garland Pictures

a Which relates to the combustion of sulfur? Why?
b Write word and symbol equations for that reaction.
c Which relates to the combustion of nitrogen? Why?
d Write word and symbol equations for that reaction.
e Which relates to the combustion of carbon? Why?
f How do you know that incomplete combustion is occurring?
g Write word and symbol equations for that reaction.
h Write word and symbol equations for the complete combustion of carbon.

7 Use the facts that nonmetal atoms have nearly full or full outer electron shells and do not have any free electrons to explain each of the following properties of nonmetal elements (ignore the exceptions).

	Property	Nonmetals	Exceptions
a	State (20 °C)	many are gases	C, P, S, Br, I
b	Lustre	dull when solid	C*
c	Melting point	mostly $<$ 20 °C	C†, C*, P, S, I
d	Boiling point	mostly $<$ 20 °C	C†, C*, P, S, Br, I
e	Density	mostly light	C†, C*, P, S, Br, I
f	Hardness	mostly soft	C* very hard
g	Electrical conductivity	mostly very poor	C† excellent
h	Thermal conductivity	mostly very poor	C* excellent C† good

C* diamond, C† graphite

8 Read the article opposite, then answer the questions below.
a Why must the water we drink and swim in be sterilised?
b Why can both chlorine and ozone be used to sterilise water?
c What does ozone do to bacteria and viruses?
d How is ozone generated at a water treatment plant?
e What are the advantages and disadvantages of using ozone to purify water?
f How is Cl_2 gas supplied to public and private swimming pools?
g Which product of reacting Cl_2 with H_2O attacks micro-organisms?
h What are the advantages and disadvantages of using chlorine?

9 The S_8 ring molecules of the element sulfur can be arranged in different ways to give the two different crystals shown in the diagram below.

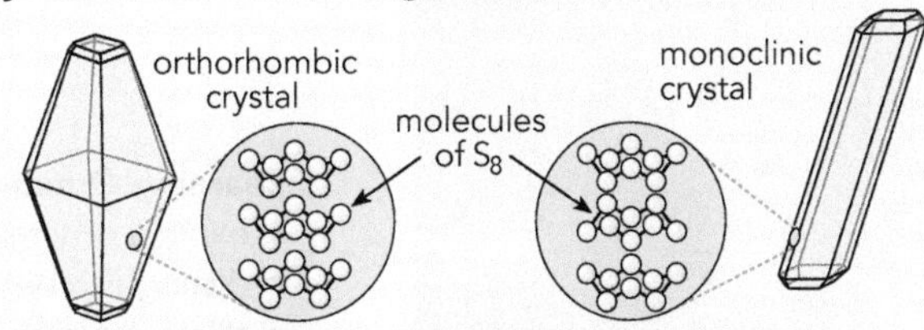

a Describe how the shapes of the two crystals differ.
b Explain how the ring molecules are arranged differently in each type of crystal.
c State the term used to describe different forms of an element.
d Identify the physical properties that are likely to differ.
e Explain why some nonmetals form allotropes.

10 The diagram opposite shows the results of inserting blue or red litmus paper into solutions consisting of an oxide dissolved in water.

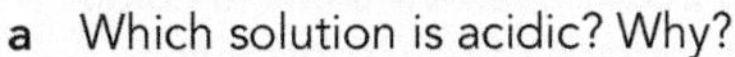
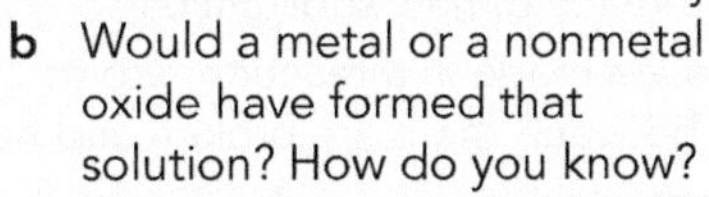

a Which solution is acidic? Why?
b Would a metal or a nonmetal oxide have formed that solution? How do you know?
c Which solution is alkaline? Why?
d Would a metal or a nonmetal oxide have formed that solution? How do you know?

11 For each reaction, describe the state of the reactants and products, then write word and symbol equations.

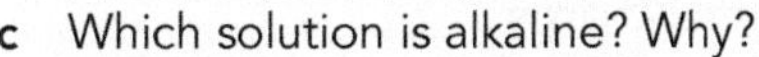
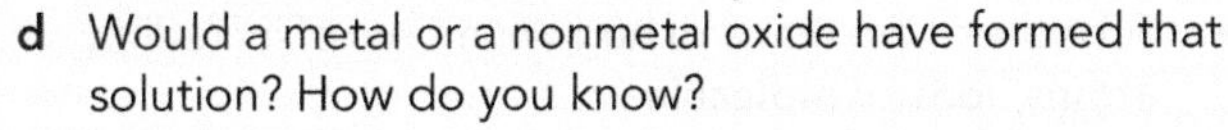

a

b

c

Disinfecting Drinking Water and Swimming Pools

It is important that our drinking water supply is kept sterile so as to prevent the spread of infectious water-borne diseases.

Ozone is a reactive chemical that is used to treat drinking water, as it destroys bacteria and deactivates viruses.

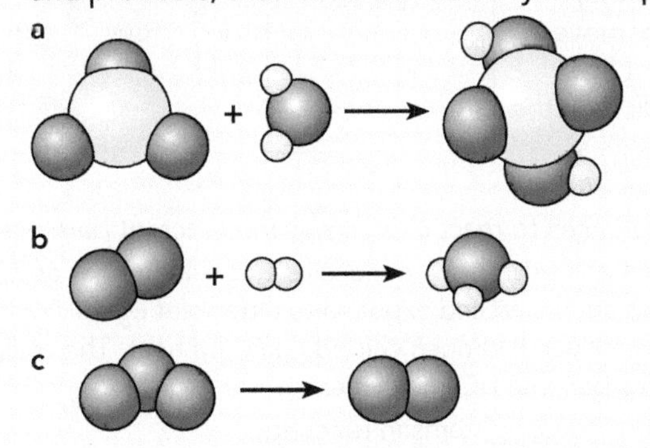

Ozone gas is made at the treatment plant by the electrolysis of water or air. Either way, no pollutants are produced.

The ozone is bubbled into the water and rapidly dissolves. It then reacts with any micro-organisms, causing the cells to burst. In the process, ozone is converted to oxygen gas.

There are no objectionable odours when water is sterilised with ozone as compared to using chlorine.

Chlorine is used to purify drinking water and to make the water in swimming and spa pools safe.

Liquefied chlorine is used in public pools but chlorine-releasing compounds are used in private pools, as handling chlorine is dangerous.

Chlorine gas readily reacts with water to form hydrochloric and hypochlorous acids. The latter acid penetrates the cell membrane of micro-organisms causing them to burst.

Only a very small amount of chlorine is needed to disinfect a pool but sometimes there may be a bleach smell resulting from chlorine reacting with urea.

Drinking water purified by chlorine may have an unappealing bleach taste.

ISBN: 9780170262316

20

Reaction Rates and Particle Collisions

Learning Outcomes - On completing this unit you should be able to:
- explain why reactions require collisions between particles
- distinguish between reactants and products in a reaction
- relate reaction rate to the number of effective collisions per second
- explain the effect of different factors on reaction rates
- *interpret and explain reaction rate curves.*

Chemical Reactions

- Substances change, eg when baking soda is added to vinegar, a gas is released causing the liquid to froth up, and the baking soda disappears. When new substances are formed, a **chemical reaction** has occurred.
- Signs that a chemical reaction is occurring include light, heat, flames, smoke, frothing, a new smell, a colour change, and the appearance of new substances.
- When new substances are made in a reaction, atoms are rearranged as **bonds** between them are broken and new ones formed.

Collisions between Atoms

- Substances are made of huge numbers of incredibly small particles, either **atoms, ions** or **molecules**.
- *Particles are in constant motion.* They vibrate, rotate or move in a straight line depending on the substance's state (see page 36).

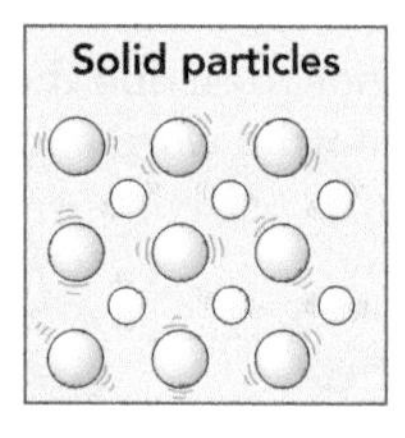

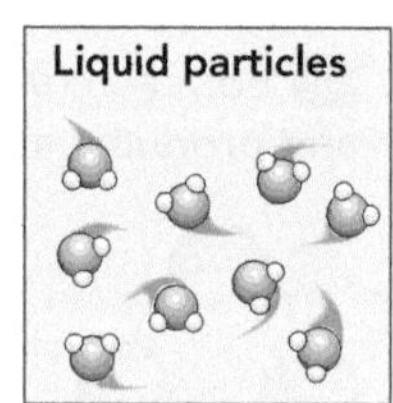

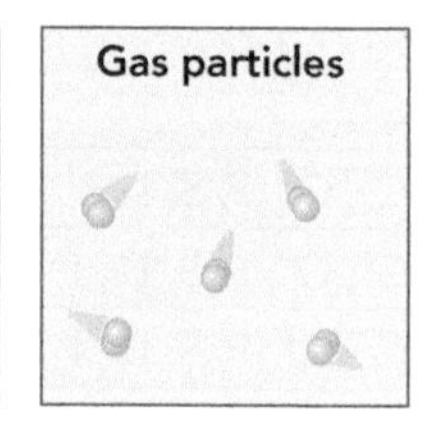

- *For substances to react chemically, the reactant particles must collide.*
- Hydrogen gas exists as H_2 molecules and chlorine gas as Cl_2 molecules. When the gases are mixed and exposed to ultraviolet light (UV), they spontaneously react to produce hydrogen chloride gas and heat. The colliding H_2 and Cl_2 molecules become HCl molecules.
- The bonds holding atoms in reacting molecules must be broken to allow the bonds in the new molecules to form.
- For a collision to be effective, it must be *sufficiently forceful.* Faster moving particles with more kinetic energy (see page 12) have more forceful collisions.
- Collisions also need to occur at the *right orientation* for the collision to be effective.

Cl_2 molecule   HCl molecule H_2 molecule

Reactants and Products

- In a chemical reaction, the original substances are called the **reactants** and the new ones the **products**.
- The change from reactants to products can be expressed as a **word equation**, with an arrow going from the reactants to the products. The reactants are put on the left and products on the right.
  Reactants → products (the arrow means 'changed into')
  eg Hydrogen gas + chlorine gas → hydrogen chloride gas
- A reaction can also be represented using a **symbol equation**:
  eg $1H_2 + 1Cl_2 \rightarrow 2HCl$ or $H_2 + Cl_2 \rightarrow 2HCl$ (as 1's are usually dropped).
- The numbers in front of formulas give the ratio of atoms, ions or molecules involved. **1** molecule of H_2 reacts with **1** molecule of Cl_2 to give **2** molecules of HCl.
- The equation is balanced, as for each type of atom the number of atoms is the same on both sides (2 H and 2 Cl   2 H and 2 Cl).

Energy in Reactions

- Most reactions need some energy to start with (**activation energy**).
- Reactions that release heat, such as the reaction between H_2 and Cl_2, are **exothermic**. The heat released will *keep them going.*
- Reactions that absorb heat, such as heating copper oxide in hydrogen gas to give copper metal, are **endothermic**. An endothermic reaction requires a source of heat both to *start the reaction and to keep it going.*

Reaction Rate

- Different reactions occur at different speeds. Rusting occurs slowly, combustion occurs rapidly, for example in the shuttle's booster rockets (see top photo).
- **Reaction rate** is the *speed at which a reaction is occurring.* The rate will change during the course of the reaction.
- The rate can be found by measuring the rate at which a *reactant is consumed* or a *product formed.*
- When magnesium ribbon is added to hydrochloric acid, it fizzes as hydrogen gas is produced and soon disappears as its atoms become ions.
- If the amount of a reactant or product is measured at regular intervals, the data can be plotted on a reaction progress graph.

H_2 HCl   Mg

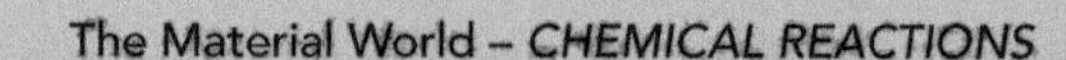

ISBN: 9780170262316

- The graph shows the volume of gas produced in the reaction: $Mg + 2HCl \rightarrow H_2 + Mg^{2+} + 2Cl^-$

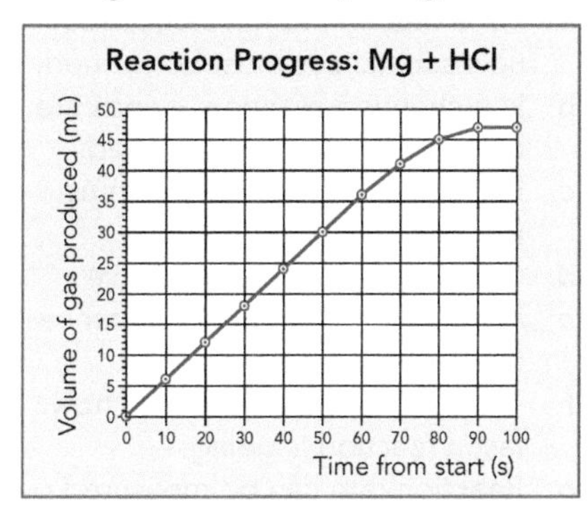

- The *gradient (slope) of the graph is a measure of the reaction rate.*
- Initially, the reaction proceeds at a steady rate, but as the Mg is consumed the reaction rate slows then eventually stops when there are no more magnesium particles.

Rate Factors

- The reaction rate depends on the number of effective collisions per second. To increase the rate, the collision *frequency and/or force must be increased.* (More frequent collisions will result in more having the right orientation.)
- Factors affecting rates are concentration, temperature, reactant state and size, and catalysts.

Concentration

- **Concentration** is the number of particles in a specific volume.
- The more reactant particles there are in a fixed volume, the *greater the frequency of collisions* and the faster the reaction rate.
- For gas reactants, particle concentration is increased by increasing the pressure. This pushes the particles closer together, so more collisions occur.
- Many reactions occur when the reactant particles are dissolved in water. A dilute solution has few particles per unit volume of water. A concentrated solution has many particles in the same volume.

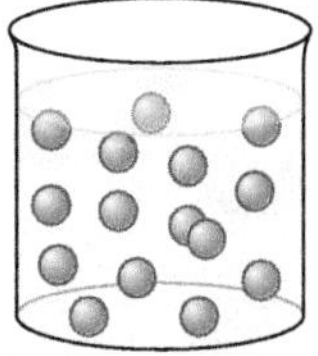
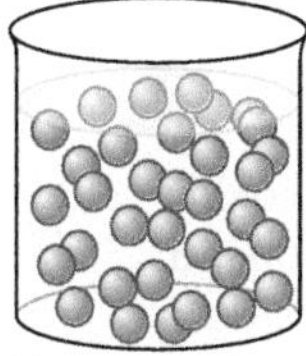
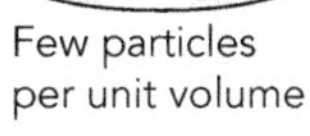

Few particles per unit volume

Many particles per unit volume

Temperature

- Some reactions are extremely slow but take place rapidly when heated. Baking soda (sodium bicarbonate, $NaHCO_3$) can be kept in the cupboard for years but breaks down into sodium carbonate (Na_2CO_3), water (H_2O) and carbon dioxide (CO_2) when heated. The CO_2 gas causes the baking to rise.
- Heated particles gain kinetic energy (see page 36). As they have more kinetic energy, they move faster and the *collision frequency increases.* More of those collisions are effective, as more of the faster-moving particles *collide with sufficient force* to break the existing bonds.
- **Temperature** is a measure of the average kinetic energy of particles (see page 36). An increase of 10 °C will often *double a reaction rate.*

Reactant State and Size

- Particles in a gas **state** are more energetic than when in a liquid state, which in turn are more energetic than when in a solid state.
- For a solid to react, other reactant particles must collide with its exposed surface particles. The more particles that are exposed, the *greater the frequency of collisions.*
- To increase the number of exposed particles, a lump is broken into smaller pieces or crushed to increase the exposed surface area.

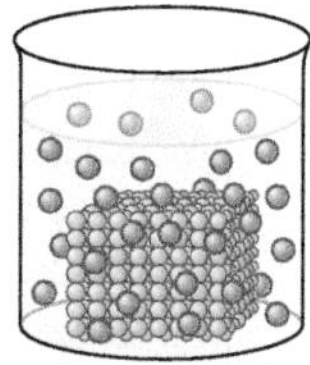

A large lump has fewer exposed surface particles

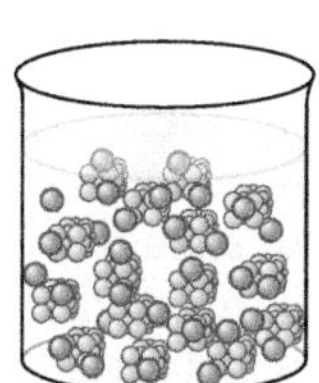

Many smaller lumps have more exposed surface particles

- Using marble chips ($CaCO_3$) rather than a large lump of marble will speed up the reaction with hydrochloric acid (HCl) that produces carbon dioxide (CO_2) gas in the reaction: $CaCO_3 + 2HCl \rightarrow Ca^{2+} + 2Cl^- + H_2O + CO_2$.

Catalysts

- A **catalyst** is a particular chemical that speeds up a specific reaction by *reducing the force needed for effective collisions.* As less activation energy is needed, the reaction occurs faster.
- As the catalyst it is not used up, it is not considered a reactant. (It is written above the equation arrow.)
- Nearly all industrial chemical processes involve a catalyst. For example, the production of sulfuric acid by reacting SO_3 with H_2O (see page 73) requires the catalyst vanadium pentoxide for it to occur rapidly.
- **Enzymes** (see page 101) are a group of catalysts made by organisms, eg rennin is a milk-curdling enzyme.

SCIENCE SKILL: Interpreting Reaction Progress Graphs

The graph shows two series of measurements for the reaction between 1 cm of magnesium ribbon and 10 ml of dilute hydrochloric acid.

Steps:

1. Decide what a line shows: the red line shows amount of gas formed.
2. Interpret the slope: the high rate of gas production is maintained during the first 70 s, then it slows until finally no more is produced after 90 s.
3. Explain the slope: initially there are plenty of colliding reactant particles to sustain the high reaction rate, but after 70 s most Mg particles have been used up, so fewer collisions occur till the reaction ceases after 90 s.

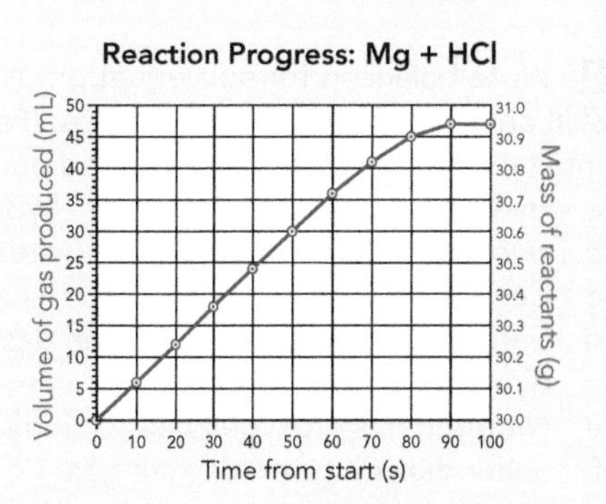

Revision Activities

1 Match up the descriptions with the terms.

Terms	Descriptions
a chemical reaction	A new substance formed in a reaction
b bond	B charged atom or molecule
c atom	C speed at which reactant particles are changed into product particles
d ion	D either a solid, liquid or gas
e molecule	E measure of average kinetic energy of particles
f reactant	F energy required to start a reaction
g product	G expressing a reaction using the names of chemicals
h word equation	H reaction that continues to absorb heat from its surroundings
i symbol equation	I catalyst produced by a living organism
j activation energy	J original substance used up in a reaction
k exothermic reaction	K occurs when new chemicals are formed
l endothermic reaction	L number of particles per unit volume
m reaction rate	M force of attraction that holds atoms together
n concentration	N very small particles that all matter is made of
o state of matter	O discrete group of covalently bonded atoms
p temperature	P expressing a reaction using the formulas of the chemicals involved
q catalyst	Q chemicals that speed up the reaction rate by lowering the activation energy
r enzyme	R reaction that releases energy to its surroundings

2 Explain the differences between the terms in each of the following:

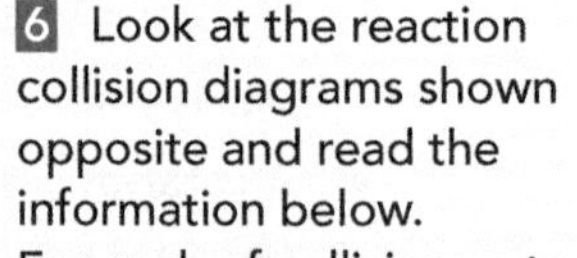

- a reactant and product particles
- b a word and a symbol equation
- c an exothermic and an endothermic reaction
- d a dilute and a concentrated solution
- e a catalyst and an enzyme.

3 For each of the reactions below, identify the reactants and products, then write word equations.

- a Hydrogen chloride gas is made by exposing a mixture of chlorine and hydrogen gases to UV light.
- b When copper oxide is heated in hydrogen gas, copper metal and steam are produced.
- c Steam is produced when hydrogen gas is combusted in oxygen gas.
- d Magnesium ribbon reacts with hydrochloric acid to give hydrogen gas and magnesium and chloride ions.
- e Sodium chloride (table salt) can be made by placing hot sodium metal in chlorine gas.

4 Write balanced formula equations (see the science skill on page 65) for these reactions. (Formulas for most of the chemicals can be found on page 73.)

- a Hydrogen gas + chlorine gas   hydrogen chloride gas
- b Hydrogen gas + oxygen gas   steam
- c Carbon powder + oxygen gas   carbon dioxide gas
- d Carbon monoxide gas + oxygen gas   carbon dioxide gas
- e Nitrogen gas + oxygen gas   nitric oxide gas
- f Sulfur dioxide gas + oxygen gas   sulfur trioxide gas
- g Sulfur trioxide gas + water   sulfuric acid

5 Decide whether the following statements are true or false. Rewrite the false ones to make them correct.

- a For a reaction to have occurred, new substances must be formed.
- b In a chemical reaction, bonds are broken and new ones formed.
- c For particles to react, they must first collide.
- d Collisions are always effective.
- e Reactants are the new substances formed in a reaction.
- f The reaction rate describes how fast a reaction is going.
- g Reaction rate can be measured by how fast a product is consumed.
- h A reaction progress graph line will often level off because one of the reactants has been used up.
- i Reaction rates are affected by concentration, temperature, reactant state, and catalysts.
- j A dilute solution has more particles in a set volume than a concentrated solution.
- k Stirring can increase reaction rates.
- l Enzymes are manufactured catalysts.

6 Look at the reaction collision diagrams shown opposite and read the information below.

For each of collisions a to d:

- i record the reactants and their states
- ii record the products and their states
- iii write a word equation that specifies the states of the reactant and product particles
- iv write a balanced symbol equation.

In reaction **a**, hydrogen iodide gas is produced by heating iodine in hydrogen gas. (Iodine becomes a gas at 184°C.)

In reaction **b**, nitric oxide gas is produced when oxygen and nitrogen molecules collide at high temperatures.

In reaction **c**, oxygen gas is produced when ozone molecules collide in atmospheric conditions.

In reaction **d**, nitrogen dioxide and oxygen gas are produced when ozone and nitric oxide molecules collide in atmospheric conditions.

a

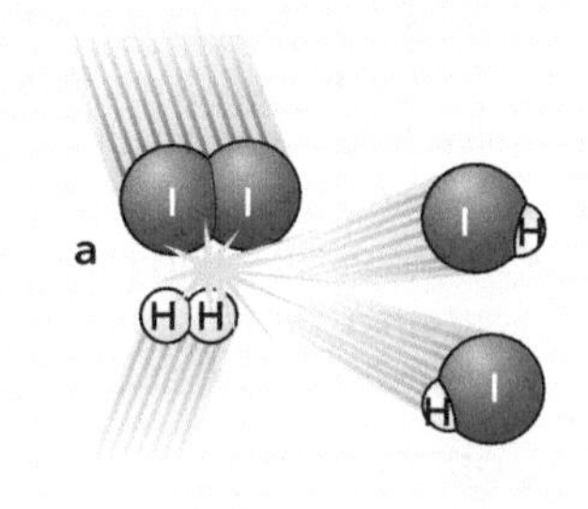

b

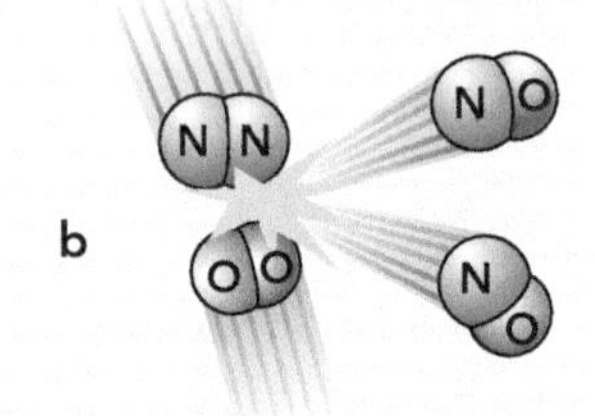

c

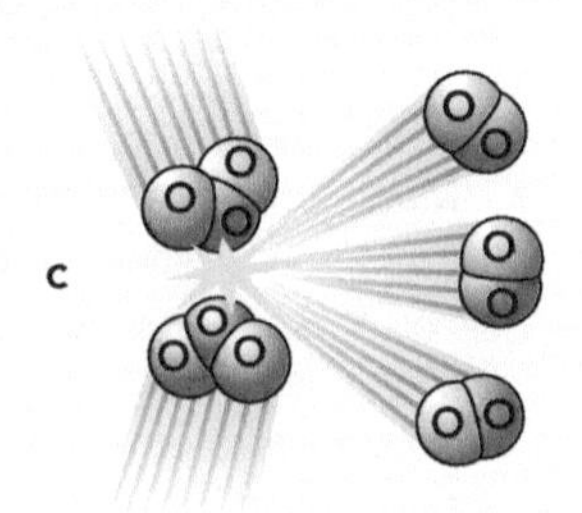

d

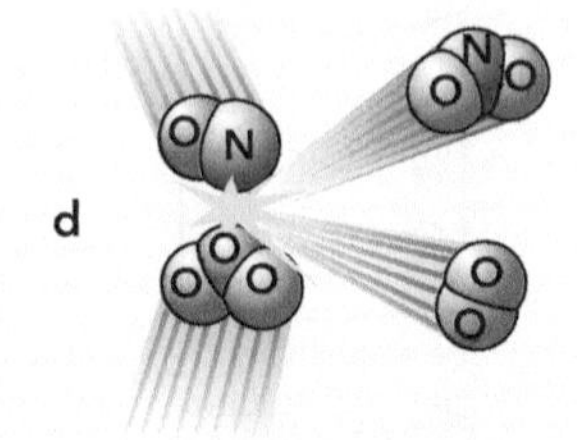

ISBN: 9780170262316

7 Hydrogen peroxide (H_2O_2) is a clear liquid that breaks down slowly to form water and oxygen gas. If a pinch of black manganese dioxide (MnO_2) powder is added, the reaction goes rapidly to completion producing steam and O_2 gas. The manganese dioxide can be recovered by evaporating the water.

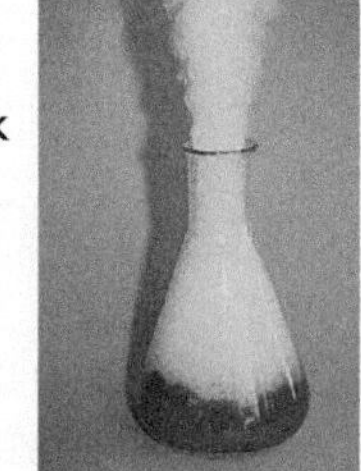

a What are the reactants and products?
b Write a word equation for the reaction.
c Write a balanced formula equation for the reaction.
d Explain the role of the manganese dioxide.
e What is the name given to such chemicals?
f Show the role of manganese dioxide on the equation.

8 Zinc reacts with hydrochloric acid to produce hydrogen gas. Predict what will happen and why in each situation:

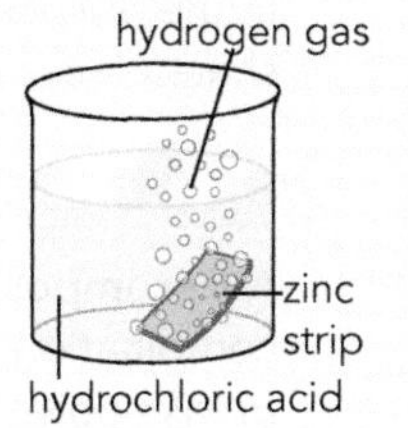

a powder is used instead of the strip
b more concentrated acid is used instead of the dilute acid
c the beaker is placed in an ice bath
d the beaker is warmed to 50°C
e an equal volume of water is added to the acid.

9 Describe how each of the following factors impacts on the number of effective collisions per second:
a increasing the concentration of reactants
b increasing the temperature of reactants
c decreasing the lump size of a solid reactant
d adding a suitable catalyst.

10 The graph relates to the reaction of Mg with HCl, which produces H_2 gas.
a What does the blue line show?
b Interpret the slope.
c Explain the slope.

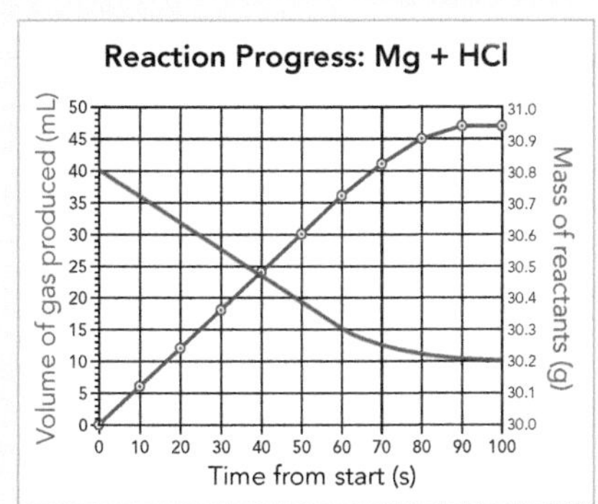

11 Read the article opposite, then answer the questions below.
a What is the normal internal temperature of the human body?
b What are enzymes?
c How many different types of enzymes are found in your body?
d What would happen to your body if enzymes ceased working?
e Describe the different roles of amylases, proteases and lipases.
f Why must food be digested before cells can use it?
g What is meant by enzymes being specific?
h List three properties of enzymes.
i Are enzymes used up in the reaction that they catalyse?

12 In an experiment, two grams of fine marble chips (calcium carbonate $CaCO_3$), are placed in excess dilute hydrochloric acid, and the gas produced by the reaction is measured using the apparatus shown.

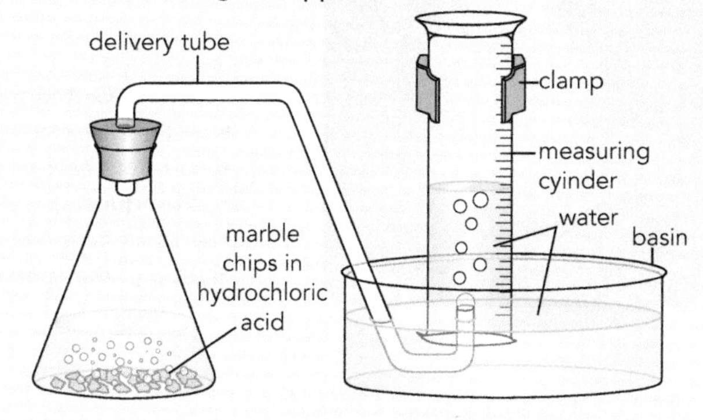

a Explain why the measuring cylinder is inverted.
b Describe how you would confirm that the gas produced is carbon dioxide.

13 The above apparatus was used to measure the gas produced over a 10-minute period. The reaction was then repeated using 2 g of powdered $CaCO_3$ and fresh dilute HCl.
a List the reactants and products.
b Write word and formula equations for the reaction.
c Plot the reaction progress for the chip and powder on the same graph.
d Describe the shapes of the two lines.
e Identify the reaction that went to completion faster.
f Explain why that reaction went faster than the other one.
g Account for why the two curves level off after a period of time.
h Suggest a reason why the amount of CO_2 produced differed.

	Volume gas (mL)	
Time (min)	Using chips	Using powder
0	0	0
1	20	70
2	60	130
3	120	200
4	190	280
5	250	370
6	310	430
7	370	460
8	420	470
9	430	470
10	430	470

Digestion and Enzymes

The internal temperature of the body is about 37°C. At this temperature the reactions that need to occur inside you, such as the digestion of food, would occur very slowly without the aid of catalysts, so slowly that you would starve to death even if you ate lots of food.

Enzymes are catalysts made by living organisms. There are several thousand different types of enzymes found inside cells of your body, each of them speeding up a particular type of reaction. Enzymes enable these reactions to occur rapidly at body temperature.

Enzymes called amylases in your saliva and small intestine aid the breakdown of starch into sugars. Enzymes called proteases in your stomach and small intestine aid the breakdown of protein into amino acids. Lipases in your small intestine are involved in the breakdown of fats into fatty acids.

The digested food molecules – sugars, amino acids and fatty acids – are small enough to pass through the lining of the gut into your blood- stream, from where they are distributed to all cells.

Enzymes have special properties. Each enzyme is specific – it catalyses only one type of reaction.

Enzymes work best at body temperature; at hotter or cooler temperatures they tend to slow down and eventually stop working.

Different enzymes work best at different pHs. Mouth enzymes work best under neutral conditions, stomach enzymes under acidic conditions, and intestinal enzymes under alkaline conditions.

An enzyme can catalyse thousands of reactions in a very short time span. It will emerge from each reaction unchanged.

ISBN: 9780170262316

Types of Chemical Reactions

Learning Outcomes - On completing this unit you should be able to:
- distinguish between four different types of reactions
- compare and contrast combination and decomposition reactions
- distinguish the effect of heat on metal oxides, carbonates and bicarbonates
- explain how dissolving and precipitation reactions are related
- *predict the formation of precipitates when two ionic solutions are mixed.*

Chemical Reactions

- For Level 1 NCEA Science and Chemistry you need to be familiar with four types of chemical reactions:

  a combination A + B → AB
  b decomposition AB → A + B
  c displacement A + BC → AC + B
  d precipitation A^+ + B^- → AB

Conventions

- Reactant and product chemicals are called **species**.
- The state of each species is shown by adding a symbol in brackets after each: *s* for solid, *l* for liquid, *g* for gas, and *aq* for chemicals dissolved in water.
- For example, when Na metal reacts with H_2O to give H_2 gas and dissolved sodium hydroxide, the symbol equation is: $2Na(s) + 2H_2O(l) \rightarrow 2NaOH(aq) + H_2(g)$

Combination Reactions

- A **combination reaction** occurs when *two substances combine to form a new substance.* At NCEA level 1 the focus is on two elements combining.
- Iron is a grey, magnetic solid. As its atoms have two loosely held valence electrons, it is a reactive metal.
- Sulfur is a brittle, nonmagnetic yellow solid. As its atoms are two electrons short of a stable full outer shell, it is a reactive nonmetal.
- When the elements are mixed in powdered form and heated, a combination reaction occurs that produces a solid, black, nonmagnetic **ionic compound** called iron sulfide ($Fe^{2+}S^{2-}$).

Fe + S
exothermic reaction
FeS

- The **ionic equation** is: $Fe(s) + S(s) \rightarrow Fe^{2+}S^{2-}(s)$.
- Each Fe atom loses two electrons to become an F^{2+} ion, and each S atom gains two electrons to become an S^{2-} ion, thus making both more stable.
- Oxygen atoms are also two electrons short of full outer shells.
- When sulfur is heated in oxygen gas, the two nonmetal elements react to produce a pungent choking gas, which is a **covalent compound** called sulfur dioxide (see page 73).

- The balanced equation for the reaction is: $S_8(s) + 8O_2(g) \rightarrow 8SO_2(g)$
  In the SO_2 molecules, the sulfur and oxygen atoms share electrons to achieve stable (full) outer shells.

Decomposition Reactions

- A **decomposition reaction** is the opposite of a combination reaction. It occurs when *a compound breaks down into simpler compounds or elements.*
- *How readily a metal compound decomposes when heated depends on its reactivity* (see page 65). The more reactive the metal, the more stable its compounds and the more difficult decomposition will be.
- Compounds can be decomposed (broken down) *by heating, electrolysis, or the action of catalysts.*

Thermal Decomposition of Hydroxides

- Metal hydroxides have positive metal ions bonded to negative hydroxide ions, OH^-, eg sodium hydroxide, Na^+OH^- and copper hydroxide, $Cu^{2+}(OH)_2^-$.
- Except for KOH and NaOH, hydroxides decompose into an oxide and water when *heated*:

Hydroxide $\xrightarrow{heat}$ oxide + water

- When blue copper hydroxide is heated it decomposes into black copper oxide and a gas that turns blue cobalt chloride paper pink, proving it is steam. The reaction is: $Cu(OH)_2(s) \rightarrow CuO(s) + H_2O(g)$.
- Thermal decomposition is carried out by heating.

Thermal Decomposition of Carbonates

- Metal carbonates have positive metal ions bonded to negative carbonate ions, CO_3^{2-}.
- The carbonates of less reactive metals such zinc ($ZnCO_3$), lead ($PbCO_3$), and copper ($CuCO_3$) readily decompose when *heated* to give an oxide and CO_2.

Carbonate $\xrightarrow{heat}$ oxide + carbon dioxide

- Heated white $ZnCO_3$ powder decomposes into solid yellow zinc oxide (ZnO) and a gas that turns limewater milky, proving that it is CO_2.
- The reaction is: $Zn(CO_3)(s) \rightarrow ZnO(s) + CO_2(g)$.
- Carbonates of the more reactive metals, K, Na and Li, do not decompose readily even when heated strongly.

Thermal Decomposition of Bicarbonates

- Metal bicarbonates have positive metal ions bonded to negative carbonate ions, HCO_3^-, eg $NaHCO_3$.

ISBN: 9780170262316

- Heated bicarbonates of reactive metals, K, Na and Li, decompose into metal carbonates, CO_2 and H_2O.

Bicarbonate $\xrightarrow{\text{heat}}$ carbonate + carbon dioxide + water

- Bicarbonates of less reactive metals exist only when dissolved in water because they undergo the same decomposition when the water evaporates.
- When white sodium bicarbonate powder is heated it decomposes to give white sodium carbonate and gases that turn blue cobalt chloride paper pink and limewater milky. The reaction is:

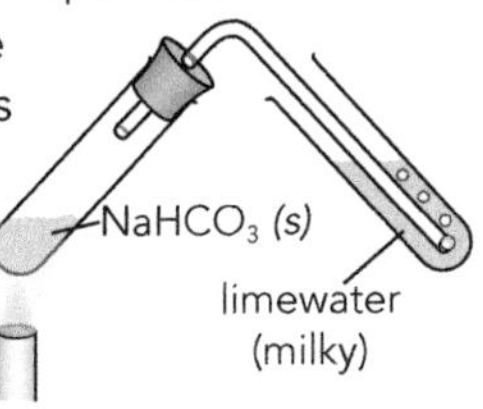

$2\mathbf{NaHCO_3}(s) \rightarrow \mathbf{Na_2CO_3}(s) + \mathbf{CO_2}(g) + \mathbf{H_2O}(g)$

Catalytic Decomposition of Hydrogen Peroxide

- Pure hydrogen peroxide (H_2O_2) is a faintly bluish, syrupy liquid that rapidly decomposes into oxygen and water when exposed to sunlight. As pure H_2O_2 is dangerous, it is dissolved in water. Dilute H_2O_2 decomposes slowly.
- When a pinch of black manganese dioxide (MnO_2) powder is added to dilute (30%) H_2O_2, rapid bubbling occurs as the liquid heats up but remains colourless.
- The gas ignites a glowing splint proving that it is O_2. The MnO_2 can be recovered by evaporating the water, which means it is a **catalyst** (see page 77).
- The reaction for this catalytic decomposition is:

$2\mathbf{H_2O_2}(aq) \xrightarrow{MnO_2} \mathbf{O_2}(g) + 2\mathbf{H_2O}(l)$.

- Catalytic decomposition is carried out by catalysts.

Displacement Reactions

- A **displacement reaction** occurs when *a free element replaces another in a compound.* The free element must be more reactive than the bonded element.
- At NCEA level 1, the focus is on reactions in which the atoms of *a more reactive metal displace the dissolved ions of a less reactive metal.* The original metal ions become metal atoms.
- In order to predict what will happen, you need to be familiar with the reactivity series (see page 65):

K > Na > Li > Ca > Mg > Al > Zn > Fe > Pb > Cu > Ag

- When an iron nail is placed in a blue copper sulfate solution [$Cu^{2+}(aq) + SO_4^{2-}(aq)$] for a day, the solution decolourises and the nail gains a thick pinky-brown coating of copper metal. The nail is seen to be thinner when cleaned.

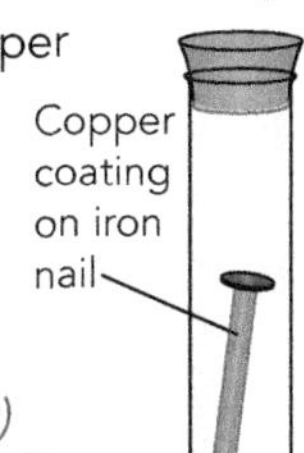

- The reaction is: $\mathbf{Fe}(s) + \mathbf{Cu^{2+}}(aq) + \mathbf{SO_4^{2-}}(aq) \rightarrow \mathbf{Cu}(s) + \mathbf{Fe^{2+}}(aq) + \mathbf{SO_4^{2-}}(aq)$

As the SO_4^{2-} ions do not participate in the reaction, they are called **spectator ions** and can be ignored thus:

$\mathbf{Fe}(s) + \mathbf{Cu^{2+}}(aq) \rightarrow \mathbf{Cu}(s) + \mathbf{Fe^{2+}}(aq)$.

- As *iron is more reactive than copper*, each Fe atom 'forces' two electrons onto a Cu^{2+} ion. The Cu^{2+} ions become Cu atoms as the Fe atoms become Fe^{2+} ions.

Dissolving

- The ions of a **soluble compound** will separate in water as they are *more strongly attracted to H_2O molecules than each other.* The ions are free to move about in the solution. An **insoluble compound** won't dissolve.
- **Dissolving** can be shown by an ionic equation. For example, when table salt (NaCl) dissolves in water: $\mathbf{NaCl}(s) \rightarrow \mathbf{Na^+}(aq) + \mathbf{Cl^-}(aq)$.

Precipitation Reactions

- A **precipitation reaction** occurs when *dissolved ions from two different solutions form an insoluble compound*: $\mathbf{A^+}(aq) + \mathbf{B^-}(aq) \rightarrow \mathbf{AB}(s)$. (An insoluble compound is one that won't dissolve in a liquid.)
- When two solutions are mixed, positive ions from one solution may be strongly attracted to negative ions from the other solution. If so, they may form strong ionic bonds that result in an insoluble compound.
- The newly formed insoluble compound forms a solid called a **precipitate**, which settles to the bottom.
- The leftover ions remain in solution as the compound they would form is soluble. They are spectator ions.
- When clear solutions of lead nitrate [$Pb^+(aq) + NO_3^-(aq)$] and sodium iodide [$Na^+(aq) + I^-(aq)$] are mixed, a yellow precipitate of lead iodide forms.
- Ignoring spectator ions (NO_3^- and Na^+) gives: $\mathbf{Pb^{2+}}(aq) + 2\mathbf{I^-}(aq) \rightarrow \mathbf{PbI_2}(s)$.

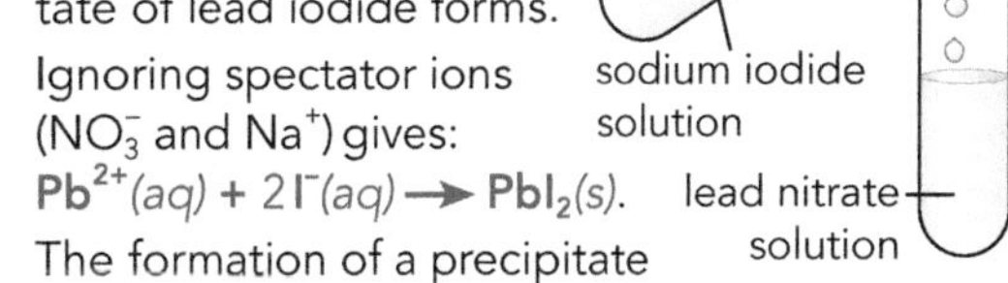

- The formation of a precipitate can be predicted using the **solubility rules** below.

Compounds	Solubility Rules
Nitrates, NO_3^-	All soluble
Chlorides, Cl^-	All soluble except AgCl, $PbCl_2$
Iodides, I^-	All soluble except AgI, PbI_2
Sulfates, SO_4^{2-}	All soluble except $CaSO_4$, $BaSO_4$, $PbSO_4$
Hydroxides, OH^-	All insoluble except KOH, NaOH
Carbonates, CO_3^{2-}	All insoluble except K_2CO_3, Na_2CO_3

SCIENCE SKILL: Predicting Precipitates

You can predict what will happen when two ionic solutions are mixed by using the solubility rules above.

Problem: Predict what happens when clear solutions of copper nitrate and sodium hydroxide are mixed.

Steps:

1. Identify the species present when the two solutions are mixed: Cu^{2+}, NO_3^-, Na^+, OH^-.
2. Identify the possible new compounds by matching up positive and negative ions: copper hydroxide (Cu^{2+} and OH^-) and sodium nitrate (Na^+ + NO_3^-).
3. Check both on the solubility rules table: only the copper hydroxide is insoluble.
4. Write an ionic equation for the precipitation reaction: $Cu^{2+}(aq) + 2OH^-(aq) \rightarrow Cu(OH)_2(s)$.
5. Predict what happens: a blue precipitate of copper hydroxide, $Cu(OH)_2$, will form.

ISBN: 9780170262316

Revision Activities

1 Match up the descriptions with the terms.

a species	A decomposition reaction that uses a catalyst
b combination reaction	B insoluble compound that comes out of solution
c ionic compound	C reaction in which a free element replaces another in a compound
d ionic equation	D symbol equation showing the charges on ions
e covalent compound	E ionic compound that dissolves in water
f decomposition reaction	F rules for which metal compounds are soluble or insoluble
g thermal decomposition	G reactant and product chemicals in a reaction
h catalytic decomposition	H compound resulting from the attraction that atomic nuclei have for shared electrons
i catalyst	I decomposition reaction that uses heat
j displacement reaction	J reaction in which ions combine to form an insoluble compound
k soluble compound	K reaction in which two substances combine
l insoluble compound	L compound that does not dissolve in water
m dissolving	M reaction in which a compound breaks down into simpler compounds or elements
n precipitation reaction	N compound formed by attraction between ions
o precipitate	O chemical that speeds up a reaction without being consumed
p solubility rules	P occurs when a compound separates into ions when mixed with water

2 Explain the differences between the terms in each item below:
- a an element and a compound
- b an ionic and a covalent compound
- c thermal and catalytic decomposition
- d spectator and participant ions
- e soluble and insoluble ionic compounds.

3 Identify the type of reaction each diagram illustrates and explain what happens in that kind of reaction.

a AB → A + B

b A + BC → AC + B

c A + B → AB

d $A^+ + B^- \rightarrow$ AB

4 Consider the components of this ionic symbol equation:

$$2Na(s) + 2H_2O(l) \rightarrow 2Na^+(aq) + 2OH^-(aq) + H_2(g)$$

then explain what each of the following means:
- a the 's' in Na(s)
- b the 'l' in $H_2O(l)$
- c the 'aq' in $OH^-(aq)$
- d the 'g' in $H_2(g)$
- e the '+' in Na^+
- f the '−' in OH^-
- g the '2' in 2Na
- h the '$_2$' in H_2O.

Next consider the overall equation.
- i What do the plus signs signify?
- j What does the arrow signify?
- k Is it a balanced symbol equation? Justify your answer.
- l Explain why is it called an ionic equation.
- m Is it a balanced ionic equation? Justify your answer.

5 Decide whether the following statements are true or false. Rewrite the false ones to make them correct.
- a The species in a reaction are the reactant chemicals.
- b Aqueous *(aq)* means dissolved in water.
- c In a combination reaction, two substances combine to form a single substance.
- d In an ionic equation, the charges on participant ions are shown.
- e Electrons are transferred in covalent bonding.
- f A decomposition reaction is the opposite of a combination reaction.
- g Thermal decomposition requires the use of heat.
- h Heating most carbonates and bicarbonates produces CO_2 gas.
- i In a displacement reaction, a more reactive metal displaces dissolved ions of a less reactive metal.
- j When an ionic compound dissolves, its ions separate.
- k A precipitate occurs when ions form an insoluble compound.

6 Identify which gas:
- a turns limewater milky
- b 'pops' when ignited in air
- c ignites a glowing splint
- d turns blue cobalt chloride pink.

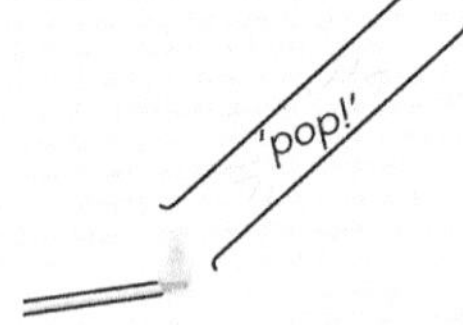

7 Write word and balanced symbol equations for these combination reactions.
- a When burning magnesium ribbon is plunged into a jar of oxygen gas, the metal burns with an intense white flame, producing a thick white powder.
- b When copper wire is heated strongly in air, it glows red-hot, then develops a black layer.
- c When hydrogen and oxygen gases are ignited, the condensation that forms turns cobalt chloride pink.
- d When heated sodium metal is plunged into chlorine gas, the sodium burns with a bright yellow flame, producing a white powder.

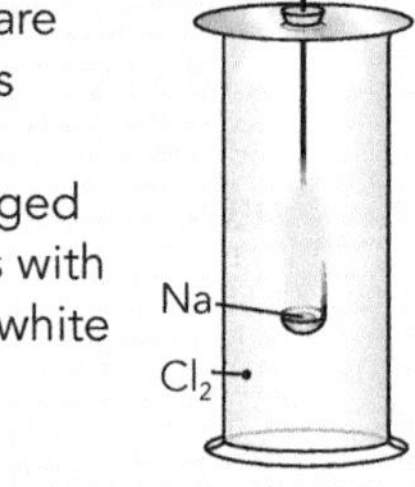

8 Write word and balanced symbol equations for these decomposition reactions.
- a When manganese dioxide (MnO_2) powder is added to dilute hydrogen peroxide, the gas produced ignites a glowing splint and the MnO_2 can be recovered after the liquid evaporates.
- b When magnesium hydroxide powder is heated, it gives off a gas that turns cobalt chloride pink and a white solid.
- c When pale green copper carbonate powder is heated, it produces a colourless gas that turns limewater milky and a black solid.

$CuCO_3$

ISBN: 9780170262316

9 **Write balanced ionic equations for these displacement reactions.**

a When a copper coil is placed in a clear silver nitrate solution, a strange sludge slowly forms on the coil and the solution turns light blue.

b When a thin piece of zinc is placed in a blue copper sulfate solution, the solution decolourises as the zinc disappears and a pinky-brown sludge forms.

c When magnesium ribbon is placed in a pale green iron sulfate solution, the solution decolourises as the Mg slowly disappears and a dark grey sludge forms.

copper wire
sludge on wire
initially $AgNO_3$ solution

10 **The solubility of an ionic compound relates to the strength of attraction that water molecules have for the ions.**

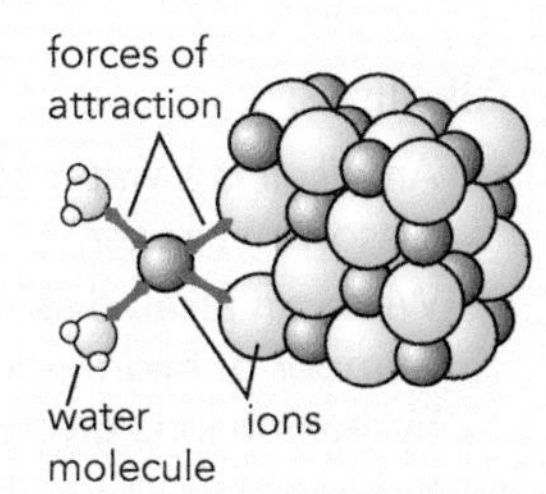

a What can be said about the attraction between water molecules and the ions making up sodium chloride?

b What can be said about insoluble silver chloride?

11 **Write balanced ionic equations for the following precipitation reactions.**

a When blue copper sulfate solution is mixed with clear sodium hydroxide solution, a blue precipitate of copper hydroxide forms as the liquid decolourises.

b When clear solutions of calcium nitrate and sodium hydroxide are mixed, a white precipitate forms.

c When clear solutions of calcium chloride and sulfuric acid are mixed, the content of the test-tube turns into a white precipitate of calcium sulfate (plaster of paris).

NaOH
$CuSO_4$ solution

12 **The two photos show a combination reaction. Describe what happens.**

Iron + sulfur

iron sulfide

13 **Read the article opposite, then answer the questions below.**

a What causes rails to expand in hot weather?

b Why did railway lines originally have gaps between rails?

c Why don't 'stretched rails' buckle on hot days?

d What is a displacement reaction?

e Why does the thermite reaction produce so much heat?

f What is displaced in this displacement reaction?

g What other metals could be produced by a displacement reaction involving Al powder?

h How does the heat produced help to separate the products?

i Why is C added to the mixture?

The Thermite Reaction

All metals expand when heated and this physical property can cause railway lines to buckle, resulting in a derailment.

The traditional solution to this problem was to have small gaps where rails met to allow for heat expansion. It was those gaps that caused the 'clickety-clack' noise.

Nowadays, the rails are stretched then welded together to prevent buckling.

When a section of rail is replaced, the thermite reaction is used to join the sections.

This is a displacement reaction between iron oxide (Fe_2O_3) and aluminium (Al). When the mixture is ignited in a crucible, the resulting exothermic reaction produces a large amount of heat, causing the temperature to rise above 2000 °C.

As aluminium is a more reactive metal than iron, the aluminium atoms 'force' the iron ions to accept three electrons. The Al atoms become Al^{3+} ions, and the Fe^{3+} ions become Fe atoms in the reaction.

At the high temperatures reached, the products are in a molten state. The heavier molten iron sinks to the bottom and the oxide and aluminium ions float on top.

The molten iron runs down from the crucible filling the mould around the rail gap. It rapidly cools into steel because of added carbon, fusing the two rails together.

The molten solution of oxide and aluminium ions run down next, overflowing the mould. As it cools, it forms solid aluminium oxide slag, which is disposed of.

14 **The equipment is used to test for CO_2 gas when a metal carbonate/bicarbonate is heated.**

a What solution is placed in the vertical test-tube?

b What happens to that solution if decomposition takes place?

Refer to the decomposition section on page 80 and the reactivity series on page 81.

c Predict which metal carbonates would not produce CO_2 gas when heated.

d Explain why those carbonates would not do so.

15 **Use the solubility rules on page 81 to predict whether a precipitate will form when these solutions are mixed. If it does, write down its name and formula.**

a Copper sulfate and sodium hydroxide solutions

b Barium chloride and sodium sulfate solutions

c Magnesium nitrate and sodium carbonate solutions

d Lead nitrate and sodium iodide solutions

e Calcium iodide and sodium carbonate solutions

f Magnesium chloride and sodium sulfate solutions

g Iron sulfate and magnesium nitrate solutions

h Lead nitrate and sodium chloride solutions

i Copper nitrate and zinc sulfate solutions

lead iodide precipitate
sodium iodide solution
lead nitrate solution

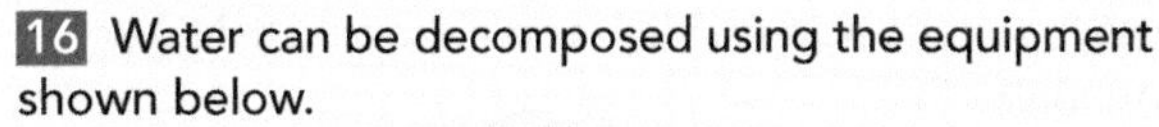

16 **Water can be decomposed using the equipment shown below.**

a Identify the type of decomposition reaction.

b Write word and symbol equations.

c Describe how you would confirm which gas is in which test-tube.

d Explain why more hydrogen than oxygen gas is produced.

e Suggest why a soluble salt is added to the water.

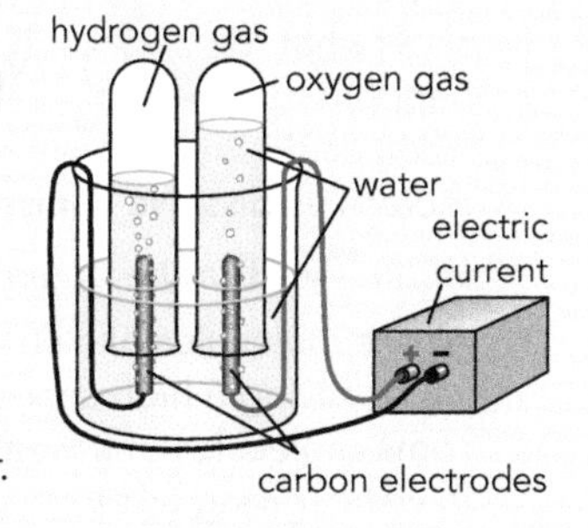

ISBN: 9780170262316

Acids, Bases and Neutralisation

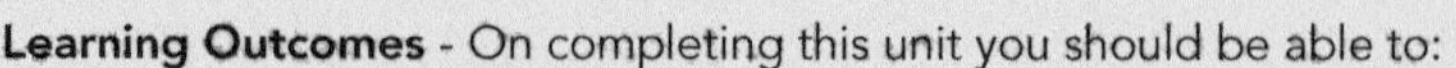

Learning Outcomes - On completing this unit you should be able to:
- compare the properties of acids and bases
- explain why acids and bases are chemical opposites
- relate the pH scale to the acidity or alkalinity of a solution
- identify the salts formed by acids reacting with metals and metal compounds
- *determine the pH of a diluted solution.*

Acids

- **Acids** are chemicals that *release hydrogen ions* (H^+) when they dissolve in water, forming an **acidic solution**.
- Weak acids, such as citric acid in lemon juice and acetic acid (CH_3COOH) in vinegar, have a sour taste.
- Strong acids, such as sulfuric acid in car batteries and hydrochloric acid in your stomach, are corrosive, attacking substances.
- When an acid dissolves in water, hydrogen ions (H^+) separate from the rest of the molecule giving the solution its acidic properties.

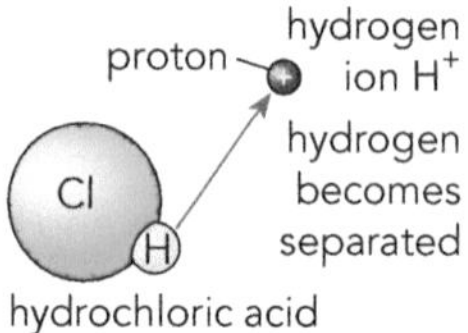

- You need to know the formulas of hydrochloric acid (HCl), sulfuric acid (H_2SO_4), and nitric acid (HNO_3).
- An ionic equation is used to show what happens when acid molecules dissolve (**ionise**) in water, for example:

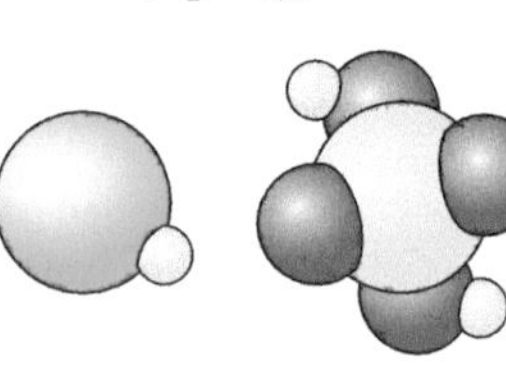
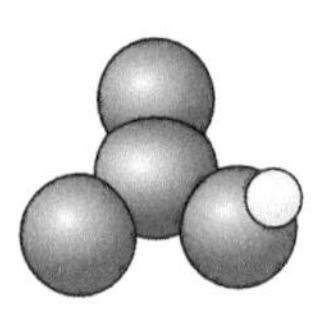

$HCl(l) \xrightarrow{water} H^+(aq) + Cl^-(aq)$.

- All the molecules of a **strong acid** like HCl ionise in water, giving a high concentration of H^+ ions. Only some molecules of a **weak acid** like acetic acid ionise in water, giving a low concentration of H^+ ions.
- Acidic solutions conduct electricity because they contain dissolved ions.
- Acidic solutions can be identified by using an **indicator**, such as blue litmus, which turns pink or red in an acidic solution.

Bases

- **Bases** are the *chemical opposites of acids* because they *remove H^+ ions from a solution.*
- **Alkalis** are bases that form **alkaline solutions** when they *dissolve in, or react with, water.*
- Weak bases, eg sodium bicarbonate in baking soda and magnesium hydroxide in antacids, have a bitter taste.
- Strong bases, eg sodium hydroxide in oven cleaners, are caustic, causing chemical burns to the skin.
- Ammonia is a base in floor cleaners.

Alkalis

- An *alkali is a water-soluble base.* (Not all bases will dissolve in water.)
- When an alkali dissolves in water it *increases the number of hydroxide ions* (OH^-) in the solution. These hydroxide ions *give the solution its alkaline properties.*
- You need to know alkali formulas such as: sodium and potassium hydroxides (NaOH and KOH), potassium carbonate (K_2CO_3), sodium bicarbonate ($KHCO_3$), and ammonia (NH_3).
- When a soluble hydroxide compound dissolves in water it **ionises**. For example, when NaOH dissolves: $NaOH(s) \xrightarrow{water} Na^+(aq) + OH^-(aq)$.
- When a carbonate dissolves, the carbonate ions react with water to give bicarbonate and hydroxide ions : $CO_3^{2-}(aq) + H_2O(l) \rightarrow HCO_3^-(aq) + OH^-(aq)$.
- When ammonia dissolves in water, some molecules pull H atoms off water molecules leaving OH^- ions: $NH_3(aq) + H_2O(l) \rightarrow NH_4^+(aq) + OH^-(aq)$.
- A **strong alkali** like NaOH fully ionises in water, giving a high concentration of OH^- ions. A **weak alkali** like ammonia only partly ionises in water, resulting in a low concentration of OH^- ions.
- Alkaline solutions conduct electricity because they contain dissolved ions.
- Alkaline solutions can be identified by using an **indicator**, such as red litmus, which turns blue in an alkaline solution.
- Insoluble bases, eg copper oxide, don't form alkaline solutions, but they do react with acids (see page 85).

Chemical Opposites

- Bases are chemical opposites of acids because they *remove H^+ ions from acidic solutions.*
- The OH^- ions produced when a base dissolves in water combine with the H^+ ions in an acidic solution to form H_2O. The reaction is: $H^+(aq) + OH^-(aq) \rightarrow H_2O(l)$
- Acids are chemical opposites of bases because they *remove OH^- ions from alkaline solutions* in the same reaction.

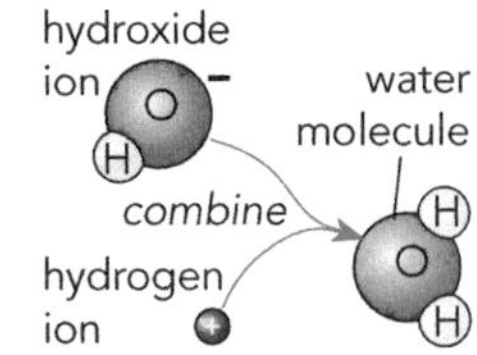

- If just the right amount of reagent is added, a **neutral solution** results, which is neither acidic nor alkaline.

ISBN: 9780170262316

The pH Scale

- A neutral solution has an *equal number of hydrogen and hydroxide ions.* The **concentration** of H^+ ions equals the concentration of OH^- ions.
- If a solution has more H^+ ions than OH^- ions, then it is acidic. If the solution has more OH^- ions than H^+ ions, then it is alkaline.
- The **pH scale** is used to indicate just how acidic or alkaline a solution is.
- The scale goes from 0 to 14, with 0 being a very acidic solution and 14 being a very alkaline solution. A neutral solution has a **pH** of 7.

0	1	2	3	4	5	6	7	8	9	10	11	12	13	14
More acidic ⟵⟶ Less acidic							neutral	Less alkaline ⟵⟶ More alkaline						

- Any solution with a pH of less than 7 is acidic. The closer the pH is to 0, the more acidic the solution. Each step down is 10 times more acidic (x10 more H^+ ions).
- Any solution with a pH above 7 is alkaline. The closer the pH is to 14, the more alkaline the solution. Each step up is 10 times more alkaline (x10 more OH^- ions).
- The pH of a solution can be measured accurately using a digital pH meter or less accurately by using drops of universal indicator solution or universal indicator paper.
- **Universal indicator** is a mixture of dyes that go a specific colour for all pHs between each pair of whole numbers.

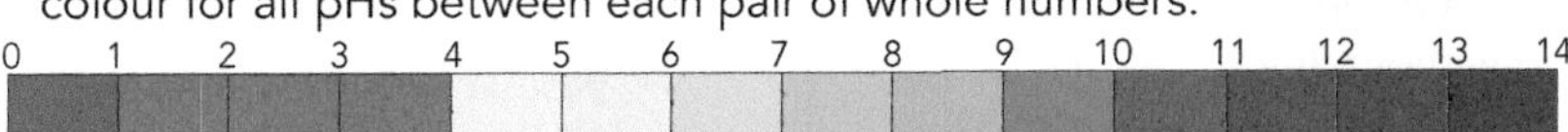

- Adding more acid to an acidic solution will reduce the pH. Adding alkali will increase the pH until the solution becomes neutral then alkaline.
- Adding more alkali to an alkaline solution will increase the pH. Adding acid will reduce the pH until the solution becomes neutral then acidic.
- Diluting with water will shift the pH of an acidic or alkaline solution *closer to 7* (see the science skill section below).

Neutralisation

- When an alkaline solution is slowly added to an acidic solution, the solution gets less acidic as the pH gets closer to 7. If the right amount is added, the pH becomes 7 and the acidic solution has been neutralised.
- Similarly, when an acidic solution is slowly added to an alkaline solution, the solution gets less alkaline as the pH gets closer to 7. If the right amount is added, the pH becomes 7 and the alkaline solution has been neutralised.
- The reaction in which an *acid reacts with a base to give a neutral solution* is called **neutralisation**. The H^+ and OH^- ions combine to form water molecules and the remaining dissolved ions will form salt crystals when the water has been evaporated. The general equation for the reaction is:

Acid + base ⟶ salt + water (+ carbon dioxide)*

* CO_2 is produced if the base is a carbonate or a **bicarbonate**.

- A **salt** is an ionic compound made of ions other than H^+ and OH^-.
- If dilute sodium hydroxide is used to neutralise dilute hydrochloric acid, the ionic equation for the reaction is:
  $H^+(aq) + Cl^-(aq) + Na^+(aq) + OH^-(aq) \rightarrow NaCl(s) + H_2O(l)$,
  or more simply: $HCl(aq) + NaOH(aq) \rightarrow NaCl(s) + H_2O(l)$.
  The salt crystals left after the water has evaporated are sodium chloride.
- For any neutralisation reaction, the salt can be identified by teaming up the positive ion of the base with the negative ion of the acid. For example, if dilute potassium hydroxide (KOH) is used to neutralise dilute sulfuric (H_2SO_4) acid, then the salt formed will be K_2SO_4, potassium sulfate.
- To know exactly when a solution has been neutralised, an indicator is used, eg universal indicator, which turns green when the solution is neutral. Near neutral point the acid or alkali must be added slowly.

Salt Formation

Acid + Metal Oxide

- Most metal oxides are insoluble except for K_2O, Na_2O and CaO. Those that react with acids to give a salt are called **basic oxides**.
- For example, insoluble black copper oxide reacts with dilute sulfuric acid to give the blue salt copper sulphate after evaporation:
  $H_2SO_4(aq) + CuO(s) \rightarrow CuSO_4(s) + H_2O(l)$

Acid + Metal Hydroxide

- Only KOH and NaOH are soluble but most *hydroxides are bases.*
- For example, insoluble white zinc hydroxide reacts with nitric acid to give colourless crystals of zinc nitrate after evaporation:
  $2\,HNO_3(aq) + Zn(OH)_2(s) \rightarrow Zn(NO_3)_2(s) + 2\,H_2O(l)$

Acid + Metal Carbonate

- Only KCO_3 and $NaCO_3$ are soluble but *all carbonates act as bases.*
- For example, an alkaline solution of sodium carbonate reacts with hydrochloric acid to produce CO_2 gas, H_2O, and white crystals of sodium chloride after evaporation:
  $2\,HCl(aq) + Na_2CO_3(aq) \rightarrow 2\,NaCl(s) + H_2O(l) + CO_2(g)$

Acid + Metal Bicarbonate

- Bicarbonates (hydrogen carbonates) of K, Na, Mg and Ca are soluble and *act as bases.*
- For example, an alkaline solution of sodium bicarbonate reacts with sulfuric acid to produce CO_2 gas, H_2O, and colourless sodium sulfate crystals after evaporation:
  $H_2SO_4(aq) + 2\,NaHCO_3(aq) \rightarrow Na_2SO_4(s) + 2\,H_2O(l) + 2\,CO_2(g)$

Acid + Metal

- Acids and reactive metals form a salt and H_2 gas (see page 65).

SKILL: Dilution and pH

Find the pH when an alkaline or acidic solution is diluted by H_2O.

Steps:

1. If diluted in 10 times as much water: add 1 to pH of acid/ subtract 1 from pH of alkali.
2. If diluted in 100 times as much water: add 2 to pH of acid/ subtract 2 from pH of alkali.

ISBN: 9780170262316

Revision Activities

1 Match up the descriptions with the terms.

a acid	A indicator that gives the pH of a solution
b acidic solution	B alkali that ionises completely in water
c strong acid	C chemical that releases H^+ ions in water
d weak acid	D scale that goes from 0 to 14 specifying the acidity or alkalinity of a solution
e indicator	E acid-base reaction that gives a neutral solution
f base	F acid that ionises completely in water
g alkaline solution	G chemical that changes colour as pH changes
h alkali	H number of ions or molecules in a fixed volume
i ionise	I metal oxide that reacts with an acid to give a salt
j strong alkali	J water-soluble base
k weak alkali	K solution with more H^+ ions than OH^- ions
l neutral solution	L solution with equal numbers of H^+ and OH^- ions
m concentration	M number that specifies how acid or alkaline a solution is
n pH scale	N acid that only partly ionises in water
o pH	O chemical opposite of an acid
p universal indicator	P another name for a hydrogen carbonate compound
q neutralisation	Q solution with more OH^- ions than H^+ ions
r salt	R ionic compound without any H^+ or OH^- ions
s bicarbonate compound	S occurs when a compound separates into ions
t basic oxide	T alkali that only partly ionises in water

2 Explain the differences between the terms in each of the items below:

a an acid and a base
b an alkali and a base
c a strong and a weak acid
d a strong and a weak alkali
e an acidic and an alkaline solution
f a hydrogen ion and a hydroxide ion
g a concentrated and a dilute aqueous solution
h a salt and any other ionic compound.

3 Copy and complete this table to compare and contrast acidic and alkaline solutions.

Property	Acidic solution	Alkaline solution
Taste		
Feel	sticky	
Strong solution	corrosive	
Ions present		
Effect on litmus		
pH		
Electricity		
Base reaction	forms	
Acid reaction		forms
Metal reaction	forms	not covered
Bi/carbonate reactions	forms	not covered

4 Describe the state of most acids and the state of most bases. Mention any exceptions you know about.

5 Decide whether the following statements are true or false. Rewrite the false ones to make them correct.

a Acids exist as molecules.
b Strong acids are corrosive and strong bases are caustic.
c When an acid dissolves, it increases the number of H^+ in solution.
d Acidic properties are due to H^+ ions.
e When a compound ionises, its ions combine.
f Both acidic and alkaline solutions conduct electricity.
g Acids are opposites of bases.
h Alkalis are insoluble bases.
i Adding an alkali increases the number of OH^- ions in solution.
j Alkaline properties are due to OH^- ions.
k Most bases are ionic compounds.
l Acids add H^+ ions to a solution and bases remove them.
m For acidic solutions the pH < 7 and for alkaline solutions the pH > 7.
n In a neutral solution the number of H^+ and OH^- ions are unequal.
o A salt is an ionic compound that doesn't have any H^+ or OH^- ions.

6 Study each of these space-filling molecular models. Identify each molecule, then:

i record its name
ii write its formula
iii decide whether it will release or remove H^+ ions.

Clue: **e** is an organic acid.

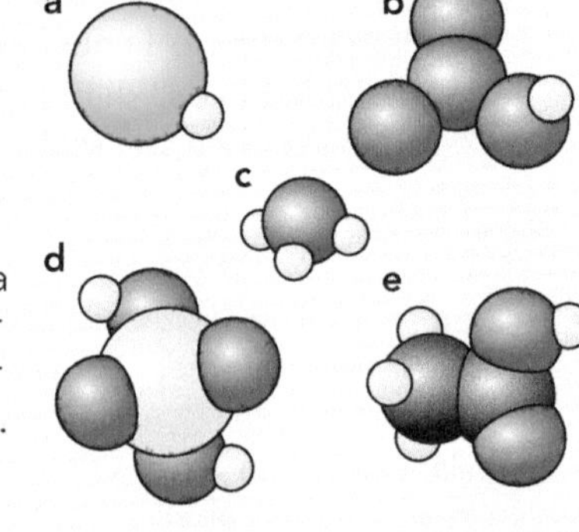

7 Look at the two signs.

a Explain why the left one would be on bottles of dilute hydrochloric acid.
b Explain why the right one would be on bottles of concentrated hydrochloric acid.

8 The photo shows some universal indicator paper after it has been dipped in a test solution.

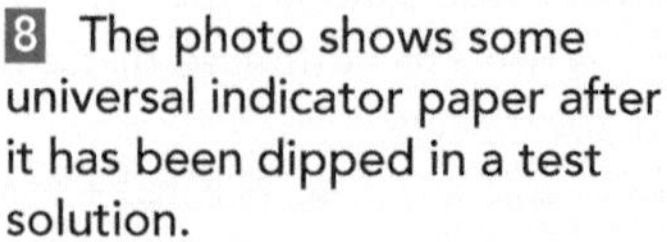

a Is the test solution acidic or alkaline?
b Is the test solution a weak or a strong solution?
c Would you add an acidic or an alkaline solution to neutralise the test solution?
d How would you know when the test solution was neutral?
e If you added excess of the solution you chose in **c**, what would happen to the test solution?

ISBN: 9780170262316

9 Decide whether each of these bases is an alkali (a water-soluble base) or not:

a calcium hydroxide, $Ca(OH)_2$
b sodium carbonate, Na_2CO_3
c calcium bicarbonate, $Ca(HCO_3)_2$
d sodium oxide, Na_2O
e zinc carbonate, $ZnCO_3$
f lead oxide, PbO
g potassium oxide, K_2O
h iron hydroxide, $Fe(OH)_2$.

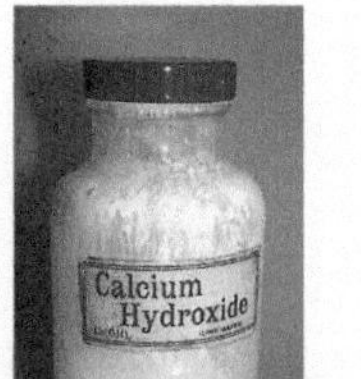

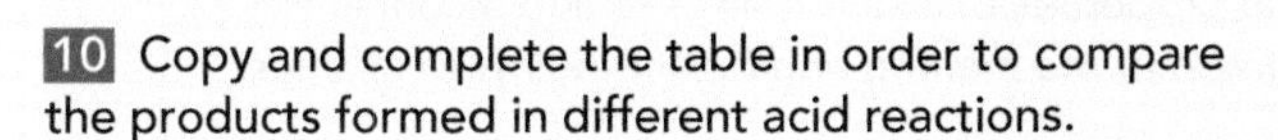

10 Copy and complete the table in order to compare the products formed in different acid reactions.

	Reaction Products		
Acid +	**Salt?**	**Water?**	**Gas?**
Metal	yes	no	H_2
Metal oxide	yes	yes	
Metal hydroxide			
Metal carbonate			
Metal bicarbonate			

11 A dilute solution of sodium hydroxide in the conical flask is to be neutralised by running some dilute hydrochloric acid solution from the tall tube into the flask. The indicator phenolphthalein is colourless in an acidic solution but turns pink just as a neutral solution is reached.

a What colour will the indicator be when first added to the acidic solution?
b When the mixture in the beaker just turns purple, what will the pH be?
c Explain why the alkaline solution should be run in slowly.
d What ions will be left in the conical flask?
e Would those ions form a soluble or an insoluble salt?
f How could that salt be recovered?
g Write a balanced ionic equation for the reaction, specifying states.

HCl
NaOH

12 Read the article opposite, then answer the questions below.

a What chemicals in your body are involved in digesting food?
b What conditions do the enzymes that digest protein in your stomach require?
c What is the pH of gastric juice?
d How is the stomach lining protected from attack by gastric juice?
e What causes a burning sensation when you vomit?
f What causes heartburn?
g What are most antacids?
h How do antacids stop heartburn?
i Is the reaction between antacid and hydrochloric acid an example of neutralisation? Explain.

13 Dilute hydrochloric acid reacts with the zinc granules in the experiment shown.

gas
water
zinc granules
hydrochloric acid

a Name the gas in the inverted test-tube.
b Describe how you would test for that gas.
c The reaction stops before the zinc granules disappear. Explain why the reaction stopped.
d Write a balanced ionic equation (ignore spectators ions).
e Identify the ions remaining in solution.
f State whether a soluble or an insoluble salt is formed.
g Describe how you would obtain a sample of that salt.
h Write a balanced symbol equation for the overall reaction, including the formation of the salt.

14 Dilute hydrochloric acid reacts with marble chips (calcium carbonate, $CaCO_3$) as shown opposite.

gas
water
marble chips
hydrochloric acid

a Name the gas in the inverted test-tube.
b Describe how you would test for that gas.
c The reaction stops before the marble chips disappear. Explain why the reaction stopped.
d Write a balanced ionic equation (ignore spectator ions).
e Identify the ions remaining in solution.
f State whether a soluble or an insoluble salt is formed.
g Describe how you would obtain a sample of that salt.
h Write a balanced symbol equation for the overall reaction, including the formation of the salt.

15 Study the diagram below, then explain why:

a a strong alkaline solution can neutralise an acidic solution of equal strength
b a strong acidic solution can neutralise an alkaline solution of equal strength
c a weak acidic solution can't neutralise a strong alkaline solution.

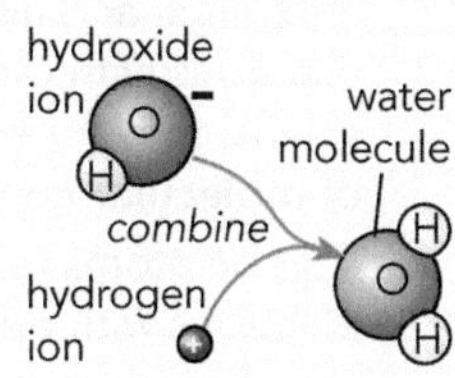

Neutralising Stomach Acid

Your food is digested by enzymes in your mouth, stomach and intestine. Enzymes in your stomach break down protein foods, but they require acidic conditions.

The stomach wall produces hydrochloric acid which helps enzymes to break down proteins molecules more rapidly.

The hydrochloric acid gives gastric juice a pH between 1.5 and 3.5. This is strong enough to make a hole in a rug, but the stomach wall is protected by a layer of mucus that prevents the acid from reaching the stomach wall.

That burning sensation in your throat when you vomit is due to acid from gastric juices irritating your oesophagus as the stomach contents are regurgitated.

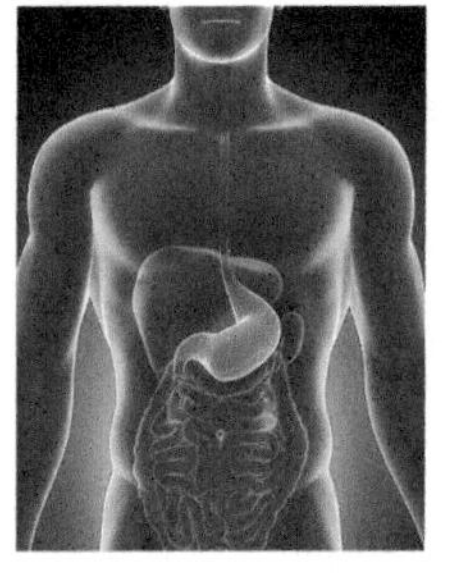

The acid in your stomach can also cause heartburn – a burning sensation in your chest that occurs after a meal. A small amount of gastric juice passes back up the oesophagus and the acid irritates the lining of the oesophagus, which is not as well protected as the stomach lining is.

Heartburn symptoms can be relieved by the use of antacid. Most antacids contain a metal compound that acts as a base when it encounters an acidic solution.

The dissolved metal hydroxide, carbonate or bicarbonate reduces the number of H^+ ions present in gastric juice, thus raising the pH.

ISBN: 9780170262316

Properties of Carbon Compounds

Learning Outcomes - On completing this unit you should be able to:
- explain why carbon atoms share four electrons, forming covalent compounds
- distinguish between alkanes, alkenes and alcohols
- relate physical properties to the structure of the carbon compounds
- compare the chemical properties of alkanes, alkenes and alcohols
- *interpret and use a graph of the melting and boiling points of alkanes.*

Carbon Bonding

- Carbon atoms are unusual as they *form bonds with up to four other atoms* in order to become stable.
- As each carbon atom has four **valence electrons** in its outermost occupied **electron shell**, the easiest way to gain a stable (full) outermost shell is by *sharing these electrons with other atoms.*
- Carbon atoms form four **covalent bonds**. In each bond a pair of electrons are shared – one from the carbon atom and the other from the other atom. This gives the carbon atom a total of eight shared electrons.

Carbon Compounds

- As carbon atoms join up in chains and rings, and also readily share electrons with other kinds of atoms, they form a huge number of **molecular compounds**.
- Carbon atoms readily share electrons with other nonmetal atoms, such as hydrogen, oxygen and nitrogen atoms, forming covalent bonds. Hydrogen atoms form one bond, oxygen atoms two, nitrogen atoms three, and carbon atoms four.
- A pair of C atoms in a molecule can form a single bond (C–C), a double bond (C=C) and even a triple bond (C≡C).
- Living things make natural carbon compounds such as proteins, carbohydrates and fats. Fossil fuels, such as coal, crude oil and natural gas, are formed from the remains of ancient living things.
- Chemists make many synthetic carbon compounds, such as plastics, paints, fuels and medicinal drugs.

Hydrocarbons

- **Hydrocarbons** are compounds that consist of *carbon and hydrogen atoms only*. For NCEA level 1 you need to know about the alkanes and alkenes.

Alkanes

Structure and Physical Properties

- **Alkane** molecules have *single bonds only between carbons* (C–C), so each C atom is bonded to four other atoms, either C or H. They are **saturated hydrocarbons** as the C atoms *cannot bond with any more atoms.*
- The C atoms that are bonded to each other form a long chain that acts as the *backbone of the molecule*. The chain may be straight or branched.
- There are many different alkanes that differ in the number of C atoms that form the backbone of the molecule and in how the backbone is structured.
- The general formula of an alkane is: C_nH_{2n+2}, where **n** = number of C atoms.
- For NCEA level 1 you need to be familiar with straight-chain molecules of the first eight alkanes, as listed in the table below.
- A **chemical formula** gives the number of each kind of atom in a molecule. A **structural formula** shows which atoms each atom is bonded to. Each line (eg C-C or C-H) is a single bond consisting of two shared electrons.
- The space-filling models show how atoms are arranged in the molecules.

Alkane	Structural Formula	Space-filling Model	MP(°C)	BP(°C)
methane CH_4	H H-C-H H		–183	–162
ethane C_2H_6	H H H-C-C-H H H	Straight-chain alkanes have a zigzag-shaped carbon backbone	–172	–88
propane C_3H_8	H H H H-C-C-C-H H H H		–188	–42
butane C_4H_{10}	H H H H H-C-C-C-C-H H H H H		–138	0
pentane C_5H_{12}	H H H H H H-C-C-C-C-C-H H H H H H		–130	36
hexane C_6H_{14}	H H H H H H H-C-C-C-C-C-C-H H H H H H H		–95	69
heptane C_7H_{16}	H H H H H H H H-C-C-C-C-C-C-C-H H H H H H H H		–91	98
octane C_8H_{18}	H H H H H H H H H-C-C-C-C-C-C-C-C-H H H H H H H H H		–57	126

ISBN: 9780170262316

- An alkane's state *depends on its melting and boiling points.* At the melting point (MP), molecules that were in a fixed array gain enough kinetic energy to move about freely in the liquid state. At the boiling point (BP), molecules that were close together in the liquid gain enough energy to separate in the gas state.

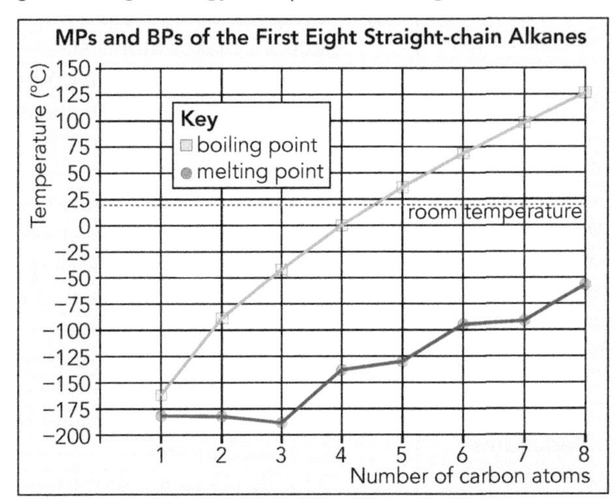

- If an alkane's BP is below room temperature (20°C), then the alkane will be a gas. If its MP is above 20°C, then it will be a solid. If its MP and BP lie on either side of 20°C, then it will be a liquid at room temperature.
- As can be seen on the graph, alkanes with 1 to 4 C atoms are gases. It has been found that those alkanes with 5 to 17 C atoms are liquids, and those with 17+ C atoms are solids at room temperature (20 °C).
- In general, as alkane molecules get larger, both MPs and BPs increase. These trends occur because *heavier molecules require more energy to move faster*, and because there are *more attractive forces between larger molecules*. More heat is needed to overcome the forces holding molecules together (see page 36).
- Alkanes are insoluble in water, as the H_2O molecules *do not attract the alkane molecules.* Liquid alkanes and water are **immiscible**, forming layers with the alkane on top as *alkanes are less dense than water.*

Chemical Properties

- Alkanes are unreactive with acids and bases but do undergo **oxidation** when ignited in oxygen gas (O_2).
- In *abundant oxygen*, alkanes burn with a strong blue flame, releasing *much heat* as **complete combustion** occurs. The products are carbon dioxide and steam.

Alkane + oxygen → carbon dioxide + water

- If the air-hole of a natural gas Bunsen burner is opened to supply lots of O_2, the reaction is: **Methane + oxygen → carbon dioxide + water**
  $CH_4(g) + 2O_2(g) \rightarrow CO_2(g) + 2H_2O(g)$
- Larger alkane molecules require more O_2.
- In *limited oxygen*, alkanes burn with a weak yellow, sooty flame, releasing *much less heat* as **incomplete combustion** occurs. The products are soot (C), carbon monoxide and water.
- If the air-hole of the Bunsen burner is closed to limit the amount of O_2 supplied, the overall reaction is: **Methane + oxygen → carbon + carbon monoxide + water.**

Alkenes

- **Alkenes** are *hydrocarbons with a double bond* between two C atoms (C=C). They are **unsaturated hydrocarbons** as they *can bond with more atoms.*
- Alkenes are named in a similar way to alkanes but with an '-ene' ending. (There is no equivalent to methane.)
- The general formula of an alkene molecule is: C_nH_{2n}, where **n** = number of C atoms.
- For NCEA level 1 you only need to know about ethene and propene gases. The ball-and-stick models show the number of bonds between pairs of atoms.

Alkene	Structural Formula	Space-filling Model	Ball-and-stick Model
ethene C_2H_4	H H / C=C / H H		
propene C_3H_6	H H / H-C-C=C / H H H		double bond; single bond

- Alkene MPs and BPs increase similarly to alkanes.
- Alkenes are unreactive with acids and bases. When combusted they produce more soot than alkanes. They are valued more for making plastics than as fuels.
- Numerous identical small molecules (**monomers**) can be *joined into long-chain molecules* (**polymers**) in a process called **polymerisation**. Plastics are polymers.
- For example, by using heat, pressure and a catalyst, thousands of ethene molecules can be joined to form polyethene (polythene) molecules as shown below:

```
H H    H H    H H    H H   heat      H H H H H H H H
C=C + C=C + C=C + C=C   pressure  -C-C-C-C-C-C-C-C-
H H    H H    H H    H H   catalyst  H H H H H H H H
```

- The double bond in each monomer is replaced by two single bonds, one of which links to the next monomer.

Alcohols

- **Alcohols** are carbon compounds that consist of a hydrocarbon chain bonded to an O atom with an attached H atom (-O-H).
- Alcohols are named in a similar way to alkanes but with an '-ol' ending. For NCEA level 1 you need to know about methanol (CH_3OH) and ethanol (C_2H_5OH).

Alcohol	Structural Formula	Space-filling Model	MP (°C)	BP (°C)
methanol CH_3OH	H / H-C-O-H / H		−98	64
ethanol C_2H_5OH	H H / H-C-C-O-H / H H		−114	78

- Both are liquids as the MP and BP of each lie on either side of 20°C. Alcohols are soluble in water, as the H_2O molecules *attract the alcohol molecules.*
- Alcohols are unreactive with acids and bases. They need less O_2 than alkanes for complete combustion, burning with a nearly invisible flame (see feature photo).

ISBN: 9780170262316

Revision Activities

1 Match up the descriptions with the terms.

a valence electron	A hydrocarbon with single bonds only
b electron shell	B combustion that occurs in abundant oxygen
c covalent bond	C hydrocarbon with a double bond between a pair of carbon atoms
d molecular compound	D electron in the outermost shell that can be shared or transferred to another atom
e hydrocarbon	E liquids that will not mix, forming layers instead
f alkane	F bond formed by a pair of atoms sharing electrons
g saturated hydrocarbon	G process of linking monomers into polymers
h structural formula	H hydrocarbon whose C atoms can't form more bonds
i immiscible liquids	I carbon compound consisting of a hydrocarbon chain bonded to an O atom with an attached H
j oxidation	J region around the nucleus in which electrons with a specific amount of energy exist
k complete combustion	K small molecules that can link to form a polymer
l incomplete combustion	L compound made of covalently bonded atoms
m alkene	M diagram showing covalent bonds between atoms
n unsaturated hydrocarbon	N reaction in which oxygen atoms combine with other atoms
o monomer	O carbon compound made of C and H atoms only
p polymer	P long molecule formed by linking monomers
q polymerisation	Q hydrocarbon capable of bonding with more atoms
r alcohol	R combustion that occurs in limited oxygen

2 Explain the differences between the terms in each of the items below:

a an alkane, an alkene and an alcohol molecule
b a saturated and unsaturated hydrocarbon
c a chemical formula and a structural formula
d complete and incomplete combustion
e monomers and a polymers.

3 Identify the space-filling molecular models shown below, then write their chemical formulas.

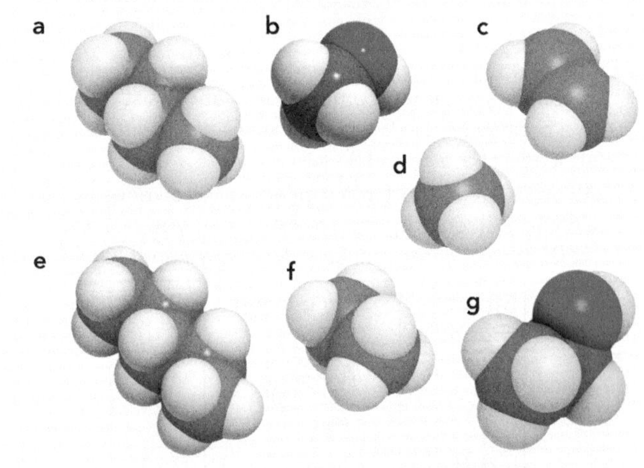

4 Use the diagrams below to explain what ball-and-stick models can tell you about the structure and bonding of carbon molecules.

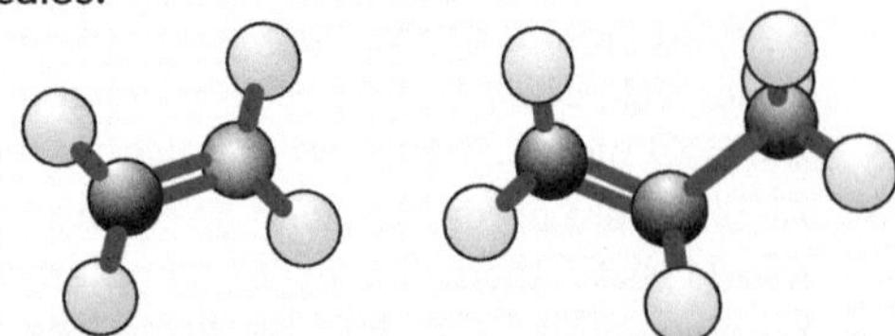

5 Decide whether the following statements are true or false. Rewrite the false ones to make them correct.

a Carbon atoms form four covalent bonds by sharing their four valence electrons.
b Having full outer shells by sharing electrons makes atoms stable.
c When two atoms form a double bond, four electrons are shared.
d Alkanes are unsaturated hydrocarbon molecules.
e Space-filling models show the number of bonds between atoms.
f Both the melting and boiling points of alkanes increase as the number of C atoms increases.
g Complete combustion requires more oxygen.
h Alkenes have twice as many H atoms as C atoms.
i The polymerisation of propene results in polypropene.
j Alcohols are hydrocarbons.
k Alcohols are soluble in water but alkanes and alkenes are not.
l Alkanes, alkanes and alcohols don't react with acids or bases.

6 Use the general formulas for alkanes and alkanes, and your knowledge of the structure of alcohols, to find the chemical formulas of:

a nonane, an alkane with 9 carbon atoms
b decane, an alkane with 10 carbon atoms
c butene **d** pentene **e** propanol **f** butanol.

7 Write word and balanced symbol equations for the complete combustion of these compounds:

a ethane (7 O_2 needed)
b propane (5 O_2 needed)
c ethene (3 O_2 needed)
d propene (9 O_2 needed)
e methanol (3 O_2 needed)
f ethanol (3 O_2 needed).

8 Write word and balanced symbol equations for the incomplete combustion of these compounds:

a methane (forming carbon monoxide and water)
b methane (forming soot [carbon] and water)
c ethene (forming carbon monoxide and water)
d ethene (forming soot [carbon] and water)
e methanol (forming carbon monoxide and water).

9 The fuel used by Bunsen burners in most schools is natural gas, which is mostly methane, CH_4.

a Explain how opening the air-hole by rotating the collar causes the complete combustion of methane to occur.
b Explain how closing the air-hole causes incomplete combustion.
c Justify why the yellow flame is called the 'safety flame'?

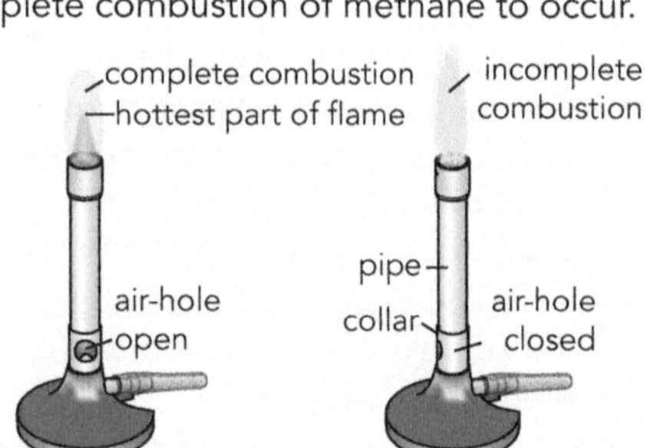

ISBN: 9780170262316

10 Study the graph below, which shows the melting and boiling points of the first 14 alkanes.

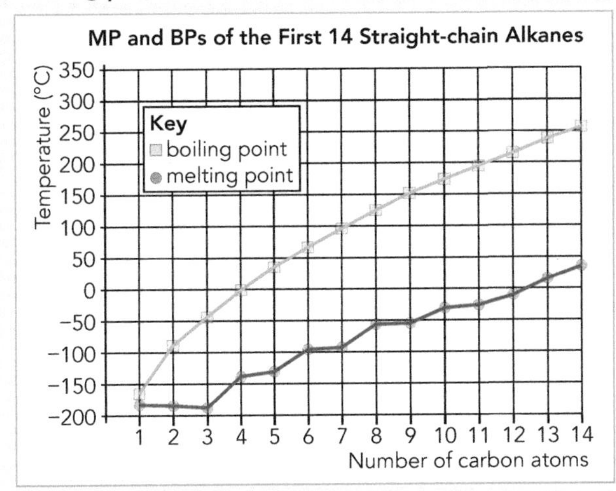

a Record the melting and boiling points of C_9H_{20}.
b Estimate the melting and boiling points of $C_{15}H_{32}$.
c At room temperature (20 °C), which of the above alkanes will be:
i gases ii liquids iii solids.
d Describe the trend shown by the boiling points and explain why that trend occurs.
e Summarise the general trend for the melting points.
f Describe the exceptions to the general trend for melting points shown by the graph line.

11 Ethanol in a spoon burns with a blue flame, while butane from a cigarette lighter burns with a yellow flame, as shown in the photos below.
a Which hydrocarbon would be completely combusted? How do you know?
b Which would be undergoing incomplete combustion? How do you know?
c Which hydrocarbon would produce the most heat? Why?

12 Read the article opposite, then answer the questions below.
a What is meths?
b What are three everyday uses of meths?
c Why are unpleasant chemicals added to meths?
d Why was the ethanol in meths diluted with methanol?
e How does methanol affect humans?
f What gives meths its purple colour and why is this done?
g Why might the NZ government have banned the use of methanol in meths?
h How is the current composition of meths meant to discourage drinking the product?

13 Marine ropes are made from a tough, water-resistant polymer called polypropene, which is formed by the polymerisation of propene (C_3H_6) molecules.

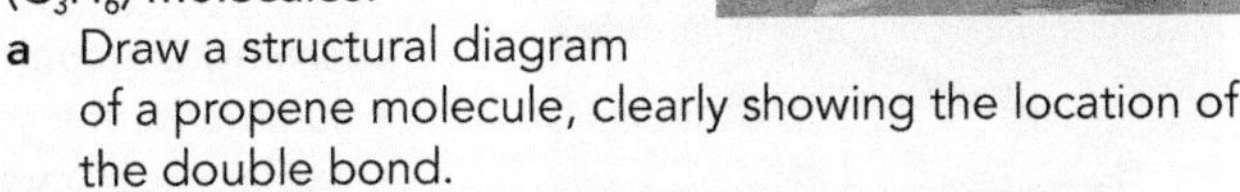

a Draw a structural diagram of a propene molecule, clearly showing the location of the double bond.
b Copy and complete this structural diagram of a polypropene molecule.

```
   CH3 H  CH3 H
   |   |   |   |   |   |   |   |
 - C - C - C - C - C - C - C - C -
   |   |   |   |   |   |   |   |
   H   H   H   H
```

c Explain what happens to the double covalent bond of each propene molecule when polymerisation occurs.
d Identify three conditions that are required for the reaction to proceed.

14 Imagine that you have been given unlabelled samples of the liquids pentane and ethanol.
a Describe a simple test based on a physical property of alkanes and alcohols that would distinguish them. (Tasting of laboratory chemicals is not permitted.)
b Suggest how they could be distinguished on the basis of a chemical property.

15 Ball-and stick molecular models are used to show the bonds between atoms as well as the orientation of the bonds and the underlying shape of the molecule.
a Identify common features of the molecules.
b Describe the distinguishing features of each molecule.
c Decide whether each is an alkane, an alkene or an alcohol.
d Write the name and chemical formula of each.
e List the molecules from lightest to heaviest.

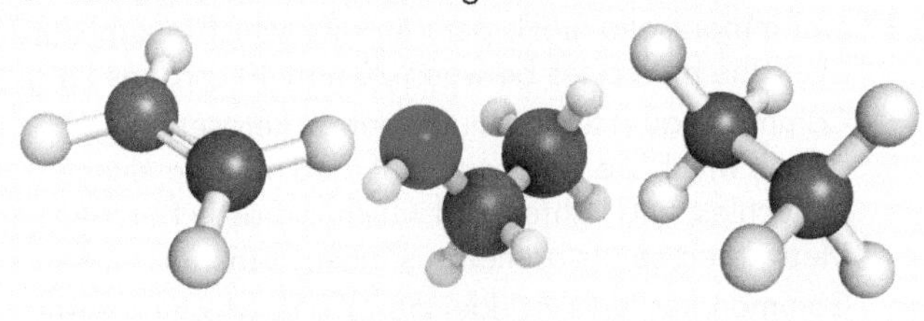

Methylated Spirits

Ethanol is the alcohol that is found in drinks such as beer, wine and spirits.

Meths (methylated spirits) is basically ethanol with other chemicals added to make it either poisonous, foul tasting, vile smelling, or nauseating.

The reason why this is done is because ethanol has a wide range of domestic uses (eg cleaning fluid), industrial uses (eg grease solvent), and recreational uses (eg in spirit burners), and governments do not want people drinking this much cheaper source of ethanol, as alcoholic beverages are taxed.

Traditionally, meths was 10% methanol, which is a toxic alcohol that can cause blindness, organ damage and death.

As methanol has a similar smell and appearance to ethanol, a purple aniline die is often added to meths to reduce incidents of people drinking meths by mistake. (Clear meths is also available.)

In 2007, the New Zealand government made it illegal to add methanol to meths. Instead, a carbon compound called denatonium benzoate is now added in very small quantities. It is the bitterest chemical known to chemists. Humans can still taste it in concentrations as low as five molecules in 10 million.

Unfortunately, there have been some incidents of alcoholics drinking large volumes despite its extremely bitter taste.

ISBN: 9780170262316

Production and Uses of Carbon Compounds

Learning Outcomes - On completing this unit you should be able to:
- summarise the sources and uses of different kinds of fuels
- explain how fractional distillation separates components of crude oil
- describe the effects of combustion products on humans and the environment
- explain how alkenes and alcohols are produced industrially or naturally
- *systematically balance complete and incomplete combustion reactions.*

Uses and Sources

- Hydrocarbons are used as fuels and for making plastics. Alcohols are used for a variety of purposes.
- **Hydrocarbon** gas fuels are obtained from **natural gas** and hydrocarbon liquid fuels from **crude oil**.
- Ethanol is produced naturally by micro-organisms but is also made by reacting ethene with water. Methanol is produced from the methane found in natural gas.
- NZ's natural gas and crude oil originated from the remains of ancient swamp forests that didn't decompose due to the lack of oxygen. Instead the remains were transformed over millions of years by heat and pressure deep underground (see page 146).

Combustion of Fuels

- **Combustion** is an **exothermic reaction** (see page 76).
- **Fuels** are substances that are combusted to *transform chemical energy into thermal energy* (see page 36), which is then used for heating, lighting, cooking, moving vehicles, and generating electricity.
- A common solid fuel is paraffin wax, which is a mixture of alkane molecules between 20 and 40 carbons long.
- Common liquid fuels include petrol, kerosene and diesel, which are all mixtures of medium-size alkane molecules, and lighter fluid which is mostly butane. Methylated spirits or meths is mostly ethanol.
- Common gas fuels include natural gas, which is mostly methane with a small amount of ethane, and petroleum gas, which is a mixture of propane and butane.
- Petroleum gas is supplied as a pressurised liquid called LPG (liquefied petroleum gas). The liquid vaporises as soon as the pressure is released, thus enabling the gases to be burnt readily.

Heat Content

- The heat content of fuels can be compared by measuring the amount of heat produced when a *standard mass of fuel undergoes complete combustion.*
- Typically, the **heat of combustion** is measured in kilojoules per gram (kJ/g).

Common Fuels	Heat of combustion (kJ/g)
paraffin wax	41.4
lighter fluid	49.5
petrol	47.0
kerosene	45.4
diesel	45.0
meths	29.7
natural gas	54.0
LPG	49.6

Producing Fuels

- Crude oil or **petroleum** is a thick, dark-coloured liquid obtained by drilling into underground reservoirs. It is a *mixture of hydrocarbons*, mostly alkanes.
- Transport fuels, such as petrol, diesel and kerosene, are produced from crude oil by separating different groups of alkanes, then purifying, refining and blending them.
- In NZ, crude oil is refined at the Marsden Point refinery.
- Crude oil is separated into *groups of similar-sized hydrocarbons* by **fractional distillation**, which uses differences in **condensation points** (temperature at which a cooling gas liquefies). [As NCEA uses boiling instead of condensing points, we will do the same.]
- *Larger hydrocarbons liquefy at higher temperatures. Smaller hydrocarbons liquefy at lower temperatures.*
- Crude oil is heated to about 400 °C and hydrocarbons with BPs below that boil. The gas-liquid mixture then enters the distillation tower where the liquids with BPs above 400 °C flow out the bottom.
- The hot gases bubble up through a series of liquid-filled trays, each of which is cooler than the previous one.
- The gases cool as they bubble through a tray, *and those with BPs higher than the tray temperature liquefy.*
- Larger molecules with high BPs liquefy as they cool in the hotter lower trays. Smaller molecules with low BPs only liquefy when they cool in the colder upper trays.
- Each tray collects **fractions** within a given range of BPs. In the tower shown below, the kerosene tray at 160 °C would liquefy hydrocarbons with BPs between 160 and 240 °C, which are those with 12 to 16 C atoms.

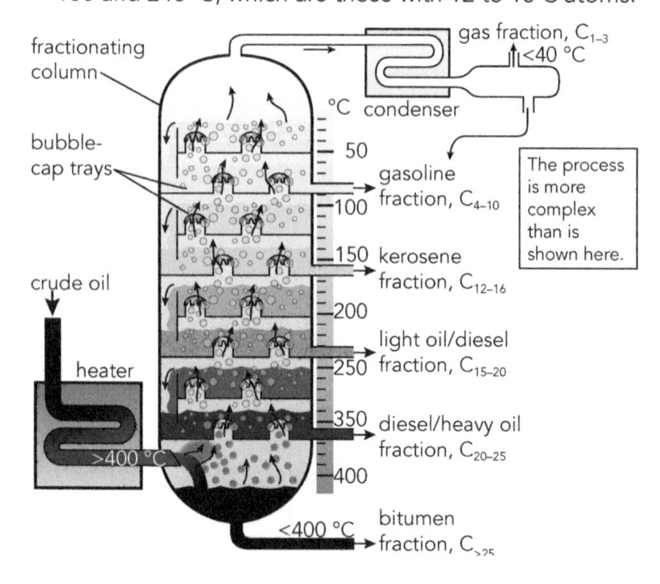

 ISBN: 9780170262316

Combustion Waste Products

- A **waste product** is an *unwanted chemical formed in the process of making something else*. As heat is the only sought after 'product' of combustion, all other chemicals formed are waste products.
- When hydrocarbons undergo **complete combustion** in abundant oxygen, the waste products are carbon dioxide gas and water vapour (see page 92).
- When hydrocarbons undergo **incomplete combustion** in limited oxygen, carbon monoxide gas and/or soot are also formed as waste products (see page 73).
- Soot consists as particles of amorphous carbon (see page 72). It is mostly produced by domestic fires and diesel trucks. The particles can blacken buildings and contribute to breathing problems such as asthma. Ongoing exposure can increase the risks of developing lung cancer and heart disease.
- Carbon monoxide is a highly toxic gas. CO molecules permanently replace the O_2 molecules that normally attach to haemoglobin in red blood cells (see page 136), starving the brain of oxygen.
- Although carbon dioxide is not a toxic gas, it is the main greenhouse gas (see page 147). CO_2 traps infrared energy from the sun in the atmosphere, thus contributing to global warming.

Cracking of Fractions

- A significant fraction of the alkanes separated in a refinery are *longer-chain alkanes that are not useful as fuels*. They are difficult to ignite because bigger molecules don't vaporise easily, and are also less efficient as fuels as they don't combust completely.
- These alkanes are broken into more smaller, more useful alkanes molecules by a process called **cracking**.
- Cracking is a **thermal decomposition** reaction (see page 80) that requires heat, pressure and a catalyst. (The heat also vapourises the liquid alkane, enabling the reaction to occur more readily in the gas state.)
- The overall reaction pattern for cracking is:

Larger alkane → smaller alkane + alkene

- For example, decane can be cracked to produce more useful octane for petrol and ethene in the reaction:

Decane → octane + ethene

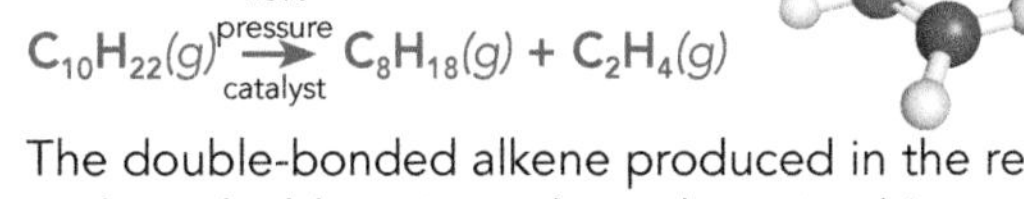

$$C_{10}H_{22}(g) \xrightarrow[\text{catalyst}]{\text{heat, pressure}} C_8H_{18}(g) + C_2H_4(g)$$

- The double-bonded alkene produced in the reaction is also valuable as it can be polymerised (see page 89), eg ethene can be made into polyethene.

Methanol Production

- Methanol (CH_3OH) is a valuable alcohol produced at Taranaki methanol plants close to the offshore Pohokura natural gas field.
- Methane (CH_4), the main component of natural gas, is the raw material of a 'two-stage' chemical process.
- In the first stage, CO and H_2 gases are produced:

$$CH_4(g) + H_2O(g) \xrightarrow[\text{catalyst}]{\text{heat}} CO(g) + 3\,H_2(g).$$

- In the second stage, methanol is made:

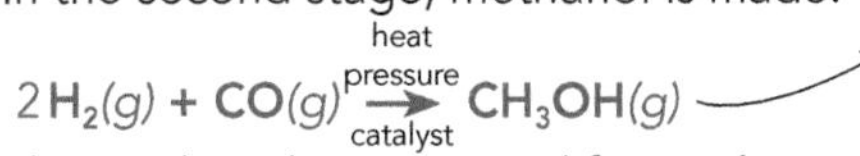

$$2\,H_2(g) + CO(g) \xrightarrow[\text{catalyst}]{\text{heat, pressure}} CH_3OH(g)$$

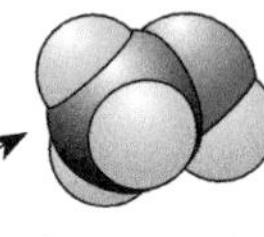

- The methanol is separated from other products and unreacted hydrocarbons by fractional distillation.
- Methanol is used as a solvent in industry, as a fuel, and as a raw material for making polymers, eg plastics.

Ethanol Production

- Ethanol (C_2H_5OH) is the alcohol found in beverages and bio-fuels. It is produced by yeast **micro-organisms**, which *transform sugars such as glucose into ethanol and carbon dioxide* in a process called **fermentation**.
- Fermentation requires warmth and an absence of oxygen, as it is an anaerobic process.
- The process involves a *complex series of reactions* that occur in solution, each of which is catalysed by a different **enzyme** (see page 101).
- The overall reaction for the fermentation of ethanol is:

Glucose → ethanol + carbon dioxide

$$C_6H_{12}O_6(aq) \xrightarrow[\text{enzymes}]{\text{warmth}} 2\,C_2H_5OH(l) + 2\,CO_2(g)$$

- Ethanol can also be made by reacting ethene gas with water at a high temperature in the catalysed reaction:

Ethene + water → ethanol

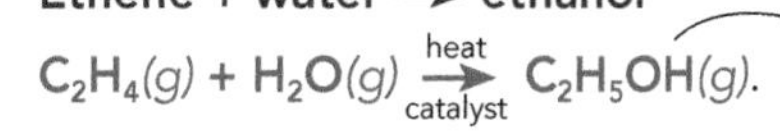

$$C_2H_4(g) + H_2O(g) \xrightarrow[\text{catalyst}]{\text{heat}} C_2H_5OH(g).$$

Polymer Production

- Many small hydrocarbon molecules can be combined to form *extremely long straight or branched molecules* by the process of **polymerisation** (see page 89).
- Plastics, such as polyethene (plastic bags), polypropylene (drinking straws), polystyrene (packaging), and PVC (pipes), are all formed by polymerisation.

SKILL: Balancing Combustion Equations

In NCEA exams you will be asked to write balanced symbol equations for combustion.

Problem: A sample of ethanol combusted in air burns with a nearly invisible blue flame. Write a balanced symbol equation for the reaction.

Steps:

1. Decide whether its complete or incomplete: blue flame and absence of smoke mean it's complete.
2. Identify formulas of reactants: C_2H_5OH and O_2.
3. Identify formulas of products: with complete combustion it will be CO_2 and H_2O.
4. Write the unbalanced symbol equation:
   $C_2H_5OH + O_2 \quad CO_2 + H_2O$.
5. First count C's: 2 on left and 1 on right side. and balance C atoms with CO_2 molecules:
   $C_2H_5OH + O_2 \quad 2\,CO_2 + H_2O$.
6. Next count H's: 6 on left and 2 on right. and balance H atoms with H_2O molecules:
   $C_2H_5OH + O_2 \quad 2\,CO_2 + 3\,H_2O$.
7. Finally count O's: 3 on left and 7 on right. and balance O atoms with O_2 molecules:
   $C_2H_5OH + 3\,O_2 \quad 2\,CO_2 + 3\,H_2O$.

Some reactions require further changes to balance.

Revision Activities

1 Match up the descriptions with the terms.

- a hydrocarbon
- b natural gas
- c crude oil
- d combustion
- e exothermic reaction
- f fuel
- g heat of combustion
- h petroleum
- i fractional distillation
- j condensation point
- k fraction
- l waste product
- m complete combustion
- n incomplete combustion
- o cracking
- p thermal decomposition
- q micro-organism
- r fermentation
- s enzyme
- t polymerisation

- A breaking into simpler chemicals using heat
- B compound made of C and H atoms only
- C occurs when all of the fuel is oxidised
- D burning a substance in oxygen
- E producing alcohol and CO_2 from glucose
- F another name for crude oil
- G substance that is combusted to generate heat
- H natural source of hydrocarbon gases
- I unwanted chemical formed in a reaction
- J organism visible only under a microscope
- K amount of heat produced when a standard amount of fuel is combusted
- L joining small molecule into long chains
- M breaking a large alkane into a smaller alkane and an alkene
- N separating substances on the basis of differing condensation points
- O occurs when the fuel is partially oxidised
- P mixture of similar-sized hydrocarbon molecules
- Q natural source of hydrocarbon liquids
- R catalyst produced by living things
- S reaction that releases heat overall
- T temperature at which a cooling gas liquefies

2 Explain the differences between the terms in each of the items below:

- a hydrocarbons and alcohols
- b diesel, petrol and LPG
- c boiling and condensing
- d kerosene and gasoline fractions
- e complete and incomplete combustion
- f cracking and polymerisation.

3 Copy and complete this table summarising the state, source, heat content, and uses of different fuels.

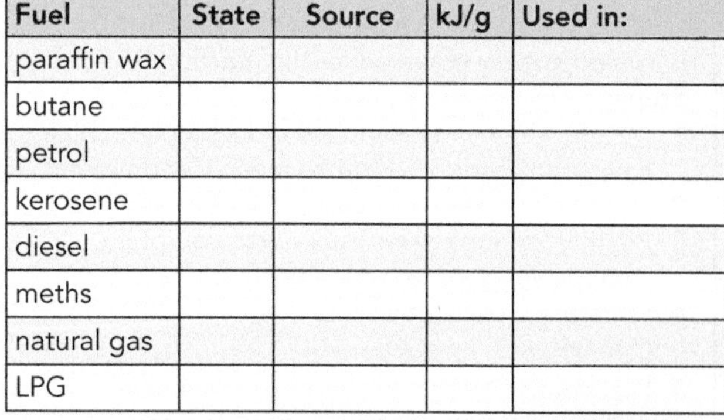

Fuel	State	Source	kJ/g	Used in:
paraffin wax				
butane				
petrol				
kerosene				
diesel				
meths				
natural gas				
LPG				

4 Identify why the fuel might be used in each of the following situations:

- a butane is used in cigarette lighters
- b LPG is used in barbecues
- c natural gas is used by stove tops
- d paraffin wax is used in candles
- e meths is used in camping stoves
- f kerosene is used to make jet fuel
- g petrol is used by most cars.

5 Decide whether the following statements are true or false. Rewrite the false ones to make them correct.

- a Most liquid fuels are produced from crude oil.
- b The energy in NZ's natural gas originally came from plants.
- c Combustion releases less heat than is required to ignite a fuel.
- d Petroleum gas can be readily liquefied.
- e Natural gas produces the most heat energy per gram.
- f Fractional distillation relies on different-sized hydrocarbons having different condensation points.
- g The temperatures in the distillation tower trays rise higher up.
- h Complete combustion produces more waste products.
- i Cracking produces a smaller alkane and an alkene.
- j Methanol is produced from the methane fraction of natural gas.
- k Fermentation is a biological process that requires oxygen.
- l A wide variety of plastics are produced by polymerisation.

6 The left photo shows a tanker bringing crude oil to the Marsden Point oil refinery near Whangarei, and the right photo shows the natural gas refinery at Kapuni.

- a Identify the fuels produced at the oil refinery.
- b Identify the fuels produced at the natural gas refinery.

7 Write balanced symbol equations for these reactions:

- a the complete combustion of methane and methanol
- b the complete combustion of ethane, ethene and ethanol
- c the incomplete combustion of propane producing soot
- d the incomplete combustion of butane producing CO.

8 A student wanted to compare the energy content of methanol and ethanol. Using the equipment shown, she measured the temperature of the water and the mass of the spirit burner before and after combusting each fuel for five minutes.

- a Identify the conditions that need to be the same for each fuel.
- b Explain how the thermal energy gained by the water can be calculated (see page 37).
- c Evaluate the accuracy of this method of measuring energy content.
- d Justify the usefulness of the results for comparing the energy in fuels.

water

spirit burner

ISBN: 9780170262316

9 The equipment is used to separate ethanol and water from a mixture of the two. Ethanol boils at 78 °C and water at 100 °C.

a Name the process that separates liquids on the basis of boiling points.
b Explain why the thermometer is placed in the neck of the flask.
c Suggest a reason why cold water runs through the outer jacket of the glass condenser.
d Describe what will happen over time as the ethanol-water mixture in the flask is heated.
e Describe how you would collect separate samples of ethanol and water.
f Explain how you would confirm that the first sample collected is ethanol and the second, water.

10 **The diagram shows the interior of a fractional distillation tower at the refinery. The colour gradient from red to blue represents a temperature gradient from nearly 400 °C down to less than 40 °C.**

a Explain why the crude oil is heated to over 400 °C before it enters the tower.
b Account for what happens to the larger, heavier hydrocarbon molecules.
c Account for what happens to the smaller, lighter hydrocarbon molecules.
d Describe how the different fractions of hydrocarbons are separated.
e List uses of the gasoline fraction.
f List uses of the heavy oil fraction.
g Describe what happens to the kerosene fraction.

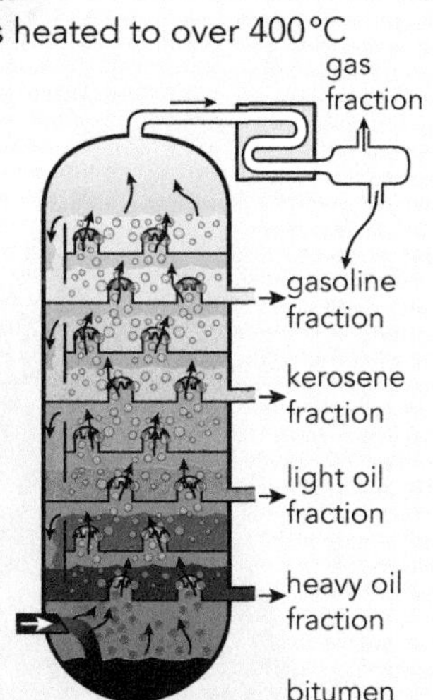

11 **Read the article opposite, then answer the questions below.**

a Why is butane normally a gas?
b Why is it a liquid in the lighter?
c What fills the space not occupied by the butane liquid?
d How is the butane allowed to escape?
e What ignites the butane?
f How do you keep the flame going? What happens to the gas?
g Why is the flame not very strong?
h How do you know that incomplete combustion is occurring?
i What are the three products?
j What is the word equation for the reaction?
k Write a balanced symbol equation for the reaction.

Butane Lighters

Many butane lighters are transparent so that you see there is a fluid inside.

As the boiling point (BP) of butane is 0 °C, it would normally be a gas at any temperature above that, but a gas can be liquefied at temperatures above its BP if it is in a pressurised container.

The pressure in the fuel container needs to be more than three times atmospheric pressure for butane to liquefy.

Inside the fuel container, butane gas fills the space not occupied by the liquid butane.

When the wheel of the lighter is flicked, a spark is caused by striking a metal against a flint. The action of flicking the wheel ends with pressing down the lighter's 'button'. This opens a valve at the top of the fuel container, which allows butane vapour in the nozzle to escape. It ignites immediately.

If the valve is kept open by your finger on the 'button', the slight pressure drop inside the fuel container causes a small amount of liquid butane to vaporise and escape through the valve.

Air mixes with butane at the top of the valve. It is not premixed as in a Bunsen burner.

Although the flame of a butane lighter is not sooty, its yellow colour indicates that combustion is not complete. The flame is between 400 and 700 °C.

Releasing the button closes the valve and the flame goes out, as there is no more fuel available for combustion.

12 **The table gives the composition of Kapuni natural gas.**

Component	Formula	%
methane	CH_4	45.8
carbon dioxide	CO_2	42.6
ethane	C_2H_6	5.5
propane	C_3H_8	3.5
butane	C_4H_{10}	1.6
pentane	C_5H_{12}	0.4
hexane	C_6H_{12}	0.3
nitrogen	N_2	0.3

a Draw a pie graph of the composition.
b Which component is not a fuel?
c Why must it be removed?
d Which alkanes are in the natural gas supplied to houses?
e Which alkanes are in LPG (liquefied petroleum gas) tanks?
f Which alkane in LPG requires pressure to liquefy?
g What is odd about butane, pentane and hexane being listed as components of natural gas (see page 88)?
h Why would they be in a liquid state before the natural gas emerges from the wells?

13 **When alkanes are cracked, a shorter alkane and an alkene are produced.**

a Write a word and balanced symbol equation for the cracking reaction that produces ethene from hexane.
b Write a word and balanced symbol equation for the cracking reaction that produces propene from pentane.

14 **Methanol is produced from methane in a two-stage process.**

a Describe what happens in the first stage, then write word and balanced symbol equations.
b Describe what happens in the second stage, then write word and balanced symbol equations.
c Explain why a fractional distillation tower is needed.

15 **Ethanol can be fermented by yeasts or produced in a chemical plant.**

a Compare the conditions needed for each to occur.
b Identify the raw materials for each process.
c Compare the products of each process.
d Write word and symbol equations for the two processes.

THE Living WORLD

Basic Biology Concepts

Learning Outcomes - On completing this unit you should be able to:
- describe the meaning of basic biological terms
- identify and explain the major concepts of biology
- relate the structure of biological objects to their function
- explain how biologists study life at different levels of organisation
- *summarise and link basic biology concepts on a concept map.*

Living Things

- Living thing are called **organisms**. They have complex internal structures and carry out complex processes, which enable them to maintain life and reproduce.
- A **species** is a group of similar organisms that are capable of interbreeding to produce fertile offspring.

Life Processes

- Organisms carry out the following **life processes**:
  - obtaining nutrients for energy and growth (**nutrition**)
  - releasing energy from food (**gas exchange** and **respiration**)
  - transporting substances internally (**circulation**)
  - disposing of cellular waste products (**excretion**)
  - holding up and moving the body or parts of it (**support** and **movement**)
  - sensing changes in the environment and responding (**sensitivity** and **co-ordination**)
  - growing larger (**growth**)
  - producing new organisms (**reproduction**).
- Different types of organisms have features that enable them to carry out life processes in different ways.

Environmental Factors

- The external **environment** of organisms includes all factors that affect them, such as the **medium** they live in, the physical conditions, and other living things.
- These **environmental factors** directly or indirectly affect how organisms carry out life processes, which impacts on their survival, growth and reproduction.
- **Physical factors** include moisture levels, temperature range, light intensity, soil or water pH, air and water currents, nutrient availability, O_2 and CO_2 levels, etc.
- Biological processes are most efficient when relevant physical factors are at **optimum levels**. Organisms can tolerate a range of conditions but if a particular factor is outside of that range it becomes a **limiting factor**.
- **Biological factors** include the activities of grazers, predators, parasites and pathogens, as well as that of competitors for the same resources.

Levels of Organisation of Life

- Although the organism is the basic unit of biological study, they exist in **populations** of organisms belonging to the same species living in the same location.
- A population exists within a **biological community**, consisting of all of the interacting plant, animal, fungi and bacteria populations found in a location.
- An **ecosystem,** consisting of a community and its environment, processes energy and chemicals.
- The **biosphere** consists of that part of the planet which is inhabited by living things.
- Biologists study lower levels of organisation as well.
- Complex organisms consist of a number of different **organ systems** (eg digestive system) each carrying out a life process (eg nutrition).
- Organ systems consists of connected **organs** (eg liver, stomach, pancreas), which are made out of **tissues** (eg nerve, muscle) consisting of masses of similar cells.
- Cells have structures called **organelles**, which are assembled out of **biomolecules** made out of **atoms**.

Biological Building Blocks

- All organisms are made out of microscopic **cells**. They are the basic building blocks of all living things.
- Cells are the simplest biological structures that can be considered to be alive, but they are highly complex.
- **Unicellular organisms** consist of a single cell that is able to carry out all of the life processes.
- **Multicellular organisms** consist of huge numbers of different kinds of cells that are organised into tissues, organs and organs systems. Specialised structures carry out different life processes.
- All cells come from pre-existing cells by a process of **cell division** or by the fusion of two cells.

The Molecules of Life

- Cells are packed with complex biomolecules that have been assembled out of simpler molecules, which either are manufactured using inorganic chemicals from the environment or come from digested food.
- **Carbohydrates** are either single sugar molecules (eg glucose), double sugar units (eg sucrose) or long chains of linked sugar units (eg starch, cellulose), forming cell structures (eg cell wall) or energy stores (eg in seeds).
- **Lipids** are fats and oils. Lipid molecules consist of three fatty acids bonded to a glycerol unit. The cell membrane is a double layer of lipid molecules. High energy lipids act as energy stores (eg in fat cells).
- **Proteins** are highly complex biomolecules that play many roles in cells. As enzymes, they catalyse and control the reactions occurring inside cells. Muscle fibres, hair and nails are made of structural proteins.
- **Nucleic acids**, such as DNA, are also highly complex molecules, which encode genetic information.

ISBN: 9780170262316

Cell Structure and Processes

- Cells are basically chemical factories. They consist of a membrane enclosing the jelly-like **cytoplasm**, where **metabolism** (the reactions of life) occurs.
- The **cell membrane** controls the exchange of chemicals between the cell and its environment. Chemicals either move passively by **diffusion** from higher to lower concentration sides, or are **actively transported** in or out by the membrane, which requires energy.
- Embedded in the cytoplasm are various organelles (eg nucleus, mitochondria, chloroplasts, ribosomes vacuoles) depending on the type of cell.
- All cells, other than bacteria, have a **nucleus** containing **chromosomes** with inherited instructions called **genes**. Cells use the genetic information to make the proteins that determine their structure and function.
- There are many different cell types, each with particular **structures** that enable it to carry out a specific **function**.
- All cells carry out **respiration** in organelles called mitochondria. Oxygen is used to release small packets of energy from glucose, which power metabolism. The waste product carbon dioxide is released into the air.
- Plant cells with chloroplasts carry out **photosynthesis**. Sunlight, water from the soil, and carbon dioxide from the air are used to make energy-rich glucose, the starting point for making all other biomolecules. The waste product oxygen is released into the air.
- Cells multiply through repeated cell division, a process by which cells enlarge then divide into two. This rapidly results in a huge number of cells, which then develop into different types through the process of **cell specialisation**.
- Organisms reproduce by developing sex cells called **gametes**. The female parent produces large, immobile egg cells, and the male parent, smaller mobile sperm cells. At **fertilisation** a sperm fuses with an egg to produce a **zygote**, the first cell of a new organism.

Adaptive Features

- Inherited features or **traits** that enable organisms to carry out life processes in their current environment are called **adaptive features** or **adaptations**.
- Three types of adaptations are:
  - **structural adaptations**, such as the parts of organisms that carry out specific functions
  - **physiological (process) adaptations**, such as plants producing toxins and animals, digestive enzymes
  - **behavioural adaptations**, such as plant responses to external stimuli and animal behaviours.
- New adaptive features arise when genetic **mutations** result in traits that suit changing environmental conditions, which are then favoured by **natural selection**.

The Diversity of Life

- On the basis of common features and shared ancestry, living things are classified into five kingdoms: Monerans, Protists, Fungi, Plants and Animals.
- **Monerans** are the simplest organisms (eg bacteria, blue-green algae). They are mostly unicellular and lack a nucleus, but they do have a circular chromosome.
- **Protists** are either unicellular or colonies of identical cells. Their cells do have a nucleus with chromosomes.
- Monerans and protists obtain food in a variety of ways.
- **Fungi** are immobile unicellular or multicellular organisms, whose cells do have a nucleus. They obtain food by digesting dead organisms or organic matter.
- **Plants** are complex, immobile multicellular organisms that are able to make food by photosynthesis.
- **Animals** are complex, mobile multicellular organisms that obtain food by consuming other organisms.

Feeding Roles and Ecological Relationships

- Different species have different feeding roles in a community. Plants are **producers**, although a few are also consumers (eg Venus fly-catcher). Animals are **consumers**, but specialise as **herbivores**, **carnivores**, **omnivores**, **parasites** and **detritivores**. Bacteria are producers, consumers or **decomposers**. Fungi are mostly decomposers or parasites.
- Feeding roles often involve relationships between species, such as **grazing**, **predation** and **parasitism**.
- Other relationships between species include:
  - **resource competition**, which harms both species (eg sheep and rabbits compete for grasses)
  - **mutualism**, which benefits both species (eg flowers being pollinated by bees)
  - **commensalism**, which benefits one species without harming the other (eg barnacles growing on whales).

Energy Flow and Nutrient Recycling

- **Food chains** show 'who eats whom' and a **food web** connects all of the food chains within a community.
- Light energy absorbed by producers is used to make food. Some of the energy in food is used to power life processes, then radiated as heat. The rest is converted into chemical energy stored in plant tissue.
- As plants get eaten by herbivores, which get eaten by carnivores, chemical energy passes along food chains. Some is used to power life processes, then radiated as heat. The rest is stored in animal tissue.
- Decomposers use the energy in dead matter to power life processes, then radiate it as heat into the environment, so **energy flows** through an ecosystem.
- Plants absorb chemicals they need to make food from the soil and air. These nutrients move along food chains from one **trophic level** to the next, eventually being released back into the environment by decomposers, so **nutrient cycling** occurs in an ecosystem.

SCIENCE SKILL: Concept Mapping

A useful way of understanding the connections between concepts is to construct a concept map.

Steps:

1. Obtain an A3 sheet and some colour felt-tip pens.
2. Identify the theme of the unit: biology concepts, and the major topics: the blue-green headings.
3. Draw a central bubble with the theme, then add the major topics in other bubbles radiating off it.
4. Identify the key concepts of each topic: for Living Things it would be organisms and species; and draw them as bubbles off the major topic bubbles.
5. Identify and link lower level concepts to the key concept bubbles, expressing them as briefly as you can, using symbols wherever possible.
6. Use arrows or colour to highlight related concepts.

ISBN: 9780170262316

Traits, Genes, Chromosomes and DNA

Learning Outcomes - On completing this unit you should be able to:
- distinguish between genetic and acquired variation
- apply the basic terminology of genetics to specific situations
- explain the connections between chromosomes, DNA, genes and alleles
- describe how the structure of DNA codes for proteins and how it is replicated
- *determine a short sequence of amino acids in a protein by decoding DNA.*

Species

- Living things (eg gannets, cats) are called **organisms**.
- All organisms carry out vital life processes (see page 98) such as moving, sensing, growing and reproducing.
- Every organism belongs to a species. A **species** is a *group of similar organisms capable of interbreeding to produce fertile offspring* (eg all cats belong to the same species, as they can interbreed).
- Every species is given Latin genus and species names (eg the domestic cat species is *Felis catus*; the dog species, *Canis lupus*).
- Every organism is part of a **population**. A population is a group of individuals belonging to the same species living in the same location at the same time (eg the gannet colony at Cape Kidnappers shown at top).

Acquired and Genetic Variation

- Members of the same species will have many features (characteristics) in common, but individuals will differ in the appearance of other features.
- Features that vary include colouration (eg skin and hair colour), form (shape and structure), functioning (eg sweating), and behaviour (eg levels of aggression).
- The features of an individual organism are influenced by:
  - **a** inherited instructions (eg hair curliness instruction)
  - **b** environmental effects (eg freckling caused by exposure to the sun)
  - **c** individual actions (eg muscle-building in the gym).

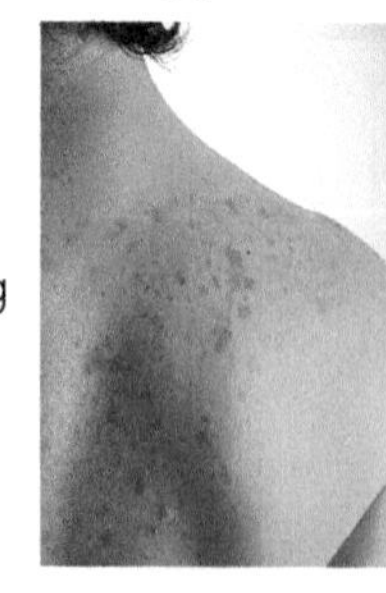

- Variation caused by an organism's environment or actions is **acquired variation**. It is usually assumed to be non-inheritable.
- Variation that is passed on from one generation to the next through reproduction is **genetic variation**.
- Genetic variation in a population may help it survive changing environmental conditions, as variation enables new adaptive features to develop (see page 109).

Traits and Genes

- A **trait** is any feature relating to the colouration, form, functioning or behaviour of an organism. At NCEA level 1 the focus is on traits that are determined solely by inherited instructions, which are called **genes**.
- Most plants and animals have thousands of genes. An organism's **genome** consists of its complete collection of genes, which enable it to grow, live and reproduce.
- Most organisms have two parents and therefore *inherit two copies of each gene*.
- A gene (eg the 'hair curliness' gene) can have different forms called **alleles** (eg 'curly hair' and 'straight hair' alleles) that cause the trait to vary (eg curly hair, straight hair). *Alleles are a source of genetic variation.*
- For a particular trait, the **genotype** of an organism is the *two alleles it has inherited*. The **phenotype** is how that genotype has been expressed.
- If an organism inherits two identical alleles (eg two 'straight hair' alleles), it is said to be **homozygous** and the phenotype is as you would expect (straight hair).
- If the organism inherits two different alleles (eg a 'straight hair' and a 'curly hair' allele), it is said to be **heterozygous**. In NCEA level 1 exams, it is assumed that only one allele will be expressed (see page 108).
- Inheritance is more complex than this and often intermediate phenotypes occur.

Chromosomes

- **Cells** are the microscopic living 'building blocks' that organisms are made out of.
- Cells have a structure called the **nucleus**, which contains thread-like objects called **chromosomes**. These coil up and become visible under the microscope just before a cell divides.
- All body cells (other than sex cells) *have an even number of chromosomes*. Those chromosomes can be sorted into pairs that are the same shape and size. *One member of each pair is inherited from each parent.*
- Individuals of a species will have the *same number of chromosomes in each of their body cells* (eg humans have 46 chromosomes and cats 38).
- Each chromosome consists of a sequence of hundreds or even thousands of different genes, each found at a specific location along the chromosome.

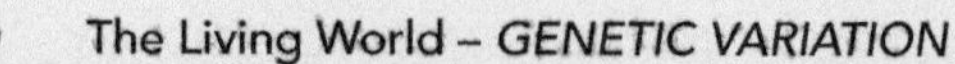

ISBN: 9780170262316

Genes and Proteins

- Genes code for proteins which then determine traits (eg the gene INS specifies the protein insulin which helps to control blood sugar levels).
- Every organism makes thousands of different proteins, each of which has a specific role.
- **Proteins** are long chains of linked **amino acids** (eg insulin consists of two linked amino acid chains).
- There are 20 different kinds of amino acids (eg serine, glycine). The sequence of amino acids in a protein causes it to fold into a unique shape that determines what it does.
- Some proteins form structures (eg keratin in hair), others act as **enzymes** (eg pepsin in the stomach), a few act as messengers called hormones (eg insulin).

DNA and Genes

- Each chromosome consists of an extremely long **DNA** (deoxyribonucleic acid) molecule wrapped around proteins that help it fit in the nucleus.
- The DNA consists of two linked strands twisted into a double helix shape.

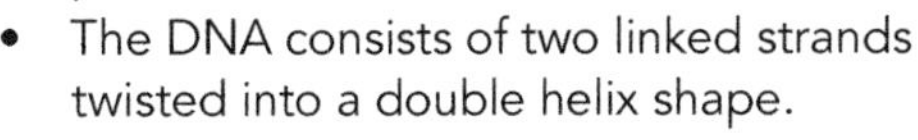

- Both strands are chains of alternating phosphate and sugar units. Each sugar has one of four **bases** (C, G, A, T) attached.
- One of the strands encodes the genes.

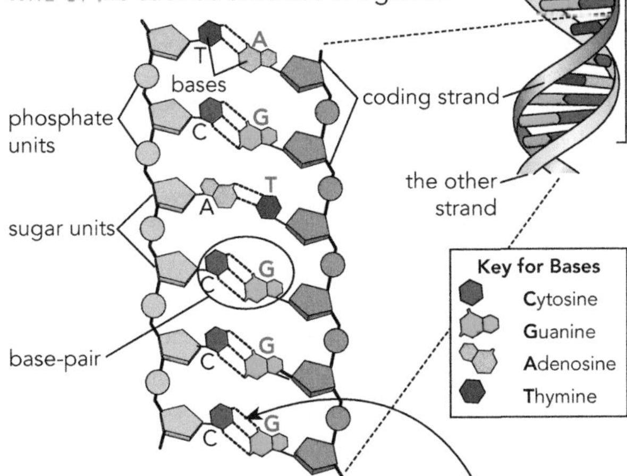

- The two strands are held together by the electrostatic attraction between particular pairs of bases. The **base-pairing rule** is: C pairs with G, and A with T.
- As can be seen, DNA is bit like a ladder: phosphate-sugar chains form the rails and base-pairs, the rungs.
- A DNA molecule consists of genes arranged in a linear fashion. A gene is a *specific section of a DNA molecule*, hundreds or thousands of base-pairs long.

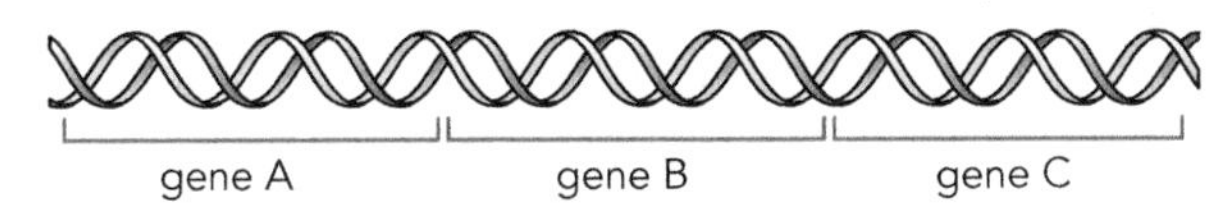

Protein Synthesis

- A cell uses particular genes found at fixed locations on specific DNA molecules to make the proteins it needs.
- Before **protein synthesis** occurs, the two DNA strands are pulled apart between the bases of each base-pair.
- Each gene is *coded for by the order of bases* (eg -**AGT-GGG**-) along a specific section of the coding strand.
- Each triplet of bases codes for one of the 20 different kinds of amino acids (eg AGT codes for serine and GGG for glycine).
- The gene's triplet sequence determines the sequence of amino acids in the protein and hence, the protein's shape and role.
- The cell reads the sequence as it assembles the protein.
- Alleles occur due to changes (see page 109) in a gene's base sequence. Eg, if **A** replaces **T** in the sequence **AGT-GGG**, then arginine is specified instead of serine, resulting in a modified protein affecting the phenotype.

DNA Replication

- The two strands of a DNA molecule encode the same genetic information, except C's and G's are swapped, as are A's and T's.
- When a cell needs to duplicate its chromosomes just before cell division (see page 104), the two DNA strands 'unzip' and new strands are assembled on the exposed bases that act as templates. This process is **DNA replication**.

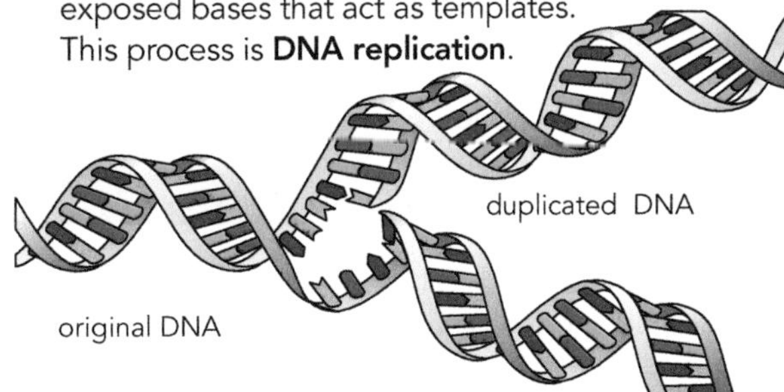

- This results in two *identical double-stranded DNA molecules* and therefore duplicated chromosomes. The duplicates are initially attached, which gives chromosomes their X-shape when the DNA coils up tightly.
- Duplicates are pulled apart when the cell divides (see page 104).

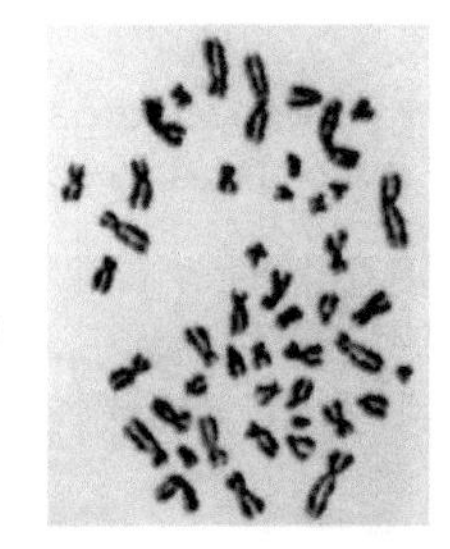

SCIENCE SKILL: Decoding DNA

To find the amino acid order of a protein, decode the DNA triplets.

Steps:

1 Locate triplet: eg ATT

2 Identify amino acid: Ile

3 Record sequence: -Ile-etc-

	T		C		A		G		
T	TTT	Phe	TCT	Ser	TAT	Tyr	TGT	Cys	T
	TTC	Phe	TCC	Ser	TAC	Tyr	TGC	Cys	C
	TTA	Leu	TCA	Ser	TAA	Stop	TGA	Stop	A
	TTG	Leu	TCG	Ser	TAG	Stop	TGG	Trp	G
C	CTT	Leu	CCT	Pro	CAT	His	CGT	Arg	T
	CTC	Leu	CCC	Pro	CAC	His	CGC	Arg	C
	CTA	Leu	CCA	Pro	CAA	Gin	CGA	Arg	A
	CTG	Leu	CCG	Pro	CAG	Gin	CGG	Arg	G
A	ATT	Ile	ACT	Thr	AAT	Asn	AGT	Ser	T
	ATC	Ile	ACC	Thr	AAC	Asn	AGC	Ser	C
	ATA	Ile	ACA	Thr	AAA	Lys	AGA	Arg	A
	ATG	Met	ACG	Thr	AAG	Lys	AGG	Arg	G
G	GTT	Val	GCY	Ala	GAT	Asp	GGT	Gly	T
	GTC	Val	GCC	Ala	GAC	Asp	GGC	Gly	C
	GTA	Val	GCA	Ala	GAA	Glu	GGA	Gly	A
	GTG	Val	GCG	Ala	GAG	Glu	GGG	Gly	G

ISBN: 9780170262316

Revision Activities

1 Match up the descriptions with the terms.

a organism	A chemicals that cells use to code for genes
b species	B possessing two identical alleles for a gene
c acquired variation	C variation caused by actions or the environment
d genetic variation	D microscopic living building blocks of organisms
e trait	E any feature of an organism
f gene	F proteins that control chemical reactions
g genome	G cell structure that contains chromosomes
h alleles	H individual living thing
i genotype	I different forms of a gene
j phenotype	J group of organisms able to interbreed successfully
k homozygous	K thread-like structure consisting of many genes
l heterozygous	L inherited biological instruction
m cell	M shorthand for deoxyribonucleic acid
n nucleus	N two alleles possessed for a trait
o chromosome	O results in the duplication of chromosomes
p proteins	P complex molecules that determine traits
q amino acids	Q possessing two different alleles for a gene
r enzymes	R building blocks of protein molecules
s DNA	S complete collection of genes of an organism
t bases	T how the genotype has been expressed
u base-pairing rule	U variation passed on from one generation to next
v protein synthesis	V making proteins using the information in genes
w DNA replication	W C and G bases pair up, as do A and T bases

2 Explain how the terms differ in each of the items below:

a acquired and inherited variation
b traits, genes and alleles
c genotype and phenotype
d homozygous and heterozygous genotypes
e a chromosome and a DNA molecule
f a base-pair and a base triplet
g a base sequence and an amino acid sequence
h protein synthesis and DNA replication.

3 For each organism shown below, describe a single trait that relates to colouration, form, function or behaviour.

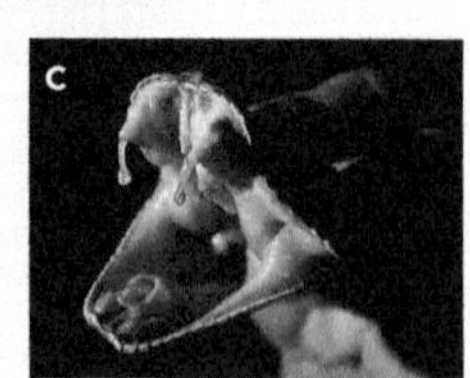

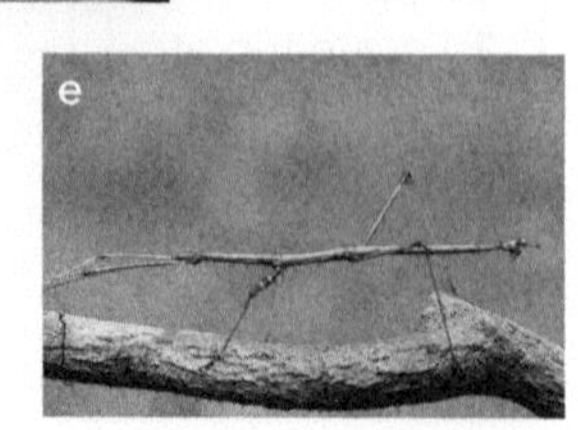

4 Decide whether the following statements are true or false. Rewrite the false ones to make them correct.

a Members of a species are capable of interbreeding successfully.
b Genetic variation is inheritable.
c Only some features of organisms are considered to be traits.
d An organism's genome usually consists of thousands of genes.
e Organisms have one copy of each gene only.
f Alternative forms of genes are called alleles.
g The phenotype is the expression of the organism's genotype.
h Heterozygous genotypes have two identical alleles.
i Chromosomes are always visible.
j Proteins are folded chains of amino acids.
k Genes code for proteins.
l Genes are located along DNA molecules.
m Genes are encoded by the order of bases along a DNA molecule.
n DNA replication results in duplicated chromosomes.

5 The diagram shows the sub-units of the hormone insulin, which is a type of protein.

a What are the sub-units?
b How are they arranged?
c What type of biological molecule is insulin?
d How do you know?
e What causes the linked chains to fold into a special shape?
f What determines the order of the amino acids in the two chains?
g Which gene is insulin expressing? Where is it located?
h How is the order of bases on that gene related to the order of amino acids in the insulin molecule?

The diagram below shows the atoms in an insulin molecule to compare with the diagram above.

i Why type of model is it?
j What atoms are in the molecule (see page 73)?
k Which model best shows how complex insulin is? Why?
l Which model best shows the overall structure of insulin? Why?

6 INS is a gene, insulin is a protein, and blood sugar level is a trait.

a What is the relationship between gene and protein?
b What is the relationship between protein and trait?
c How many INS genes does each person have?
d What are different forms of the INS gene called?
e What might happen to the insulin produced if two defective INS alleles were inherited?
f How might the DNA of a defective allele differ from that of a normal allele?

ISBN: 9780170262316

7 Hair shape is determined by the inheritance of a hair curliness gene that has two alleles: the 'curly hair' allele and the 'straight hair' allele.

a What is the phenotype of the girl?
b What is the phenotype of the boy?
c If the girl is homozygous, what is her genotype for the trait?
d If the boy is homozygous, what would his genotype be?
e If their baby sister was heterozygous, what would her genotype be?
f What could the baby sister's phenotype be?

8 Onion cells have four chromosomes. The photo shows root-tip cells at a stage when chromosomes are visible.

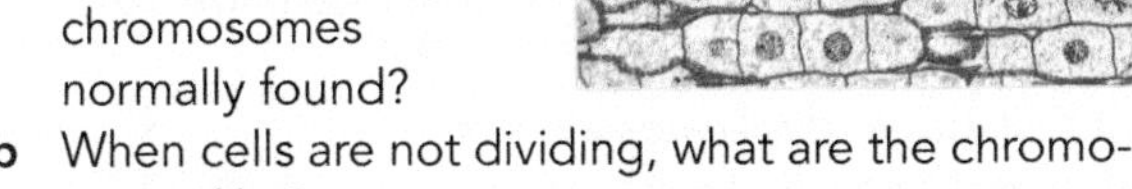

a Where are the chromosomes normally found?
b When cells are not dividing, what are the chromosomes like?
c What happens to the chromosomes to make them visible under a microscope when the cells are dividing?
d What will have happened to each onion chromosome just before cell division occurs?
e How many chromosomes will cells have at that stage?
f How many 'daughter' cells result from each onion cell division?
g How many chromosomes will each 'daughter' onion cell have after division has occurred?

9 Cubic water melons are grown in Japan by placing the fruits inside boxes, which forces them to grow into an unusual shape.

a What type of variation is this? How do you know?
b Why wouldn't the seeds of a cubic watermelon produce cubic watermelons naturally?

10 Read the article opposite, then answer the questions below.

a Why is the ability to taste bitter chemicals important?
b Why do some plants produce toxic chemicals?
c Why is it that we are able to detect a wide range of tastes?
d Name the trait, protein and gene involved.
e What are the two common alleles?
f How many alleles do we inherit?
g Describe the phenotypes of individuals who are homozygous and heterozygous for the two alleles.
h Why is the ability to taste PTC odd?
i How might being a supertaster be an adaptive feature in a human population?

Supertasters

The ability to taste different chemicals is clearly important for enjoying foods but it is also important to be able to detect foods that might be toxic.

Many plants produce bitter chemicals to prevent herbivores from eating them. Humans have many different taste receptors on the tongue that can detect a wide variety of bitter tastes.

Receptors have proteins that are able to detect particular chemicals.

The protein 'Taste receptor 2 member 38' or TAS2R38 for short enables people to taste a bitter chemical called PTC.

The protein is coded for by the TAS2R38 gene found on chromosome number 7 (see above image).

There are two common alleles for the gene: the 'taster' allele and the 'non-taster' allele.

Those people with two 'taster' alleles find strips soaked in PTC intensely bitter. Those people with two 'non-taster' alleles usually can't taste PTC at all. Those with a 'taster' allele and a 'non-taster' allele find PTC moderately bitter.

What's odd about this gene is that PTC is an industrial chemical not found in nature.

A 'supertaster' tastes far more intensely than others, which may help in detecting potentially toxic plants. Scientists believe that being a supertaster is related to the presence of TAS2R38 'taster' alleles.

11 Imagine that the section of a DNA molecule shown opposite represents a single gene. The red coding strand is read from the top to the bottom.

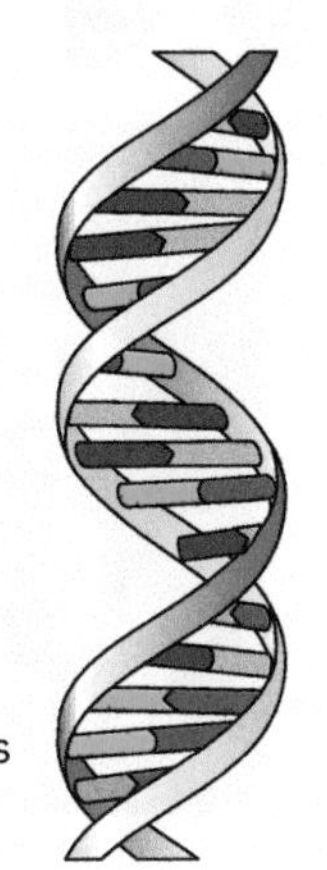

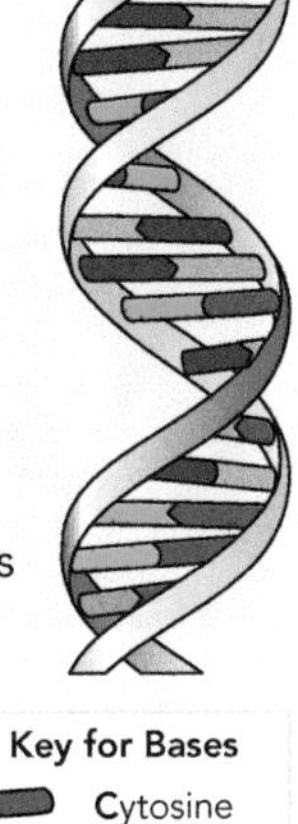

a How many bases long is the gene?
b What is the sequence of bases on the coding strand?
c How many base triplets are there in this gene?
d How many amino acids will the protein that is synthesised have?
e What will be the sequence of amino acids in that protein? (Refer to the table in the science skill panel on page 101).
f What is the base sequence on the other strand? (Use the pairing rule.)
g Why is this sequence important?
h What happens when DNA replication occurs? What does it result in?

12 The karyotype image shows the chromosomes found in the body cells of a male human. They have been arranged into pairs of similar size and shape.

a How many chromosomes do human body cells have?
b How many pairs can the chromosomes be arranged into?
c What can be said about the origin of the members of each pair?
d What is odd about the pair in the bottom right corner?
e What are those two chromosomes usually called?
f If the karyotype had been of the chromosomes of a human female, what would the last pair be like?
g If the man inherited one copy of the INS gene from his father on chromosome 11 as shown, where would the copy of the INS gene he inherited from his mother be?

ISBN: 9780170262316

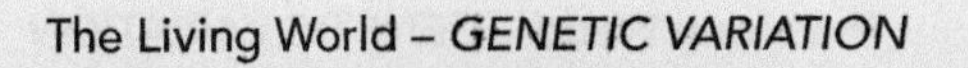

Reproduction, Growth and Variation

Learning Outcomes - On completing this unit you should be able to:
- explain the advantages and disadvantages of asexual and sexual reproduction
- outline the stages of reproduction and the functions of meiosis and fertilisation
- compare and contrast mitosis and meiosis cell division
- explain how sexual reproduction leads to variation within a population
- *identify and interpret cell division and karyotype diagrams.*

Asexual and Sexual Reproduction

- In some species (eg aphids), a single organism can produce offspring. This is called **asexual reproduction**, and can result in a rapid increase in population numbers. The *offspring are genetically identical* to the parent and each other, as the parent passes on its entire collection of **alleles**.
- In most species, a male and a female produce off-spring. This is called **sexual reproduction** and the *offspring are genetically different* (except for identical siblings), as alleles from two organisms are combined.
- In sexual reproduction, the parent organisms produce sex cell called **gametes**. The male parent produces mobile **sperm** and the female parent, immobile **eggs**.

Stages of Reproduction

- The term **body cells** refers to all cells in an organism except for reproductive (sex) cells. The body cells of all members of a species have the same even number of **chromosomes** (eg human body cells have 46).
- Reproductive cells have *half the normal number of chromosomes* (eg human sperm and eggs have 23).
- At **fertilisation**, the nucleus of a sperm enters an egg to form the first cell of a new organism, which is called a **zygote**.
- The sperm and egg nuclei fuse, resulting in the zygote having a nucleus with a full set of chromo-somes (eg 46 in human zygotes).

23 chromosomes
sperm
egg
fertilisation
zygote
46

- The zygote then divides repeatedly, rapidly becoming a multicellular organism.
- The stages of reproduction shown are:
  1. gamete production
  2. fertilisation to give a zygote
  3. cell division and speciali-sation to give an embryo
  4. growth of the embryo into a foetus.

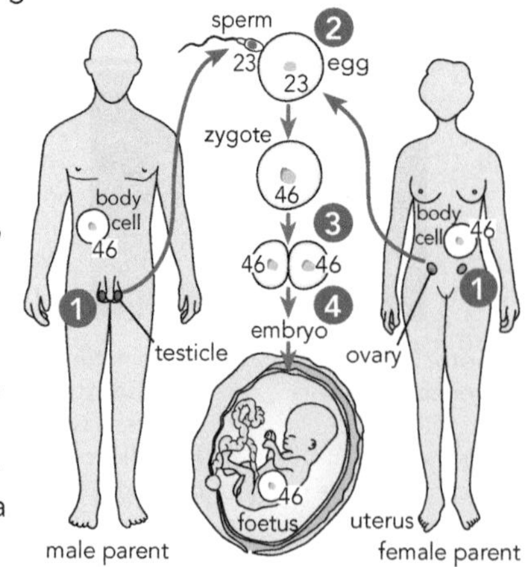

Mitosis and Growth

- New body cells are produced from existing body cells by a process of cell division called **mitosis**.
- Just before mitosis occurs, the cell duplicates all of its chromosomes by **DNA replication** (see page 101), (eg human cells would then have 92).
- At that stage, duplicates are still attached to each at other at the part called the **centromere**.

centromere
duplicates

- As the original cell divides into two, the *duplicated chromosomes are separated* as shown below. For simplicity, the diagram shows an organism that has just four chromosomes (two matching pairs - see opposite page) in its body cells.

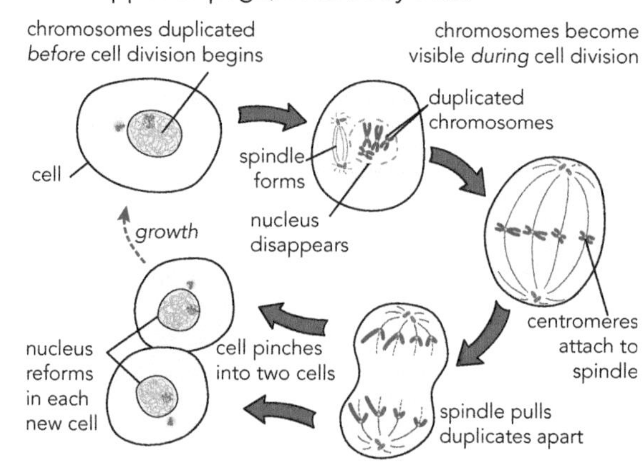

- The new cells then grow to the size of the original cell.
- By repeated cell divisions, a single-celled zygote can grow into a multi-cellular organism consisting of huge numbers of cells. It takes just 20 repeated divisions to reach over a million cells (humans have trillions).
- Body cells are genetically identical but there are many different kinds of body cells (eg over 300 in humans).
- When groups of cells stop dividing, a process called **cell specialisation** occurs in which particular *sets of genes get expressed.* This results in different cell types having different structures and functions.

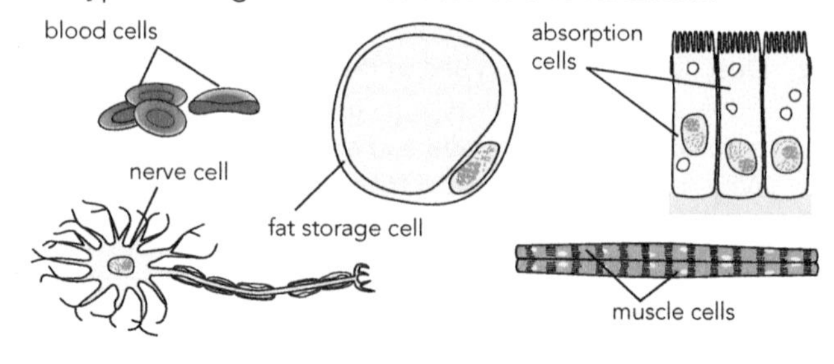

ISBN: 9780170262316

Homologous Chromosomes

- Body cell chromosomes can be *matched in pairs*, one member of each pair from each parent. The members of a matching pair are **homologous chromosomes**.
- Fluorescent probes can be used to identify homologous chromosomes. Digital images of individual chromosomes are arranged in pairs of decreasing size. The resulting image is called a **karyotype**.
- This image shows the karyotype of a human female.

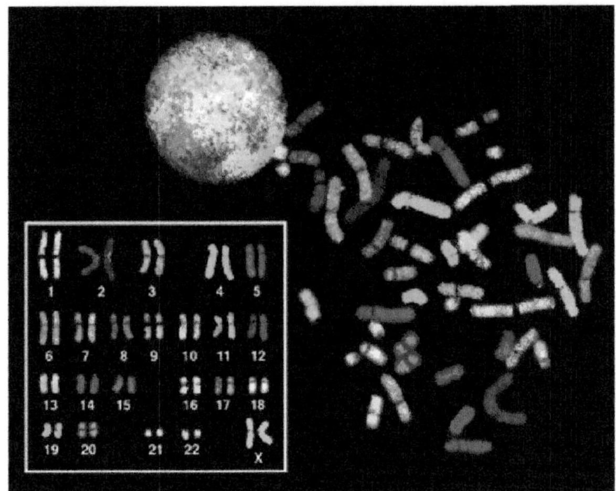

- Each homologous pair consists of a paternal and a maternal chromosome inherited from the respective parents. The two alleles (see page 100) inherited for a particular trait are found at the *same location on homologous chromosomes*.

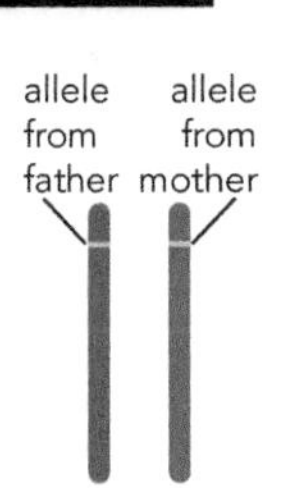

Meiosis Cell Division

- Gametes are produced from body cells, so the number of chromosomes must be halved (eg from 46 to 23 in humans). This occurs in a cell division called **meiosis**.
- In meiosis, body cells *divide twice but the chromosomes are duplicated just once*. This results in gametes with half the number of chromosomes found in body cells.
- Just before meiosis occurs, the cell duplicates all of its chromosomes. They stay attached at the centromere.
- The diagram below shows an organism that has just four chromosomes (two matching pairs) in its body cells.

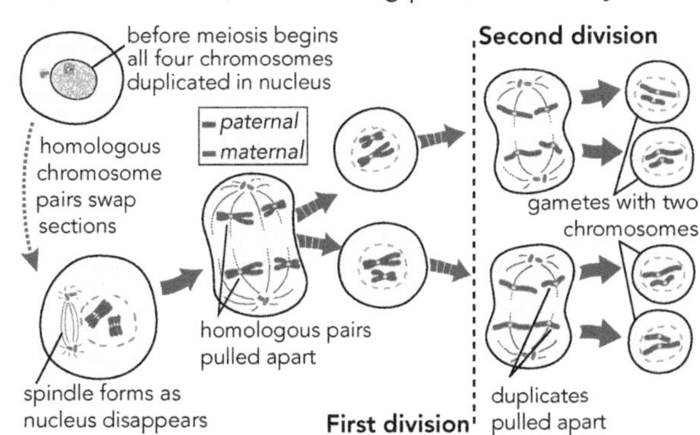

- In the first division, the cell randomly separates homologous chromosomes, thus *reshuffling maternal and paternal chromosomes* and also *separating alleles*.
- In the second, the cells separate the chromosome duplicates, thus *halving the chromosome number*.

Variation in a Population

- At NCEA level 1, each trait is considered to be determined by a single gene with just two alleles. In reality the variation in most traits is influenced by several different genes, often with multiple alleles.
- It is *mixing of the alleles of several different genes* that generates **phenotype** variation in a population.
- Sexual reproduction mixes alleles of different genes in three steps:
  1. Parts of homologous chromosomes are *swapped before duplication*. This is called **crossing-over**.
  2. Homologous chromosomes randomly separate in meiosis, resulting in *each gamete having a different collection of chromosomes* and therefore alleles.
  3. Gametes randomly combine at fertilisation, resulting in offspring having unique collections of chromosomes, and therefore *different combinations of the alleles of several genes affecting a trait.*

Sex Determination in Humans

- In many species, the sex of an organism is *determined by a single pair of chromosomes.* For humans it is the X and Y chromosomes, which are considered to be homologous even though they are different sizes.
- Human males have an X and a Y chromosome (XY). As paired chromosomes separate in meiosis, half of the sperm produced in a male's testicles will have an X chromosome and half a Y chromosome.
- Human females have two X chromosomes (XX). When eggs are produced by meiosis in a female's ovaries, all will have an X chromosome.
- The possible combinations of X and Y chromosomes that offspring might have are shown in the blue boxes:

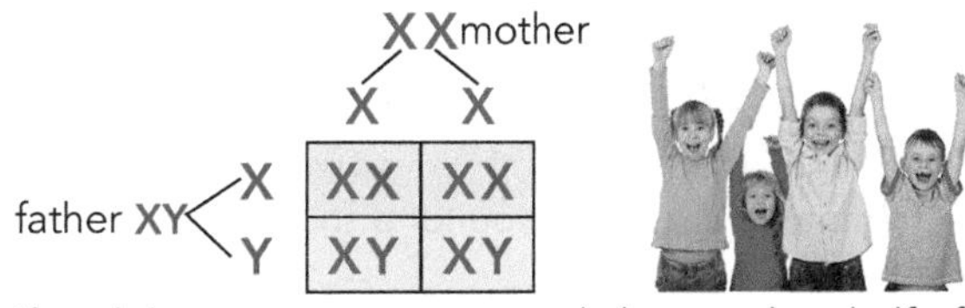

- If each box represents an equal chance, then half of the possible offspring will be boys and half girls, depending on *which type of sperm fertilises the egg.*

SCIENCE SKILL: Identifying Cell Division

You should be able to distinguish between mitosis and meiosis cell division diagrams.

Problem: What type of cell division is shown below?

Steps:

1. Identify how many divisions are shown: two.
2. Count the number of new cells formed: four.
3. Check out whether the number of chromosomes has changed: the original cell had four but the four new cells have only two.
4. Is it unchanged, as occurs in mitosis? No, so it isn't mitosis.
5. Has it been halved, as occurs in meiosis? Yes, so it must be meiosis.

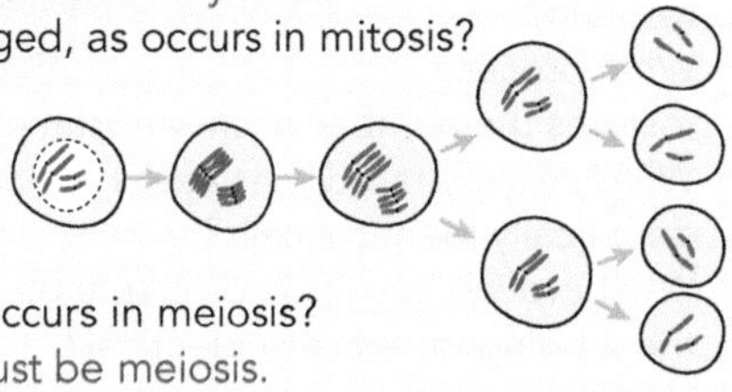

ISBN: 9780170262316

Revision Activities

1 Match up the descriptions with the terms.

a asexual reproduction	A all cells in organism other than reproductive cells
b alleles	B results in the duplication of chromosomes
c sexual reproduction	C first cell of an organism, resulting from fertilisation
d gametes	D alternative forms of a specific gene
e sperm	E reproductive cells, either sperm or eggs
f eggs	F image showing paired chromosomes
g body cells	G reproductive cells produced by the male parent
h chromosomes	H cell division that produces gametes with half the normal number of chromosomes
i fertilisation	I the two alleles possessed for a specific trait
j zygote	J nucleus of a sperm enters an egg to form a zygote
k mitosis	K single organism produces offspring
l DNA replication	L occurs when different types of cells are formed
m centromere	M thread-like structures bearing many genes
n cell specialisation	N cell division that produces two genetically identical 'daughter' cells
o homologous chromosomes	O physical appearance or state of a trait
p karyotype	P part where chromosome duplicates are attached
q meiosis	Q chromosomes with the same sequence of genes
r phenotype	R reproductive cells produced by the female parent
s genotype	S occurs when parts of homologous chromosomes are swapped
t crossing-over	T two organisms produce varying offspring

2 Explain the differences between the terms in each of the items below:

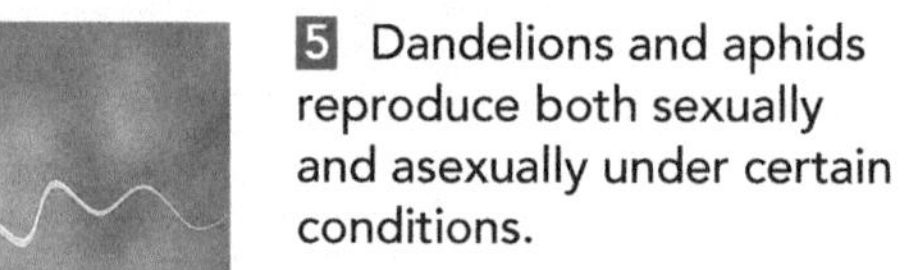

a sexual and asexual reproduction
b gametes and body cells
c sperm and eggs
d mitosis and meiosis
e maternal and paternal chromosomes
f phenotype and genotype for a specific trait.

3 The diagram below shows some of the stages of human reproduction. Describe what is happening at each of the points labelled **a** to **f**, then answer the questions.

g What process halves the number of chromosomes?
h Why does this need to occur?
i What processes are involved in creating new combinations of chromosomes?
j What does this result in?
k What process restores the number of chromosomes?
l What process rapidly increases the number of genetically identical cells?
m What process results in diverse types of cells despite them being genetically identical?
n How many cells and cells types are found in humans?

4 Decide whether the following statements are true or false. Rewrite the false ones to make them correct.

a Asexually reproduced siblings are genetically identical.
b Gametes have half the number of chromosomes as body cells.
c All human cells have 46 chromosomes.
d Fertilisation results in a zygote with the normal chromosome number.
e Mitosis produces genetically identical daughter cells.
f The spindle pulls duplicated chromosomes apart.
g Cell specialisation results in cells whose genotypes differ.
h Homologous chromosomes have identical alleles.
i Meiosis randomly separates chromosomes into gametes, which randomly combine at fertilisation.
j Meiosis combines pairs of alleles.
k Sexual reproduction results in new combinations of alleles for different genes.
l A child's sex depends on the sex chromosome carried by the egg.

5 Dandelions and aphids reproduce both sexually and asexually under certain conditions.

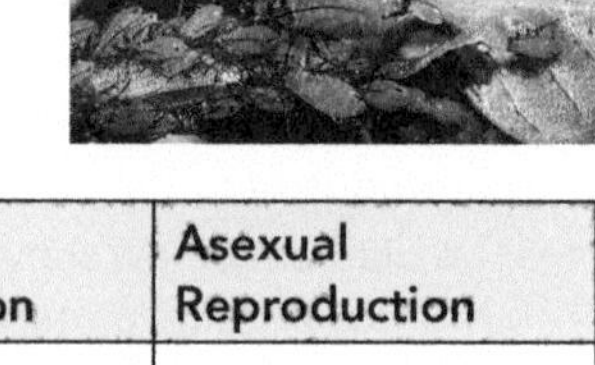

Copy the table below and use it to compare and contrast the advantages and disadvantages of sexual and asexual reproduction. Consider issues such as the ease of producing offspring as well as the value of genetically identical and genetically diverse offspring.

	Sexual Reproduction	Asexual Reproduction
Advantages		
Disadvantages		

6 Study the diagram below of a cell division process.

a How many cell divisions are shown?
b How many new cells are formed?
c The original cell had four chromosomes. How many do the new cells have?
d Is it a diagram of mitosis or meiosis?
e Justify your answer.
f Are the new cells genetically identical or different?
g Describe what happens at stages **i** to **v**.

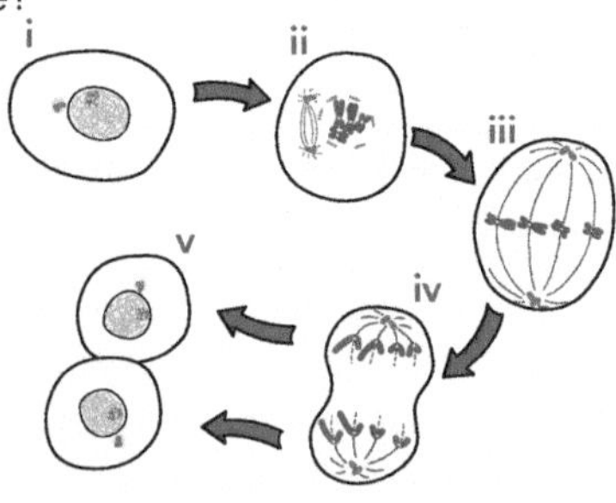

ISBN: 9780170262316

7 Study the diagram below of a cell division process.

a How many cell divisions are shown?

b How many new cells are formed?

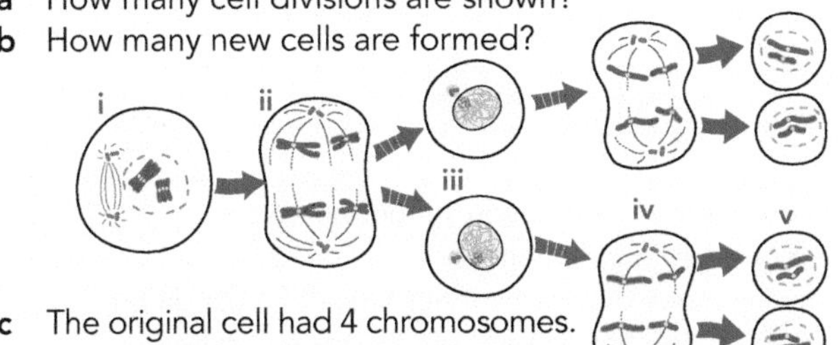

c The original cell had 4 chromosomes. How many do the new cells have?

d Is it a diagram of mitosis or meiosis? Why?

e Are the new cells genetically identical or different?

f Describe what happens at stages **i** to **v**.

8 Copy the table below, and use it compare and contrast the two types of cell division.

Type / Feature	Mitosis Cell Division	Meiosis Cell Division
Purpose of division	growth	
Location of division		testicle/ovary
Original cell type	body cell	
No. of divisions		two
No. of cells formed	two	
Type of new cells		gametes
Chromosome no.	unchanged	
Chromosome sets		different
Homologous pairs	remain together	
Allele pairs		separate
Subsequent event		fertilisation

9 Genetic variation in a family occurs when individuals inherit different sets of chromosomes.

Explain how each of the following contributes to variation:

a crossing-over

b meiosis

c fertilisation.

10 Read the article opposite, then answer the questions below.

a What facial features distinguish the boy from the two girls?

b What facial features do the two girls have in common?

c How many chromosomes will each baby have?

d Why does the boy have a different set of chromosomes?

e How did the identical twins arise?

f How many eggs were fertilised inside the mother?

g Why is the boy so different from the girls?

h Why will the twins have identical alleles?

i What might cause their phenotypes to differ later in life?

11 The paired chromosomes on the karyotype shown below have been arranged on the basis of decreasing size and the location of centromeres.

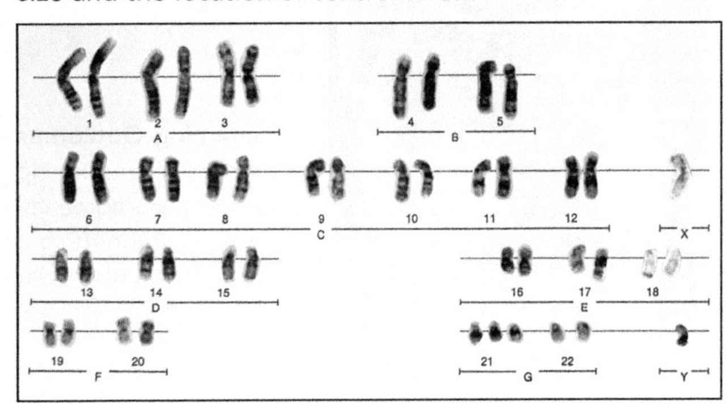

a How many chromosomes does the organism have?

b Can you assume that the organism is definitely human?

c What is odd about the pairings shown above?

d Assuming it is a human, what is the sex of the organism? How do you know?

e Given that the Y chromosome is much shorter that the X chromosome, what can be said about the difference between males and females in terms of having two alleles for every gene?

12 The diagram below shows the stages of reproduction of a kowhai tree. Compare the diagram with the stages of human reproduction on the opposite page.

a How does the plant's gender differ from humans?

b Which part makes female gametes?

c Which part makes male gametes?

d What are they called?

e How do they reach the eggs?

f Where does fertilisation occur?

g Place the red stages in order.

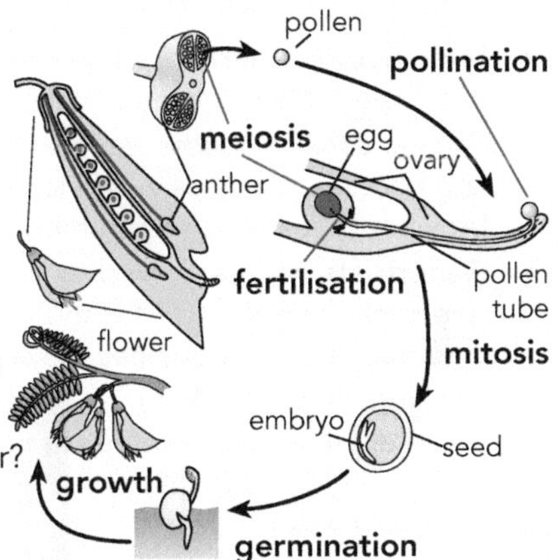

Identical and Non-identical Siblings

The triplets shown in the photo were all born to the same mother on the same day.

The blue-eyed baby is a boy and the two brown-eyed babies are girls. The girls are identical twins.

The boy started life when a single sperm fertilised a single egg. The resulting zygote underwent repeated cell divisions as it developed into an embryo, then a foetus and eventually into a male baby.

The two girls also started life when a single sperm fertilised a single egg to form a zygote. The zygote divided several times to form a ball of cells, which then split into two for some reason. The two smaller balls of cells developed into embryos, then foetuses and eventually into the two female babies.

The sisters have identical sets of chromosomes, as they originated from the same zygote.

The brother inherited quite a different set of chromosomes. In particular, he inherited a Y chromosome from his father and an X from his mother. His sisters inherited an X chromosome from both of their parents.

As the girls have identical chromosomes they have identical collections of alleles. This means that to the extent to which genes alone determine a particular trait, they will have identical phenotypes.

ISBN: 9780170262316

Patterns of Inheritance and Adaptive Features

Learning Outcomes - On completing this unit you should be able to:
- explain the principles of monohybrid inheritance
- use a pedigree chart or family tree to study the inheritance of a condition
- use a Punnett square to determine phenotype probabilities and ratios
- describe how adaptive features and new alleles arise
- *determine the genotypes of individuals expressing a dominant allele.*

Genes and Alleles

- Sexually produced organisms inherit two sets of chromosomes – one from the male parent and the other from the female parent. They therefore *inherit two copies of each gene.*
- The two copies are found at a specific location on homologous chromosomes (see page 105).
- Different alleles exist for most genes, which may result in different phenotypes (eg straight, wavy and curly hair are caused by different combinations of alleles of the trichohyalin gene on chromosome 1).

gene from father | gene from mother

Monohybrid Inheritance

- **Monohybrid inheritance** is about the inheritance and expression of a single gene considered in isolation.
- The ear canal produces wax that traps dust, dead cells and bacteria. It comes in two forms: sticky yellow-brown 'wet' earwax, and crumbly grey 'dry' earwax.
- **Geneticists** have recently discovered that a single gene, labelled ABCC11 for short and also found on chromosome 1, determines the state of the wax.
- The gene has two alleles. When expressed, one allele causes wet wax and the other, dry wax.
- As there are two alleles, there are three possible genotypes: two 'wet' alleles, two 'dry' alleles, and one of each.
- The phenotypes of the two homozygous genotypes are as expected, but all individuals with heterozygous genotypes have wet wax.
- An allele that is *expressed if just one copy is inherited* is said to be **dominant**. An allele that is *expressed only if two copies are inherited* is said to be **recessive**.
- A dominant allele is notated by a capital letter (eg W for the 'wet' allele) and a recessive allele by the same letter in lower case (eg w for the 'dry' allele).

Complete Dominance

- When *one allele completely masks the expression of another*, this known as **complete dominance**.
- Earwax is one of the very few confirmed examples of complete dominance in human features. Better known examples of complete dominance relate to features of pea plants, such as plant height, flower position and colour, pod and seed colour, etc.

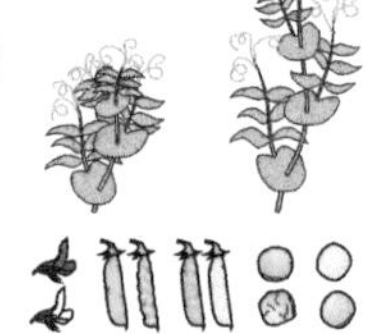

- In NCEA level 1 exams, any trait used as an example is assumed to be determined by a single gene, with just two alleles, one of which shows complete dominance.

Pedigree Charts & Family Trees

- A **pedigree chart** is used to track the inheritance of a condition when breeding plants or animals.
- A **family tree** is used to study the inheritance of a condition over several generations. It is used in genetic counselling to explain the risk of passing on a **genetic disease**, such as cystic fibrosis, which results in the lungs producing a very thick, sticky mucus.
- The tree below shows the inheritance of cystic fibrosis. The squares represent males; circles, females; line joining males and female, matings; and branches beneath, offspring. Affected individuals are coloured.
- The pattern of inheritance over generations shows whether the *condition is due to a dominant or a recessive allele.*

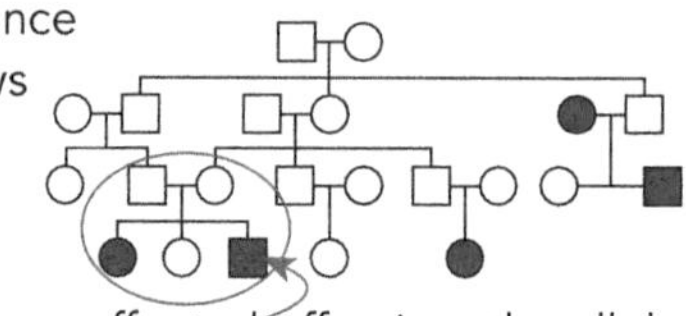

- If normal parents produce affected offspring, the allele involved is recessive (which is true for cystic fibrosis). The normal state must be due to a dominant allele.
- If affected parents produce normal offspring, the allele involved must be dominant.
- If the status of the alleles is known, then some genotypes can be determined using two simple rules:
  1. any individual expressing the recessive allele must have two recessive alleles
  2. any individual expressing the dominant allele must have at least one dominant allele.
- It may be possible to determine what the other allele will be by looking closely at what alleles the parents and/or offspring possess (see the science skill).

Punnett Squares

- When the status of alleles and parental genotypes are known, it is possible to work out the *chances of an offspring having a particular phenotype* and to *predict the ratio of phenotypes amongst offspring.*
- *Alleles separate when gametes form and combine randomly when zygotes form* (see page 105).

ISBN: 9780170262316

- A **Punnett square** gives the genotypes of all possible **zygotes**, and the resulting phenotype probabilities and ratios.
- In pea plants, the allele P for purple flowers is completely dominant over the allele p for white flowers.
- The genotypes of all possible zygotes for a cross (mating) between two heterozygous (Pp) purple-flowered plants are shown below. The colours in the squares indicate what the phenotype flower colour will be.
- As each square is equally likely, the probability of getting a purple-flowered plant (PP or Pp) is 3 out of 4 or ¾. The probability of getting a white-flowered plant (pp) is 1 out of 4 or ¼.

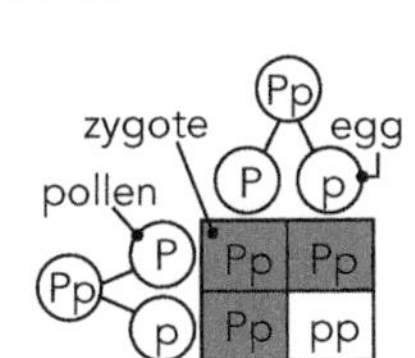

Expected Phenotype Ratios

- A Punnett square is also used to predict the **phenotype ratio** of offspring. As the expected genotype ratio for the above cross is 1 PP : 2 Pp : 1 pp, the expected phenotype ratio will be 3 purple-flowered plants to every 1 white-flowered plant. This means that 75% of the peas produced should grow into purple-flowered plants and 25% into white-flowered plants.
- Predicted phenotype ratios are only reliable if large numbers of offspring are involved because of *chance effects with small numbers.*

Test Crosses

- To check whether a plant or animal is **pure-breeding** (homozygous) for a dominant allele, it is mated with a homozygous recessive individual in a **test cross**.
- For example, in cats, the allele S for short hair is completely dominant over the allele s for long hair. A short-haired cat could have the genotype SS or Ss.
- If the short-haired cat is homozygous (SS), then all offspring of a test cross with a long-haired (ss) cat will be short-haired, as shown in the top Punnett square.
- If the short-haired cat is heterozygous (Ss), then you would expect that 50% of offspring would be short-haired and 50% long-haired, as in the bottom Punnett square.

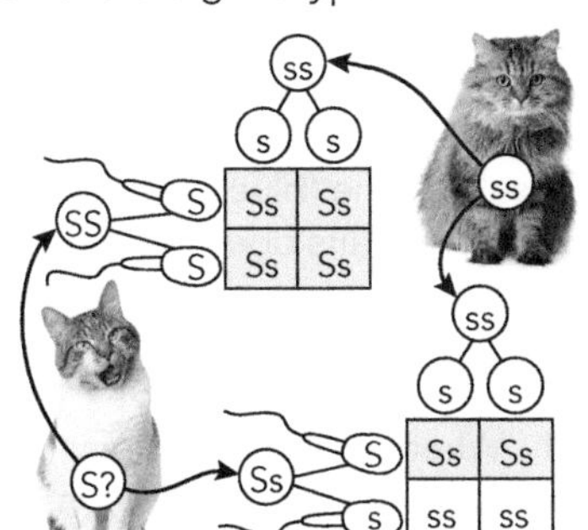

- A significant number of offspring are needed to avoid chance effects, but if any offspring have the recessive phenotype, then the parent is not pure-breeding.

Adaptive Features

- Members of a species have traits (eg the sow's multiple teats) that enable them to *survive and reproduce in their environment.* These traits are called **adaptive features** (or adaptations).
- The **environment** includes the physical surroundings and conditions, as well as other organisms such as food sources, competitors, predators and diseases.

An Example of an Adaptive Feature

- All mammals feed their young on milk, which is a rich source of lactose sugar. Lactose has to be digested first by the enzyme lact<u>ase</u> before the body can use it.
- The lactase gene that codes for the lactase enzyme usually gets 'switched off' late in infancy, after weaning.
- Some humans produce lactase beyond infancy. This phenotype, known as lactase persistence, is due to a dominant allele that keeps the gene 'switched on'.

The Origin of Adaptive Features

- The environment of a population (eg northern Europeans 5000 years ago) may change adversely (eg food shortage) or beneficially (eg domestication of cattle).
- **Genetic variation** *occurs in sexually reproducing populations* (see page 105). Some individuals may have an allele (eg lactase persistence allele) that results in a phenotype (eg lactase persistence) which enables them to cope with an adverse change or take advantage of a beneficial one (eg abundant cow's milk).
- *Environmental changes affect survival and reproduction.* Individuals with a favourable phenotype (eg lactase persistence) survive better (eg because of the dairy food source) and produce more offspring. Those offspring inherit the allele, which over many generations will increase in frequency in the population.
- The process by which *nature 'selects' favourable phenotypes* is called **natural selection**. It results in a population evolving to become adapted to a changed environment (eg most adults of northern European origin can now digest an abundant food source).

The Origin of New Alleles

- A **mutation** is a *random change in the base sequence of a gene* (see page 101). It may be due to radiation, chemicals or a DNA copying error (see page 101).
- Most mutations are harmful, as they result in proteins that do not work properly. A few mutations convey some benefit (eg lactase persistence).
- Mutations that occur in body cells cannot be inherited but those that occur in gametes can be. *Inherited mutations are the ultimate source of genetic variation* in a population, as they may give rise to useful alleles.
- When an environmental change (eg drought, disease) threatens to wipe out a population or even the whole species, genetic variation may mean that a few individuals have phenotypes (eg drought/disease resistance) that *ensure the population or species survives.*

SCIENCE SKILL: Determining Genotypes

To find the genotype of an individual expressing a dominant allele, check the parents and offspring.

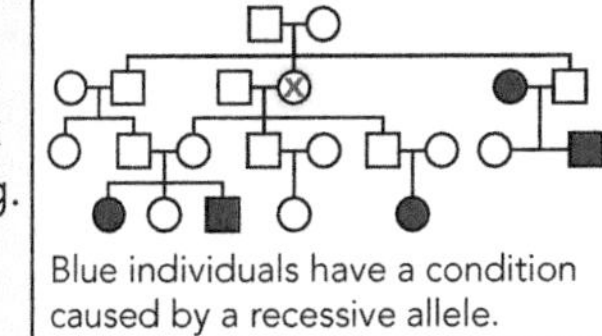
Blue individuals have a condition caused by a recessive allele.

Problem: What is X's genotype?

1. Specify the allele the individual must have: the dominant allele.
2. Check parents: either could have a recessive allele.
3. Check offspring: two of the three offspring have children with the condition, so X must have a recessive allele. She has a heterozygous genotype.

ISBN: 9780170262316

Revision Activities

1 Match up the descriptions with the terms.

a monohybrid inheritance	A environment 'selects' favourable phenotypes
b genetics	B chart showing phenotypes of several generations
c dominant allele	C variation that can be inherited through reproduction
d recessive allele	D inheritance involving a single gene only
e complete dominance	E disease caused by the inheritance of defective alleles
f pedigree chart	F cell that results from a sperm fertilising an egg
g family tree	G mating with a homozygous recessive individual
h genetic disease	H study of how genes are inherited and expressed
i Punnett square	I random change in the base sequence of a gene
j zygote	J feature that enables individuals to survive and reproduce
k phenotype ratio	K allele expressed when just one copy is present
l pure-breeding	L expected frequency of different phenotypes
m test cross	M chart showing phenotypes of a human family
n adaptive feature	N allele expressed only when two copies are present
o environment	O expression of one allele completely masks the expression of another allele
p genetic variation	P diagram used to predict phenotype probabilities and ratios
q natural selection	Q individual homozygous for a particular trait
r mutation	R physical and biological factors affecting a population

2 Explain the differences between the terms in each of the items below:

a genes and alleles
b genotype and phenotype
c a homozygous and a heterozygous genotype
d gametes and zygotes
e dominant and recessive alleles
f a pedigree chart and a family tree
g phenotype probabilities and ratios.

3 Cats have coats of many colours but a few cats are completely white. This is due to the presence of just one 'white masking' allele W, which prevents other colour genes from being expressed. The normal allele w allows coat colours to be expressed if two are present.

a What are the two phenotypes?
b How many alleles are there for the gene that controls the expression of coat colour?
c Which allele is dominant and how do you know?
d Which allele is recessive and how do you know?
e What are the three possible genotypes?
f Which genotype is homozygous dominant, homozygous recessive, and heterozygous?
g What is the phenotype of each of the three genotypes?
h What is the genotype of the tan-coloured cat?
i What is the genotype of the white cat?
j How could you confirm whether the white cat was pure breeding or not?
k What event would confirm that the white cat had a recessive allele for expressing coat colour?

4 Decide whether the following statements are true or false. Rewrite the false ones to make them correct.

a Sexually produced organisms have one copy of each gene.
b In a monohybrid cross, one gene is considered in isolation.
c A dominant allele will always be expressed.
d You need two copies of a recessive allele for it to be fully expressed.
e Carriers are usually affected by the condition in some way.
f A family tree is used to explain the risk of a genetic disease being passed on.
g Punnett squares are used to predict what proportion of offspring will have a phenotype.
h When heterozygotes mate, the expected phenotype ratio is 2 expressing the dominant allele to 1 expressing the recessive allele.
i A test cross involves a mating with a homozygous dominant individual.
j An adaptive feature arises when an allele results in a phenotype that is selected for by the environment.

5 White coats in cats are caused by a mutation to the gene that controls the expression of other coat colour genes. The 'white masking' allele is dominant and prevents pigment-producing cells from developing as the embryo grows. This results in an absence of colour, as no pigments are produced. Some cats with the mutant allele are deaf and lack pigmentation in their irises, which results in the 'blue' eyes shown above.

a What traits does the 'white masking' allele affect?
b What is unusual about the gene involved?
c Why would the white coat condition be considered to be a genetic disease?
d Where would the original mutation have occurred?
e Why might the condition persist despite the defects associated with it?

6 Pedigree charts like the one below are used by breeders to keep track of phenotypes and to identify genotypes. If the condition is 'white coat' (see questions 3 and 5), then the 'wild type' is coloured coat.

a What are the phenotypes of the first generation?
b What are their genotypes? How do you know?
c How many kittens were in their litter?
d What allele will all kittens in that litter have? Why?

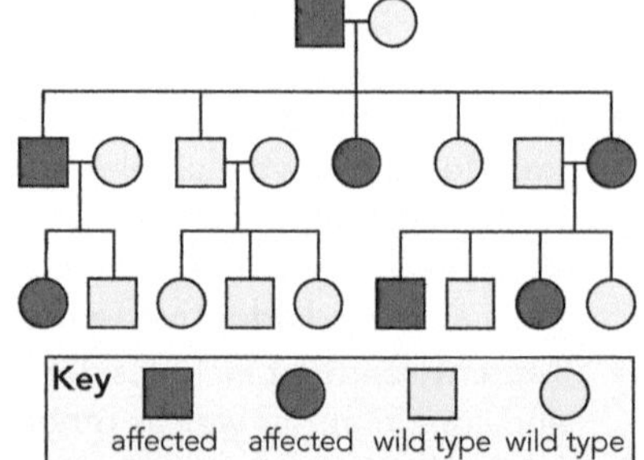

ISBN: 9780170262316

7 Cystic fibrosis is one of the commonest genetic diseases, occurring in about 1 in every 2000 people. The disease is caused by a mutated form of a gene called CFTR on chromosome 7. The mutated allele is recessive, and the normal allele shows complete dominance.

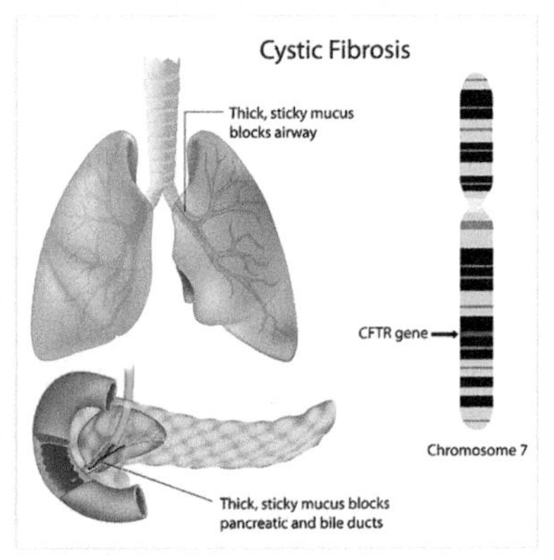

Usually, an affected person has two normal parents, each of whom is a 'carrier', having 1 copy of the mutated allele.

a What would be the genotype of a child who suffers from cystic fibrosis?
b What would be the genotypes of 'normal' parents who have a child with cystic fibrosis?
c Why are the parents called 'carriers'?
d Why is it unlikely that the parents would know they are carriers before they had a child with cystic fibrosis?
e What were the chances that the child inherited cystic fibrosis? Draw a Punnett square to support your answer.
f What is the probability that their next child has cystic fibrosis as well?
g What is the probability that their next child is a carrier?
h What are the chances their next child would be normal?
i What would you advise the parents to do?

8 All human babies can digest lactose sugar in milk but only some adults can digest the lactose in cow's milk.

a Why is milk production an adaptive feature?
b Why is lactase production an adaptive feature?
c Why does lactase production cease?
d How did lactase persistence originate?
e How did it become an adaptive trait?

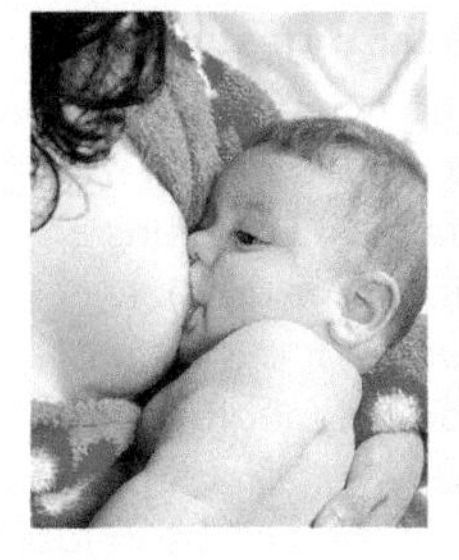

9 Read the article opposite, then answer the questions below.

a What is the function of haemoglobin molecules?
b Why is it critical that red blood cells are flexible?
c How does the mutated haemoglobin A gene result in sickling?
d How does having sickle cells result in anaemia?
e What are the chances that two carriers will have a child with sickle cell anaemia?
f If they already have one child with anaemia, what are the chances of having another?
g Why is the inheritance of the gene called incomplete dominance?
h Why is the allele common in Africa?

10 A genetically healthy population has a large amount of genetic variation. Endangered species, such as the native saddleback, often have very limited variation.

a What factors caused many NZ bird species to become endangered?
b How might being endangered cause limited variation?
c How might breeding programmes reduce variation?
d How could breeding programmes increase variation?
e How might a new virus put an endangered species at risk of extinction?
f Explain why genetic variation in a population is often crucial for the survival of a species.

11 **In the nineteenth century, an Austrian monk named Gregor Mendel discovered the basic laws of inheritance. He also found that the seven traits of pea plants shown below were all examples of complete dominance. Use a Punnett square to find the expected phenotype ratio of offspring from each of the following crosses.**

a a pure-breeding tall and a pure-breeding short plant
b two plants that are heterozygous for seed shape
c two plant that produced green seeds
d a heterozygous pink- and a white-flowered plant
e two plants that produce pinched pods
f two plants both of which produced green pods
g two heterozygous plants with axial flowers.

	Plant height	Flower position	Flower colour	Pod shape	Pod colour	Seed shape	Seed colour
Phenotype when dominant allele is expressed	Tall	Axial	Purple	Inflated	Green	Smooth	Yellow
Phenotype when recessive allele is expressed	Short	Terminal	White	Pinched	Yellow	Wrinkled	Green

Sickle Cell Anaemia

Anaemia occurs when insufficient oxygen is reaching body cells. Red blood cells transport oxygen from the lungs to all cells in the body. A complex molecule called haemoglobin holds the oxygen molecules.

Cells with normal haemoglobin molecules are doughnut-shaped and quite elastic, readily deforming in shape when they have to pass through narrow capillaries.

Haemoglobin consists of two kinds of protein (A and B), each coded for by a different gene. A mutated allele for the haemoglobin A gene results in haemoglobin that crystallises readily in low-oxygen conditions.

The crystallised molecules deform the blood cells giving them a rigid sickle (curved) shape, which means capillaries often get blocked causing problems for organs. The sickle cells don't last as long as normal cells (10-20 as compared to 90-120 days), which results in a shortage of red blood cells, ie in anaemia.

The mutated allele is recessive and therefore two alleles are required for the condition. The normal allele is not completely dominant as individuals with heterozygous genotypes produce both normal and abnormal haemoglobin but this does not result in sickle cells.

Heterozygous individuals are less affected by the malaria parasite, so the mutated allele is more frequent in Africa.

ISBN: 9780170262316

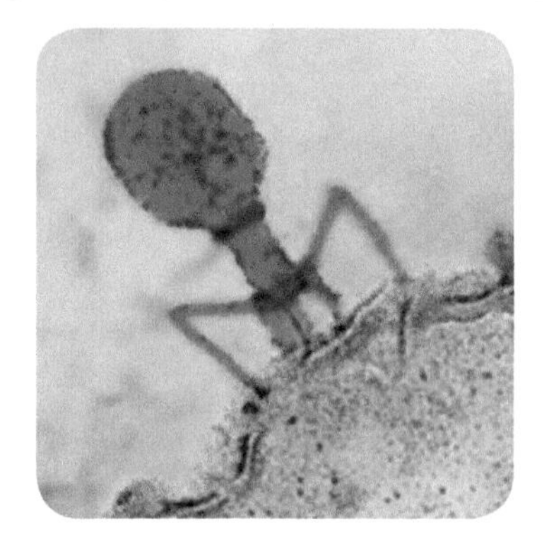

Bacteria, Fungi and Viruses

Learning Outcomes - On completing this unit you should be able to:
- explain why micro-organisms such as bacteria and fungi are considered to be alive
- describe the functions of the components of typical bacteria, fungi and viruses
- describe how bacteria and fungi carry out specific life processes
- explain why viruses are not alive and yet are able to reproduce
- *culture microbes and distinguish between bacterial and fungal colonies.*

Micro-organisms

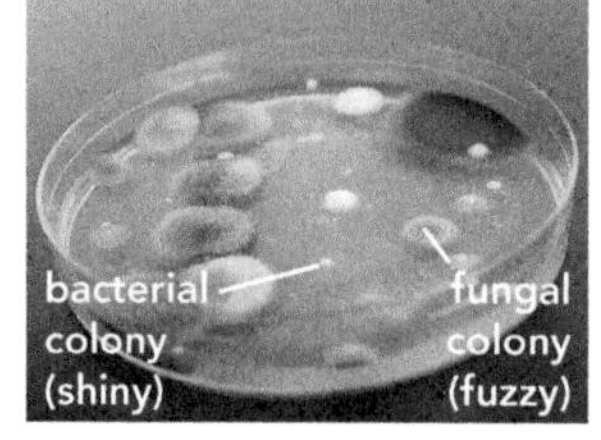

- **Micro-organisms** (or microbes) are microscopic organisms. Individual microbes are usually only visible under powerful microscopes. Colonies of microbes can sometimes be seen with the naked eye, as in the photo of a plastic Petri plate shown above.
- Microbes that exist as single cells are called unicellular organisms.
- Even though micro-organisms are not easily seen, the results of their activities are significant for humans. Harmful microbes spoil food, damage crops, and cause many diseases. Helpful microbes recycle nutrients, dispose of sewage, and produce foodstuffs.
- Like other organisms, micro-organisms must carry out life processes (see page 98) such as nutrition, respiration, excretion, growth and reproduction.
- Microbes that feed on other organisms are consumers. Those that live on or in larger organisms, feeding off them, are **parasites**. If they feed on dead organisms or foodstuffs, they are **saprophytes** (decomposers). If they make food using raw materials, they are producers.
- For NCEA level 1, you need to know about bacteria, fungi and viruses, although *viruses are not alive*.

Bacteria

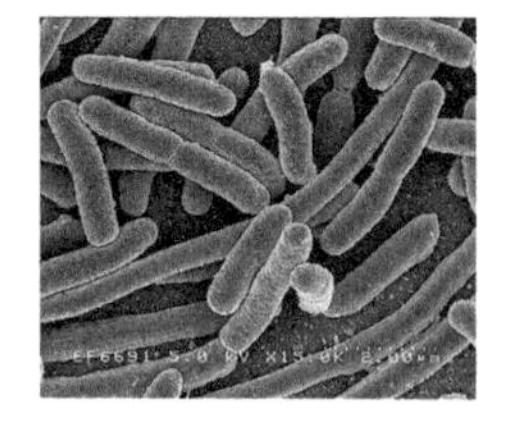

- **Bacteria** are very small organisms, less than $\frac{1}{10}$ the size of animal cells. Thousands of different kinds of bacteria have been identified. Most bacteria *grow and reproduce in moist, warm conditions.*
- Bacteria are named after their shape: rod-shaped bacteria are called bacilli; spherical-shaped, cocci; spiral-shaped, spirilli; and bent rod-shaped, vibrio.

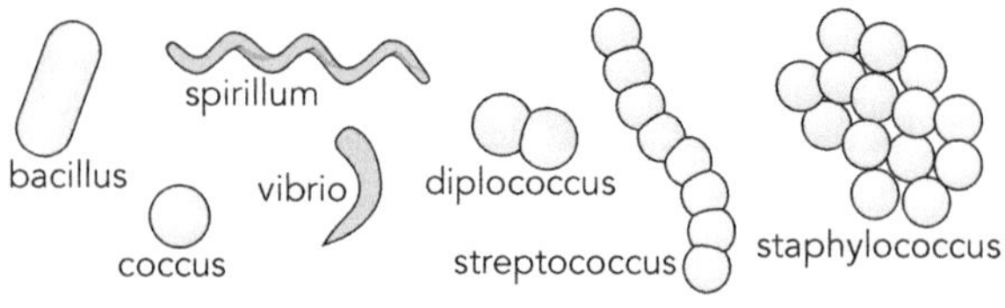

- Bacteria are also named by how they group together: paired bacteria are given the prefix 'diplo'; chained, 'strepto'; and clustered, 'staphylo'.

Structure

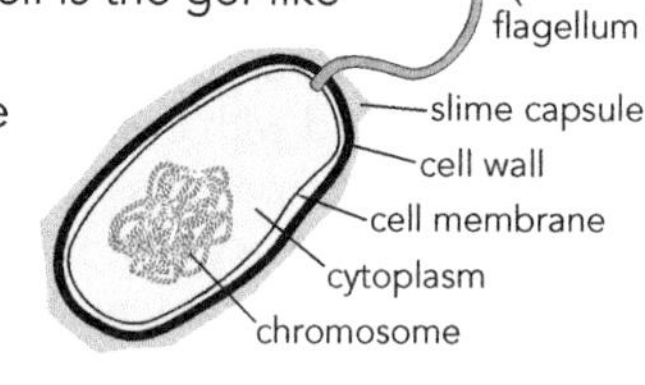

- The bulk of a bacterial cell is the gel-like **cytoplasm**, in which the chemical reactions of life (**metabolism**) occur. It is enclosed by the cell membrane, which controls the entry and exit of chemicals, such as nutrients, ions, water, and oxygen and carbon dioxide gases.
- Outside the membrane is a stiff supporting cell wall, which is enclosed in a protective slime capsule.
- Bacterial cells lack a **nucleus**. Their genes are mostly located on a single circular **chromosome** found in the cytoplasm. Bacteria have between 500 and 7500 genes.
- Some bacteria have flagella that spin like propellers, rapidly pushing the bacteria forward.

Life Processes

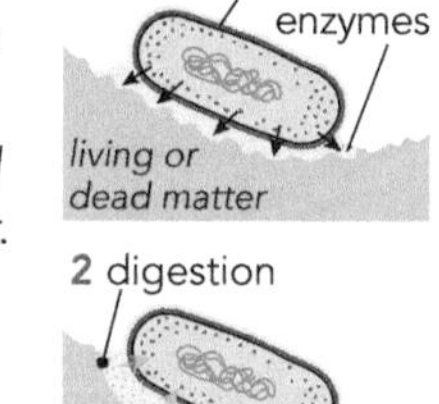

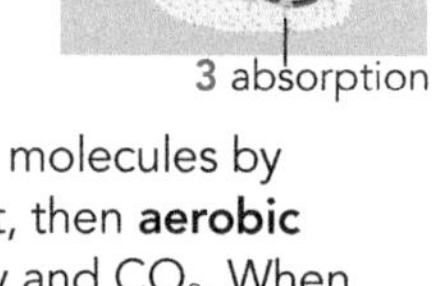

- Consumer bacteria obtain nutrients by **extra-cellular digestion**:
  1. Digestive enzymes are *secreted* onto living tissue or dead matter.
  2. Starches, fats and proteins are digested into small molecules.
  3. Small, soluble food molecules are *absorbed* through the cell membrane into the cytoplasm.
- Bacteria release the energy in food molecules by respiration. If oxygen gas is present, then **aerobic respiration** produces lots of energy and CO_2. When oxygen is absent, bacteria survive **anaerobically** but *get less energy from their food* and produce waste gases such as methane (see page 88).
- Waste products are **excreted** *by diffusing out of the cell.*
- As bacteria feed, they grow then reproduce asexually (see page 104) by **binary fission**:
  1. The chromosomal DNA is replicated (see page 101).
  2. The cell pinches in half.
  3. Two separate, genetically identical cells are formed.

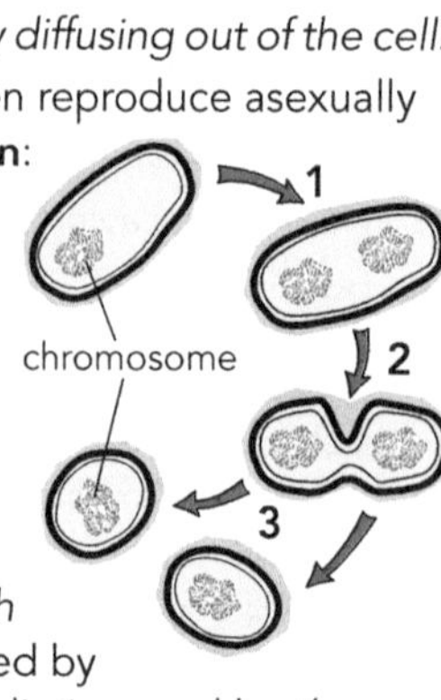

- Some bacteria form resistant **spores** in adverse conditions.
- Genetic *variation arises through mutations* (see page 109) caused by copying errors, chemicals or radiation, and by *the transfer of genes* between different strains.

ISBN: 9780170262316

Fungi

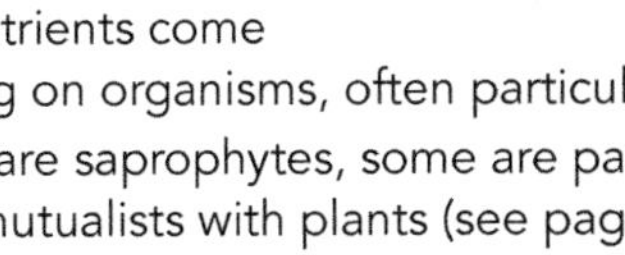

- **Fungi** include unicellular yeasts and multicellular moulds.
- *Fungi are immobile, but cannot make food as plants do*. Their nutrients come from feeding on organisms, often particular species.
- Many fungi are saprophytes, some are parasites, others are mutualists with plants (see page 121). The fungus provides the plant with mineral nutrients while the plant feeds the fungus with sugars.

Structure

- Fungi cells have cell walls and nuclei.
- Moulds consist of a network of **hyphae**, which are tiny tube-like filaments that *invade the host organism or foodstuff*.

Life Processes

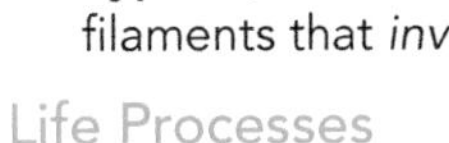

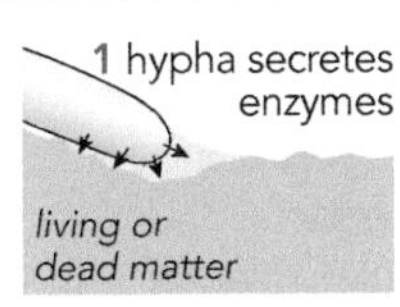

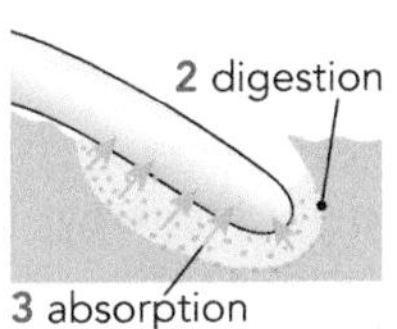

- Moulds feed like bacteria. Their hyphae secrete digestive enzymes onto food sources, then absorb the resulting small, soluble molecules.
- Fungi release energy from food molecules in aerobic or anaerobic conditions. Yeast cells live anaerobically, *fermenting alcohol* (see page 93) and carbon dioxide.
- Like bacteria, waste products are excreted by diffusion.
- Hyphae grow rapidly as the cells undergo mitosis cell divisions (see page 104), then elongate.
- Bread mould reproduces asexually by forming tiny, *genetically identical spores* in sporangia (spore cases).
- The steps are:
  1. Spore cases grow on upright hyphae.
  2. Spore cases split, releasing mature spores.
  3. Light spores disperse in air.
  4. Spores germinate if they land on suitable surfaces.

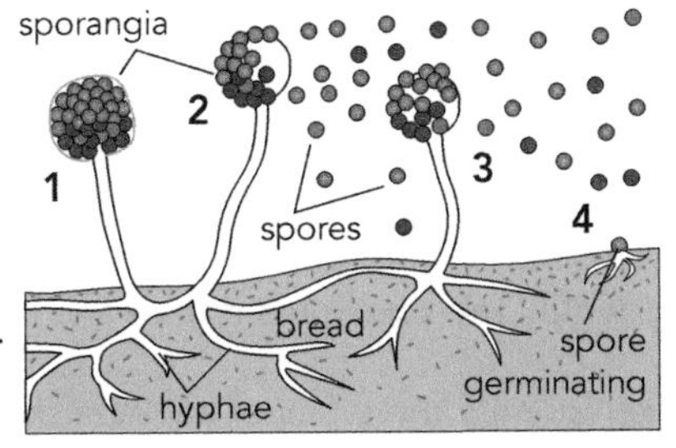

- When spores germinate on a specific dead or living moist food source (eg bread), they send out hyphae that rapidly start digesting the food.
- Mould also reproduce sexually, forming highly resistant spores that *remain dormant in unfavourable conditions*.
- Unicellular yeast cells reproduce asexually by budding off new cells.

Viruses

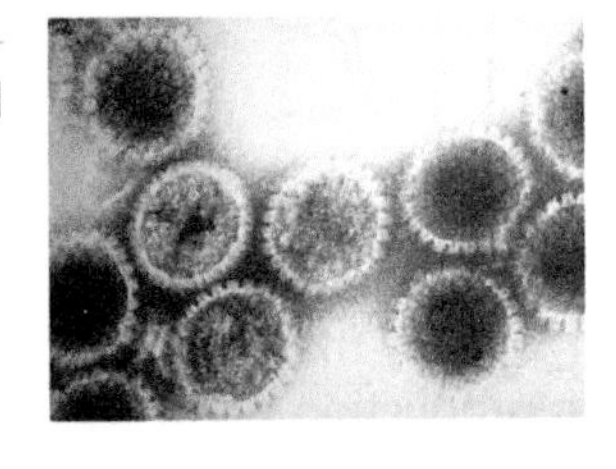

- **Viruses** are extremely small objects, *much smaller than bacteria*. They are not considered to be alive because they are not cells and cannot carry out any life processes.
- The image is of herpes viruses that cause cold sores.
- Viruses *depend entirely on the cells of host organisms*, such as humans, but are classed as pathogens (disease-causing organisms, (see page 116) rather than parasites, as they do not feed on the host cells.

Structure

- A virus consists of a DNA molecule, or RNA (ribonucleic acid) molecules, found inside a protective protein coat.
- The herpes virus has at least 75 genes along its DNA molecule.

Life Process

- As viruses are not cells, they do not move, feed, respire, excrete or grow, *but they can replicate*. To do that, each type of *virus takes over cells* of a specific type in a particular host species.
- Viruses reproduce by **viral replication**:
  1. Virus attaches to cell surface.
  2. Virus inserts its genes into cell.
  3. Viral genes take over cell, making thousands of identical viruses.
  4. Cell releases new viruses, which spread to other cells or organisms.

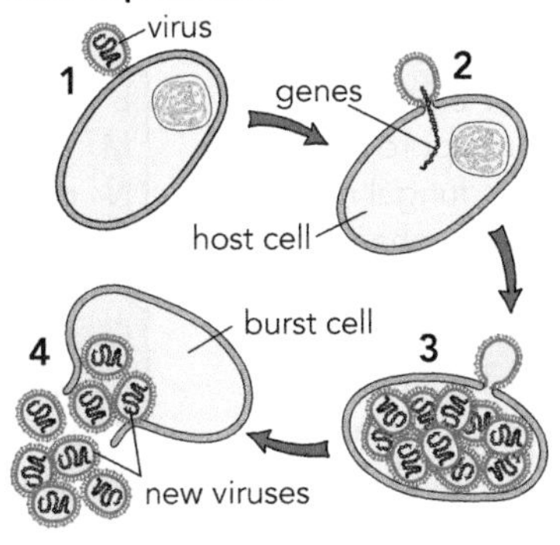

- In the lab, viruses can be *cultured in hen egg cells*.
- The genes of viruses *readily mutate*, which is why new strains of viruses appear regularly.

Environmental Factors

- Like all organisms, the life processes of all microbes are affected by **environmental factors**.
- Physical factors that affect microbes include temperature, moisture and pH (see page 85), nutrient and oxygen availability, as well as chemicals such as toxins, disinfectants and antibiotics (see page 117).
- Physical factors need to be within a *range for optimum growth* (eg bread mould grows best from 15 to 30 °C).
- Biological factors include resource competition with other microbes and the availability of host organisms.

SCIENCE SKILL: Culturing Microbes

Microbes are **cultured** (grown) by **inoculating** (infecting) then **incubating** (keeping warm) nutrient agar.

Culturing bacteria and fungi:

1. As instructed by your teacher, lightly wipe a cotton bud over surfaces to collect microbes.
2. Lift the lid off the plate slightly and inoculate the sterile agar by gently brushing the bud tip over the surface. Quickly replace the lid.
3. Turn the dish upside down to prevent water condensing on the agar, and seal it with tape.
4. Incubate at 25 °C for several days. *Do not reopen!*
5. Shiny spots are bacterial colonies; furry areas are fungal colonies. (Viruses only multiply inside cells.)

ISBN: 9780170262316

Revision Activities

1 Match up the descriptions with the terms.

Terms	Descriptions
a micro-organism	A internal cell structure containing chromosomes
b parasite	B occurs when a bacterium divides into two genetically identical cells
c saprophyte	C without oxygen
d bacteria	D unicellular micro-organisms that lack a nucleus
e cytoplasm	E immobile micro-organisms with nuclei, which are either parasites or saprophytes
f metabolism	F growing microbe colonies in the laboratory
g nucleus	G organisms usually only visible under a microscope
h chromosome	H fine feeding filaments of moulds
i extra-cellular digestion	I complex structure that has genes along its DNA
j aerobic respiration	J organism that feeds off another living organism
k anaerobic	K culturing microbes at an optimal temperature
l excretion	L interior of a cell, where metabolism occurs
m binary fission	M getting rid of wastes produced by metabolism
n spore	N non-living objects that use cells to reproduce
o fungal microbes	O transfer microbes or spores onto agar
p hyphae	P organism feeding on dead or organic matter
q virus	Q reproductive cells used for dispersal
r viral replication	R when digestion occurs outside of the cell
s culture	S making multiple genetically identical viruses
t inoculate	T respiration with oxygen producing much energy
u incubate	U all the chemical reactions that occur in a cell

2 Explain the differences between the terms in each of the items below.

a producers, consumers and decomposers
b parasites and saprophytes
c bacterial and fungal cells
d secretion and absorption
e aerobic and anaerobic conditions
f asexual and sexual reproduction
g binary fission and budding
h inoculation and incubation.

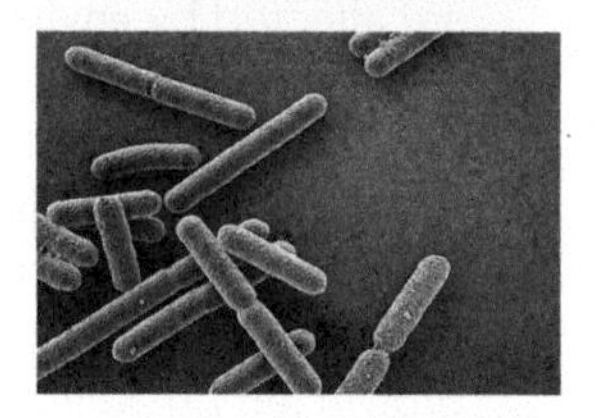

3 The photo shows some cultured micro-organisms. Identify objects **a** to **d**.

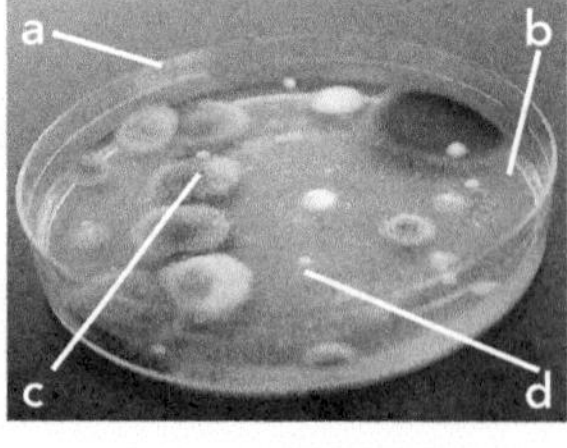

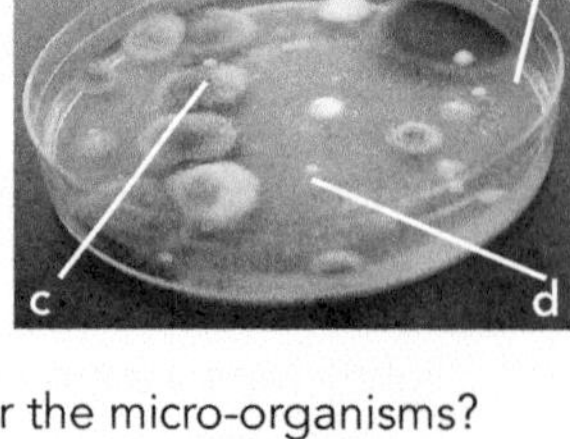

e When are micro-organisms visible to the naked eye?
f Why do bacterial and fungal colonies look different?
g What does the jelly-like agar in the plate provide for the micro-organisms?
h Why are there no virus colonies growing on the agar?
i Under what conditions would culturing have occurred?
j What should not have occurred with the plate above?

4 The photo shows a bacterial colony and a fungal mould called penicillin.

a Which colony is which?
b Why is one colony a zigzag shape?
c How has the penicillin affected the growth of the bacterial colony?

5 Decide whether the following statements are true or false. Rewrite the false ones to make them correct.

a It is possible to see colonies of micro-organisms.
b Living objects are made of cells and carry out life processes.
c Parasites feed on living organisms, saprophytes feed on dead ones.
d Bacteria are much larger than body cells but have no nuclei.
e Bacteria digest food internally.
f Binary fission occurs rapidly and results in identical offspring.
g Spores allow bacteria and fungi to survive until conditions improve.
h Anaerobic metabolism is more energy efficient.
i Fungal microbes exists as single cells or as networks of fine threads.
j Bread mould spores are dispersed by water.
k Viruses are non-living as they are not made of cells.
l Viruses never destroy host cells.
m Viruses can only reproduce in cells.
n Microbes rapidly multiply when nutrients are readily available.

6 Identify the types of biological objects illustrated in boxes **a**, **b** and **c**.

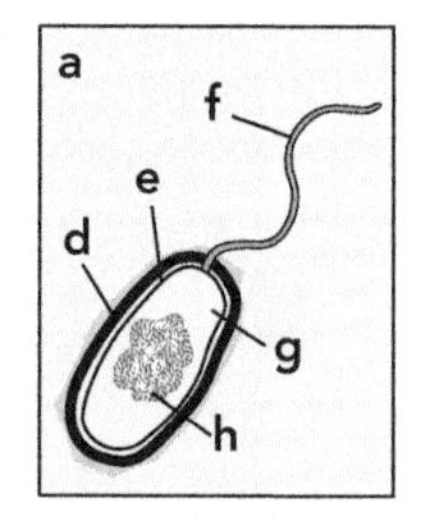

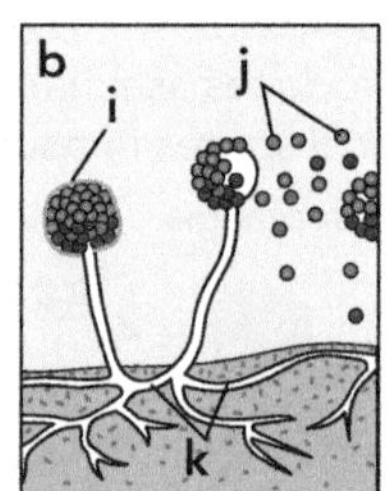

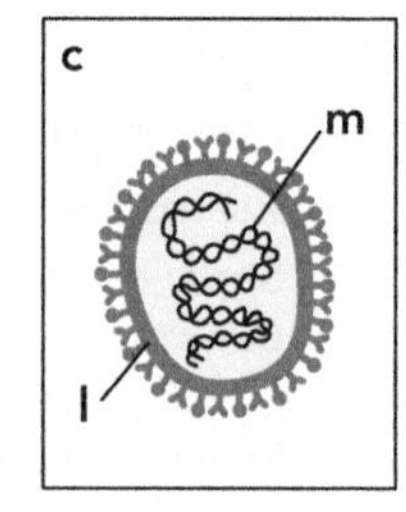

Identify the structures **d** to **m** and state their functions.

7 Identify the types of bacteria shown in the sketches below. Use singular and plural terms, as well as prefixes where needed.

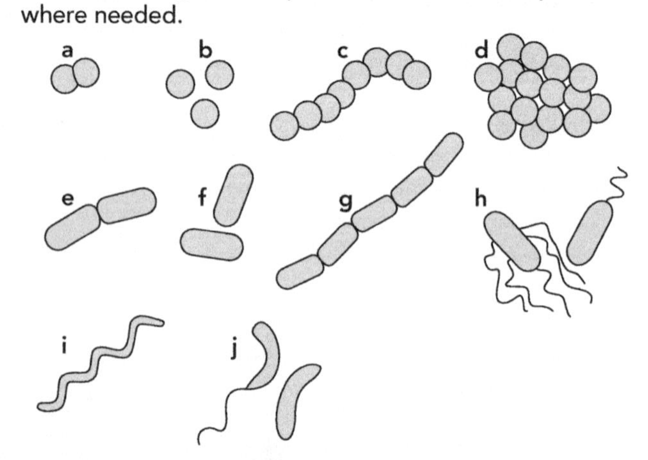

8 Study the diagram of bacterial reproduction shown opposite.

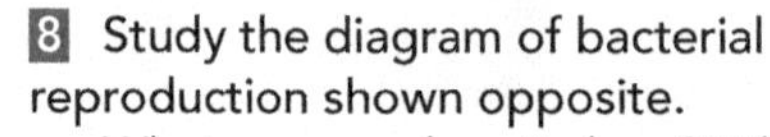

a What process does it show? Why?
b What will the process result in?

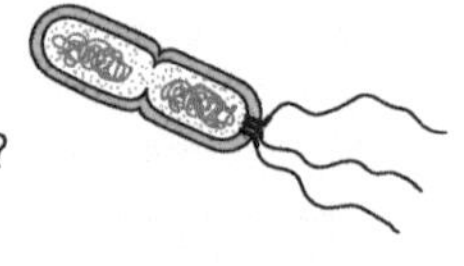

ISBN: 9780170262316

9 Copy and complete the table and use it to compare and contrast the features of bacteria, fungi and viruses.

Feature	Bacteria	Fungi	Viruses
Living			
Made of cells			
Cell nucleus			–
Life processes			one only
Feeding roles	wide variety		–
Method of digestion			–
Excretion			–
Anaerobic metabolism	produces methane		–
Asexual reproduction method			
Dormant spores			–
Source of genetic variation			

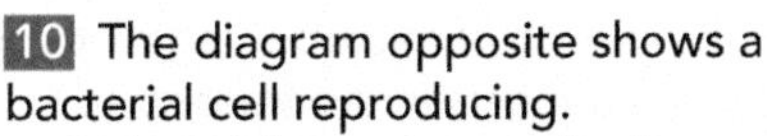

10 The diagram opposite shows a bacterial cell reproducing.

a What is the process called?
b What type of reproduction is it?
c What happens at each stage?
d How is the bacteria's single chromosome duplicated?
e If the process takes 20 minutes, how many bacteria would there be after 6 hours?
f What is the overall significance of the process?

11 A Petri plate was opened then incubated.

a Explain how many fungal spores landed on it.
b Explain how many bacterial spores landed on it.

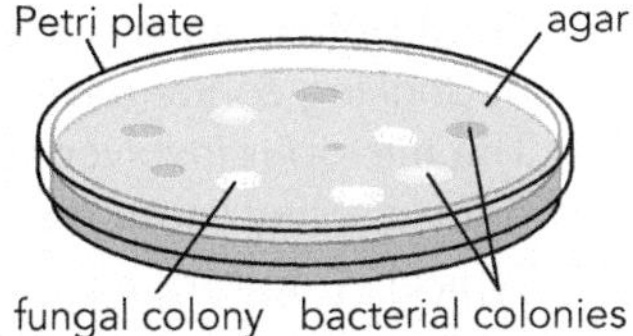
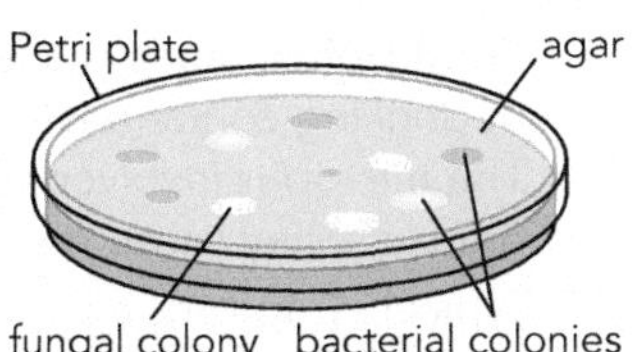

12 Read the article opposite, then answer the questions below.

a Which types of microbes are used to make sourdough bread?
b What is meant by the term 'optimum growth'?
c Why is the sourdough left in a warm place for several hours?
d What is the name given to anaerobic metabolism by yeast?
e State the temperature range over which yeast and lactic acid bacteria grow and the optimal values.
f Which microbe has a wider temperature tolerance?
g Why would the bacteria be able to tolerate a low pH?
h Why might being able to survive in salt solutions be important?

13 The left side of the diagram shows how bacteria digest food sources and the right side, how fungi do it.

a What is the process called?
b Why are enzymes secreted onto the food source?
c What happens to the food source?
d How do digested food molecules enter the microbes?
e Are there any significant differences between how bacteria and fungi digest their food sources?

14 The diagram shows a fungal mould reproducing.

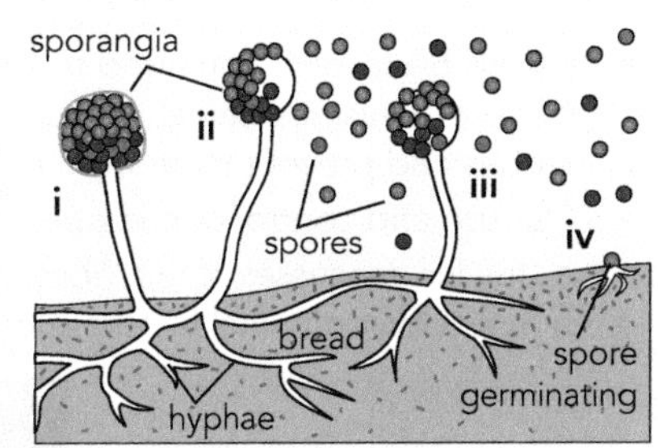

a What is the process called?
b What type of reproduction is it?
c What happens at each stage?
d What will each spore have in its nucleus?
e How does the process help ensure the mould spreads to new food sources?
f What is the significance of the process for the species?
g Why do sexually produced spores have resistant coats?

15 The diagram shows a virus using a cell to reproduce.

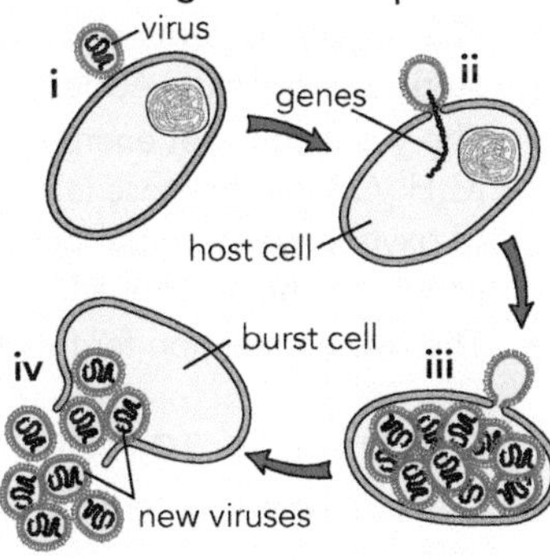

a What is the process called?
b What type of reproduction is it?
c What happens at each stage?
d What will each new virus have inside it?
e What is the significance of the process for the virus?

Effects of Environmental Factors

Micro-organisms play an important role in bread-making. Sourdough bread requires the activity of yeast (a unicellular fungus) and lactic acid bacteria.

The growth and reproduction of micro-organisms are affected by environmental conditions, such as temperature, moisture and pH, as well as oxygen and nutrient levels.

Micro-organisms grow within a certain range for each environmental factor, and optimum growth occurs at a specific value.

The yeast and lactic acid bacteria are mixed in with flour, milk and other ingredients to form the sourdough, which is then left in a warm place to rise.

In the oxygen-free conditions inside the dough, bacteria and yeast carry out anaerobic metabolism, resulting in the production of lactic acid by the bacteria, and ethanol and CO_2 gas by the yeast. The gas causes the sourdough to rise. (The alcohol evaporates during baking.)

The graph shows the relative growth rates for the two microbes over a range of temperatures.

Bacterial growth occurred in a pH range of 3.9 to 6.7, with optimal growth at about 5.5. Yeast growth was unchanged across that pH range.

The lactic acid bacteria tolerated up to a 4% salt solution, while the yeast tolerated up to an 8% salt solution.

ISBN: 9780170262316

Interactions between Humans and Micro-organisms

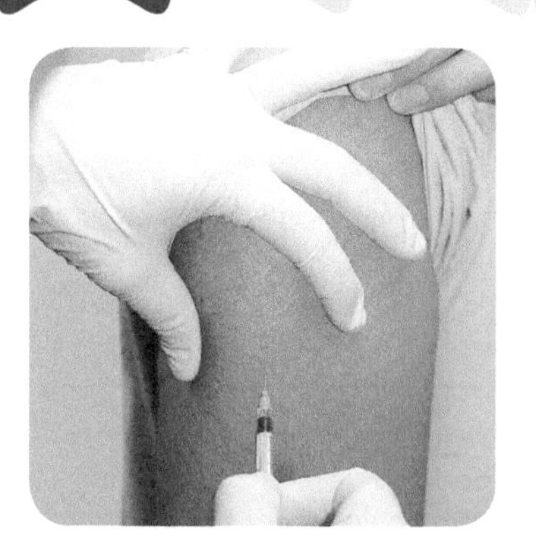

Learning Outcomes - On completing this unit you should be able to:
- outline how microbes are used to make wine and yoghurt
- describe how methods of food preservation manipulate environmental factors
- explain how the human body defends itself against pathogens
- summarise the role microbes play in recycling dead matter and organic wastes
- *identify the conditions required for fair testing of antiseptics.*

Using and Controlling Microbes

- For NCEA level 1, you need to know how humans and micro-organisms interact, in particular how we use microbes and prevent them from harming us.
- Our use and control of microbes is based upon:
  a the life processes of microbes (see page 112)
  b how environmental factors affect microbes.

Producing Food

- Microbes are used to produce many foodstuffs, for example: fungi are used to make bread, beer and wine; bacteria are used to make cheese and yoghurt.

Wine-making

- The sugary contents of crushed grapes are placed in sterile, airless vats. Added yeast cells rapidly multiply.
- The yeast cells get energy from two sugars – glucose ($C_6H_{12}O_6$) and fructose (also $C_6H_{12}O_6$). In the absence of oxygen, yeast metabolism results in two waste products, the alcohol ethanol and carbon dioxide gas.
- The overall reaction for this **fermentation** process is:
  Glucose (or fructose) → ethanol + carbon dioxide
  $C_6H_{12}O_6(aq) \rightarrow 2C_2H_5OH(l) + 2CO_2(g)$
- Different wines are fermented at different temperatures.

Yoghurt-making

- The milk is **pasteurised** (heated at 72 °C for 15 s) *to kill all pathogenic bacteria.*
- The milk is inoculated with *Lactobacillus* bacteria as it cools, which rapidly multiply as the culture is incubated at about 45 °C for 6 hours.
- In the absence of oxygen, the bacteria get energy from the lactose sugar in milk by fermentation, a process which produces lactic acid in this case.
- The lactic acid changes the proteins in milk so that it thickens into yoghurt and gains that sour taste. The lactic acid also lowers the pH, helping to preserve the yoghurt by preventing other microbes from growing.

Food Preservation and Safety

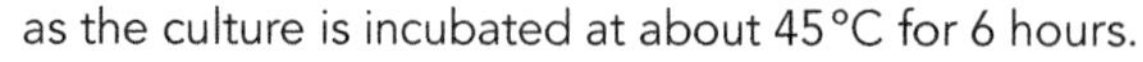

- **Saprophytic microbes** cause foodstuffs to decay; pathogenic microbes cause food poisoning.
- Foodstuffs are preserved and kept safe by methods that either *kill microbes or limit growth rates*, such as:
  a sterilising by heat, radiation or chemicals, to kill microbes (eg pasteurisation, SO_2 in wine, bottling fruit)
  b sealing, to remove the oxygen that aerobic microbes need (eg vacuum packing seafood, canning fruit)
  c dehydrating, to remove the moisture that microbes need to grow (eg drying milk into a powdered form)
  d refrigerating, to slow the growth of microbes (eg chilling fresh meat and cooked food in the fridge)
  e freezing, to stop the growth of microbes (eg placing vegetables and instant dinners in the freezer)
  f acidifying, to make the pH less than the level that microbes can tolerate (eg pickling in vinegar)
  g salting, to increase the amount of dissolved salt beyond the level microbes tolerate (eg salting fish).
- Often **sterilisation** is combined with another method.
- Some of the above methods inhibit microbe growth, but dormant microbes or airborne spores will quickly grow if the food is exposed to normal conditions.

Food Poisoning

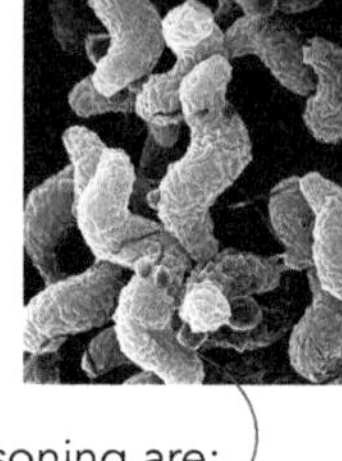

- **Food poisoning** is caused by eating or drinking contaminated foodstuffs. The microbes involved are **pathogens**, as they cause disease.
- Typical symptoms include: nausea, cramps, diarrhoea, vomiting, fever.
- Some pathogens that cause food poisoning are:
  - the bacteria *Salmonella*, *Listeria* and *Campylobacter*
  - the viruses Norovirus, Rotavirus and Hepatitis A.
- The microbes either attack human cells or produce chemicals called **toxins** that poison or damage cells.

Controlling Pathogens

- Microbes that harm living plants, animals or humans by causing disease are all classed as pathogens. All viruses are pathogens, as are many bacteria and fungi.
- Some human diseases caused by different types of pathogenic 'microbes' are listed below.

Bacterial Diseases	Fungal Diseases	Viral Diseases
tetanus, polio, tuberculosis, meningitis, typhoid, cholera	athlete's foot, thrush, tinea, 'jock itch', ring worm	colds, influenza, mumps, measles, chickenpox, HIV, smallpox, herpes

ISBN: 9780170262316

- Different types of pathogenic diseases are shown below.

Type	Bacterial	Fungal	Viral
Disease	gastroenteritis	athlete's foot	herpes
Microbe name	*Campylobacter species*	*Trichophyton* species	*Herpes simplex*
Microbe structure			
Infection	infected food	shower floors	kissing
Symptoms	nausea, vomiting, diarrhoea		

- **Infection** is the *invasion of body tissues by pathogens.* It occurs through: contact (cold sores), stepping on old nails in the soil (tetanus); drinking infected water (cholera); breathing air droplets (influenza); eating infected food (gastroenteritis); and body fluid exchange (HIV).
- **Epidemics** are pathogenic diseases (eg whooping cough) that spread rapidly in a population (eg NZ in 2011).
- The transfer of pathogens can be prevented or at least minimised by washing hands thoroughly, practising kitchen hygiene, cooking food well, isolating infected individuals, and coughing into sleeves.
- **Disinfectants** (eg antibacterial wipes) are powerful chemicals applied to surfaces to kill microbes.
- **Antiseptics** (eg surgical alcohol) are milder chemicals applied to cuts to kill bacteria but not human cells.

The Body's Defences against Pathogens

- The body has several lines of defence:
  1. Dead layers of skin cells act as a physical barrier keeping pathogens away from living cells beneath.
  2. The enzyme lysozyme in saliva, nose mucus, tears and urine kills some microbes. Stomach acid kills most of the bacteria living on or in food.
  3. If pathogens reach body tissues, then white blood cells known as **phagocytes** engulf and digest the microbes.

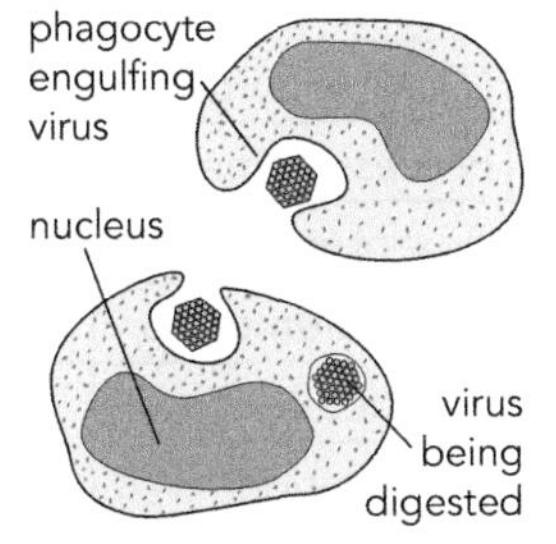

  4. On the outer surface of each pathogen species are unique marker chemicals called **antigens**. White blood cells called **lymphocytes** absorb escaped antigens and use them to create **antibodies** that are released into the blood. These custom-made proteins recognise microbes with a specific antigen and either tag them for destruction or disable them.

Natural and Artificial Immunity

- When first infected by a particular pathogen, the microbes rapidly multiply causing the symptoms. As antibody numbers build up, the pathogens are killed.
- The body then has **natural immunity** to further infection from that pathogen, as there are large numbers of the specific antibody in the blood.
- **Artificial immunity** to a disease is gained through **vaccination**. Weakened or dead pathogens are injected into the blood, which *stimulates lymphocytes to make antibodies* in preparation for a real infection.
- Vaccination can provide immunity to bacterial and viral diseases, but some viruses (eg influenza) *mutate frequently* thus overcoming previous immunity.

Antibiotics and Resistance

- **Antibiotics** are medicines, often derived from natural products, used to fight bacterial diseases. They kill bacteria or inhibit bacterial growth but don't affect viruses.
- A mutation (see page 109) may give a bacterium **resistance** to a particular antibiotic. That microbe multiplies, *passing on its resistant genes*. As the bacteria affected by the antibiotic are killed off, the population gradually changes into a resistant strain. This is an example of natural selection (see page 109).

Nutrient Recycling

- Many bacteria and fungi are decomposers, *breaking down dead organisms as well as organic wastes*. These microbes release the nutrients in carbohydrates, fats and proteins in a form that plants can then absorb.
- Microbes play a vital role in **nutrient cycles** by recycling chemical elements back into the environment.
- The carbon atoms in organic matter are recycled by aerobic decomposer respiration back into the air as part of carbon dioxide molecules for plants to absorb.

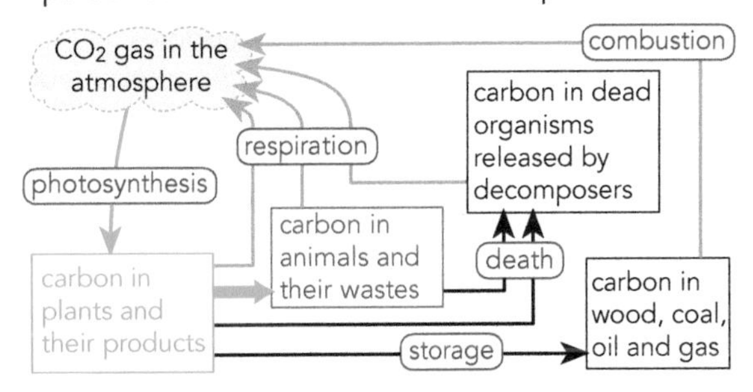

- The nitrogen atoms in organic matter are recycled by decomposer nutrition (see page 113) back into the soil in the form of nitrate ions (NO_3^-). The nitrate ions are absorbed by the roots of plants to make proteins.

Composting and Sewage Treatment

- Gardeners use aerobic bacteria and fungi to digest plants and food wastes to form nutrient-rich compost.
- Aerobic bacteria in sewage treatment plants break down human wastes so the water can be discharged safely.

SCIENCE SKILL: Fair Testing

Fair testing requires identical conditions.

Compare the effectiveness of three antiseptics:

1. Obtain a sterile Petri plate with nutrient agar.
2. Make up a standard solution of each antiseptic.
3. Dip a standard-sized, labelled filter paper disc in each solution, and one disc in water as a control.
4. Wipe a cotton bud on a source of bacteria, then dip it into sterile water. Pour the water onto the agar to just cover the surface.
5. Place the disks as shown.
6. Seal the plate and incubate for three days.
7. Observe the size of clear areas around each disc. Decide which antiseptic is most effective.

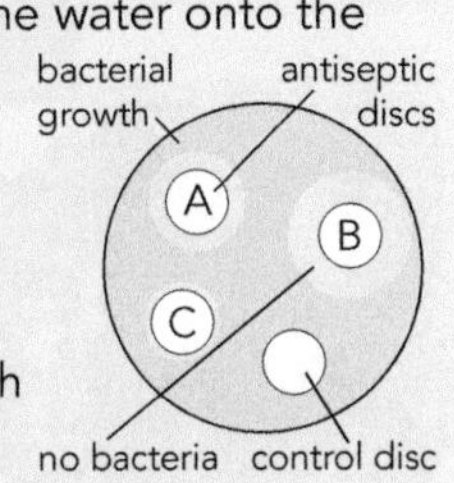

ISBN: 9780170262316

Revision Activities

1 Match up the descriptions with the terms.

a fermentation	A blood cells that engulf then digest microbes
b pasteurising	B killing microbes using heat, radiation or chemicals
c saprophytes	C powerful chemicals that kill microbes on surfaces
d sterilisation	D immunity created by a previous infection
e food poisoning	E anaerobic metabolism by bacteria or fungi
f pathogens	F bacterial strain no longer affected by antibiotic
g toxins	G injecting with dead or weakened microbes
h infection	H special proteins that recognise specific microbes
i epidemic	I killing microbes in foodstuffs by briefly heating
j disinfectants	J immunity created by vaccination
k antiseptics	K invasion of tissues by micro-organisms
l phagocytes	L infection of the digestive system caused by eating contaminated food
m antigens	M medicines designed to kill or inhibit bacteria
n lymphocytes	N recycling of chemical elements in nature
o antibodies	O specific marker chemicals on pathogens
p natural immunity	P bacteria or fungi that break down organic matter
q artificial immunity	Q disease-causing micro-organisms
r vaccination	R chemical applied to cuts to kill bacteria cells
s antibiotics	S blood cells that absorb antigens and use them to make antibodies
t antibiotic resistance	T poisonous chemicals made by pathogens
u nutrient cycles	U outbreak of a pathogenic disease in a population

2 Explain the differences between the terms in each of the items below.

- **a** aerobic and anaerobic conditions
- **b** pasteurisation and fermentation
- **c** pathogens and saprophytes
- **d** sterilising and inoculating
- **e** disinfectants and antiseptics
- **f** antigens and antibodies
- **g** natural and artificial immunity.

3 Identify preservation method(s) being used and the environmental factor that kills or inhibits microbes.

4 Decide whether the following statements are true or false. Rewrite the false ones to make them correct.

- **a** Microbes are both helpful and harmful to humans.
- **b** Wine-making relies on fungi and yoghurt-making on bacteria.
- **c** Fermentation always produces the alcohol ethanol.
- **d** Pasteurisation involves boiling milk.
- **e** Preserving food involves making environmental conditions difficult for microbes.
- **f** Either pathogens attack body cells or their toxins poison them.
- **g** Disinfectants can be safely applied to wounds.
- **h** Phagocytes hunt down microbes.
- **i** Antibodies recognise specific pathogens and help destroy them.
- **j** Vaccination gives natural immunity.
- **k** Antibiotics will destroy viruses.
- **l** Antibiotic resistance arises because bacteria mutate.
- **m** Decomposers release chemical elements into the environment.
- **n** Nitrogen atoms are recycled as nitrate ions released in the soil.

5 Copy the table and use it to compare and contrast wine-making with yoghurt-making.

Feature	Wine-making	Yoghurt-making
Raw materials		
Sugars involved		
Sterilisation used		
Type of microbes		
Microbe name		
Process name		
Type of metabolism		
Incubation temp.		
Products		
Preservation	SO_2 added	

6 *Campylobacter* bacteria are the most common cause of food poisoning from under-cooked chicken. Their rate of reproduction is affected by temperature (see graph).

- **a** What's the best temperature for the bacteria?
- **b** How does that relate to body temperature?
- **c** Above and below what temperatures does reproduction cease?
- **d** Would the bacteria be killed at 20 °C? At 50 °C?
- **e** What would chilling and freezing chicken do to *Campylobacter* bacteria?
- **f** What would cooking chicken do to the bacteria?
- **g** Why is it important to thoroughly cook chicken?

ISBN: 9780170262316

7 Mould forms on wallpaper, bags, shoes and clothing.

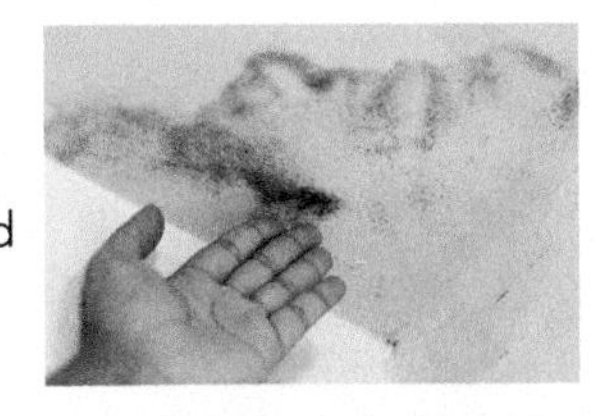

a Why is this a problem?
b What conditions are needed for moulds to form?
c How could mould be prevented from growing?

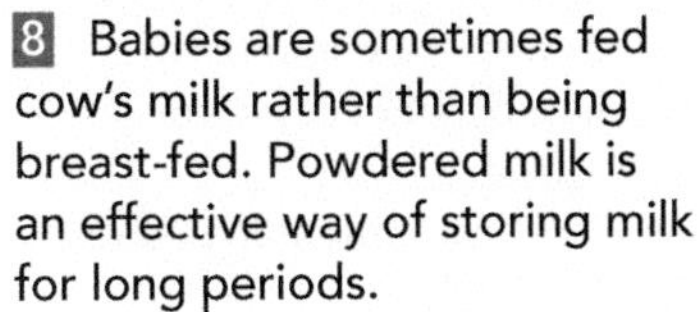

8 Babies are sometimes fed cow's milk rather than being breast-fed. Powdered milk is an effective way of storing milk for long periods.

a Why does powdered milk have a long shelf-life?
b Why don't microbes grow on the milk powder after the container has been opened then resealed?
c Should an open container be refrigerated? Why?
d How would the situation differ using fresh milk?

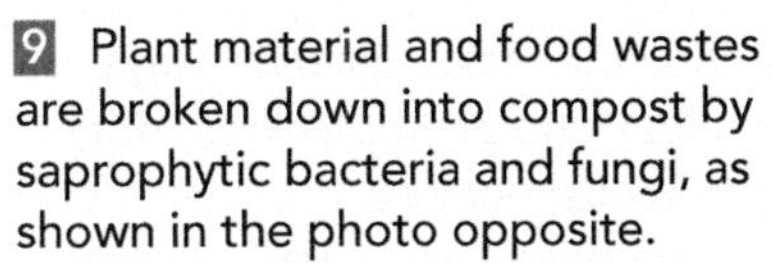

9 Plant material and food wastes are broken down into compost by saprophytic bacteria and fungi, as shown in the photo opposite.
a Why are they called saprophytes?
b What conditions inside the compost bin would result in the rapid breakdown of organic matter?
c What would happen if air was not circulating in the bin?
d How would the action of saprophytes aid plants?

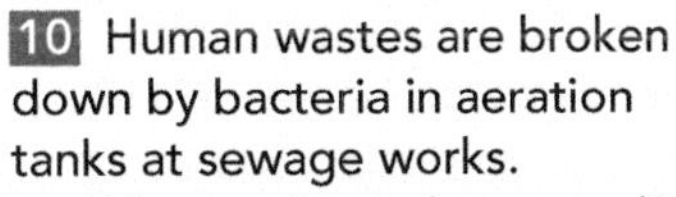

10 Human wastes are broken down by bacteria in aeration tanks at sewage works.
a Why are the tanks aerated?
b What other conditions do the bacteria need to function?
c Which three life processes are significant for the bacteria's role as decomposers?
d What are the products of the aerobic bacterial action?
e What happens to the treated water?
f What dissolved chemical could cause problems?

11 Read the article opposite, then answer the questions below.
a What do the acronyms AIDS and HIV stand for?
b What pathogen type causes AIDS?
c HIV has crossed a 'species barrier'. What does this mean?
d How do AIDS infections occur?
e How can the spread of the disease be reduced?
f What does it mean when a person is found to be HIV positive?
g Why is there usually a long delay between the initial infection and the development of AIDs?
h What are 'opportunistic infections'?
i Why is it difficult to develop an effective vaccine?
j Why is AIDs a 'pandemic'?

12 The diagram shows the recycling of nitrogen atoms in an ecosystem.
a Where are most N atoms found?
b How do most plants get N atoms?
c How do decomposers increase the amount of nitrate ions in the soil?
d Why do organisms need nitrogen atoms?

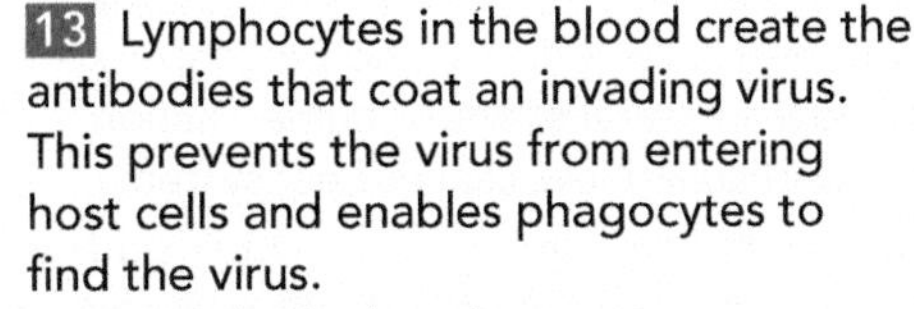

13 Lymphocytes in the blood create the antibodies that coat an invading virus. This prevents the virus from entering host cells and enables phagocytes to find the virus.

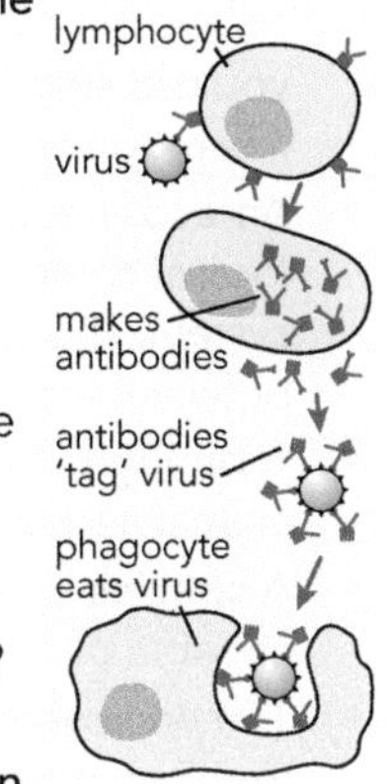

a How do the lymphocytes create antibodies?
b What do antibodies recognise on the viruses?
c How does the antibody coating ensure that the virus doesn't replicate?
d How do phagocytes destroy viruses?

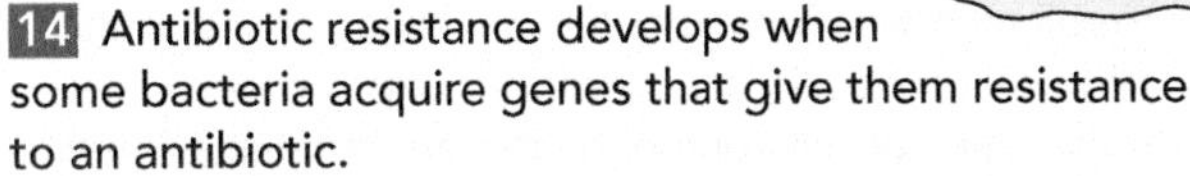

14 Antibiotic resistance develops when some bacteria acquire genes that give them resistance to an antibiotic.
a How might a single bacterium have acquired resistance?
b How would all the bacteria become resistant over time?
c What do scientists do to overcome resistance?

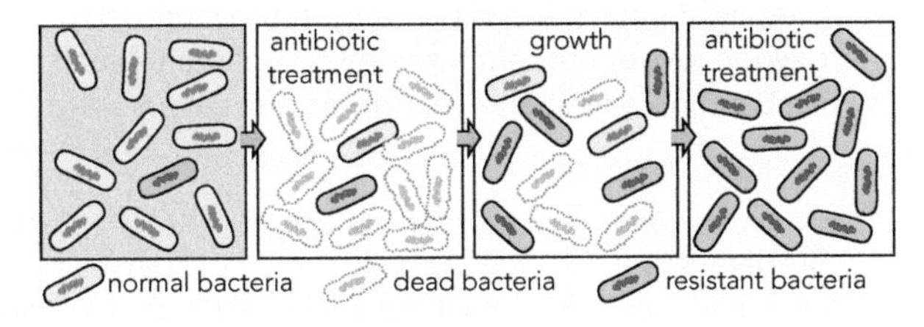

HIV and the AIDS Pandemic

The disease AIDS (Acquired Immune Deficiency Syndrome) is caused by the HIV virus (Human Immunodeficiency Virus).

The disease was first observed in 1981 in USA but the virus originally came from a monkey species in Africa.

The virus is transferred from person to person in body fluids (semen, blood, breast milk). Infection can occur through having sex, blood transfusions, shared syringes, birth and breast feeding. It is not caught by touching or coughing.

HIV (green) on a lymphocyte

Once the virus enters the body, it invades lymphocytes that fight diseases.

The viruses can remain dormant in the lymphocytes for up to 10 years. When the viruses become active, they take over the lymphocyte cells and make them produce many more copies of the virus, which escape to invade other lymphocytes.

The body's immune system is badly affected and the person becomes vulnerable to many opportunistic infections. These include diseases such as thrush, diarrhoea, tuberculosis, pneumonia and some cancers. When these symptoms occur, the person then has full-blown AIDS. There may be times of recovery followed by relapse. Eventually health deteriorates and an infection proves fatal.

Currently, there is no cure for AIDS. Medical drugs have turned it into a chronic rather than a fatal disease. Vaccines are being investigated, but the virus mutates rapidly.

The AIDS pandemic has killed 30 million and 34 million more are infected worldwide.

ISBN: 9780170262316

Ecological Features and Processes

Learning Outcomes - On completing this unit you should be able to:
- explain how species in a community interact with each other and the environment
- distinguish different feeding and other ecological roles within a community
- compare and contrast energy flow and nutrient recycling by an ecosystem
- distinguish between species density, distribution and diversity
- *use quadrats along a transect line to study the effect of an environmental gradient.*

Ecosystems

- For NCEA level 1 Science you need to know about ecological features (characteristics) and processes so you can study how a natural event or human action has modified a particular New Zealand ecosystem.
- An **ecosystem** (eg mudflat, stream, grassland) is a *biological community and its environment interacting as a system*. The interactions occur within the community and between the community and the environment.
- A **biological community** consists of all of the plant, animal and micro-organism species in a given location.
- A **species** is a group of similar organisms that are capable of interbreeding to produce fertile offspring.
- The **environment** of a community consists of the medium in which it exists (eg air, water, soil) and the *physical and chemical conditions* experienced. The main types of environments are terrestrial, marine and freshwater, each with very different conditions.
- Depending on the type of environment, environmental factors that affect a community include: temperature, light intensity and pH ranges; moisture and oxygen levels; nutrient availability; pollutant levels; wind action and shelter; water movement and clarity.

Species in Ecosystems

- **Biodiversity** is the *number and variety of species* present in an ecosystem. Natural ecosystems tend to have high biodiversity unless affected by human actions, such as allowing effluent to enter a stream or introducing a species that undergoes a population explosion.
- Managed ecosystems, such as forest plantations and farms, usually have low biodiversity.
- An ecosystem may have a **key species**, which forms the physical environment in which other species live (eg mangrove trees in a mangrove swamp), or has a major influence over the lives of many other species present (eg introduced trout negatively affect native fish, crayfish and insect populations).
- **Indicator species** can *indicate healthy or degraded environments.* For example, the presence of stoneflies indicates that a stream has a high level of dissolved oxygen, which is a healthy environment.

Feeding Relationships

- Organisms get the energy and nutrients needed for life and growth by the process of **nutrition** (see page 132).

Modes of Nutrition

- Species get energy and nutrients in different ways:
  - producers (plants, algae and phytoplankton) obtain them directly from the environment
  - consumers (grazers, predators and parasites) obtain them by feeding on living organisms
  - decomposers (saprophytes and detritivores) obtain them by feeding on organic matter.
- **Phytoplankton** are floating producer micro-organisms.
- Parasites are organisms that live on or in larger organisms, feeding off them without killing them.
- Saprophytes are bacteria and fungi that decompose organic matter (dead organisms and wastes) by carrying out extra-cellular digestion (see page 112).
- **Detritivores** (eg worms) feed on similar food sources (called detritus) as decomposers, but they eat lumps of food and digest it internally.

Food Chains and Webs

- **Food chains** show who is eaten by whom, in this order: Producer → herbivore → primary carnivore → secondary carnivore.
- Food chains start with a producer and should end with a saprophyte. Each arrow means 'is eaten by'.
- Food chains connect to form a **food web**, eg the pond food web shown.

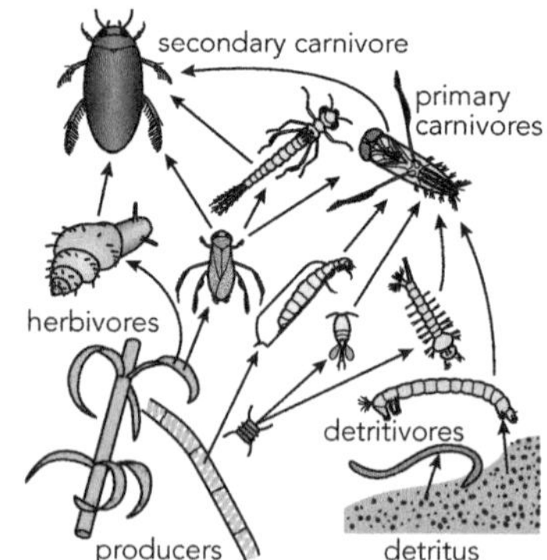

Trophic Levels

- **Trophic levels** (see page 121) are the steps along food chains. Producers are on the first trophic level; herbivores, on the second level; primary carnivores, on the third level; and so on. Animals are placed on the highest trophic level from which they feed.
- Decomposers are put on a separate trophic level because *their food comes from all the other levels.*
- Some food chains start with detritus, which is often placed on the first level alongside producers.

ISBN: 9780170262316

Other Ecological Relationships

- A relationship in which *both species benefits* is called **mutualism**, eg nitrogen-fixing bacteria living in the root nodules of legume plants gain food from the plant and provide the plant with nitrate ions.
- A relationship in which *one species benefits without harming the other* is called **commensalism**, eg plants growing in the forks of kauri trees receive more light.
- **Competition** occurs when two species need the same resource, eg different sea anemone species in an aquarium ecosystem compete for attachment sites. This *negatively affects both* and may result in one species being eliminated from the ecosystem.

Energy Flow

- **Energy flows** through an ecosystem *as species get eaten by other species.*
- Energy enters most biological communities through producers that absorb light energy from the sun and use it to manufacture energy-rich sugars in the process called photosynthesis (see page 129). The sugar molecules have stored **chemical energy**.
- As producers get eaten by herbivores, which get eaten by carnivores, etc, chemical energy is passed along food chains through the trophic levels.
- At each trophic level much of the energy input is lost through respiration, wastes and death. Only a fraction gets passed on to the next trophic level, as shown by the size of the boxes symbolising trophic levels below.

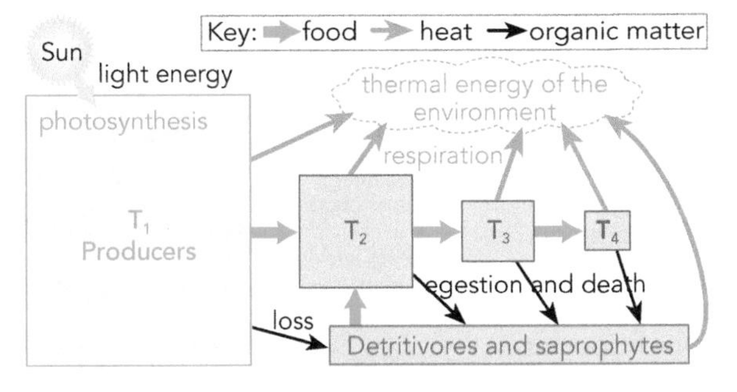

- The chemical energy in food is transformed into heat as producers, consumers and decomposers carry out the process of respiration (see page 137). This heat warms the environment. As plants use light rather than heat, *energy is not recycled.*

Nutrient Cycles

- Organisms need *atoms of different elements* (eg C, O, H, N, S, P, K, Mg, Ca) to live, grow and reproduce.
- Consumers and decomposers get them by digesting other organisms; producers, by absorbing **nutrients** from the environment (eg CO_2 gas and NO_3^- ions).
- Ecosystems recycle elements through **nutrient cycles**.
- Nutrients absorbed by producers from the environment are used to make the **biomolecules** (proteins, carbohydrates, lipids) that pass along food chains.
- As they digest organic matter, decomposers play a vital role in *recycling nutrients to the environment for producers to use*, particularly mineral nutrients (ions) in the soil.
- Nutrient shortage can reduce productivity. Nutrient overload can reduce biodiversity (see page 123).

- The diagram shows how mineral nutrients are recycled.

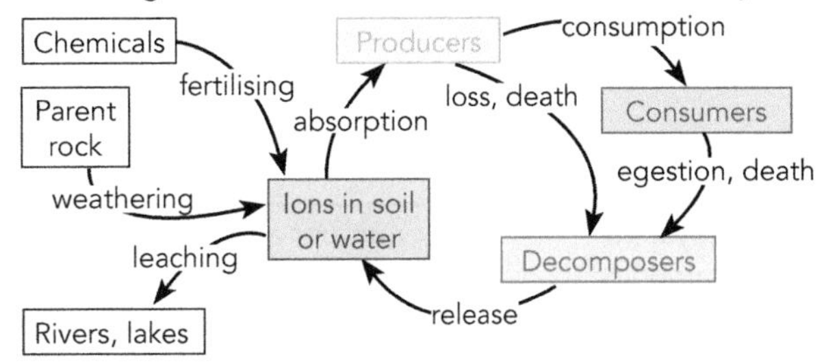

- Cycles you need to know about are the carbon (see page 117), nitrogen (see page 119) and water cycles.

Water Cycle

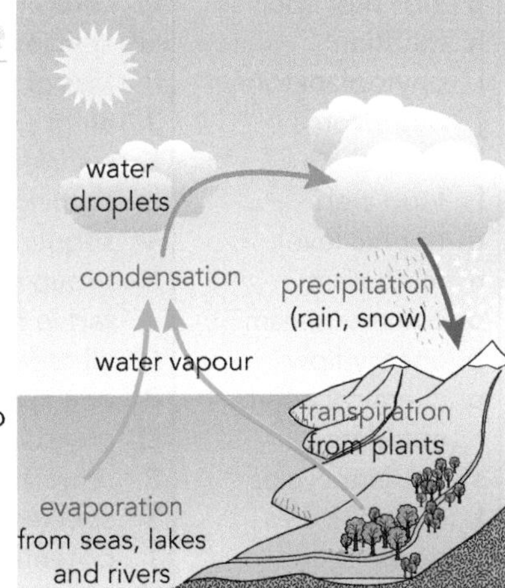

- Water is a vital substance, often forming the main constituent of organisms. It provides support for cells, and enables the reactions of life to occur in solution.
- The water cycle involves physical processes, such as evaporation, condensation and precipitation, and biological processes, such as absorption and **transpiration** (see page 129) by plants.

Distribution, Density & Diversity

- The distribution of species in an ecosystem may occur in response to an **environmental gradient**, such as light intensity at different depths in a pond or tidal exposure on a mudflat. In a forest, distribution can also be determined by the growth forms of plant species.
- The density or abundance of a species is measured by the number of animals per unit area or percentage ground cover for small plants. Sampling frames are used to study how the density of immobile species changes along a gradient, such as tidal exposure.
- The diversity and abundance of invertebrate species in two freshwater ecosystems can be compared by using a net to collect samples in a standardised way.

Impact of Events

- Events such as floods, landslides, fires and urban development can dramatically alter the physical surroundings of a biological community.
- Other events, such as drought, nutrient overload and pollution, *gradually alter environmental factors,* resulting in the composition of the community changing over time as different species tolerate different environmental conditions.

- Some events directly change the composition of a community, such as eliminating pests (eg possums in wildlife sanctuaries).

ISBN: 9780170262316

Revision Activities

1 Match up the descriptions with the terms.

a ecosystem	A number and variety of species in an ecosystem
b community	B organisms that digest organic matter internally
c species	C both species benefit from the relationship
d environment	D chemicals required for organisms to live
e biodiversity	E community and environment acting as a system
f key species	F linked-up food chains in a community
g indicator species	G carbohydrates, proteins and lipids
h nutrition	H species that indicate environmental quality
i phytoplankton	I loss of water by evaporation of water from leaves
j detritivores	J all of the species in a given location
k food chain	K transfer of energy through trophic levels
l food web	L floating micro-organisms that are producers
m trophic level	M sequence of species, each one eaten by the next
n mutualism	N group of similar organisms capable of producing fertile offspring
o commensalism	O all organisms at a food chain step of a community
p energy flow	P one species benefits but the other is unaffected
q chemical energy	Q physical surroundings and conditions
r nutrients	R recycling of chemical elements in an ecosystem
s nutrient cycles	S species with a major influence over other species
t biomolecules	T potential energy stored in the bonds of molecules
u transpiration	U obtaining and processing nutrients and energy
v environmental gradient	V gradual change in a factor across a location

4 Decide whether the following statements are true or false. Rewrite the false ones to make them correct.

a An ecosystem consists of a community and its environment.
b Different environmental factors are important in different ecosystems.
c Biodiversity is higher in managed ecosystems than natural ones.
d Indictor species provide useful information on environmental quality.
e Saprophytes and detritivores eat similar foods in the same way.
f The arrows in a food web show how nutrients and energy move.
g Resource competition may result in one species displacing another.
h Plants use thermal energy to make food.
i Nutrients exit and re-enter communities, but energy only exits.
j Nutrients mostly enter a community through producers.
k Animals play a major role in the water cycle.
l Environmental gradients affect where species live in ecosystems.
m Exotic species alter communities.

2 Explain the differences between the terms in each of the items below:

a key and indicator species
b predators and parasites
c saprophytes and detritivores
d food chains and food webs
e mutualism and commensalism
f energy flow and nutrient cycling
g distribution, density and diversity.

3 Identify each ecological relationship or feeding role.

5 Natural ecosystems have many species. The diagram shows a simplified food web of a sheltered estuary.

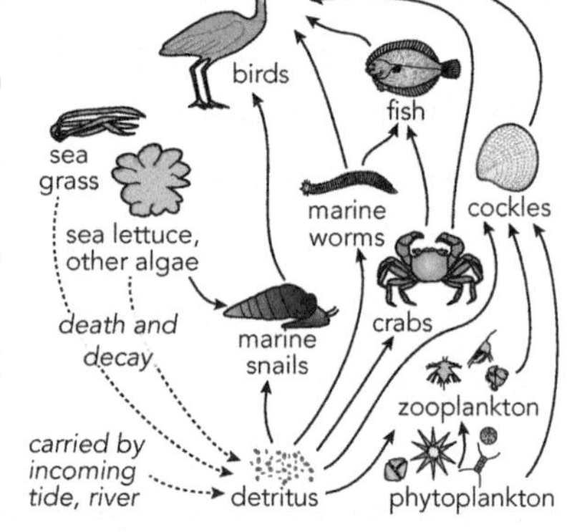

a Which species are the producers?
b Which species are the herbivores?
c Which species are primary carnivores?
d Which species are secondary carnivores?
e What is detritus and why is it placed separately?
f What are phytoplankton and zooplankton?
g What natural events could impact on the food web?
h What human actions could affect the food web?

6 A group of students netted samples of aquatic invertebrates found in two streams.

a Which stream type has the greatest number of species?
b Which stream type has the greatest diversity of species?
c Which stream is likely to have poorer quality water? Why?
d What natural events or human actions might affect each type of stream?

	No. of species	
Stream type	**Rural**	**Urban**
beetles	4	2
caddis flies	5	3
crustaceans	2	0
damsel flies	2	0
flies	0	1
mayflies	3	0
midges	0	2
snails	1	1
stoneflies	3	0
worms	2	4

ISBN: 9780170262316

7 Environmental factors can be more or less important in ecosystems.

a Rate the importance of factors as: 3 = high 2 = moderate 1 = little 0 = no.

b What natural events might affect each ecosystem significantly?

c What about human actions?

	Ecosystem Type		
Factor	**Field**	**Mudflat**	**Stream**
dissolved O_2			
humidity			
light intensity			
moisture			
nutrients			
pH			
pollutants			
salinity			
soil layer			
temperature			
tide exposure			
water clarity			
water flow			
wave action			
wind action			

8 The diagram shows energy flow and nutrient cycling in an ecosystem. The orange arrows show the transfer of energy and the black arrows, the transfer of nutrients.

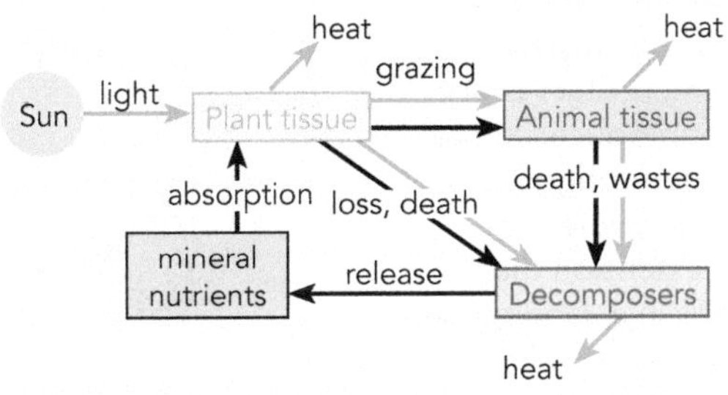

a Describe how energy flows through an ecosystem.

b Describe how nutrients are recycled by an ecosystem.

9 Describe how each event would impact on nutrient cycles in the ecosystem:

a fertilising a field

b the silting up of a lake

c heavy rainfall on a hill farm

d clearing bush

e planting legumes

f effluent entering a stream.

10 Read the article opposite, then answer the questions below.

a What is an aquatic ecosystem?

b What does an algal bloom look like?

c Why is phosphorus a limiting factor in aquatic ecosystems?

d What causes an algal bloom?

e What two negative effects does the decomposition of the bloom have on water quality?

f Why might fish die after an algal bloom?

g How can overloading an ecosystem lead to reduced biodiversity?

h How is it that algal blooms sometimes result in human deaths?

i What are the possible sources of excess nutrients?

11 Describe how each event would impact on particular species in a stream and affect the ecosystem as a whole:

a a lengthy drought

b silting from urbanisation

c increased farm effluent

d the arrival of didymo algae.

12 Mud flats are sheltered shallow beaches with areas of exposed mud as the tide goes out. To find out how the distribution, density and diversity of mudflat species change, a quadrat can be placed on the mud every 5 m along a transect line from high- to low-tide marks. The types of species and their numbers inside the quadrat are recorded each time. The data is then displayed on a kite graph, where kite width indicates abundance.

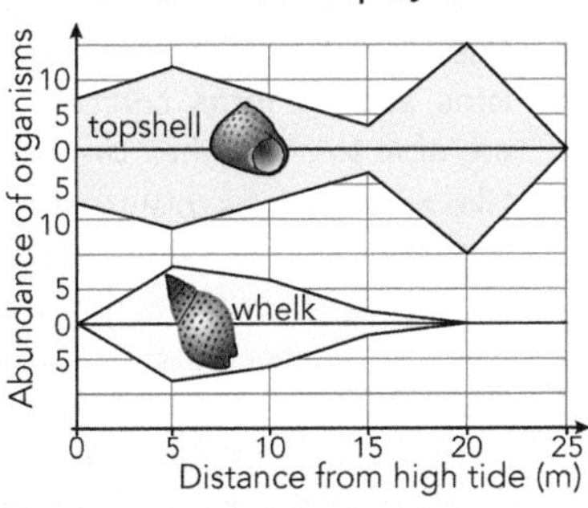

a What is a quadrat and how is it used?

b What kinds of organisms would be recorded?

c What kinds of organisms would be missed ?

d What is a transect line and how is it used?

e What environmental gradient is being studied when a transect line is run from high- to low-tide marks?

f What other environmental factors would change along that environmental gradient?

g How would a kite graph of the mudflat show:

i the distribution of species?

ii the density of species?

iii the diversity of species?

13 Mudflat snails are sensitive to 'heavy metal' pollution.

a What would happen on a polluted mudflat?

b Why are mudflat snails called an indicator species?

Overloaded Ecosystems

Aquatic ecosystems, such as lakes, rivers, streams and harbours, sometimes experience algal blooms. Large masses of algae rapidly grow, often covering the surface of the water. The algae may be multicellular plants or microscopic producers called phytoplankton.

The chemical element phosphorus is often the environmental factor that limits the growth of aquatic producers. Blooms occur when large amounts of phosphate enter the water.

The blooming algal species absorb large amounts of other nutrients, leaving little for other plant species.

When the algal bloom eventually dies and sinks to the bottom, decomposers digest the organic matter but the nutrients they release remain on the bottom, unavailable to top-dwelling species.

The decomposers also use up most of the oxygen in the water as they release energy from the abundant food source by aerobic respiration.

The lack of oxygen kills some animal species in the water, such as the invertebrates that fish feed on, as well as the fish themselves.

Algal blooms also produce toxins that shellfish, such as mussels and oysters, absorb, making them toxic to humans.

The excess phosphate that causes the problem can come from farm effluent, fertiliser run-off, or treated sewage.

ISBN: 9780170262316

The Life Cycle of Flowering Plants

Learning Outcomes - On completing this unit you should be able to:
- compare and contrast sexual and asexual reproduction in flowering plants
- relate reproductive structures to their functions
- describe the different stages in the life cycle of flowering plants
- relate features of flowers, fruits and seeds to pollinating and dispersal agents
- *identify optimal conditions by interpreting graphs with multiple lines.*

Plant Life Processes

- **Vascular plants** are large, complex, multicellular organisms, such as ferns, conifers and flowering plants, that are *able to make their own food.*
- Like all organisms, plants carry out the life processes of nutrition, gas exchange, excretion, transport, support, sensitivity, co-ordination, growth and reproduction.
- Vascular plants consist of leaves supported by a shoot system anchored by a root system, all supplied with water and nutrients by the **vascular system** (see page 128).

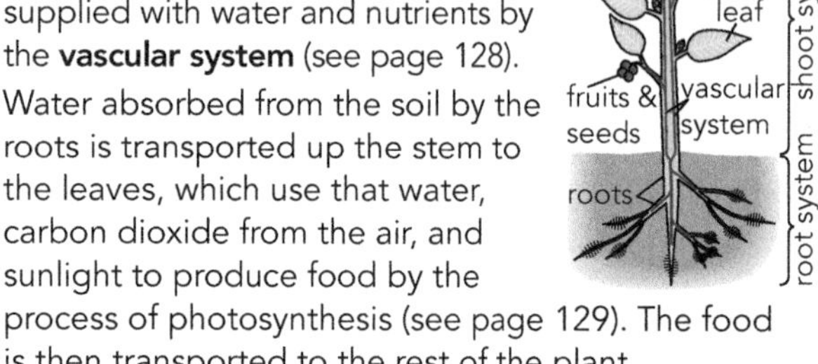

- Water absorbed from the soil by the roots is transported up the stem to the leaves, which use that water, carbon dioxide from the air, and sunlight to produce food by the process of photosynthesis (see page 129). The food is then transported to the rest of the plant.
- Flowering plants are vascular plants with complex reproductive organs that *produce seeds inside fruits.*

Asexual Reproduction

- Asexual reproduction (see page 104) occurs when a single plant produces *genetically identical offspring* called **clones**. Natural methods include runners (eg strawberry plants), bulbs (eg onions), tubers (eg potatoes) and stolons (eg pingao). Plant nurseries use cuttings to rapidly produce large numbers of a variety.
- Asexual reproduction enables plants with inherited variations that suit current conditions *to rapidly spread and multiply.* But clones lack variations that might enable a population to survive changing conditions.

Sexual Reproduction

- Sexual reproduction (see page 104) involves combining the genes of two parents from the same or closely related species. This results in *genetically different offspring.*
- The genetic variation may be expressed in phenotypes (see page 105) that enable some plants in a population to *thrive when the environment changes.* These plants pass on their advantageous gene combinations to offspring (see page 109).
- Sexual reproduction does require plants to produce gametes and to evolve features to ensure male gametes are transported from one plant to another.

Flower Structure and Function

- Most flowering plants are *bisexual*, as their flowers have both male and female sex organs.
- The male organ, called a **stamen**, consists of an **anther** perched on top of a long filament. Multiple anthers in each flower produce microscopic **pollen grains**, which contain the male gametes (sperm).
- The female organ, called a **carpel**, consists of an **ovary** that produces female gametes (eggs) inside **ovules**, and an extension that ends in a 'landing pad' for pollen, called the **stigma**.

stamen: anther, filament; carpel: stigma, style, ovary; petal; ovule

- Some flowers have a single carpel; others have several fused carpels.

Pollination

- For sexual reproduction to *generate significant variation*, pollen must be transferred from the male organ of one plant to the female organ of another plant of the same species. This is known as **cross-pollination**.
- As plants can't move, they must use a pollinating agent, such as insects, birds or the wind. Plants have specific **adaptations** for different agents, for example:
  - **a** insect-pollinated flowers (eg manuka by bees) are white, scented, produce nectar, and have anthers and stigmas arranged to aid pollen transfer
  - **b** bird-pollinated flowers (eg flax by tui) are brightly coloured, unscented, produce lots of nectar, and have sex organs arranged to aid pollen transfer
  - **c** wind-pollinated flowers (eg grass) are drab with no scent or nectar, produce abundant pollen in exposed anthers, and have feathery or silky stigmas.

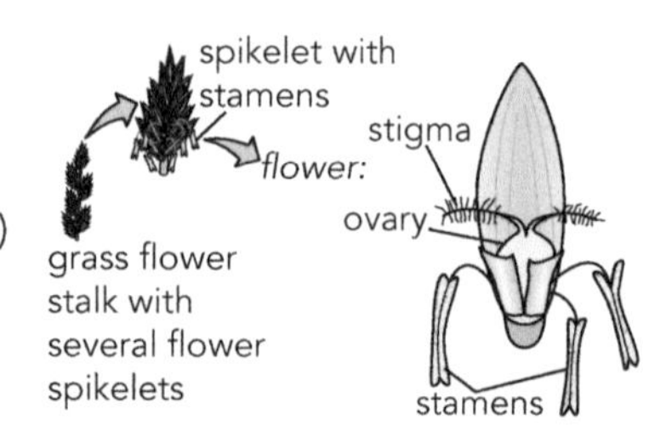

- To avoid **self-pollination**, some species have male and female plants (eg kiwifruit), or separate male and female flowers (eg corn). The sex organs of bisexual flowers mature at different times or are well separated to reduce the chances of self-pollination occurring.

ISBN: 9780170262316

Life Cycle

Flowering

- Flowering is a major event in the life cycle of a plant. It needs to occur at a time which is favourable for pollination and the development of fruits and seeds.
- The timing of flowering is an inherited feature initiated by the plant's response to a changing environment.
- Away from the tropics, seasonal temperature trends have a significant influence on the timing of flowering, with most species flowering in the spring.
- The **photoperiod** is another environmental factor involved in determining when a species will flower. Plants are able to *sense the relative lengths of day and night* and respond by producing hormones.
- Biologists believe that a yet-to-be-identified chemical, which they call florigen, is produced in leaves when various conditions are favourable. This **hormone** is transported to certain leaf buds, which then switch to being flower buds. Cell specialisation (see page 104) results in tissues that develop into flower structures.

Meiosis, Fertilisation and Seed Development

- In the anthers and ovules, cells undergo meiosis cell division (see page 105), resulting in pollen or eggs, each with *half the number of chromosomes*.
- When pollen from another plant of the same species lands on the stigma, they develop tubes that grow down the style to reach the ovules. Sperm swim down the tubes to fertilise the egg cells inside the ovules.

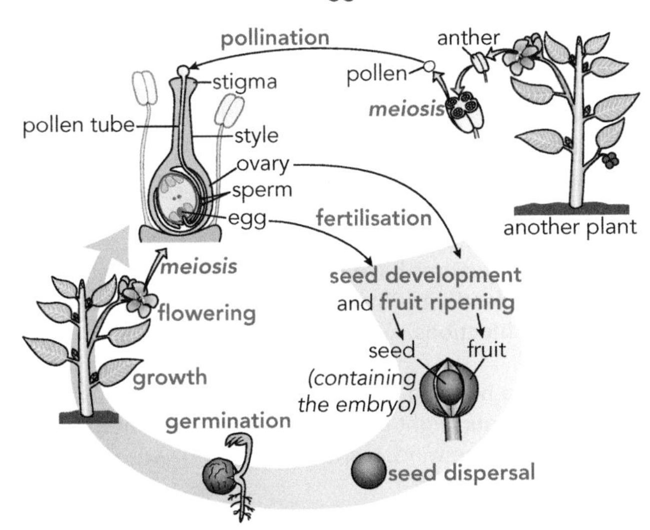

- Each fertilised egg (zygote) grows into an embryo by repeated mitosis cell divisions (see page 104) followed by cell specialisation. Each ovule becomes a **seed** with an embryo inside. The ovary enclosing the seed(s) develops into a **fruit**, which can be dry (eg gorse and pea pods) or fleshy (eg grape, pumpkin).

Seed Structure and Function

- Most seeds have a thick protective seed coat, which helps to ensure that it lasts until conditions are favourable for germination.
- The embryo consists of a **radicle** (first root), **plumule** (first shoot), and **cotyledons**, which supply the germinating seed with food *till it makes enough by photosynthesis*.

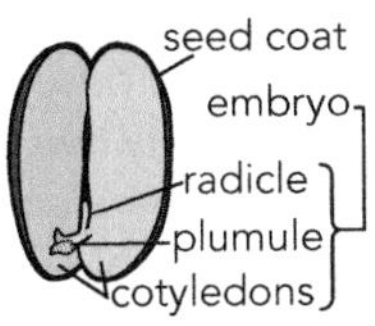

seed coat
embryo
radicle
plumule
cotyledons

Seed Dispersal

- Plants have adaptations to disperse seeds, otherwise offspring would *compete with each other or the parent plant* for space, light, water and nutrients.
- Plants use a variety of dispersal agents including:
  - **a** wind - seeds are light (eg pohutukawa), have 'wings' (eg sycamore) or 'parachutes' (dandelion)
  - **b** water - seeds float in the sea (eg mangrove seeds) or pods float in streams (eg kowhai pods)
  - **c** animals - sweet juicy fruits (eg berries) are eaten by animals and the tough intact seeds pass through the gut to be deposited elsewhere in droppings
  - **d** self-dispersal - pods (eg gorse) dry out and suddenly split open, throwing the seeds some distance.

Seed Germination

- Seeds are living organisms that are in a state of **dormancy**, with no **metabolism** or growth occurring.
- Different types of seeds require specific soil temperatures and moisture levels before **germination** occurs. Water is required for cell metabolism to start and warmth for rapid growth to occur.
- Oxygen is also required for the process of respiration to *release energy from the food stored in the seed.*
- As seeds absorb water, they swell up and the seed coat splits, letting the radicle and plumule emerge.
- As the radicle is sensitive to gravity it grows down toward moisture, becoming the first root of the plant.

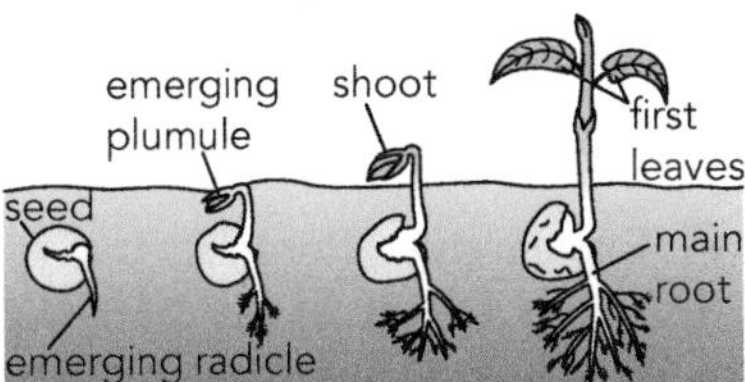

- The plumule is also sensitive to gravity but grows upward instead. Once above ground, it becomes the first shoot, growing toward the light source.

Growth

- Once a flowering plant has germinated, growth occurs at the root and shoot tips that develop, and also from buds. This life process is explained in the next unit.

SKILL: Interpreting Multiple Line Graphs

The most successful germination of seeds occurs when all environmental factors are at optimum levels. To determine the level for a factor, seeds are germinated over a range of values and rates are compared. The graph shows germination rates of canola seeds.

Steps:

1. Identify the environmental factor that is being modified: soil temperature.
2. Determine the values used: 2, 3, 4, 6 and 8 °C.
3. Identify how success is being measured: the % germinated by specified numbers of days.
4. Identify the optimal level: 8 °C, as 100% germination occurs within the least number of days (6 days).

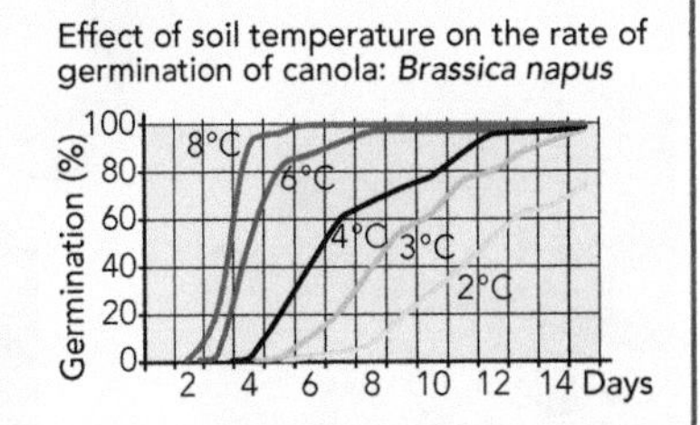

ISBN: 9780170262316

Revision Activities

1 Match up the descriptions with the terms.

a vascular plants	A female sex organs of flowering plants
b vascular system	B plant embryo enclosed in a protective seed coat
c clones	C 'landing pad' for pollen grains
d stamens	D part of the stamen that produces pollen
e anther	E chemical messenger transported around organism
f pollen grains	F occurs when pollen comes from the same plant
g carpels	G system that transports water, minerals and food
h ovary	H part of the embryo that develops into the root
i ovule	I part of a female organ that develops into a seed
j stigma	J food stores that are part of the plant embryo
k cross-pollination	K plants that are genetically identical to the parent
l adaptation	L part of a female organ that develops into a fruit
m self-pollination	M period of time during which a plant is illuminated
n photoperiod	N microscopic objects bearing the male gametes
o hormone	O inherited feature aiding survival or reproduction
p seed	P all of the chemical processes necessary for life
q fruit	Q male sex organs of flowering plants
r radicle	R part of the embryo that develops into the shoot
s plumule	S occurs when pollen comes from another plant
t cotyledons	T stage at which seeds begin to grow and develop
u dormancy	U ripened ovary of a flower
v metabolism	V state in which no metabolism or growth occurs
w germination	W plants that have an internal transport system

2 Explain the differences between the terms in each of the items below:

a sexual and asexual reproduction
b cross- and self-pollination
c meiosis and fertilisation
d seeds and fruits
e radicle and plumule
f dormancy and germination.

3 Copy and complete the table.

Reproduction	Advantage	Disadvantage
Sexual		
Asexual		

4 Identify the methods of asexual reproduction shown below. Choose from: bulb, runner, tuber, cutting, stolon.

5 Decide whether the following statements are true or false. Rewrite the false ones to make them correct.

a Plants carry out different life processes to animals.
b The tubes of the vascular system transport water and nutrients.
c Many plant varieties sold by nurseries are clones.
d Asexual reproduction results in diversity and sexual reproduction in uniformity.
e Anthers release pollen and stigmas receive it.
f Wind-pollinated flowers produce less pollen than insect-pollinated.
g Self-pollination generates no significant genetic variation.
h Both photoperiod and temperature affect the timing of flowering.
i Both flowering plants and mammals have swimming sperm.
j After fertilisation, ovules become fruits and ovaries, seeds.
k Seeds are dispersed to reduce competition for resources.
l Seedlings can detect gravity, water and light.

6 Use the diagram to answer the questions.

a What life process are flowers involved in?
b What is the overall purpose of that life process?
c What is the particular function of flowers?

Biological structures have particular functions. Identify the structures labelled **d** to **j**, and state their functions.

7 Describe how the reproductive organs of manuka, flax and corn plants might be adapted to be pollinated by bees, tui and the wind, respectively.

8 The life cycle of a plant consists of a series of stages.

a Arrange these stages in order starting with a seed in the soil: dispersal, pollination, germination, growth, seed development, ripening, fertilisation, meiosis.
b Briefly describe what happens at each stage.
c Identify two stages at which plants rely on other agents.
d Compare and contrast the role of pollination and seed dispersal.
e What physical factors affect flowering and germination?

ISBN: 9780170262316

9 Copy and complete the table to make generalisations about insect-, bird- and wind-pollinated flowers.

	Pollinating Agent		
Feature	**Insect**	**Bird**	**Wind**
flower size			
petals			
petal colour			
scent			
nectar			
anthers			
pollen			
pollen amount			
pollen size			
stigma			
stigma surface			

10 The diagram summarises the kowhai life cycle.

a Identify the stages labelled **i** to **viii**.

b Decide whether kowhai flowers are likely to be pollinated by insects, birds or the wind.

c State your reasons.

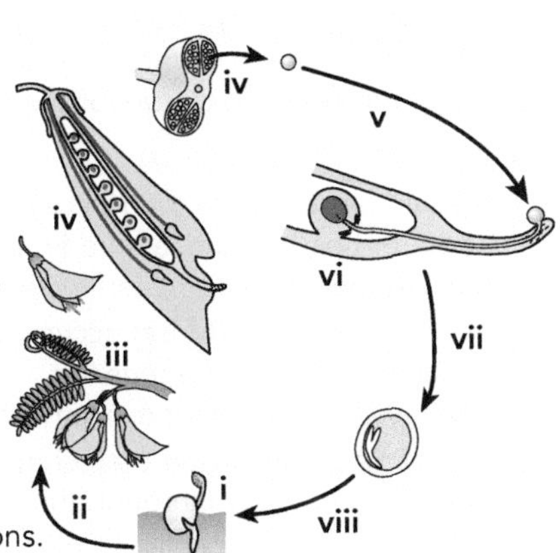

11 The diagram below shows the internal structures of a bean seed.

a What life process do seeds play a key role in?

b What is the particular function of seeds in that process?

c
d
e
f

Identify the structures labelled c to f, and state their functions.

g Which three structures form the embryo?

h How does the seed protect the embryo?

i How does the seed feed the embryo?

12 Read the article opposite, then answer the questions below.

a When you eat a cob, what biological objects are you consuming?

b Where are the male and female flowers located?

c How do they differ?

d Why would the male and female flowers mature at different times?

e What is the pollinating agent?

f What adaptations do corn flowers have to aid pollination?

g Why would fertilisation occur several hours after pollination?

h Why is it that animals can no longer act as dispersal agents?

i How have humans become a necessary part of the life cycle of modern varieties of corn?

13 Identify each plant, its dispersal agent, and state one adaptation that would assist dispersal by that agent.

i Explain why plants need to disperse their seeds.

14 The graph shows the effect of temperature on the germination of seeds of a different species of canola.

Effect of soil temperature on the rate of germination of canola: *Brassica rapa*

Germination (%)
0
20
40
60
80
100
2 4 6 8 10 12 14 16 18 20 Days
2 °C
3 °C
4 °C
5 °C
6 °C
7 °C

a At what soil temperature do the seeds germinate most rapidly?

b At what temperatures is germination close to 100%?

c Based on the data, what is the optimum temperature?

d Why might that not necessarily be the optimum?

Pollinating and Dispersal Agents

Most people look forward to eating corn cobs over summer. A cob is basically a ripened female reproductive organ and the kernels are ripe seeds.

Corn plants have separate male and female flowers. The tassels at the top of the stalk consist of many small green male flowers. The female flowers are the green 'ears' with silky threads, found growing lower down.

The male and female flowers of a plant mature at different times.

The male flowers have dangling stamens that release lots of very light pollen when mature. The pollen is blown by the wind to mature female flowers on other plants.

Those silky threads on the female flowers are actually stigmas. When pollen grains land on stigmas, they grow long tubes down the stigmas to the ovules, so that sperm can fertilise the egg cells.

The resulting zygotes develop into embryos and the ovules containing the embryos become the seeds.

Originally, corn seeds had much tougher coats that could survive passing through the gut of animals that ate the seeds.

Modern varieties are unable to survive in the wild. They are dependent on humans for seed dispersal.

ISBN: 9780170262316

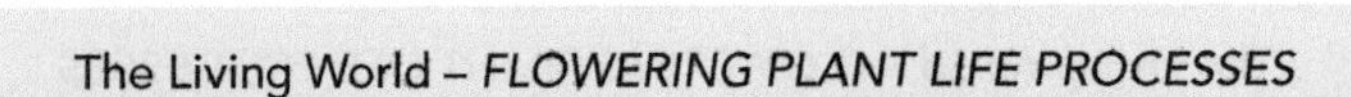

Plant Processes that Maintain Life

Learning Outcomes - On completing this unit you should be able to:

- identify the processes that plants carry out in order to live
- distinguish between cells, tissues, organs and organ systems
- compare and contrast the roles of primary and secondary growth
- explain how different parts of a plant are involved in making food
- *interpret photos and diagrams of the cellular structure of stems and leaves.*

Non-reproductive Life Processes

- For plants to remain alive, they must be able to:
  - obtain energy and nutrients to make food (nutrition)
  - circulate water and food to all parts (internal transport)
  - get energy from food (gas exchange and respiration)
  - dispose of the waste products of cells (excretion)
  - grow larger and form organs (growth & development)
  - elevate leaves and anchor in the soil (support)
  - sense changes (sensitivity) and respond (co-ordination).
- *Plant structures enable them to carry out life processes.*

Plant Structures

- Plants cells have a **vacuole** and a semi-rigid, external **cell wall** made of cellulose. It constrains the cell when the vacuole inflates with water.
  The resulting pressure helps support soft tissue, eg leaf.
- Plants have a variety of cell types, each with *features or structures that enable it to perform a particular function.*

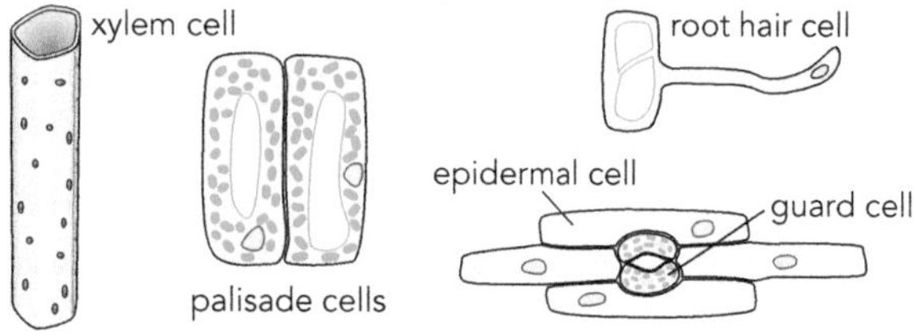

- Tissues (eg palisade, epidermis) are large masses of similar cells. Organs (eg leaf, vascular bundle) are constructed out of different tissues. Organ systems (eg vascular system) consist of organs that *work together to carry out a life process* (eg transport).

Internal Transport

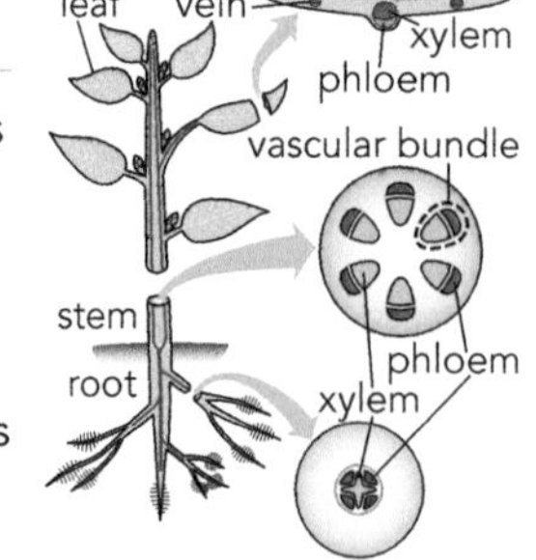

- The **vascular system** consists of a network of conducting tissues that run throughout the length of the plant.
- The root vascular system divides into separate bundles that ascend the stem and connect to leaf veins.
- In each bundle, thick-walled hollow **xylem vessels** lift absorbed water and minerals up to leaves and flowers, while living **phloem tubes** transport dissolved sugar (sucrose) up or down the plant to wherever needed.

Growth and Development

- Growth and development occurs through cell division, enlargement and specialisation. The zones of plants where cell division occurs are called **meristems**. They *retain the ability to divide* for the life of the plant.

Primary Growth

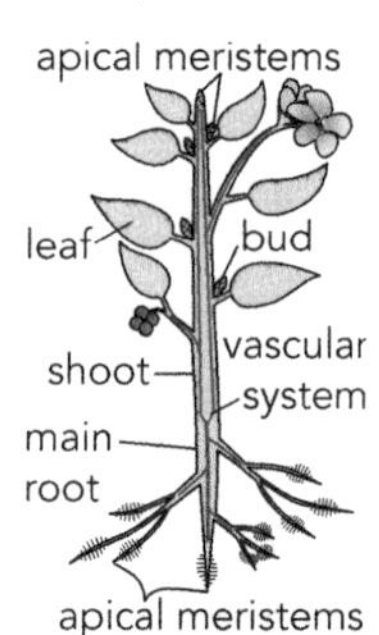

- Apical meristems are found in shoot and root tips, and in leaf and floral buds at the base of leaves.
- The **primary growth** that occurs in these meristems *lengthens the plant and forms organs* (eg leaves).
- In root tips, cells divide then elongate, forcing the tip through the soil.
- Numerous root hair cells develop on the surface just behind the tip. These penetrate the soil, providing *a large surface area for absorbing water and mineral nutrients.*

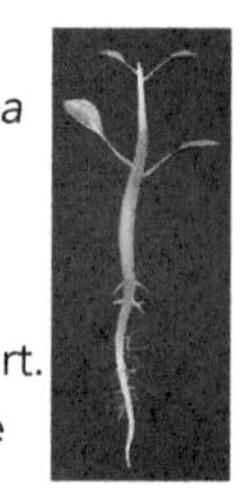

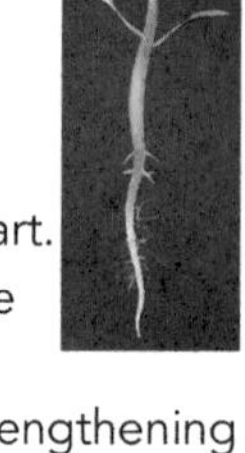

- As the main root grows down through the soil, side roots develop, which help anchor the plant and support the above-ground part.
- Behind the tip, elongated cells in the centre of the root become new vascular tissue.
- In shoot tips, newly divided cells elongate, lengthening the tip. The inner elongated cells behind the meristem specialise to become vascular or other stem tissues.
- Cells on the outside of shoot tips specialise to form **leaf primordia**, each of which enlarges to form a leaf. The leaf vein connects with a stem vascular bundle.
- As the shoot grows taller, the leaves that are formed are exposed to more light.

Secondary Growth

- **Secondary growth** *increases the vascular supply* and the girth of stems and roots.
- When plants become woody, the **cambium** (meristem zone) becomes a continuous cylinder and divides to form xylem on the inside and phloem on the outside.
- As new xylem forms inside the cambium and new phloem outside, the shrub or tree's girth increases.
- Older layers of xylem nearer the centre of the stem or root stop transporting water. The hollow tubes become the strong, flexible woody support system of the plant.

ISBN: 9780170262316

- Each growth ring of a tree trunk represents one year's secondary growth. The wider lighter bands are caused by rapid growth in spring and summer; the thinner darker bands by slower growth in the cooling autumn.

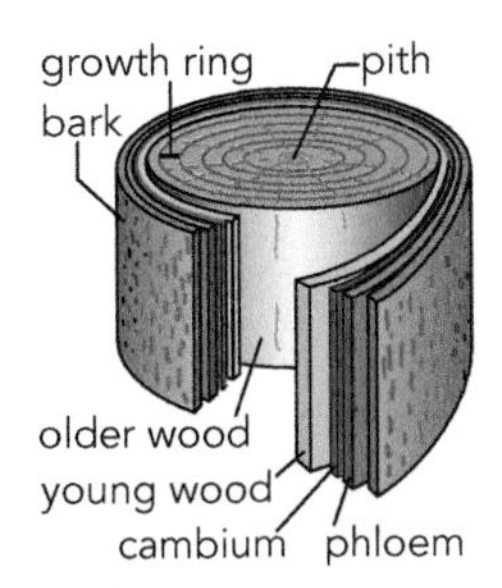

Nutrition and Gas Exchange

- Plants require *energy and raw materials* (eg CO_2, H_2O and mineral nutrients) from the environment to make the biomolecules that cells are constructed out of. Energy is also needed for other life processes.
- The process of **photosynthesis** enables plants to make glucose sugar using H_2O from the soil, CO_2 from the air, and sunlight.

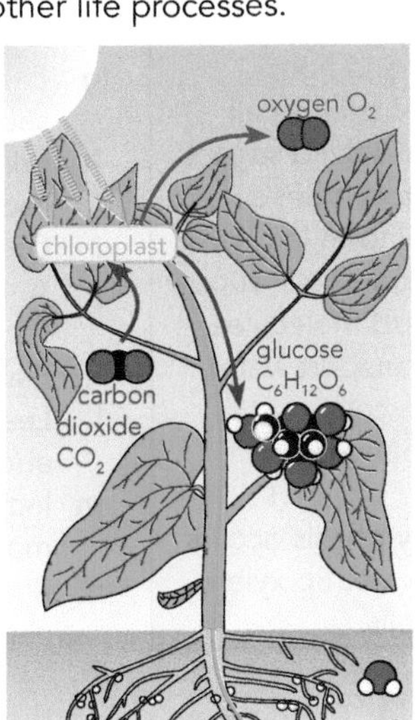

- Glucose is then used to make carbohydrates, fats, proteins and DNA. The latter two require atoms of the elements N, K, Ca, Mg, P and S as well.
- The process also produces oxygen gas, which is the *main source of the oxygen in the atmosphere* that organisms require for respiration.
- Photosynthesis occurs in the green parts of plants, in special cell structures called **chloroplasts**. They contain **chlorophyll**, a green pigment, that *absorbs light energy and transforms it* into chemical potential energy stored in the bonds of molecules.

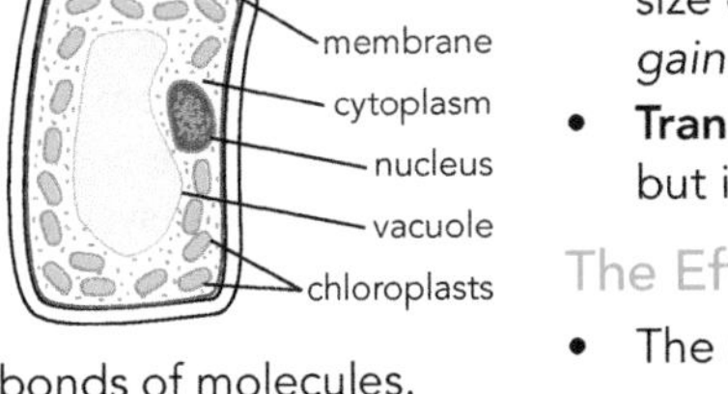

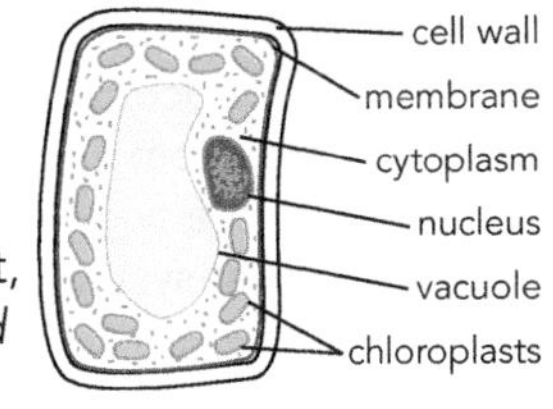

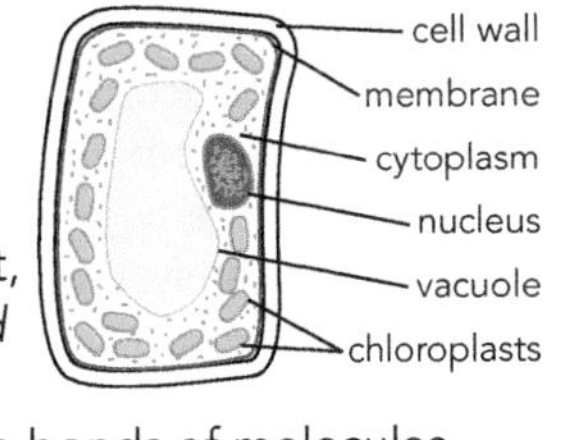

- Although complex, the process can be summarised as:

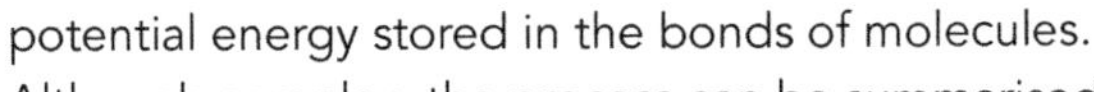

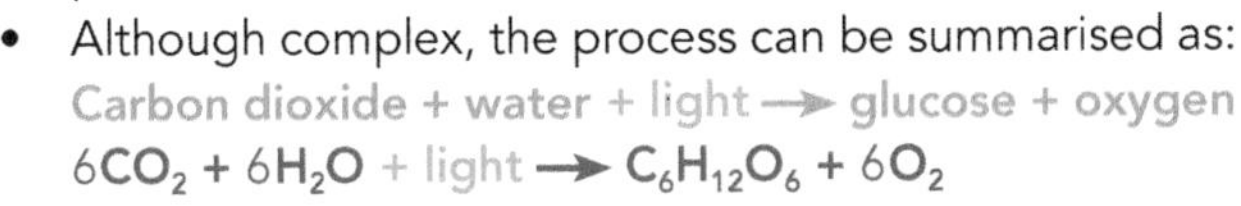

Carbon dioxide + water + light → glucose + oxygen

$6CO_2 + 6H_2O$ + light → $C_6H_{12}O_6 + 6O_2$

Leaf Structure & Function

- Leaves have adaptations that enable them to *photosynthesise efficiently without losing too much water.*
- Their large surface area enables them to absorb more sunlight, and their thinness allows the sunlight to penetrate all layers.
- The petiole (stalk) holds the blade of the leaf up to the light. Its vascular bundle becomes the network of veins that supplies all leaf cells. The midrib and other major veins help to support the leaf and keep it flat.
- The epidermis makes a transparent waxy cuticle that transmits light but limits the loss of water vapour from the leaf. Microscopic pores (**stomata**) in the epidermis open to allow CO_2 to enter but close on hot, dry or windy days to prevent water loss causing wilting when the cells deflate.

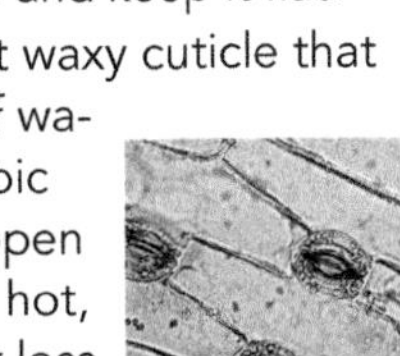

Internal Structure of Leaves

- The structure is in cross-section.

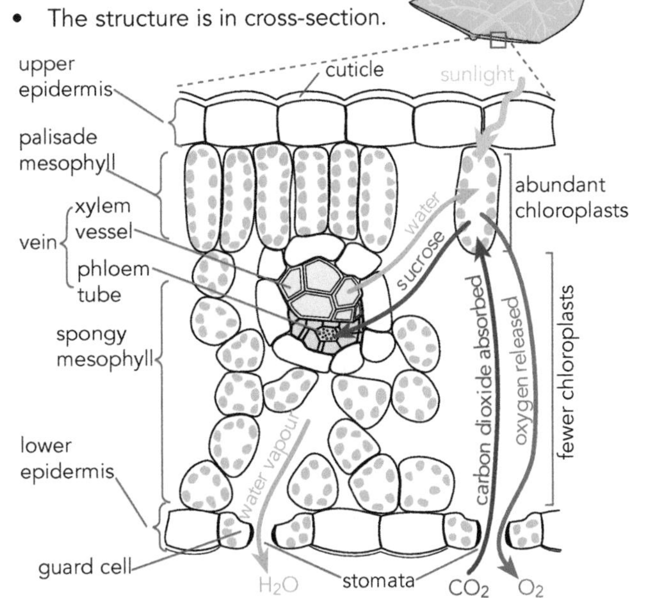

- H_2O from xylem vessels and CO_2 from stomata diffuse through the moist cell membranes of **mesophyll** cells.
- Chloroplasts absorb and use light energy to convert CO_2 and H_2O into energy-rich glucose and O_2 gas.
- Closely packed **palisade cells** maximise light absorption; loosely packed **spongy cells** maximise CO_2 diffusion.
- Glucose is converted into starch for storage or sucrose for transport in the phloem. O_2 diffuses out of the leaf.
- In the life process of **gas exchange**, the plant controls the exchange of CO_2 and O_2 between its internal and external environments. **Guard cells**, which alter the size of the stomata by *inflating or deflating as they gain or lose water*, play a key role in this process.
- **Transpiration** is the loss of water vapour via stomata but it pulls more water up the xylem from the roots.

The Effect of Environmental Factors

- The rate of photosynthesis is increased by:
  - greater light intensity overhead in the summer
  - higher CO_2 levels in the air during daytime
  - higher moisture levels in the soil after watering
  - warming air and soil temperatures in spring
  - higher soil nutrient levels from added fertiliser.
- Increasing the level of an environmental factor will boost the rate of photosynthesis to an optimum level, after which it *no longer rises and may even fall.*

Respiration and Excretion

- Phloem tubes transport sugars to tissues needing energy for metabolism and growth. The energy in the sugar is released by the life process of **respiration** (see page 137), which *occurs in all cells all of the time.*
- Efficient respiration requires O_2 gas, which is made in leaves during the day or diffuses into organs from the air (21% oxygen) or soil during the day and night.
- Although complex, the process can be summarised as:

Glucose + oxygen → carbon dioxide + water + energy

$C_6H_{12}O_6 + 6O_2 \rightarrow 6CO_2 + 6H_2O$ + energy

- Overall, plants absorb CO_2 and release O_2 in the day. The *reverse happens at night* as only respiration occurs.
- Plant cells produce few wastes that need to be **excreted**.

ISBN: 9780170262316

Revision Activities

1 Match up the descriptions with the terms.

a vacuole	A zone of cells that produce vascular tissue in stems
b cell wall	B green pigment that absorbs light energy
c vascular system	C zones of plants where cell division occurs
d xylem vessels	D upright, closely packed photosynthetic cells
e phloem tubes	E network of conducting tissues supplying organs
f meristems	F microscopic pores in epidermis of leaves
g primary growth	G occurs at shoot and root tips, lengthening plant
h leaf primordia	H structure that fills with water inflating the cell
i secondary growth	I pairs of cells that alter the size of stomata
j cambium	J making food using raw materials and light
k photosynthesis	K external layer of cellulose that constrains cell
l chloroplasts	L conducting tissue transporting dissolved sugars
m chlorophyll	M cell structures that carry out photosynthesis
n stomata	N masses of immature cells that will form leaves
o mesophyll	O occurs in stem and root, increasing vascular supply
p palisade cells	P releasing the energy stored in food molecules
q spongy cells	Q thick walled, hollow tubes that transport water
r gas exchange	R disposing of waste products from cells
s guard cells	S exchange of CO_2 and O_2 gases between an organism and its environment
t transpiration	T cell layers in a leaf where photosynthesis occurs
u respiration	U loss of water vapour, which lifts water up xylem
v excretion	V loosely packed photosynthetic cells

2 Explain the differences between the terms in each of the following items:

a cells, tissues and organs
b xylem vessels and phloem tubes
c primary and secondary growth
d photosynthesis and respiration
e palisade and spongy mesophyll
f gas exchange and excretion.

3 Make a generalisation about each leaf feature listed and explain how it might help leaves to carry out photosynthesis:

a colour
b thickness
c flatness
d surface area
e veins
f stalk length.

4 For each plant cell type illustrated below:

i state its name
ii describe the cell's function
iii explain one feature of the cell that enables it to carry out that function.

a
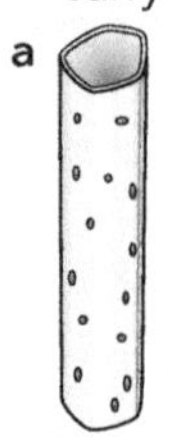

b
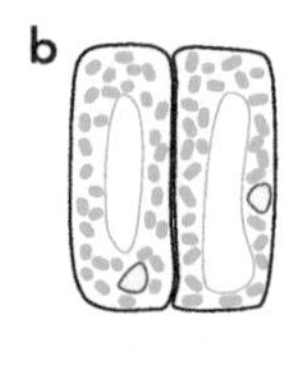

c
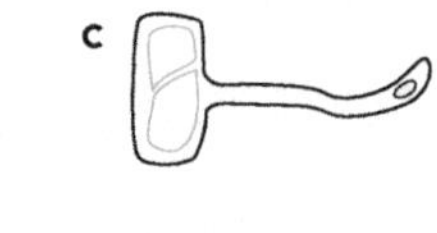

d
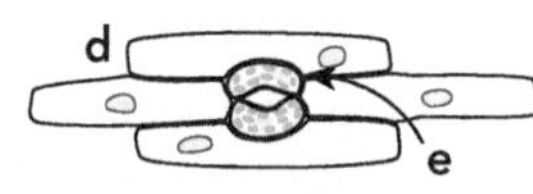
e

5 Decide whether the following statements are true or false. Rewrite the false ones to make them correct.

a Plants carry out the same life processes as animals, only differently.
b The pressure of water in vacuoles helps make soft tissues more rigid.
c Organs are made of cells, which are made of tissues.
d Xylem vessels transport water and phloem tubes, dissolved sugars.
e Sugar is always transported downward, and water upward.
f Meristems cease dividing.
g Primary growth lengthens and secondary growth widens plants.
h Primary growth gets more light; secondary growth, more water.
i Growth rings reveal a tree's age.
j Animals ultimately rely on plants for water and oxygen.
k Plants transform light into heat.
l Leaves maximise photosynthesis and minimise transpiration.
m Increasing an environmental factor may sometimes slow a process.
n Respiration reverses the gases exchanged in photosynthesis.

6 The diagram shows the vascular system and cross-sections of a young plant.

a What life process does the vascular system carry out?
b What are the two functions of the vascular system.
c Identify the red and blue vascular tissue?
d What materials are transported in each tissue?
e What directions do substances move in each tissue?
f Compare vascular tissue in the roots, stems and leaves.

7 Copy and complete the table to compare and contrast features of primary and secondary growth in plants.

Feature	Primary Growth	Secondary Growth
Type of plant		
Location		
Meristem name		
Height of plant		
Girth of plant		
Organs formed		
Vascular supply		
Main function		

8 Estimate the age of the tree trunk shown below and explain how you are able to do this.

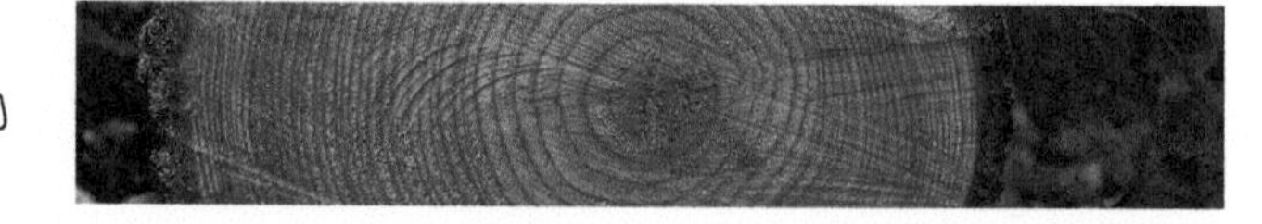

ISBN: 9780170262316

9 Nutrition is a life process that is carried out by the whole plant.

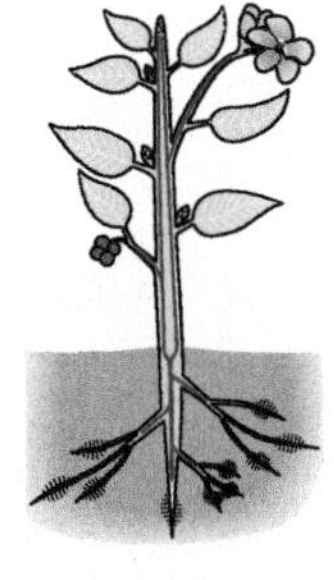

a What is the purpose of plant nutrition?

b Describe the role of each of the following in carrying out nutrition:
- i the root system
- ii the stem system
- iii the leaves.

c Explain how roots, stems and leaves work together to carry out nutrition.

10 The diagram shows the internal structure of a typical leaf.

For structures a to j:
- i write its name
- ii state its function
- iii describe how its structure helps it to function.

Explain how these structures work together to optimise photosynthesis:

k structures **a**, **b**, **c** and **j**

l structures **f** and **d**

m structures **g**, **h** and **i**.

n Explain how structures **b**, **e**, **a** and **f** work together to minimise transpiration.

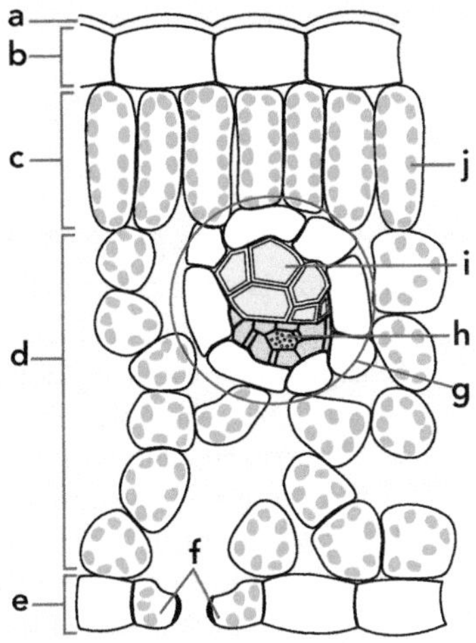

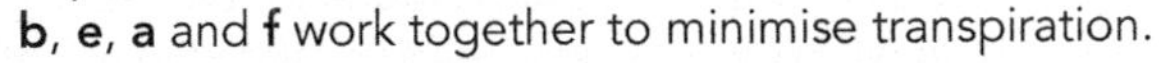

11 Leaf cells convert the glucose they have made into insoluble starch. Iodine solution turns starch blue-black. After a green-and-white-leaved plant had been in the dark for 24 hours, a black strip was attached to one leaf and a clear plastic bag with CO_2 absorber sealed another. The plant was placed in sunlight and after 24 hours the leaves tested for starch using iodine.

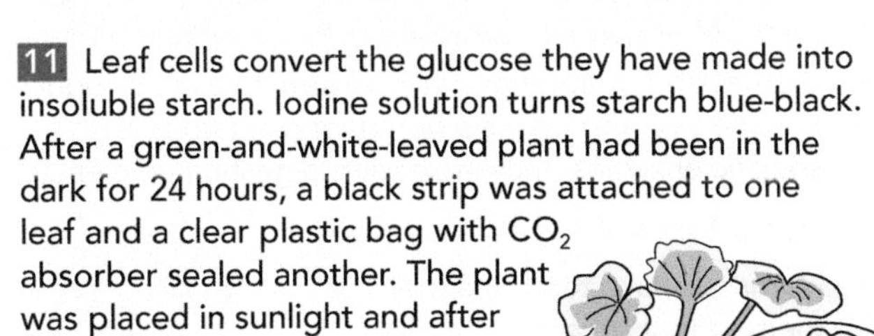

a What colours would parts of the two leaves go?

b What would this show about the factors needed for photosynthesis?

c What other factor is required?

12 Read the article opposite, then answer the questions below.

a The levels of which two dissolved gases were being measured?

b Which line shows which gas?

c What is a 'diurnal pattern'?

d Describe the pattern for each gas.

e Explain what causes the CO_2 level to change during the day.

f Explain what causes the O_2 level to change during the day.

g Describe the gas exchange pattern associated with plants.

h Describe the gas exchange pattern associated with microbes.

i How do you know that the CO_2 pattern is related to organisms?

j What would happen to gas levels if aquatic plants were removed?

13 A water weed produces bubbles of gas when it is exposed to light and sodium bicarbonate has been added to the water to supply dissolved CO_2.

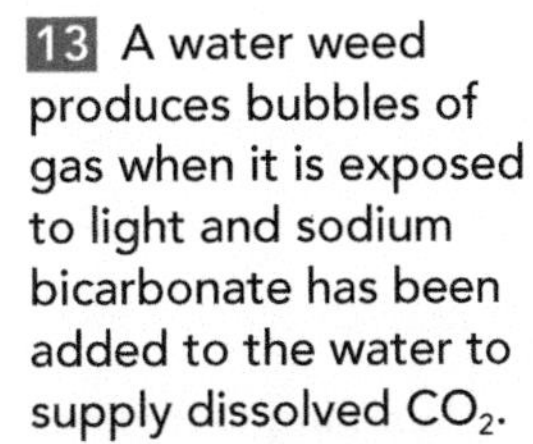

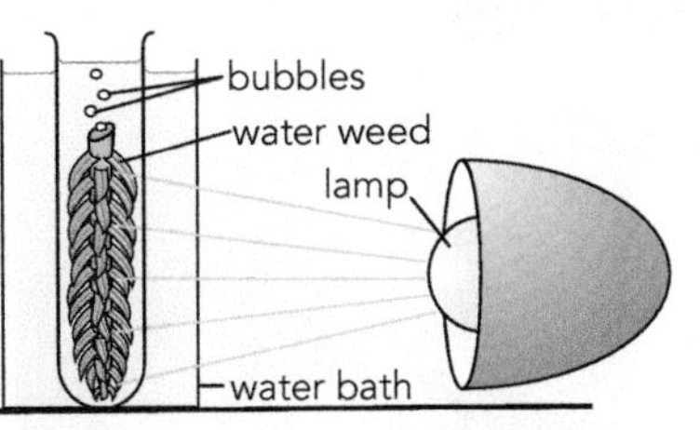

The gas ignites a glowing splint, proving that it is O_2. The number of bubbles produced per minute indicates the rate of photosynthesis.

a Explain why counting bubbles indicates the rate of photosynthesis.

Three environmental factors that will affect the rate of photosynthesis in the water weed are:

b the light intensity

c the amount of dissolved CO_2 in the water

d the temperature of the water.

For each environmental factor:
- i Explain how the equipment could be used to investigate how the factor affects the rate of photosynthesis.
- ii Describe how you would increase the factor.
- iii Predict how the rate of photosynthesis would change.

14 The images are very thin slices, called cross-sections, of two plant organs. The cells are stained with special dyes.

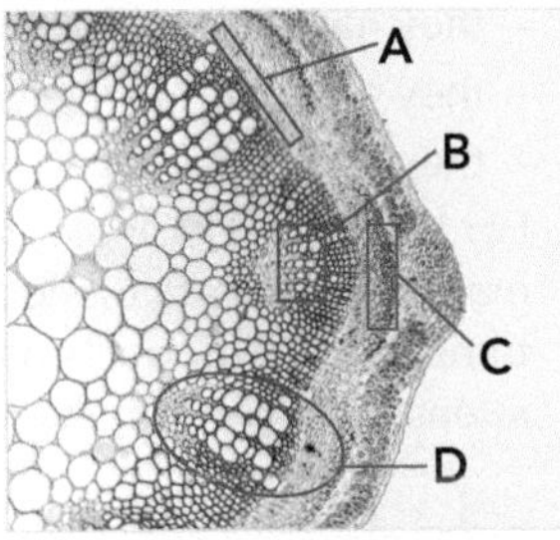

a What is the top organ?

b How do you know?

c Identify tissues **A** to **C**, and structure **D**.

d Where would secondary growth occur?

e What is the bottom organ?

f How do you know?

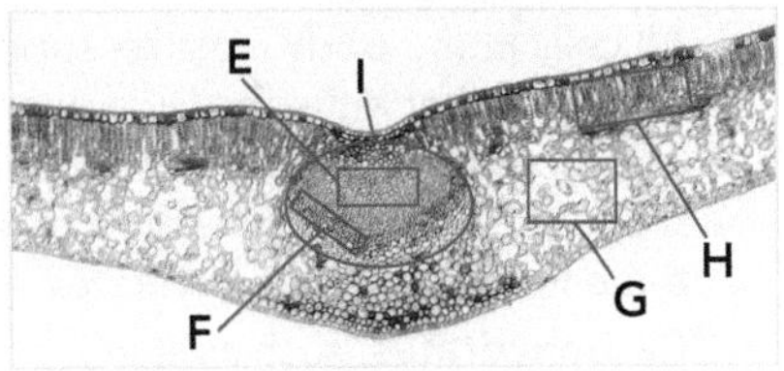

g Identify tissues **E** to **H**, and structure **I**.

h What is the connection between structures **D** and **E**?

Gas Exchange in a Pond

The levels of dissolved oxygen and carbon dioxide gases in a pond colonised by aquatic plants were measured every three hours over a period of six days. The results have been plotted in the graph.

As the graph shows, the CO_2 level followed a diurnal pattern with the level being high near dawn, then falling to a minimum late in the day.

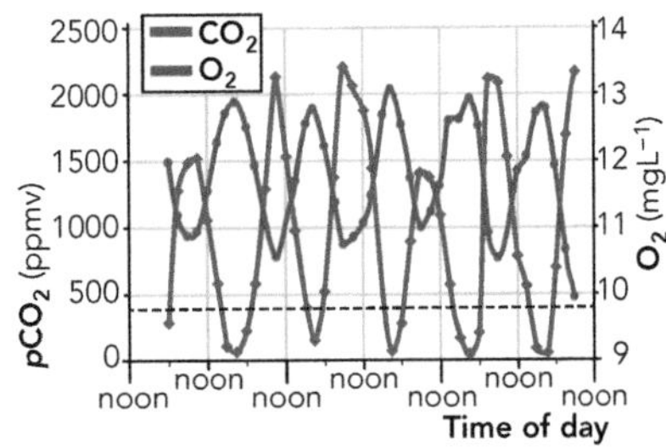

Similarly the O_2 level followed a diurnal pattern, but with level being high late in the day and falling to a minimum early in the morning.

The falling CO_2 level and rising O_2 level during the day is due to photosynthesis being carried out by the plants.

The rising CO_2 level and falling O_2 level during the night are caused by plants and microbe respiration in the pond.

The amount of dissolved gases in the pond is also affected by the partial pressures of gases in the air above the pond.

The dotted line shows the concentration of CO_2 that would have occurred if no biological processes were happening in the pond. As you can see, photosynthesis is a significant process.

ISBN: 9780170262316

Food and the Digestive System

Learning Outcomes - On completing this unit you should be able to:

- list the organic nutrients that mammals need to carry out life processes
- distinguish the different stages of nutrition that occur within mammals
- relate the adaptations of herbivores, carnivores and omnivores to their diets
- describe the structures that are part of the digestive system and their functions
- explain how the different parts of the digestive system work together.

Consumers

- All organisms need *energy and nutrients from their environment to carry out life processes* such as sensitivity and co-ordination, movement, excretion, growth and development, and reproduction.
- All animals are consumers, as they obtain the energy and nutrients they need by eating other organisms.
- Mammals are animals that have the following features:
  - they are 'warm-blooded' vertebrates
  - they have fur or hair on their bodies
  - they give birth to live young (rather than lay eggs)
  - they feed their young milk from mammary glands.
- For NCEA level 1 Biology, you need to know how mammals process the food they eat (nutrition), transport the products of digestion (circulation), and release energy from those products (respiration).

Organic Nutrients

- All cells in the body must be supplied with **organic nutrients** that either provide energy or are used to assemble the molecules needed to carry out life processes. The nutrients are obtained by digesting three food types: carbohydrates, lipids and proteins.
- **Carbohydrates**, such as starch and cellulose in plants and lactose in milk, consist of linked sugar units. When digested, the products are single sugar molecules.

- Single sugars, such as glucose, provide an immediate source of energy for activities like muscle movement and sending messages along nerves.
- **Lipids**, such as plant oils and animal fat, consist of three fatty acid units attached to a glycerol unit. When digested, the products are separate fatty acid and glycerol molecules.

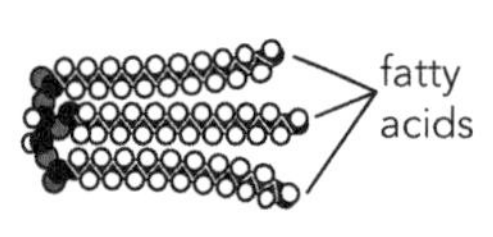

- **Proteins**, such as casein in milk, consist of large numbers of different amino acid units bonded together (see page 101). When digested, the products are individual amino acid molecules.

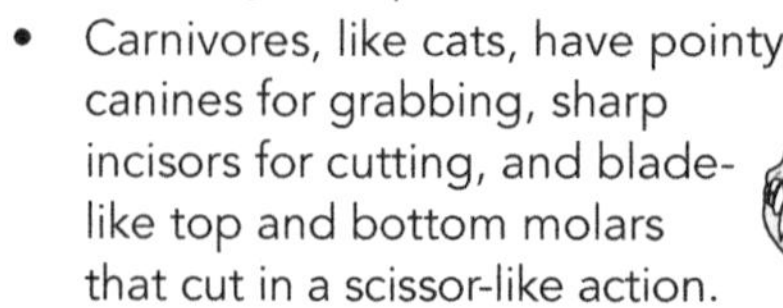

- The digestion of carbohydrates, lipids and proteins results in *small soluble molecules* (single sugars, fatty and amino acids) that can be *absorbed into the blood.*
- Cells need **inorganic nutrients** (minerals), eg Ca^{2+}, K^{+}, Na^{+}, Mg^{2+}. These are absorbed directly into the blood.

The Digestive System

- Nutrition occurs primarily in the digestive system (gut).
- Nutrition is a life process that consists of five stages:
  1. **ingestion** - getting food into the gut
  2. **digestion** - breaking down large lumps of food into small, soluble molecules
  3. **absorption** - moving the products of digestion into the circulatory system
  4. **egestion** - getting rid of undigestible food wastes
  5. **assimilation** - ensuring the products of digestion become part of cell structures or an energy source.
- Mammals have specific **adaptations** for processing foods. In NCEA exams, you need to relate the adaptations of herbivores, carnivores and omnivores to diets.
- Diet-related adaptations involve structures that *reduce particle size and provide optimum conditions for specific enzymes,* thus increasing digestion efficiency.

Ingestion and Physical Digestion

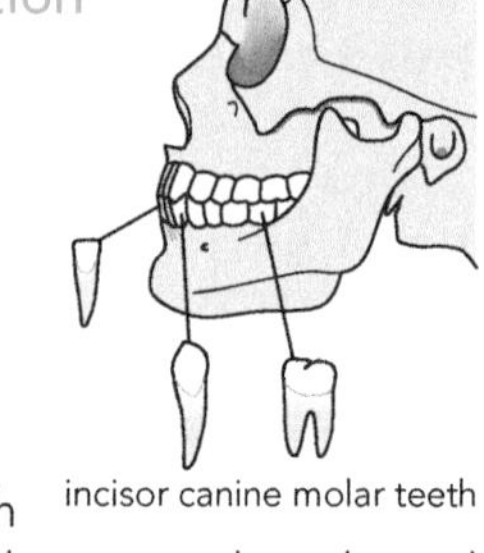

- Mammals have teeth that help them to tear off pieces of plant or animal matter and then cut or grind them into smaller bits that can be readily swallowed.
- Omnivores, like humans, have pointed canine teeth to tear food, chisel-shaped incisor teeth to cut food, and flat-topped molars to crush and grind food into smaller pieces that are more easily digested.
- Herbivores, like sheep, have a bottom row of incisor and canine teeth that press leaves against an upper pad, enabling them to be torn off. Numerous large flat-topped molars grind up the leaves.

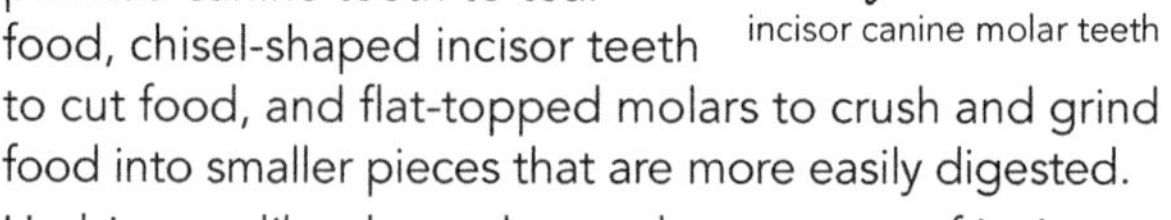

- Carnivores, like cats, have pointy canines for grabbing, sharp incisors for cutting, and blade-like top and bottom molars that cut in a scissor-like action.

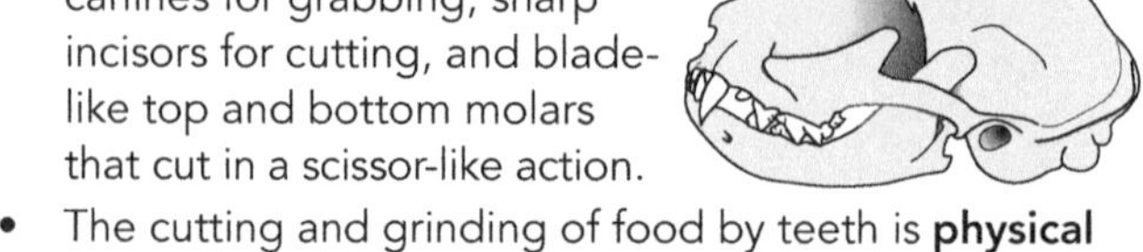

- The cutting and grinding of food by teeth is **physical digestion**. It reduces food particle size, *increasing the surface area exposed for the next stage of digestion.*
- Salivary glands secrete saliva, which lubricates food before it is swallowed into the oesophagus. Waves of muscle contractions, called **peristalsis**, squeeze the food along the oesophagus and into the stomach.

ISBN: 9780170262316

Herbivore and Carnivore Digestive Systems

- The digestive system is a series of connected organs: mouth, oesophagus, stomach, caecum and intestines.
- Rabbits mostly eat leaves, which are *difficult to digest* because of the cellulose cell walls.

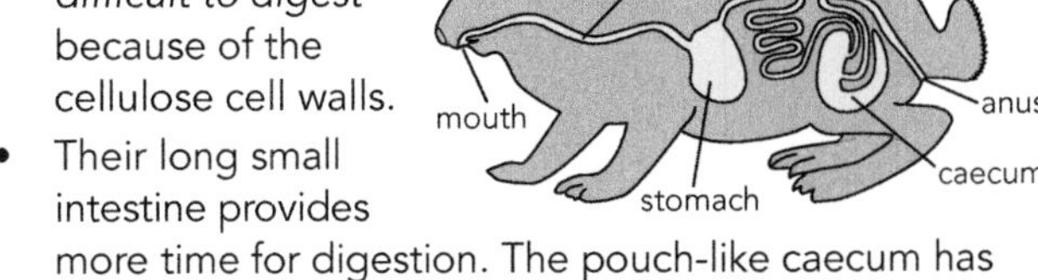

- Their long small intestine provides more time for digestion. The pouch-like caecum has bacteria that aid the digestion of cellulose into sugar.
- As carnivores, dogs eat flesh, which is *easier to digest* as it's protein and fat. Their intestines are therefore relatively short and the caecum is tiny.

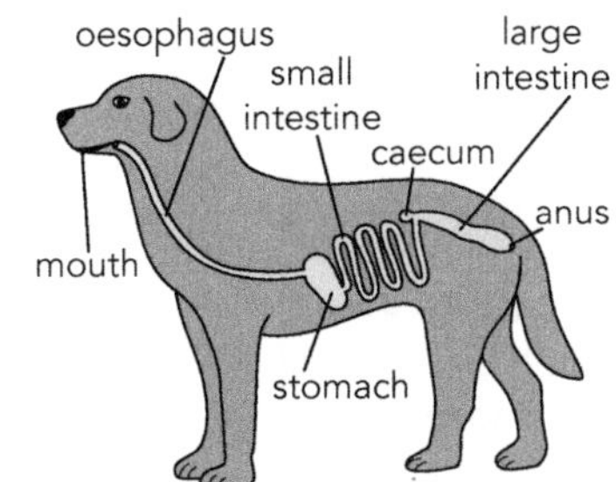

Chemical Digestion

- **Chemical digestion** is a *process involving splitting* complex carbohydrates, lipids and proteins into small soluble molecules that can be absorbed.
- **Digestive enzymes** are catalysts (see page 77) that *specialise in splitting specific food types or molecules.* Each works best at its *optimum level of acidity or alkalinity*, so different organs have different pHs (see page 85). Most work best at body temperature (37 °C).
- The table details some of the enzymes that digest food in parts of the human gut with different pHs.

Part	pH	Enzyme	Food	Product	Source
mouth	6.5-7.5	amylase	starch	maltose and glucose	salivary glands
stomach	1.5-4.0	pepsin	proteins	peptides	stomach
small intestine	7.5-8.5	lipase	lipids	fatty acids and glycerol	pancreas
		amylase	glycogen	glucose	
		trypsin	peptides	amino acids	
		sucrase	sucrose	glucose, plus other sugars	small intestine
		maltase	maltose		

Human Digestive System

- When food reaches the stomach, it is churned up with gastric juice containing hydrochloric acid and enzymes. Pepsin is an enzyme that splits proteins into **peptides** (short chains of amino acids) most efficiently under high acidity.
- The stomach then releases its contents in small amounts into the small intestine.
- The gall bladder stores **bile** from the liver and secretes it into the small intestine when food is present.
- Bile **emulsifies** blobs of fat into tiny oil droplets. The *increased surface area makes digestion more efficient.*

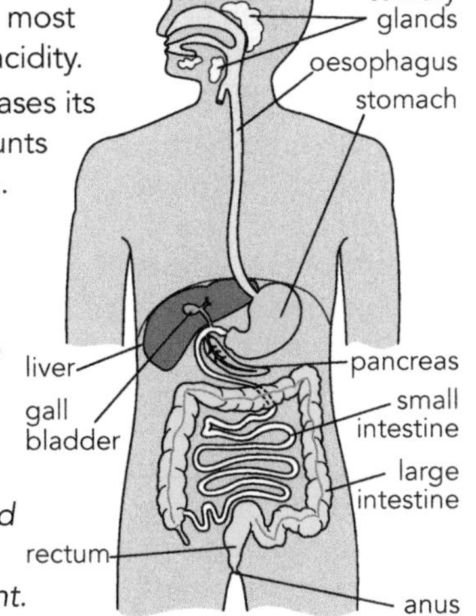

- The pancreas secretes pancreatic juice into the small intestine, which has enzymes that digest all food types.
- Bicarbonate ions in bile and pancreatic juice neutralise stomach acid making the gut pH slightly alkaline.
- The small intestine secretes enzymes, such as sucrase which digests the double sugar sucrose (table sugar).

Digestive Enzyme Action

- Each enzyme has a special *active site into which part of a particular food molecule fits before it is split*, eg:

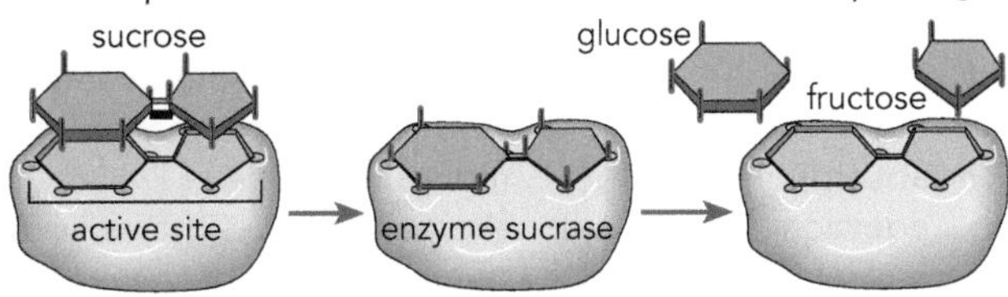

- If the pH or temperature is not right, the **denatured enzyme** *loses its special shape* and no longer works.

Absorption

- The products of digestion are absorbed by the small intestine. It is lined with tiny extensions, called **villi**, that protrude into the contents of the intestine. The villi have thousands of microvilli, *greatly increasing the surface area for absorbing nutrients.*
- Different nutrients are *transferred to different structures in the villi* by a process called **active transport**.
- Sugars and amino acids are transported into blood vessels called **capillaries**, and are carried to the liver dissolved in the liquid part of blood (see page 137).
- Fatty acids and glycerol are transported into lymph vessels called **lacteals**. They enter the blood when lymph fluid drains into a vein.

Egestion

- Peristalsis pushes undigestible food into the large intestine, where water is reabsorbed *otherwise the mammal would dehydrate.* The resulting faeces are stored in the rectum, then expelled through the anus.

Processing and Assimilation

- Assimilation occurs when cells use absorbed digestion products *to assemble molecules or supply energy.*
- Digestion products circulating in the blood can supply all cells directly, but *supply needs to match demand* otherwise cells get damaged by excess or starve.
- The liver processes, stores and releases nutrients according to the body's current needs.
- Cells *get energy* from absorbed glucose (see page 137).
- The liver and pancreas work together *to control blood sugar levels.* Some excess glucose is converted into glycogen in the liver and muscles. It acts as a rapid-access energy store. The rest is changed into fatty acids.
- Cells *form membranes* out of absorbed fatty acids.
- Excess fatty acids are stored in fats cells found throughout the body. They act as a long-term energy store.
- Glycogen and fat are changed into glucose as needed.
- Cells *assemble proteins* (see page 101) out of absorbed amino acids.
- Excess amino acids cannot be stored, so the liver breaks them down. The resulting urea is transported in the blood to the kidneys to be excreted in urine.

ISBN: 9780170262316

Revision Activities

1 Match up the descriptions with the terms.

a organic nutrients	A breaking down food into usable organic nutrients
b carbohydrates	B cells transporting the products of digestion
c lipids	C liquid produced by liver that helps digest fats
d proteins	D cutting and grinding of food into smaller lumps
e inorganic nutrients	E simple molecules derived from other organisms
f ingestion	F ions of other elements required by cells
g digestion	G inherited features that aid survival or reproduction
h absorption	H molecules made of linked sugar units
i egestion	I short amino acid chains, usually less than 50 long
j assimilation	J getting rid of undigested food wastes
k adaptations	K muscle contractions pushing food through gut
l physical digestion	L lymph vessels that drain villi
m peristalsis	M molecules made of fatty acids linked to glycerol
n chemical digestion	N enzyme that doesn't work as its shape has changed
o digestive enzymes	O molecules made of various linked amino acids
p peptides	P fine blood vessels that supply cells directly
q bile	Q splitting large molecules into small soluble ones
r emulsify	R selectively moving molecules into cells
s denatured enzyme	S getting food into the gut
t villi	T tiny extensions protruding into the gut
u active transport	U converting blobs of fat into tiny oil droplets
v capillaries	V enzymes that split large molecules into small ones
w lacteals	W getting digestion products into cells

2 Explain the differences between the terms in each of the following items:

a consumers and producers
b ingestion and digestion
c absorption and assimilation
d herbivores, carnivores and omnivores
e physical and chemical digestion
f acidic, neutral and alkaline conditions.

3 The diagram below summarises the digestion of a starch, lipid and protein molecule. Consult the table on the previous page, then for each of the reactions labelled a to e identify:

i the enzyme involve
ii the source of the enzyme
iii the location of the digestion reaction
iv the pH of the location
v the product of the digestion reaction.

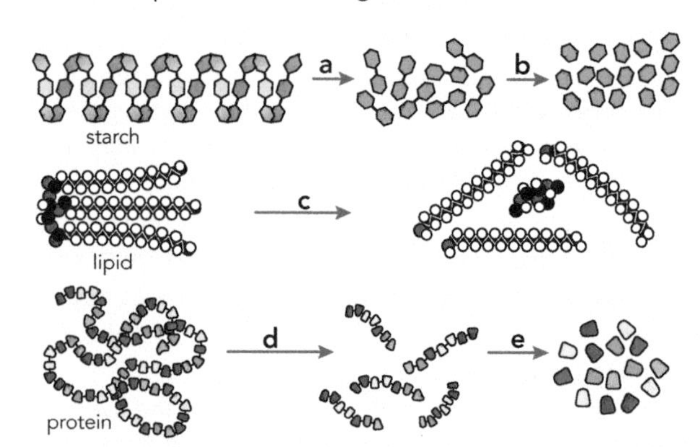

4 Decide whether the following statements are true or false. Rewrite the false ones to make them correct.

a Consumers rely ultimately on producers for energy and nutrients.
b Organisms assemble molecules.
c Required organic nutrients include sugars, fatty acids and amino acids.
d Digestion results in small insoluble molecules.
e Nutritional adaptations are often related to diet.
f Leaves digest quicker than meat.
g Human canines tear, incisors cut, and molars grind food.
h Physical digestion is necessary for efficient chemical digestion.
i Muscles move food along the gut.
j Bacteria help herbivores to digest the starch in plant cell walls.
k Optimum chemical digestion depends on pH only.
l Bile dissolves fats and raises the pH.
m Denatured enzymes no longer have active sites.
n Passive transport transfers particular kinds of molecules.
o The liver processes digested food.

5 Study the dog, rabbit and chimp skulls, then copy and complete the table.

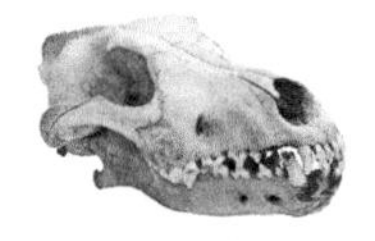

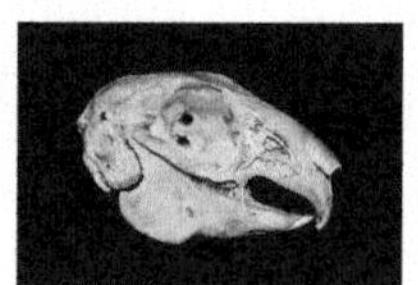

Feature	Dog	Rabbit	Chimp
Feeding role			omnivore
Food			fruit, meat
Canine shape			pointy, long
Canine role			tear food
Incisor shape			chisel shape
Incisor role			cut food
Molar shape			flat-topped
Molar role			grind food

6 The diagram below can be used to compare and contrast the digestive systems of a carnivore and herbivore.

Name the common organs labelled a to d.

e Which is the herbivore?
f How do you know?
g How are its guts adapted for efficient digestion of its diet?
h Which is the carnivore?
i How do you know?
j How are its guts adapted for efficient digestion of its diet?
k Which gut has a mutualistic relationship?

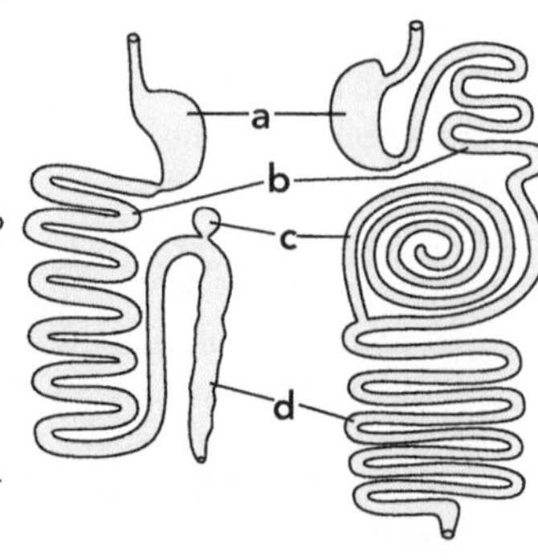

ISBN: 9780170262316

7 The diagram shows the structures that are part of the digestive system.

Identify the structures labelled **a** to **m** and describe how it carries out its digestive function.

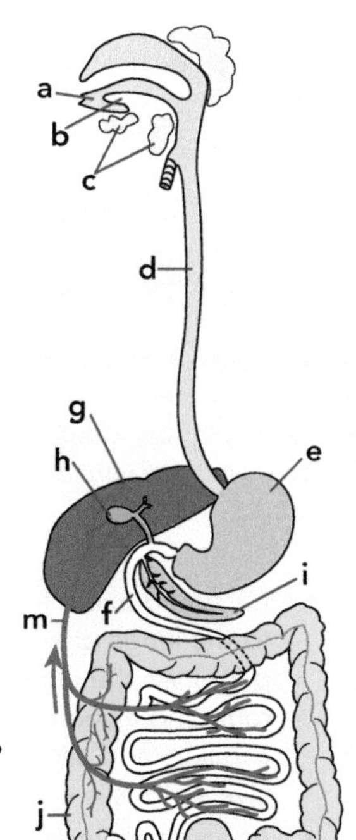

n Which structure produces saliva and what is its function?

o Which structure produces gastric juice and what is the function of the juice?

p Which structure produces greenish-yellow bile and what is the function of bile?

q Which structure produces pancreatic juice and what is the function of the juice?

r What is the circled object called and what is its function?

8 Bile and pancreatic juice work together to digest lipids.

a Describe how bile affects the pH of the intestinal contents.

b Explain how bile affects fat blobs in the small intestine.

c Explain how pancreatic juice digests lipids.

d Explain how the efficiency of lipid digestion depends on the effect of bile on the intestinal contents.

9 The diagram shows a microscopic section of the inner lining of the small intestine.

Identify structures labelled **a** to **d**.

e What role of the small intestine is the diagram relevant to?

f How do structures **a** and **b** enable the small intestine to carry out that role efficiently?

g What is the function of network **c** and how is its structure adapted to that function?

h What is the function of vessel **d** and how is its structure adapted to that function?

10 Read the article opposite, then answer the questions below.

a What food types does milk supply?

b Why must carbohydrates, lipids and proteins be digested?

c Why does milk taste a bit sweet?

d Describe the structure of lactose.

e What does lactase do to lactose?

f What happens to the products?

g Describe the structure of a typical milkfat.

h What does lipase do to a milkfat?

i What happens to the products?

j Describe the structure of a casein molecule.

k What does pepsin do to casein?

l What does trypsin do to the resulting products?

m What happens to the products?

Digesting Milk

Milk contains carbohydrates, lipids and proteins as well as mineral ions (Ca^{2+}, K^+, Na^+, Mg^{2+}) and vitamins A, B_1, B_2 and E.

The digestive system must break down the carbohydrate, lipid and protein molecules into smaller soluble molecules before they can be absorbed into the blood to supply cells.

The main carbohydrate is the double-sugar lactose, consisting of bonded galactose and glucose units.

The pancreas makes the enzyme lactase and secretes it into the small intestine. Lactase splits lactose into glucose and galactose molecules, which are absorbed into the blood by the villi.

Milk contains a variety of lipids, which are collectively called milkfat. Each lipid consists of three fatty acid units bonded to a single glycerol unit.

The pancreas makes pancreatic lipase and secretes it into the small intestine. The enzyme splits milk fats into glycerol and fatty acid molecules, which are absorbed by the lymph system before entering the blood.

The main protein in milk is casein, which is a complex molecule consisting of a large number of different linked amino acids.

The stomach wall produces the enzyme pepsin, which splits casein into shorter peptides. These are then digested into amino acids by trypsin and other enzymes secreted by the pancreas. The amino acids are absorbed into the blood.

11 The diagram shows the internal structure of a tooth and the shape and size of human teeth.

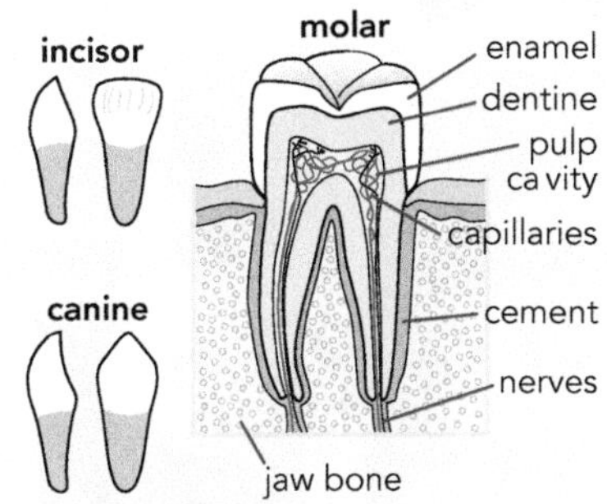

a What is the hardest part of a tooth?

b Why are teeth supplied with blood vessels?

c Why are teeth supplied with nerves?

d What holds teeth securely in the jaw bone?

e Draw up a table to contrast the size, roots, shape, sharpness, role and action of different types of teeth.

12 The graphs shows the rate at which salivary amylase digests starch molecules at certain temperatures.

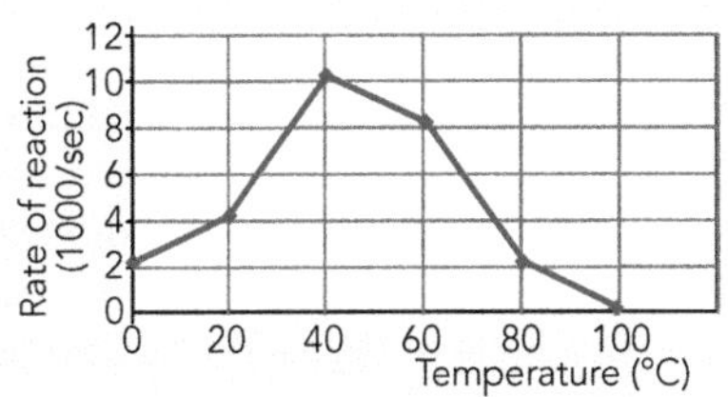

a Describe the trend.

b What is the optimum temperature for digestion by this enzyme?

c Over what temperature range does digestion occur?

d Explain why it would work over such a wide range.

13 The diagram shows the mouth of a human being.

Identify the structures labelled **a** to **g** and briefly describe how each is adapted to carry out a nutritional function.

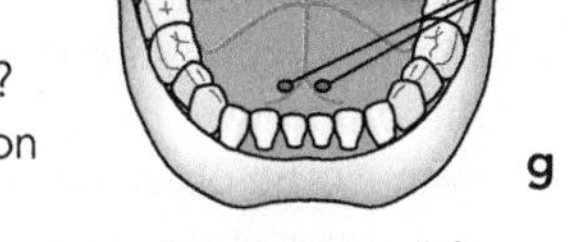

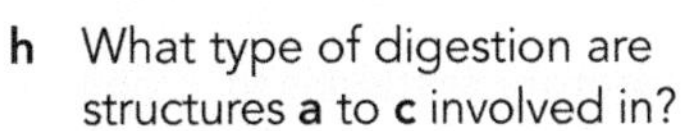

h What type of digestion are structures **a** to **c** involved in?

i What effect does this have on food particles?

j What type of digestion does the saliva secreted from structure **f** carry out?

k What effect does this have on starch molecules?

l Explain how the efficiency of chemical digestion depends upon the effectiveness of physical digestion.

ISBN: 9780170262316

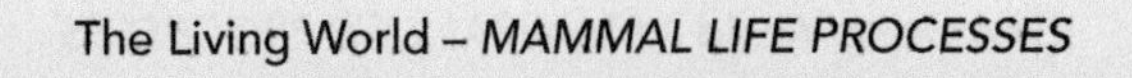

Respiratory, Circulatory and Excretory Systems

Learning Outcomes - On completing this unit you should be able to:
- describe how the respiratory, circulatory and excretory systems work
- relate the structure of organs in an organ system to their particular functions
- compare and contrast what happens in aerobic and anaerobic metabolism
- explain how the digestive, respiratory, circulatory and excretory systems cooperate
- *use a triple-axis graph to show how variables with different units change.*

Maintaining Life

- Mammals require organic nutrients, such as amino acids, fatty acids and simple sugars, for their cells to assemble the *complex molecules that are needed to carry out life processes*. The required nutrients are the products of digestion (see page 132).
- Cells use glucose molecules to provide the *energy needed by the chemical reactions that maintain life.* These reactions are collectively known as **metabolism**.
- Releasing all the energy in glucose molecules requires oxygen gas, which enters the body through the lungs. They also dispose of the carbon dioxide produced.
- Mammals are able to use the products of digestion to stay alive, keep warm, grow and reproduce, because the digestive (see Unit 31), respiratory, circulatory and excretory systems work together.

The Respiratory System

- The function of the **respiratory system** is to *supply the blood with oxygen and to remove carbon dioxide* from it. The process is called **gas exchange**, as *respiratory gases are exchanged with the environment.*
- The respiratory system consists of nasal passages, trachea and **lungs**.
- When diaphragm and rib muscles contract, the chest cavity enlarges reducing the pressure inside. Atmospheric pressure forces air through the nasal passages, down the **trachea**, into the bronchi, along ever finer bronchioles, to inflate **alveoli**.
- Their balloon-like shape greatly *increases their surface area, which optimises gas exchange.*
- The alveoli are enclosed by capillaries carrying red blood cells, whose job is to transport O_2.
- Oxygen molecules in inhaled air *dissolve* in the moist lining of the alveoli, then *diffuse* through the capillary walls into red blood cells. These contain the oxygen-carrying chemical called **haemoglobin**, which turns from dark red to bright red when loaded with O_2 molecules.
- Carbon dioxide from cellular respiration is transported to the lungs dissolved in the plasma (see opposite page). As the blood flows through capillaries surrounding alveoli, the CO_2 molecules diffuse through capillary and alveoli walls and escape into the air in alveoli. This is exhaled when ribs and diaphragm relax.
- As alveoli linings must be kept moist for gas exchange to occur, alveoli air is saturated with water vapour, some of which is lost in exhaling. Having the gas exchange surface inside the body *minimises water loss.*

The Circulatory System

- The function of the **circulatory system** is to *transport materials to all body cells and to remove their wastes.*
- The circulatory system consists of the heart, which pumps blood through **arteries** to organs. The arteries divide into **capillaries** passing cells, which reconnect to form **veins** that return the blood.
- As the blood circulates, it transports oxygen from the lungs and nutrients from the gut and liver to all cells.
- It also transports CO_2 to the lungs and other cellular wastes to the kidneys for excretion.

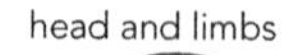
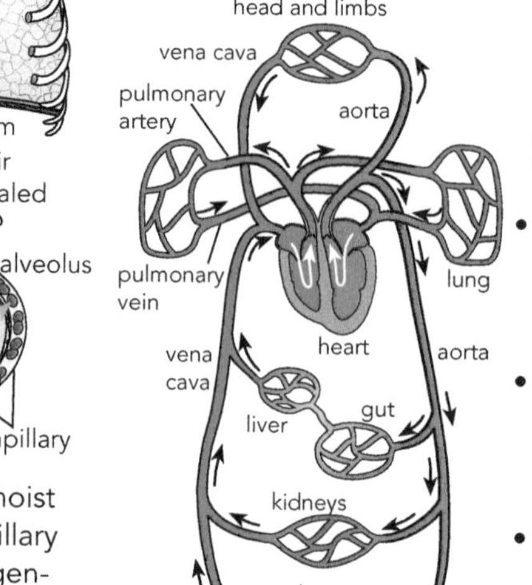
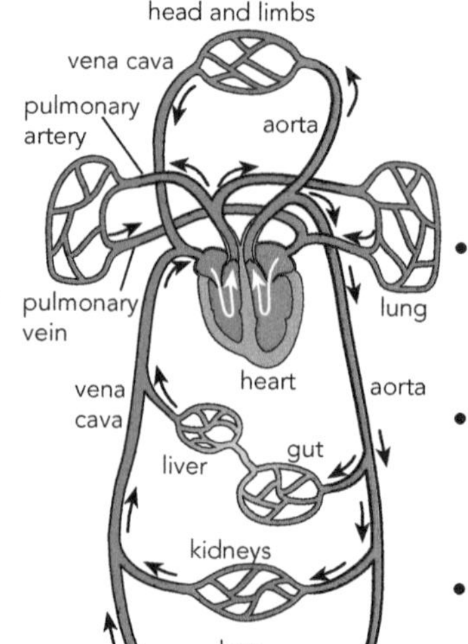

- Oxygenated blood leaving the lungs is pumped via the heart to all other organs.
- Deoxygenated blood leaving body organs is pumped via the heart back to the lungs.
- Mammals have a *double circulatory system*: one supplies the lungs, the other, all other organs.

ISBN: 9780170262316

- The **heart** has two pumps. The smaller one pumps deoxygenated blood to the lungs. The larger one pumps oxygenated blood to the body. The double system ensures the *pressure is high enough to return the blood.*

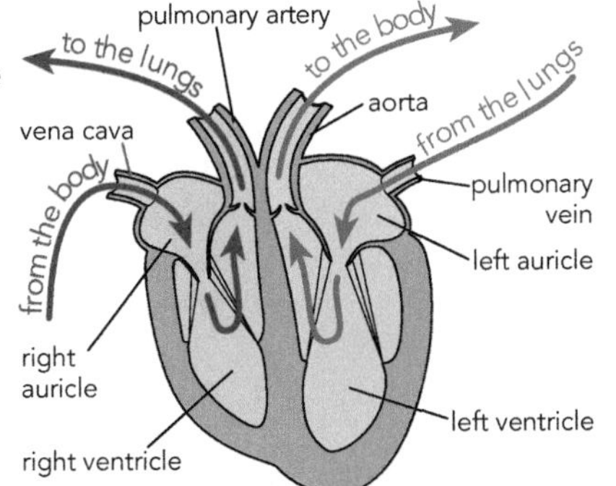

- Blood consists of red and white cells (see page 117) floating in **plasma** (water with dissolved substances).

(The names of parts of organ are not needed for NCEA.)

Exchanging Chemicals

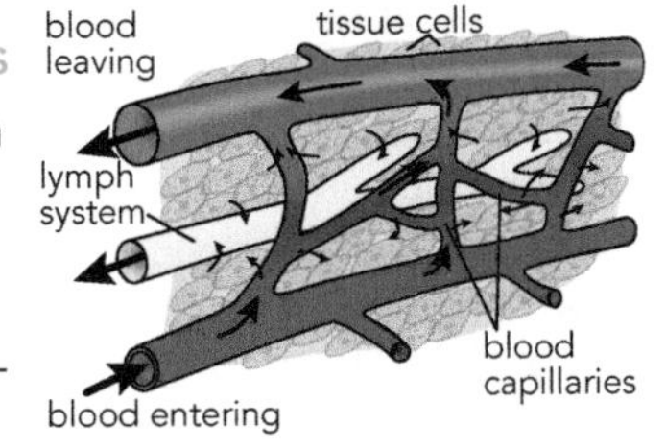

- As blood flows through capillaries, nutrients escape, and O_2 **diffuses**, through the walls into the surrounding cells. CO_2 and other wastes diffuse from cells into the blood plasma.
- The lymph system drains escaped fluid into the blood.

Animal Cells

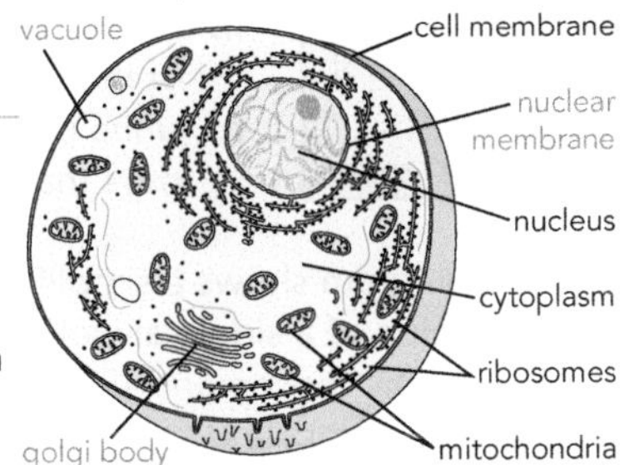

- The diagram shows the structure of a typical cell. (You do not need to know about the parts with the grey labels.)
- The cell membrane, consisting of two layers of lipids, encloses the cell and controls the entry and exit of chemicals.
- The nucleus contains the chromosomes bearing genes.
- The cytoplasm is the jelly-like interior of the cell where metabolism occurs. The ribosomes assemble proteins and the **mitochondria** carry out respiration.

Cellular Respiration

- Cells need energy in small amounts, but each glucose molecule ($C_6H_{12}O_6$) contains a large amount of energy.
- Energy is released from glucose molecules by the process of **cellular respiration**. In a series of steps controlled by respiratory enzymes, *small amounts of energy are transferred* to energy carrier molecules called **ATP** (adenosinetriphosphate).
- ATP energises all chemical reactions occurring in cells.
- The first stage of getting energy out of glucose occurs in the cell cytoplasm and does not require oxygen. In a process called **glycolysis**, each glucose molecule is split into two molecules of pyruvic acid by respiratory enzymes, but only two molecules of ATP are formed.
- What happens next depends on whether *cells are getting enough oxygen.* If not (eg in muscle cells during sprinting), the pyruvic acid is changed into lactic acid in a process called **anaerobic metabolism**. No more ATP is formed. Lactic acid causes cramps and O_2 is needed later to convert it back to pyruvic acid.

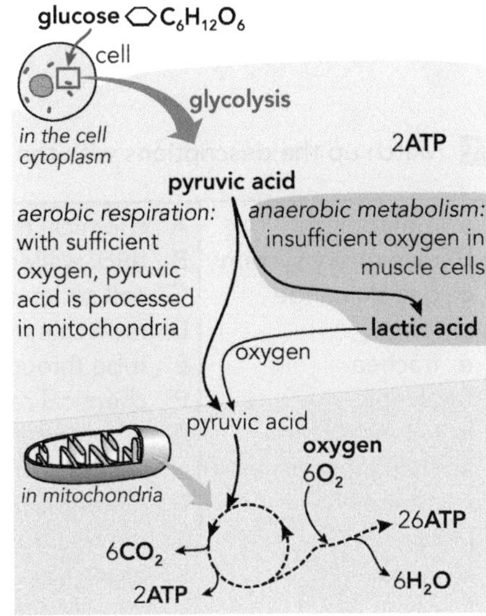

- If oxygen is plentiful, then the *rest of the energy in pyruvic acid is released* through the process of **aerobic respiration**.
- This occurs in mitochondria, producing about 28 more ATP molecules.
- Waste products CO_2 and H_2O diffuse into the blood. CO_2 is taken to the lungs for disposal. Excess H_2O is lost by urinating, sweating and exhaling.
- Although complex, the process can be summarised as:

Glucose + oxygen ⟶ carbon dioxide + water + energy

$C_6H_{12}O_6 + 6O_2 \longrightarrow 6CO_2 + 6H_2O$ + 30 ATPs

- For NCEA, you only need to know the overall steps.

The Excretory System

- The function of the **excretory system** is to *dispose of waste products of cell metabolism*, which would otherwise be toxic for cells.
- Amino acids are also used as an energy source for respiration, but ammonia is produced. This toxic chemical is transported to the liver, which converts it into urea. The blood then transports urea to the **kidneys**. Its millions of *filtering units extract urea and other soluble wastes.* They are excreted in watery urine.

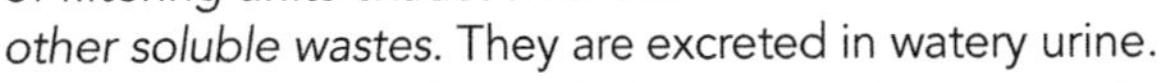

- The water content of blood plasma must kept around 92% to prevent cells being damaged. Water content rises by drinking and falls through excretion, sweating and exhaling. The kidneys play a key role in keeping the water content of the blood within safe limits by increasing or decreasing the amount of water in urine.

SCIENCE SKILL: Using a Triple-axis Graph

Three axes can be used for three different quantities.

Steps:

1 Identify the set or controlled quantity: time.

2 Identify quantities that aren't set: heart rate and volume inhaled.

3 Mark a scale for the set quantity on the horizontal axis, and scales for the other two on left and right vertical axes. Plot the data using a key.

Time (min)	Heart rate (beats)	Volume of air inhaled (litres)
1st	64	6
2nd	60	7
3rd	61	6
4th	Intense exercise	
5th	128	29
6th	124	27
7th	115	21
8th	78	15
9th	66	9
10th	62	7

ISBN: 9780170262316

Revision Activities

1 Match up the descriptions with the terms.

a metabolism	A chemical that transports oxygen in blood
b respiratory system	B thick-walled tubes carrying blood away from heart
c gas exchange	C cell structures that carry out aerobic respiration
d lungs	D molecules that carry small amounts of energy
e trachea	E tube through which air moves in and out of lungs
f alveoli	F chemical reactions in cells that maintain life
g haemoglobin	G organ system that disposes of cell waste products
h circulatory system	H very fine tubes carrying blood past cells
i arteries	I metabolism that occurs when oxygen is absent
j capillaries	J moving from higher to lower concentration areas
k veins	K organ system that carries out gas exchange
l heart	L microscopic balloon-like structures inside lungs
m plasma	M excretory organs that filter out waste products
n diffusion	N muscular pump that pushes blood around body
o mitochondria	O organs that contain the gas exchange surface
p cellular respiration	P splitting glucose to give pyruvic acid and 2 ATP
q ATP	Q organ system that transports substances to and from all organs of the body
r glycolysis	R thin-walled tubes carrying blood to heart
s anaerobic metabolism	S movement of O_2 and CO_2 in and out of organism
t aerobic respiration	T respiration that occurs when oxygen is abundant
u excretory system	U step-by-step release of energy from glucose
v kidneys	V liquid part of blood carrying dissolved substances

2 Explain the differences between the terms in each item:

a respiration and gas exchange
b inhalation and exhalation
c arteries, capillaries and veins
d diffusion and active transport
e secretion and excretion
f oxygenated and deoxygenated blood
g aerobic and anaerobic conditions.

3 Explain why nutrients are needed.

4 Name the respiratory structures labelled a to h and describe how each is adapted to carry out its function.

i Describe the main function of the respiratory system.
j Identify the gas exchange surface found inside the lungs.
k Explain why the gas exchange surface must be kept moist.
l Explain the advantage of an internal surface.
m Explain how the efficiency of gas exchange is increased.
n Describe how gases move across the surface.

5 Decide whether the following statements are true or false. Rewrite the false ones to make them correct.

a Digestion products become cell nutrients.
b Metabolism is the reactions of life.
c Gas exchange occurs in the nose.
d Alveoli inflate when the pressure in the chest cavity falls.
e O_2 travels in red blood cells and CO_2 in the blood plasma.
f Capillaries link arteries to veins.
g All arteries carry oxygenated blood.
h One side of the heart pumps deoxygenated blood and the other side, oxygenated blood.
i Chemicals enter or exit cells by diffusion or active transport.
j Ribosomes assemble proteins; mitochondria dismantle glucose.
k Haemoglobin transports O_2, and adenosinetriphosphate, energy.
l Anaerobic metabolism is 15 times more efficient than aerobic.
m Ammonia is a toxic waste product of the respiration of fatty acids.
n Urine is less concentrated on cold days than on hot days.

6 The diagram shows a simplified cross-section of a single alveolus.

a What is air movement A?
b What is molecule B?
c What is process C?
d What is blood D?
e What is blood E?
f What is process F?
g What is molecule G?
h What is air movement H?
i Why have the red blood cells changed colour between E and D?

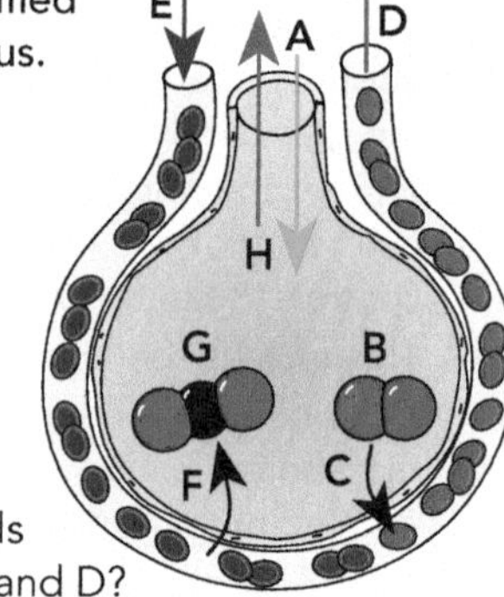

7 A PE teacher recorded the heart rate and blood pressure of a student just after a 100 m race.

Time (min)	Heart rate (beats/min)	Blood pressure (mm Hg)
1st	132	184
2nd	130	178
3rd	128	142
4th	124	120
5th	85	108
6th	72	104
7th	68	106
8th	65	102

a Draw a triple-axis graph.
b Describe the trend for each variable.
c How are heart rate and blood pressure lines related?
d Explain why the sprinter's heart beat and blood pressure took some time to return to normal.

8 The equipment shown is used to confirm that exhaled air contains CO_2 gas while inhaled air doesn't.

a How does the equipment work?
b Explain what will happen to the limewater in each test-tube.

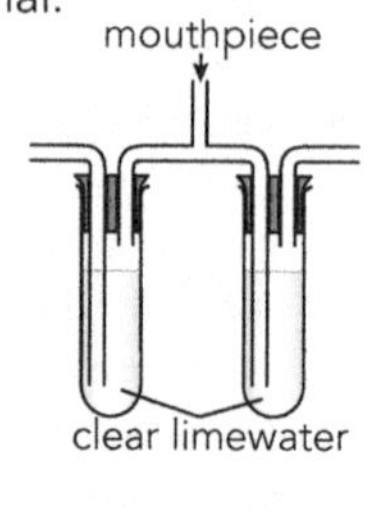

ISBN: 9780170262316

9 **The networks in the diagram represent capillaries supplying the cells within organ(s). As blood flows through a network, substances are exchanged with the organ.**

For networks labelled a to e:

- i list the substances the capillary would supply
- ii list the substances the capillary would receive.

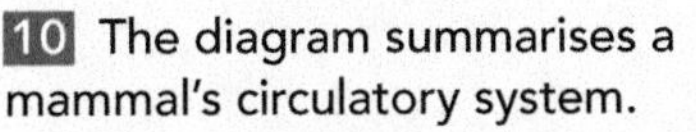

- f Describe the main function of the circulatory system.
- g Identify the four main parts of the circulatory system and describe the function of each.
- h Explain why blood in the diagram is coloured blue and red.

10 **The diagram summarises a mammal's circulatory system.**

- a What is organ **A** and why does it have two chambers?
- b Identify the organs that capillary network **B** is supplying, and justify your choice.
- c Identify the organs that capillary network **C** is supplying, and justify your choice.
- d Decide whether the blood vessels labelled **D** are arteries or veins, and justify your choice.
- e Decide whether the blood vessels labelled **E** are arteries or veins, and justify your choice.

11 **Mammals have a 'double circulatory' system. One circuit goes from the heart to the lungs and back again; the other goes from the heart to all other body organs and back again.**

- a Describe the function of each system.
- b Explain why pressure is vital for effective circulation.
- c Describe how the heart creates this pressure.
- d Predict what might happen if it was a single system.
- e Explain the advantage of having a double system.

12 **Read the article opposite, then answer the questions below.**

- a What process requires glucose and oxygen?
- b Why does a runner heat up?
- c What happens to the body if the temperature rises too much?
- d What are two ways of losing heat?
- e How does sweating cool the body?
- f What happens to the body if too much water is lost?
- g How do the kidneys help keep the body hydrated?
- h How do we know that exercise causes us to lose salt?
- i What happens to the body if too much salt is lost?
- j How do runners keep up water and salt levels in their blood?

13 **Respiration is a process that occurs in all organisms.**

- a What is the overall function of respiration?
- b Why is respiration important for all life processes?
- c What is the source of energy for respiration?
- d What gas is needed to maximise energy release?
- e Which chemical receives small amounts of energy?
- f Why is it called the 'energy carrier'?
- g What are the two waste products of respiration?
- h Write a summary word equation for respiration.
- i Write a symbol equation for respiration (see below).

14 **The diagram summarises what happens when oxygen is present and absent in muscles.**

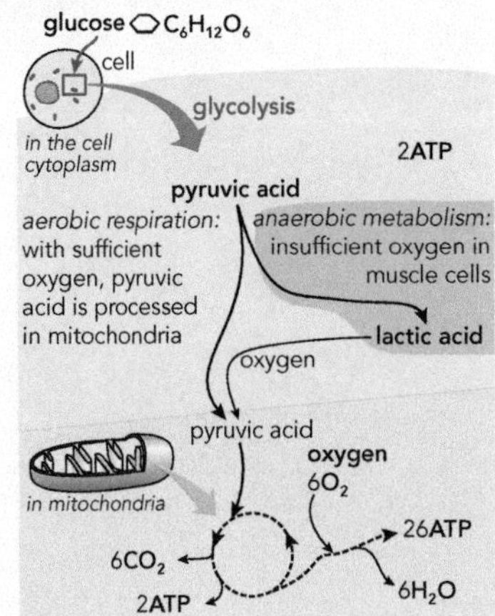

- a Compare products of aerobic and anaerobic metabolism.
- b Explain why metabolism is more energy efficient in aerobic conditions.
- c Why might anaerobic metabolism be important to sprinters?
- d Explain why it results in an O_2 debt.

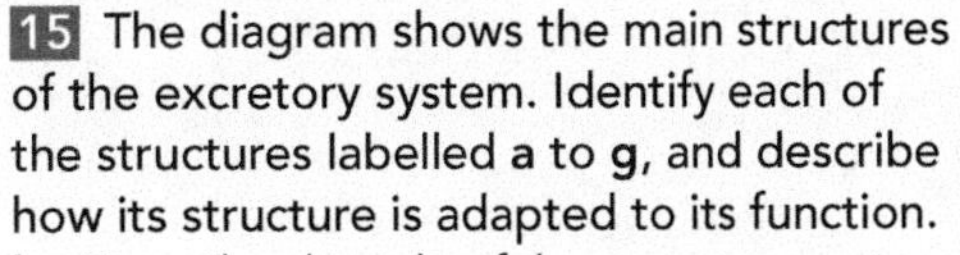

15 **The diagram shows the main structures of the excretory system. Identify each of the structures labelled a to g, and describe how its structure is adapted to its function.**

- h Describe the role of the excretory system.
- i List three chemicals that the blood in structure **a** is taking to the kidney.
- j Explain how waste chemicals are separated from useful ones in the kidneys.
- k What happens in kidney failure?
- l Outline another role of the kidneys.
- m Explain why this role is so important for cells.

16 **Write a paragraph to explain how the digestive, respiratory, circulatory and excretory systems work together to meet the needs of cells.**

Controlling Water and Salt Levels

Long distance running requires a lot of glucose and oxygen to keep the leg muscles moving, but the action of muscles also generates a lot of heat.

The temperature of the body must kept within strict limits otherwise cells and tissues will not function well, and heat exhaustion may occur.

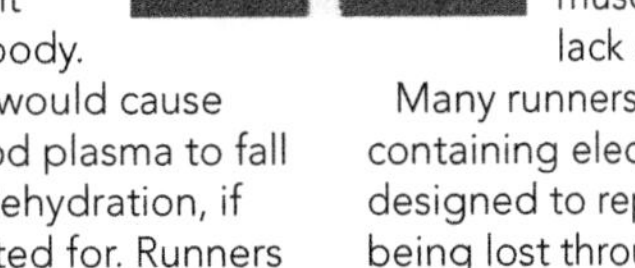

The body radiates heat, but most excess heat is lost through sweating. Water secreted by sweat glands onto the skin absorbs heat energy from the body as it evaporates, cooling the body.

Excessive loss of water would cause the water content of blood plasma to fall below 92%, resulting in dehydration, if the loss is not compensated for. Runners need to drink sufficient water to offset the loss. The kidneys also play a key role in keeping the body hydrated by reducing the amount of water excreted in urine. The urine becomes much yellower in colour, as the urea is more concentrated.

Runners also lose water in the form of water vapour in the air they exhale.

Sweat is salty in taste, which means that salt is being lost from the blood. If too much salt is lost, this can result in muscle cramps, dizziness and lack of co-ordination.

Many runners consume special drinks containing electrolytes. These are designed to replace a range of salts being lost through sweating.

ISBN: 9780170262316

PLANET Earth and BEYOND

Effects of Astronomical Cycles on Planet Earth

Learning Outcomes - On completing this unit you should be able to:
- distinguish between the real and apparent motion of heavenly bodies
- describe Earth's spin, curvature, orbit and tilt, as well as the Moon's orbit and spin
- explain the effects of each astronomical cycle on our planet
- explain how particular cycles and effects interact to cause seasons and tides
- *interpret diagrams of astronomical cycles and related events.*

Apparent Motion

- The Sun and Moon appear to rise in the east, travel westward in an arc across the sky, then set in the west. At nigh-time, the stars appear to move in circular paths, as shown in the time lapse photo opposite.
- The appearance of motion may be due to a heavenly body actually moving or *an illusion due to the fact that the planet we stand on is moving*. The latter is **apparent motion**.
- The Sun and Moon appear to move westward because Earth actually spins eastward.

Earth's Spin and Curvature

- Earth *spins eastwards* around an **axis** passing through the geographic poles.
- It has rotated for billions of years and will keep on spinning for a long time, as there are no significant forces to stop it spinning.
- The **period** for its rotation is nearly 24 hours, which defines the length of an Earth day. New Zealand lies between 34° and 47° South, and at that latitude we are spinning eastward at about 350 metres per second!
- The surface of the half of the planet facing the Sun is bathed in **solar radiation** (see page 41), consisting of ultraviolet, light and infrared radiation. Absorbed radiation heats the surface during the day, but is radiated into space during the night, cooling the surface.
- As the planet is a sphere, the higher the **latitude** north or south of the equator, the *less radiation reaches the surface because of Earth's curvature.*
- The amount of radiation received depends on the angle of the Sun, which varies according to the latitude, time of the year, and hour of the day. *The shallower the angle, the less radiation received.* At an angle of 45°, each square metre of the surface receives about 70% of the radiation when the Sun is directly overhead.

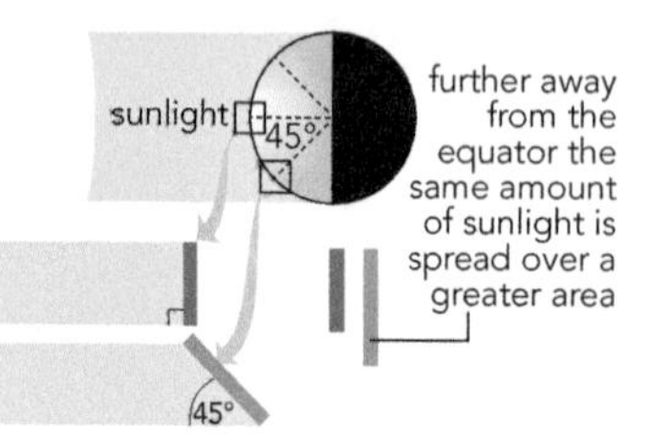

- As the Sun 'arcs' across the sky, more radiation hits the surface around the middle of the day when it is 'higher' in the sky. More radiation reaches the surface during summer, when the Sun's path is more overhead.

Earth's Orbit around the Sun

- Earth is a **planet**, which means that it orbits a **star**, the Sun. Its orbital period is approximately 365.36 days.
- As its orbit is **elliptical** rather than circular, the distance between Earth and the Sun varies between 147 and 152 million km. When Earth is closest in early January, it *receives 6.5% more radiation than when it is furthest away* in early July, but Earth's tilt (see below) has a much greater effect on temperatures.

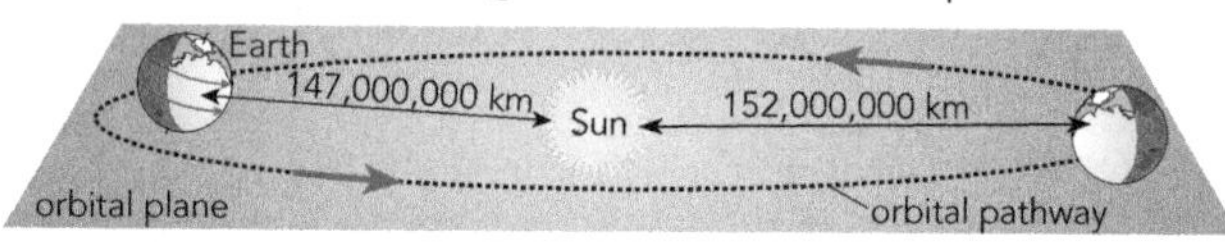

- Earth travels 940 million km around the Sun each year, at a speed of about 30 000 metres per second. We don't notice this as everything around us travels at the same speed and we have *no fixed reference point.*
- Living things can only carry out the *reactions that sustain life in solution*, but water is in a liquid state only when it is warmer than 0 °C and cooler than 100 °C.
- A planet orbiting at a distance of 150 million km gets enough solar radiation to have an average surface temperature of –19 °C. Fortunately, Earth's atmosphere *traps enough heat energy* to raise the average surface temperature to 14 °C (see page 147).
- Earth's **inertia** (see page 8) and the gravitational pull of the Sun keep the planet orbiting at a life-friendly distance. If we were closer, the planet would become too hot for life, and if further, it would freeze.
- As the planet passes through the vacuum of outer space it does not slow at all as *there is no friction.*

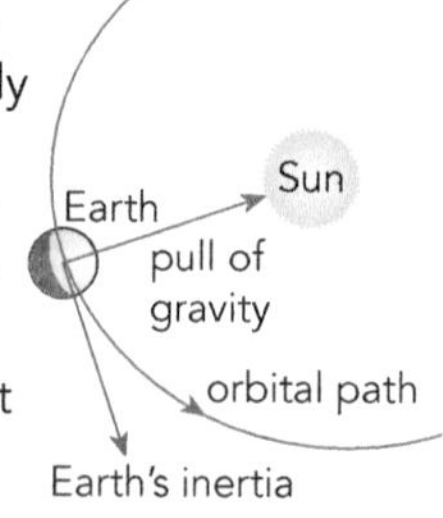

Earth's Tilted Spin

- Around 4.5 billions years ago, a smaller planet collided with Earth. The massive impact flung debris way out into space but gravity eventually pulled the debris together to form the Moon.
- The *impact tipped Earth so that it spins at an angle*. The polar axis is tilted at 23.5° to the plane in which Earth orbits the Sun, which has far-reaching consequences for the weather over each year.

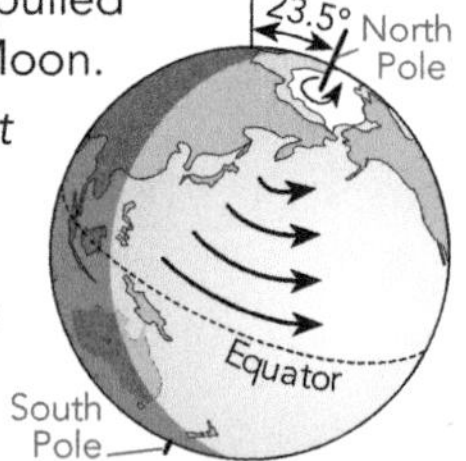

ISBN: 9780170262316

- During our summer, the Sun shines more directly onto the southern hemisphere. Temperatures are hotter as there are more hours of daylight and the Sun is higher in the sky.
- During our winter, the Sun shines more directly onto the northern hemisphere. NZ's colder temperatures result from shorter days and the Sun being lower in the sky.

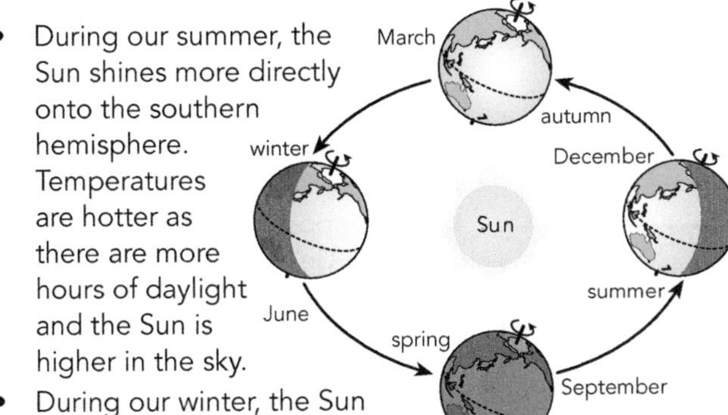

The Equator, Tropics and Polar Circles

- The **equator** is a line equally distant from both poles.
- At **summer solstice** (December 22), the Sun 'crosses' the sky directly above the Tropic of Capricorn, at 23.5° south, which is its southernmost latitude. Inside the Antarctic Circle, the surface is exposed to solar radiation all day long.

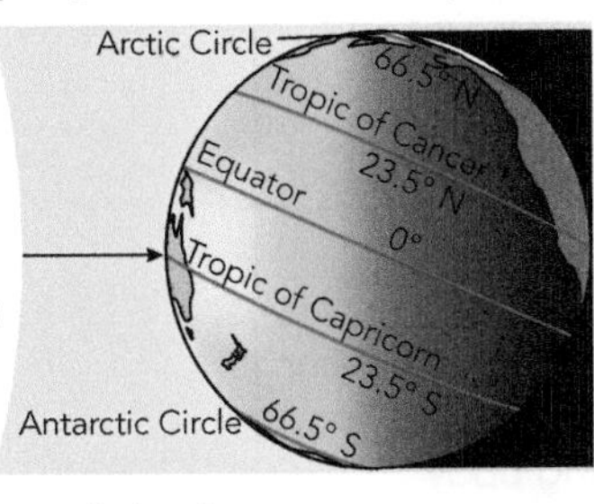

- At **winter solstice** (June 21), the Sun 'crosses' the sky directly above the Tropic of Cancer at 23.5° north, which is its northernmost latitude. Inside the Antarctic Circle, the Sun doesn't shine.

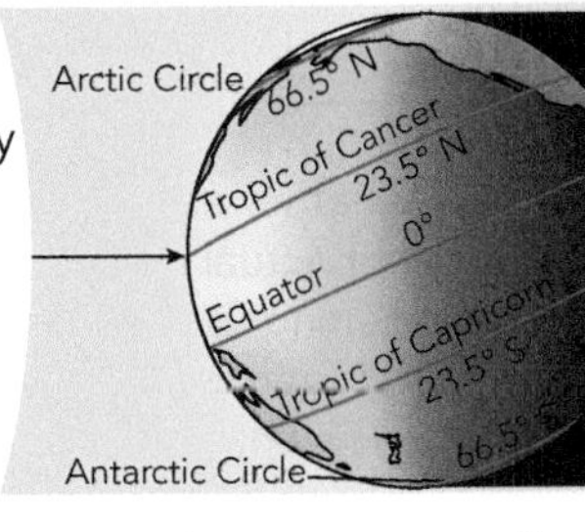

- The region between Cancer and Capricorn is called the **tropics**. It does not experience the four seasons that the regions outside do. The tropics have hot, humid weather most of the year, and a wet season once a year when the Sun is more directly overhead.

Global Winds & Ocean Currents

- The Sun *heats tropical air more than subtropical air*. The warmed tropical air expands and rises, lowering the air pressure at sea level near the equator.
- Subtropical air is cooler and denser so air pressure is higher. The *pressure difference* causes cooler surface air to flow toward the equator, creating **trade winds**.
- As Earth spins eastwards, the trade winds end up blowing from the southeast in the southern hemisphere and from the northeast in the northern hemisphere. This is known as the **Coriolis effect**.
- Higher in the atmosphere, the warmer tropical air flows away from the equator to replace the cooler subtropical surface air flowing toward the equator. This results in **convection currents** (see page 40), which *distribute heat*.

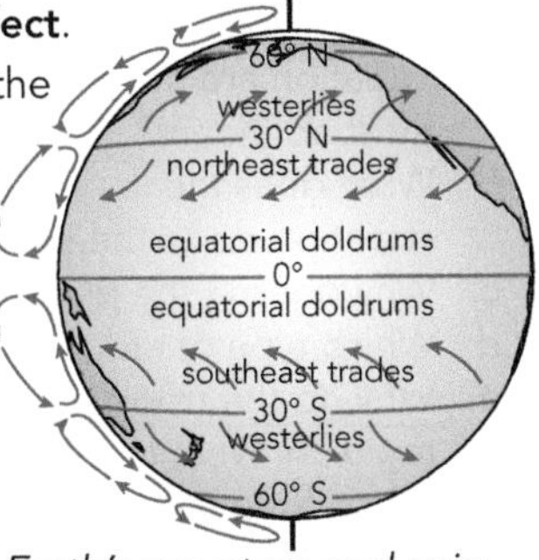

- **Global** patterns are *due to Earth's curvature and spin*.

- The chart shows the directions in which surface ocean currents move. The red arrows show warm currents and the blue, cold currents.

- Due to global wind patterns (resulting from convection and Earth's rotation), most *southern hemisphere surface currents are anti-clockwise* and northern, clockwise.
- In the South Pacific, southeast trade winds blowing across the surface cause a current that moves colder water along the coast of South America toward the equator. In response, warmer tropical water moves southward in the mid-Pacific and east of New Zealand, *distributing heat away from the tropics*.
- In the Southern Ocean surrounding Antarctica, the prevailing westerly winds create an uninterrupted circumpolar current that flows in an easterly direction.

The Moon's Orbit around Earth

- The **Moon** orbits Earth in the *same direction* in which the planet spins and orbits.

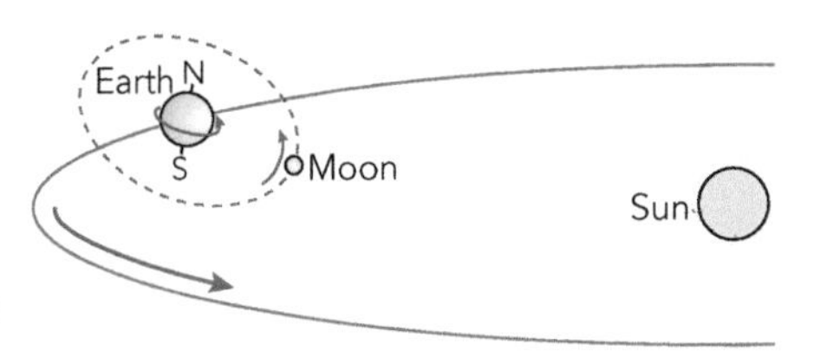

- The Moon's **orbital plane** is at an angle to Earth's, which is why the easterly direction in which it 'rises' varies.
- The same side of the Moon always faces Earth. This means it *rotates once every time it completes an orbit*, compared to Earth which rotates 365 times in an orbit.
- The moon appears in the eastern sky and 'moves' westward because Earth is actually spinning eastward.
- The period of the Moon's orbit is 27.3 Earth days.
- Each day it moves a bit further eastward around Earth, so it 'rises' on average *about 50 minutes later each day*.
- As the Moon is not a source of light, moonlight is reflected sunlight. The shape of the illuminated area of the Moon changes over a month because the *position of the Moon relative to Earth changes*.
- The chart shows the different **phases** of the Moon over a **lunar month**. It takes 29.5 days to get back to the same phase.

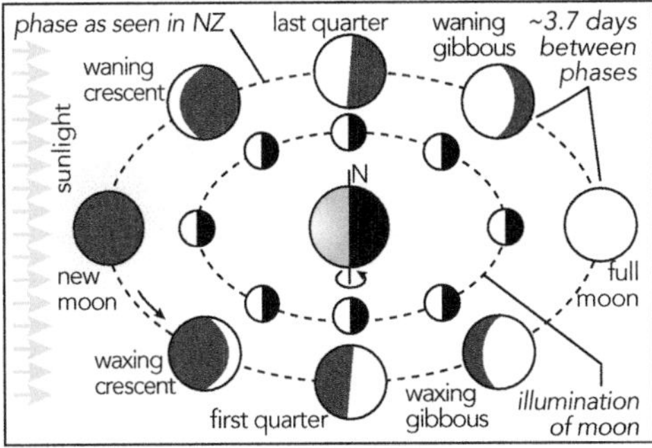

- As Earth rotates, the *gravitational pull of the Moon on the sea* causes twice-daily high (and low) tides. Each high tide occurs on average about 12 h 25 min after the previous one. The position of the Moon relative to the Sun contributes to twice-monthly spring and neap tides (see page 145).

Revision Activities

1 Match up the descriptions with the terms.

a apparent motion	A large heavenly body orbiting a star
b polar axis	B line around Earth equidistant from the poles
c latitude	C angular distance north or south of the equator
d period	D deflection of trade winds to the west
e solar radiation	E appearance of the Moon during a lunar month
f planet	F day of the year with the most hours of daylight
g star	G illusion of motion due to the observer moving
h elliptical orbit	H cooler winds moving toward the equator
i inertia	I major directions in which surface air moves
j equator	J line between poles that Earth rotates about
k summer solstice	K bulk flow of air or water due to a pressure difference caused by heating
l winter solstice	L large heavenly body orbiting a planet
m tropics	M length of time a rotation or orbit takes
n trade winds	N period of time between successive new moons
o Coriolis effect	O day of the year with the least daylight hours
p convection current	P spectrum of radiation emitted by the Sun
q global winds	Q region between Tropics of Capricorn and Cancer
r moon	R resistance of an object to any change in motion
s orbital plane	S oval-shaped pathway of an orbiting body
t Moon phases	T huge gas sphere emitting radiation from nuclear reactions
u lunar month	U imaginary flat surface in which an orbit occurs

2 Explain the differences between the terms in each item:

a apparent and actual motion
b a planet, a star and a moon
c summer and winter solstice
d a lunar day and an Earth day
e waxing and waning moons.

3 Decide whether the 'effects' shown in the photos are related to the spin, curvature, orbit or tilt of Earth, or the orbit of the Moon. (More than one may be involved.)

a

b

c

4 The table shows the maximum possible solar radiation at midday at different latitudes when the Sun is over the equator.

Latitude	Radiation (W/m^2)
0°	1030
30° S	900
45° S	730
60° S	510
90° S	0

a Identify the units.
b Graph the data.
c Describe the trend shown by the graph.
d Explain why the amount of radiation changes as it does.

5 Decide whether the following statements are true or false. Rewrite the false ones to make them correct.

a Sun and Moon travel westward.
b Stars appear to rotate around the South Celestial Pole each night.
c The Moon is only visible at night.
d The alternation of day and night is caused by Earth's rotation.
e Solar radiation warms the closest side as Earth rotates.
f Antarctica gets more radiation than Australia due to Earth's curvature.
g It takes a year for Earth to return to the same point in its orbit.
h Friction will cause Earth to slow.
i A planetary collision created the Moon and tilted planet Earth.
j The equator separates hemispheres.
k Seasons are caused by Earth's spin.
l Convection currents distribute tropical heat to subtropical areas.
m Ocean currents are indirectly affected by Earth's rotation.
n Lunar days are shorter than ours.
o The phases of the Moon are caused by the Moon orbiting Earth.
p There are two low tides each day.

6 Decide whether each of the following is an example of apparent or actual motion:

a the Sun setting
b changing Moon phases
c the Moon rising
d circular star tracks
e an eclipse of the Sun
f the Sun rising in the sky.

7 The diagram shows the location of several imaginary lines on the globe.

a What does the dotted line represent?
b How is that imaginary line defined?
c What do the solid lines represent?
d How are those two locations defined?
e Where is the polar axis located?
f In which direction does Earth spin around that axis?
g How long does a single rotation take?
h Why does it keep spinning?
i Why doesn't friction slow the spin down?

8 The diagram illustrates the composition of solar radiation reaching Earth's surface.

UV 5% | visible 46% | near infrared 49%
Wavelength relative intensity
300 600 900 1200 1500 1800 2100 2400
Wavelength (nanometres)

a Which types of radiation are the most abundant?
b What does the radiation do to the surface of the planet?
c Why doesn't the planet get hotter and hotter?
d Which type of radiation can damage living things?
e Why do the tropics get more radiation than other regions? What has this to do with Earth's curvature?
f What three factors affect how much solar radiation a particular location receives?

ISBN: 9780170262316

9 The diagram shows the Sun, Earth and Moon.

a Which object is a star?
b How is a moon different from a planet?
c What shape is Earth's orbit?
d What does this mean in terms of solar radiation?
e What force holds planets and moons in orbit?
f How long does it take for Earth to orbit the Sun? What is the significance of this period of time?
g How fast is Earth orbiting the Sun? Why are we unaware of how fast Earth is going?
h How long does it take for the Moon to orbit Earth? What is the significance of this period of time?
i Explain how the Sun's gravitational pull and Earth's inertia keep Earth in orbit around the Sun.
j Explain what keeps the Moon in orbit around Earth.

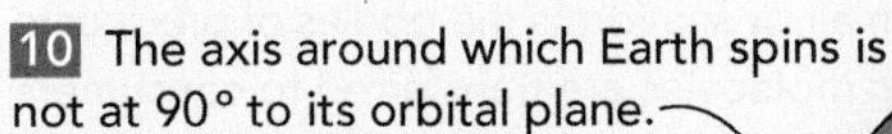

10 The axis around which Earth spins is not at 90° to its orbital plane.

a What event tilted the axis?
b What are the consequences of having a tilted axis?
c If the axis was at 90° to the orbital plane, how would conditions on Earth be different?

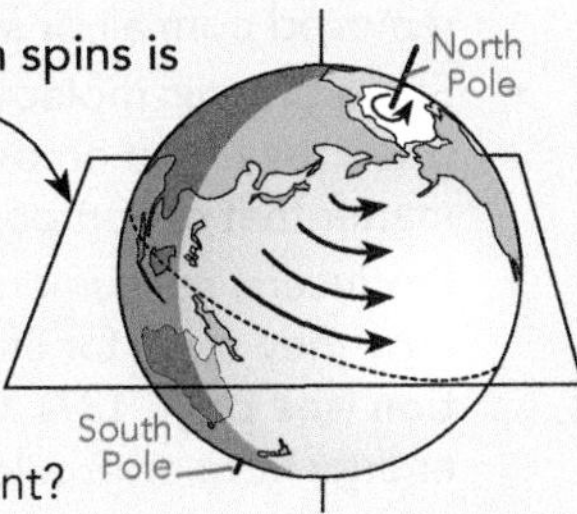

11 Earth's tilted axis results in four seasons each year in regions outside of the tropics.

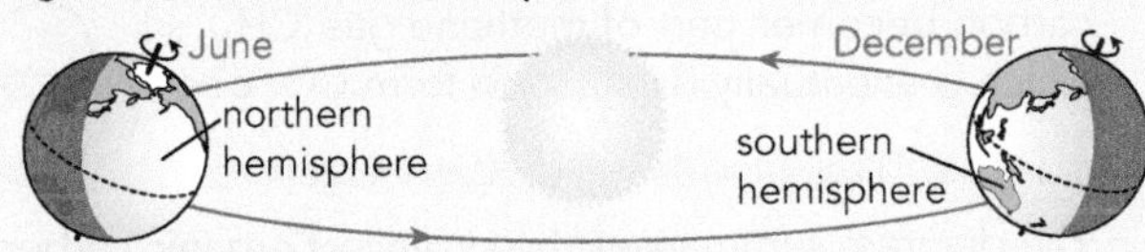

a What season is it in June in the northern hemisphere?
b Why is that season occurring?
c What season is it during December in the southern hemisphere?
d Why is that season occurring?
e What causes winter in June in the southern hemisphere?
f Why don't the tropics experience the four seasons?
g What season does occur in the tropics and when?

12 Read the article opposite, then answer the questions below.

a What forces affect Earth's tides?
b Why is the Moon's pull stronger than the Sun's despite the Sun being much more massive?
c What causes the highest tide of the day?
d What causes the two low tides?
e What causes the lowest high tide of the day?
f How does the pull of the Sun and Moon cause a spring tide?
g Why do two spring tides occur each lunar month?
h How does the pull of the Sun and Moon cause a neap tide?
i How does Earth's spin influence tides?

13 The photo was taken at midnight in Antarctica.

a What time of the year would it have been?
b Explain how it could be light at midnight by referring to astronomical cycles.

14 The diagram shows the global pattern of surface winds (white arrows) and the currents higher in the atmosphere (dark blue arrows).

a What heat transfer process causes the dark blue air flows?
b How are flows affected by Earth's curvature?
c What causes surface winds to flow from the subtropics toward the equator?
d What causes them to curve toward the west?

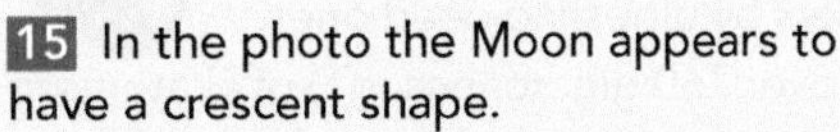

15 In the photo the Moon appears to have a crescent shape.

a Why is part of the moon illuminated and part in deep shadow?
b What astronomical cycle causes the appearance of the Moon to change?

16 The summary diagram is for the southern hemisphere.

a Why do new and full moons rise when they do?
b How are first and last quarter moons different?

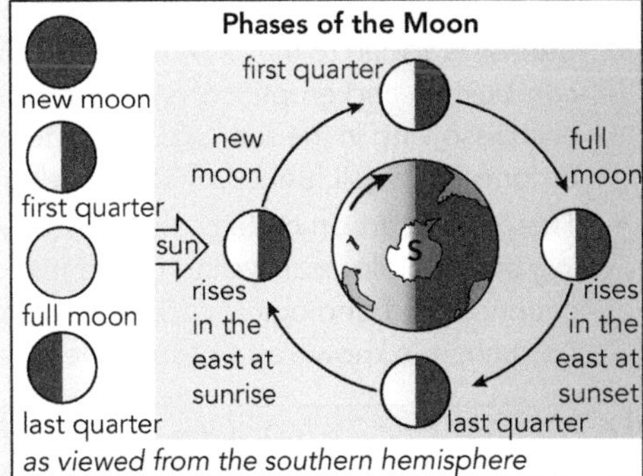

High, Low, Spring and Neap Tides

Tides are caused by the gravitational pull of the Moon and Sun changing as Earth spins. The pull of the Moon is about twice as strong as the Sun's.

As Earth rotates so that the Moon is overhead, the sea bulges in the highest high tide of the day. When the Moon rises or sets, a low tide occurs because the Moon exerts the least pull on the sea in your location.

When the opposite side of Earth is facing the Moon, a second lesser high tide of the day occurs at your location. This is due to a centrifugal effect that 'pushes' the sea outward on the side facing away from the Moon.

As the Moon orbits Earth, its position relative to the Sun changes.

Around new moon, when the Moon is on the same side of Earth as the Sun, the gravitational pull of the Sun and Moon combine to cause some very high tides each lunar month, the highest of which is called a spring tide.

Around full moon, when the Moon is on the opposite side to the Sun, the gravitational pull of the Sun and Moon interact to cause another spring tide.

When the Moon is overhead around first and last quarter, the sideway pull of the Sun to some extent flattens the upwardly bulging sea caused by the Moon's pull. This results in very low high tides, the lowest of which is called a neap tide.

Tidal times and height are predictable.

Recycling Carbon Atoms

Learning Outcomes - On completing this unit you should be able to:
- explain why carbons atoms needed to be recycled
- outline the roles played by different kinds of organisms in the cycling of carbon
- identify and describe the processes involved in organic and inorganic cycles
- compare the time scales associated with different processes of the carbon cycle
- *interpret diagrams showing the amount of stored and transferred carbon.*

Biogeochemical Cycling

- Life would not be possible without atoms of the element carbon as they *form the 'backbone'* of the complex **organic molecules** that all organisms are made of.
- Producers obtain carbon directly from the environment in the form of carbon dioxide molecules (CO_2). Consumers get C by eating living organisms and decomposers by digesting dead ones.
- There are now over 760 gigatonnes (gT) of C atoms in CO_2 molecules circulating in the atmosphere, and a similar amount (740 gT) in dissolved CO_2 molecules circulating in the oceans. ('Giga' means a billion.)
- The carbon in atmospheric CO_2 is used by plants to make organic molecules, but are returned to the atmosphere as CO_2 molecules. On average, carbon atoms in the air are *recycled once every 12.7 years.*
- Carbon is added to the air by respiration, decomposition, combustion and eruptions; removed by photosynthesis and dissolving in the sea; and stored for short (eg forest) or long (eg fossil fuels, limestone) periods of time.
- The global carbon cycle is *a complex system* consisting of the biological cycling of organic carbon and chemical and geological cycling of inorganic carbon. As such, it is known as a biogeochemical cycle.

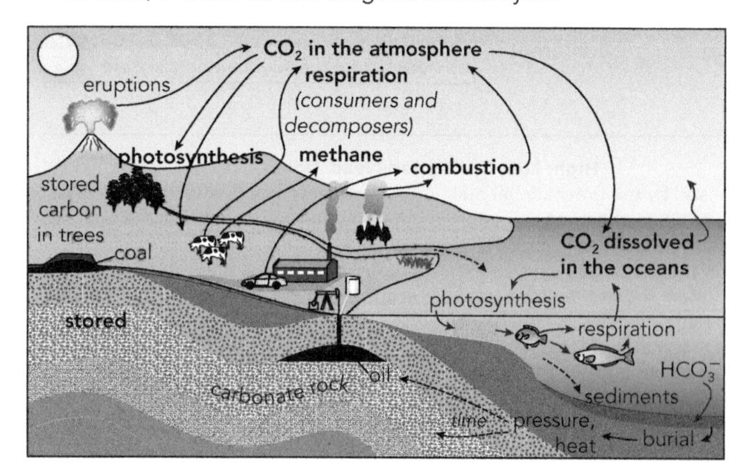

The Organic Carbon Cycle

- The biological cycling of **organic carbon** involves processes such as photosynthesis and respiration, which occur on daily and seasonal time scales, and decomposition, which can occur over days or over much longer time scales of many years.
- The key step involves producers using carbon in CO_2 gas to make organic molecules through the process of **photosynthesis** (see page 129). Much of the carbon removed from air or sea forms the bodies of producers.
- Some organic molecules are transferred to consumers when they feed on producers, others form the organic matter that decomposers feed on when organisms die.
- Producers, consumers and decomposers get the energy they need for life processes by **aerobic respiration** (see page 137). Oxygen (O_2) is used to release energy from molecules such as glucose and, in the process, *carbon is returned to the air or sea* in CO_2.
- Some decomposers get energy from organic molecules in the absence of O_2 by **anaerobic metabolism**. The carbon becomes part of methane gas (CH_4), which is eventually oxidised to form CO_2 gas.

The Terrestrial Organic Carbon Cycle

- The diagram summarises the cycling of organic carbon on a global scale, excluding fossil fuel combustion. The blue numbers show how many gigatonnes of C atoms exist in a trophic level (see page 121) or are stored in a reservoir. The red numbers show how many gigatonnes of C atoms are transferred annually.

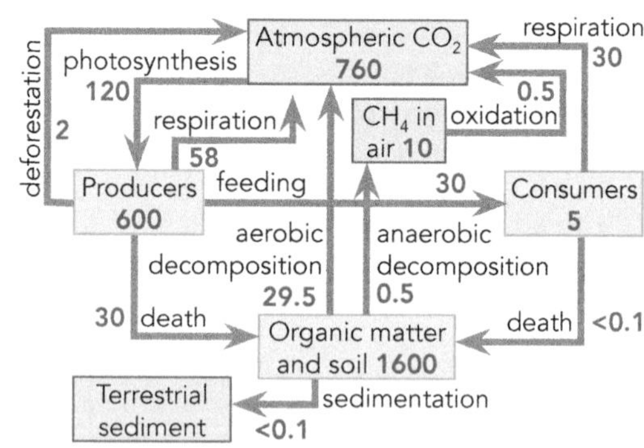

- Some carbon is stored for years in forests. **Deforestation** returns carbon to the air.
- Other carbon is removed from circulation when undecayed plant matter forms layers in swamps and bogs. Over millions of years the **sediment** gets buried and may be transformed by heat and pressure into coal or gas.
- The **combustion** of the hydrocarbons (see page 92) in **fossil fuels** adds carbon to the air in CO_2 molecules.
- The terrestrial cycle *does not occur in isolation* from the marine cycle. Depending on the location, the sea absorbs CO_2 from the air or releases it into the air.

ISBN: 9780170262316

The Marine Organic Carbon Cycle

- Marine producers are free-floating photosynthetic micro-organisms, called **phytoplankton**, that live at depths that light penetrates near the sea's surface.
- They use the carbon in dissolved CO_2 to make organic molecules by photosynthesis, just as plants do.
- Phytoplankton is consumed by **zooplankton** – free-floating micro-organisms and small invertebrates.
- As dead plankton sink to the sea floor, decomposition by bacteria transfers CO_2 and CH_4 to deep water, in *a fast process* called the **biological pump**. This CO_2 cannot be used by the surface-dwelling phytoplankton.
- Fortunately, ocean currents, called **thermohaline circulation**, return CO_2 to the surface for reuse, but this slow process *takes decades or even centuries.*
- The diagram below summarises the marine recycling of organic carbon on a global scale.

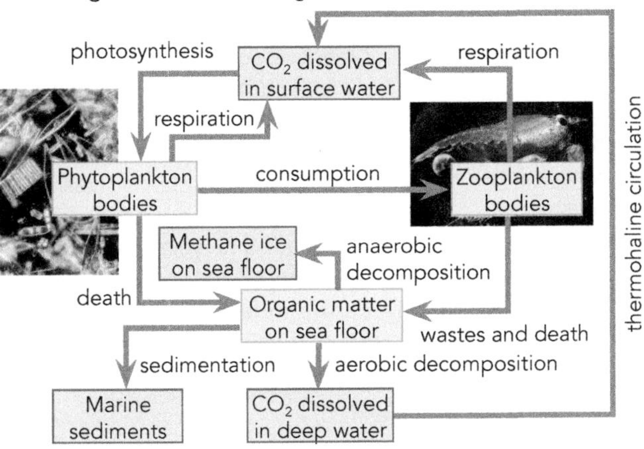

- Not all organic matter that sinks to the sea floor decomposes. Some gets buried in layers, forming marine sediments. This removes carbon from circulation.
- Over millions of years some sediments get *transformed by heat and pressure* into oil or natural gas deep underground. When these fossil fuels are combusted, carbon returns to the cycle as atmospheric CO_2.
- About 3300 gT of carbon is stored in coal, oil and gas.

The Inorganic Carbon Cycle

- Carbon is also recycled in *inorganic carbonate ions and compounds*. Some of the processes involved are chemical reactions and others are geological processes taking millions of years.
- The diagram shows the **inorganic carbon** cycle. Red and blue numbers are gigatonnes of carbon.

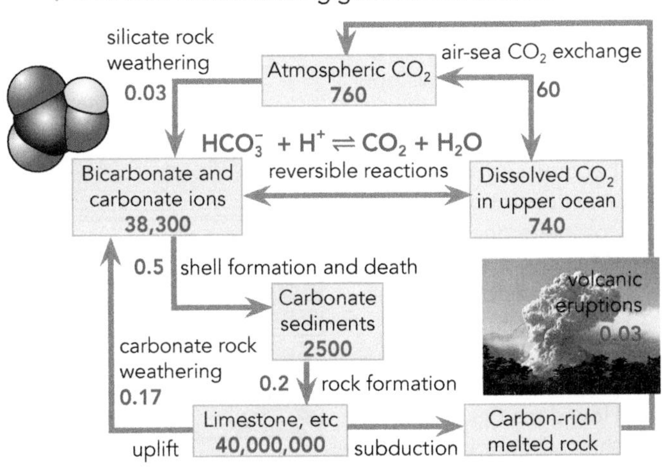

- The atmosphere and ocean surface exchange CO_2. Dissolved CO_2 reacts with H_2O to form bicarbonate (HCO_3^{2-}) and hydrogen (H^+) ions. This enables *the seas to absorb more CO_2 from the air*, resulting in the oceans having 60 times more C than the atmosphere.
- Excess H^+ ions (see page 85) cause **ocean acidification**.
- Rainwater, which is slightly acidic due to dissolved CO_2, chemically **weathers** silicate rocks. In the reaction, carbon becomes part of dissolved bicarbonate ions carried by streams and rivers into the ocean.
- Marine organisms such as shellfish and diatoms combine the bicarbonate ions with dissolved calcium ions (Ca^{2+}) to make calcium carbonate ($CaCO_3$) shells.
- When these organisms die, their shells sink to the ocean floor and get buried in layers. The carbonate sediments are *transformed by heat and pressure into limestone* (eg chalk) over millions of years.
- The annual transfer of C atoms from seawater to sediments is very small, but as it occurs over geological time spans, limestone formations now contain huge amounts of C atoms in the form of carbonate ions.
- The limestone may eventually be uplifted and exposed. Chemical weathering of this carbonate rock returns carbon to the oceans in the form of HCO_3^{2-} ions.
- Limestone may be forced deep into Earth's crust by a process called **subduction** (see page 151). When it melts into magma, CO_2 gas is formed. The pressurised gas escapes in eruptions, returning carbon to the air.

Greenhouse Gases and Effect

- CO_2 and CH_4 are the main **greenhouse gases** responsible for *global warming and climate change.*
- Global temperatures are determined by how much sunlight reaches Earth's surface, how much is reflected, and the greenhouse effect of its atmosphere.
- Earth is heated by visible **solar radiation** (see page 142) and cools by radiating infrared energy into space.

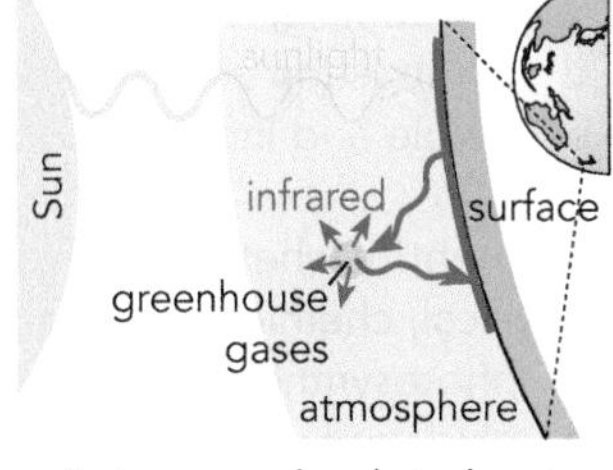

- Atmospheric gases, such as CO_2 and water vapour, absorb outgoing infrared radiation then *re-radiate some back to heat the surface more.* This **greenhouse effect** has kept the average global surface temperature around 15 °C.
- But atmospheric CO_2 passed 400 ppm (parts per million) in 2013 as a result of fossil fuel combustion and deforestation.

Atmospheric CO_2 (ppm) — graph, 320–400 ppm, Year 1970–2010

- 8.6 gT of combusted carbon was added to the atmosphere in 2012. About 4.9 gT remained in the air; the rest was absorbed by the sea *increasing its acidity*, weakening shells and corals thus affecting food chains.
- Depending on what action is or isn't taken to reduce CO_2 emissions, climate scientists predict that *global surface temperatures will rise* between 1.8 and 4.2 °C by 2100, with serious consequences for all countries.

ISBN: 9780170262316

Revision Activities

1 Match up the descriptions with the terms.

Term	Description
a organic molecules	A current that brings deep water up to the surface
b organic carbon	B feeding level within a community, eg producers
c photosynthesis	C edge of a tectonic plate is forced under another
d aerobic respiration	D releasing all the energy from glucose using oxygen
e anaerobic metabolism	E floating photosynthetic micro-organisms
f trophic level	F fuel formed from the remains of ancient organisms
g sediment	G breaking down rock through chemical reactions
h combustion	H carbon found in organic molecules
i fossil fuel	I burning a substance in oxygen to produce heat
j deforestation	J getting energy from glucose in the absence of O_2
k phytoplankton	K particles deposited on a solid surface, eg sea floor
l zooplankton	L gas that absorbs and emits infrared radiation
m biological pump	M fast transfer of carbon dioxide to the sea floor
n thermohaline circulation	N using light energy and inorganic chemicals to make organic molecules
o inorganic carbon	O removal of forest cover, often through burning
p ocean acidification	P radiation emitted by the surface of the Sun
q weathering	Q complex molecules assembled by living things
r subduction	R small floating marine consumer organisms
s greenhouse gas	S trapping heat energy within the atmosphere
t solar radiation	T carbon that is not part of an organic molecule
u greenhouse effect	U increasing amount of dissolved CO_2 in oceans releases excess H^+ ions, lowering the pH

2 Explain the differences between the terms in each of the items below:

- a producers, consumers and decomposers
- b photosynthesis and respiration
- c aerobic and anaerobic metabolism
- d combustion and respiration
- e visible and infrared radiation.

3 Identify whether each of the following is essentially a physical, chemical, biological or geological process:

- a photosynthesis
- b oxidation
- c sedimentation
- d weathering
- e dissolving
- f combustion
- g respiration
- h ocean current
- i forming ions.

4 Decide whether each event adds carbon to, or removes carbon from, the atmosphere.

5 Decide whether the following statements are true or false. Rewrite the false ones to make them correct.

- a Carbon atoms are an essential part of all organic molecules.
- b Only decomposers get carbon from the physical environment.
- c A gigatonne is a million tonnes.
- d Inorganic carbon compounds have not been produced by organisms.
- e Photosynthesis adds carbon to the atmosphere or sea.
- f Anaerobic metabolism by bacteria adds methane to the environment.
- g Photosynthesis and respiration affect the cycle in similar ways.
- h Fossil fuel and forests store carbon on different time scales.
- i Marine food chains transfer surface carbon to the ocean floor.
- j Organic sediments occur when decomposition is complete.
- k The inorganic cycle involves carbonate compounds.
- l Weathering recycles carbon.
- m Greenhouse gases cool the Earth.
- n Earth's surface absorbs light but emits infrared radiation.

6 The chart shows the movement of organic carbon through the terrestrial global cycle. Combustion is not shown. Identify each of the processes labelled a to k.

Atmospheric CO_2 760; CH_4 in air 10; Producers 600; Consumers 5; Organic matter 1600; Organic sediment; a (120); b (58); c (30); d (30); e (2); f (30); g (<0.1); h (29.5); i (0.5); j (0.5); k (<0.1)

- l What do the orange and pink boxes represent?
- m What do the blue and green numbers represent?
- n How many gigatonnes of carbon are being added to and removed from the atmosphere?
- o What does this suggest about the part of the carbon of cycle shown above?
- p What impact are human activities having on the part of the organic carbon cycle shown above?

7 The diagram shows how coal is formed.

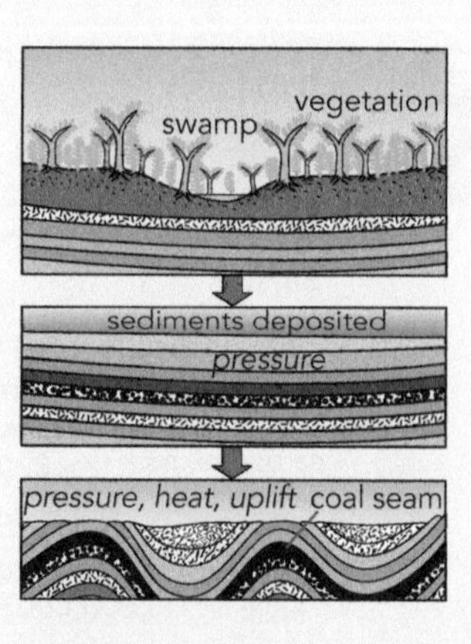

- a How does organic matter become sediment?
- b What conditions prevent complete decomposition?
- c What conditions are needed to transform organic sediment into coal?
- d How long does this process take?
- e How does this impact on the carbon cycle?

ISBN: 9780170262316

8 The diagram shows the movement of organic carbon through the marine part of the global carbon cycle.

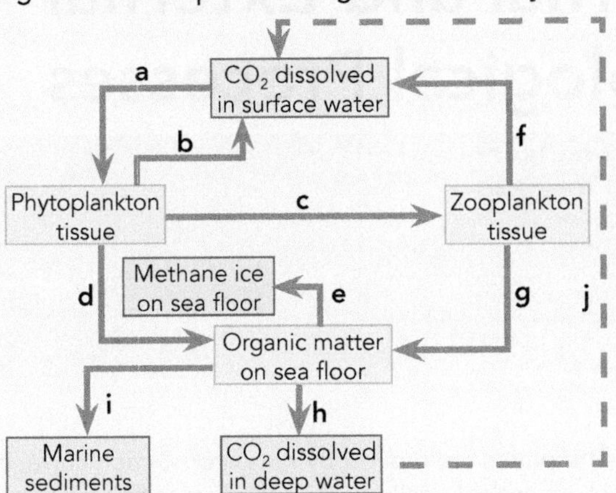

Identify each of the processes labelled a to j.

- **k** What are the three main biological processes?
- **l** Which of the processes are physical processes?
- **m** What happens to the products of aerobic and anaerobic decomposition?
- **n** Why is the CO_2 produced by aerobic decomposition not immediately available to phytoplankton?
- **o** What process makes it available and on what time scale?
- **p** How would the burning of fossil fuels impact on the part of the carbon cycle shown in the diagram?

9 The diagram shows how oil and natural gas are formed.

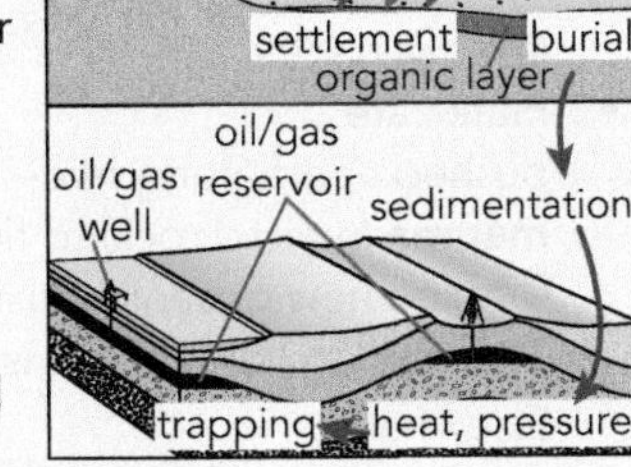

- **a** How does organic matter become sediment?
- **b** What conditions prevent its decomposition?
- **c** What conditions are needed to transform the organic sediment into oil and gas?
- **d** How long does this process take?
- **e** How does this impact on the carbon cycle?
- **f** How does the combustion of oil and gas by humans impact on the carbon cycle?
- **g** Why is the time scale for combustion so problematic?

10 Read the article opposite, then answer the questions below.

- **a** What distinguishes ruminants?
- **b** Why is plant matter hard to digest?
- **c** What is the function of cellulase?
- **d** How do ruminants and bacteria both benefit?
- **e** What kind of metabolism occurs in the rumen? Why?
- **f** What are the products of bacterial fermentation?
- **g** How much more methane does a cow produce than a sheep?
- **h** How many tonnes of CO_2 is 1 tonne of methane equivalent to?
- **i** Why is it that NZ has the highest proportion of methane in its greenhouse gas emissions?
- **j** How can it be reduced?

11 The diagram shows the movement of inorganic carbon through the global carbon cycle.

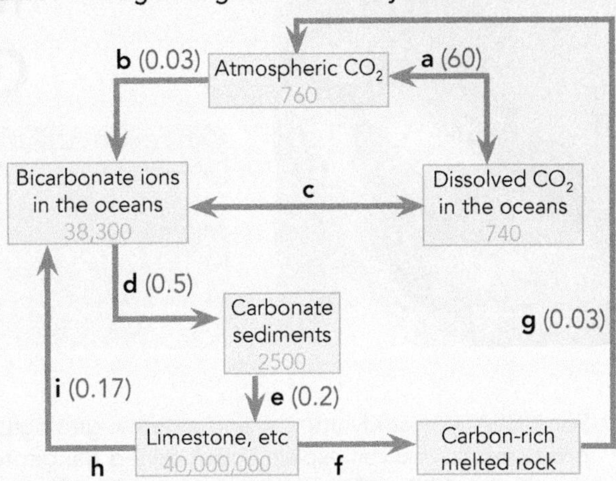

Identify each of the processes labelled a to i.

- **j** Which is a physical process?
- **k** Which two are chemical processes?
- **l** How does **c** affect ocean acidity and marine life?
- **m** Which are biological processes and what time scales do they occur in?
- **n** Which are geological processes and what time scales do they occur in?
- **o** Which processes remove CO_2 from the atmosphere?
- **p** Which processes add CO_2 to the atmosphere?

12 The diagram shows how the greenhouse effect heats the planet.

Describe each of the events labelled a to e.

- **f** Outline the three factors that determine average global temperatures.
- **g** State two greenhouse gases that are increasing.
- **h** Compare the radiation emitted by the Sun and Earth.
- **i** Explain the importance of the greenhouse effect.
- **j** Describe the processes carried out by humans that are increasing the amount of carbon dioxide in the air.
- **k** State an activity carried out by humans that is increasing the amount of methane in the air (see below).

Methane Production down on the Farm

Ruminants are animals like cows and sheep that have complex stomachs consisting of four compartments, one of which is called the rumen.

Cows and sheep eat large amounts of grass, but they lack enzymes that digest the cellulose cell walls of plant matter. They do, however, host species of bacteria in their rumens, which produce an enzyme called cellulase that digests cellulose into simple sugars. These sugars are then fermented by bacteria into fatty acids and large amounts of methane gas (CH_4).

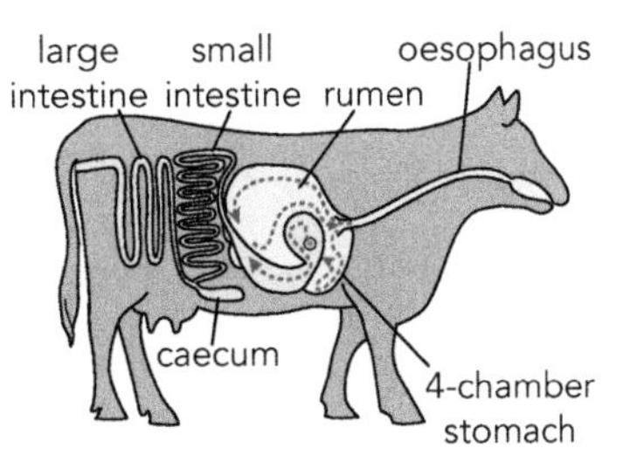

Contrary to popular opinion, most of the methane is burped rather than farted, up to 280 litres per day for a cow and around 25 litres for a sheep.

Methane may not be as abundant in the atmosphere as CO_2, but it is 22 times more potent as a greenhouse gas.

As there are about 45 million sheep and 10 million cows in NZ, potent CH_4 accounts for about a third of our contribution to global warming.

Strategies to reduce the methane produced include breeding efficient feeders, adding urea to their diet, feeding them on tannin-rich plants, and using antibiotics to control anaerobic bacteria.

ISBN: 9780170262316

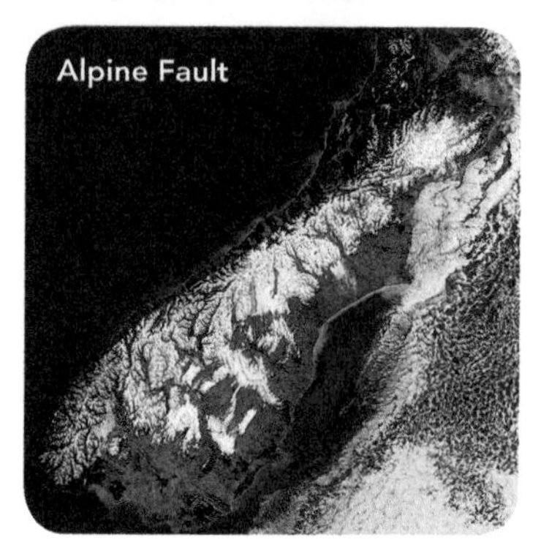

Internal and External Geological Processes

Learning Outcomes - On completing this unit you should be able to:
- describe the internal and tectonic plate structure of Earth
- explain how convection currents in the mantle move tectonic plates
- compare and contrast what happens at different kinds of plate boundaries
- explain the essential features of volcanic eruptions and earthquakes
- *interpret diagrams showing links between internal and external processes.*

- For NCEA, you study internal and external geological processes, so you can explain how local (eg Rangitoto, Lake Taupo, Milford Sound) or national surface features (eg Southern Alps, North Island ranges) are formed.

Internal Geological Processes

Earth's Internal Structure

- The 'solid' part of Planet Earth consists of four layers: the crust, mantle, outer core and inner core.
- The **crust** is the thin, cooler, outer rocky layer, which is *solid but brittle*.
- Beneath the crust is a much thicker, hotter, denser layer of rock, called the **mantle**. The rock's *semi-molten state enables it to flow extremely slowly*.

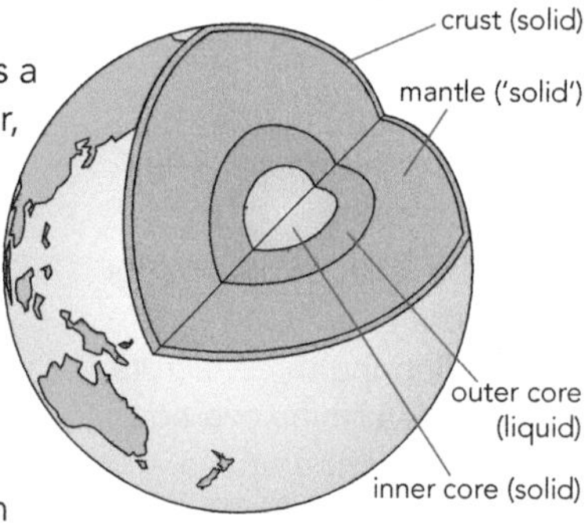

crust (solid)
mantle ('solid')
outer core (liquid)
inner core (solid)

- Nuclear reactions in the mantle produce most of the internal heat, which reaches the surface by convection currents in the mantle and conduction through the crust (see page 40).

Oceanic and Continental Crust

- The sea-floor connects with continental land masses to form Earth's continuous crust.
- Lighter **continental crust** is about 35 km thick on average, while denser **oceanic crust** is only about 5 km thick.

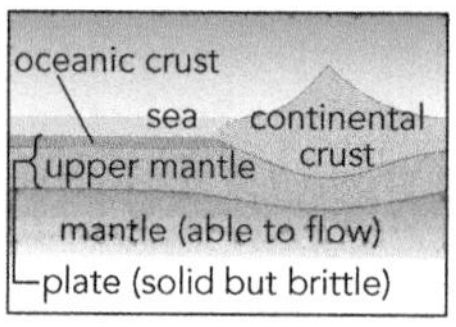

oceanic crust
sea
continental crust
upper mantle
mantle (able to flow)
plate (solid but brittle)

- The *crust is fused to the upper mantle layer,* and together they form the **lithosphere**.

Tectonic Plates

- The lithosphere is not seamless but *cracked into huge irregularly shaped slabs,* called **tectonic plates**. NZ sits astride two such plates.

Pacific Plate
Indo-Australian Plate

- The crust of some plates (eg Pacific) is almost entirely oceanic, but other plates have both. 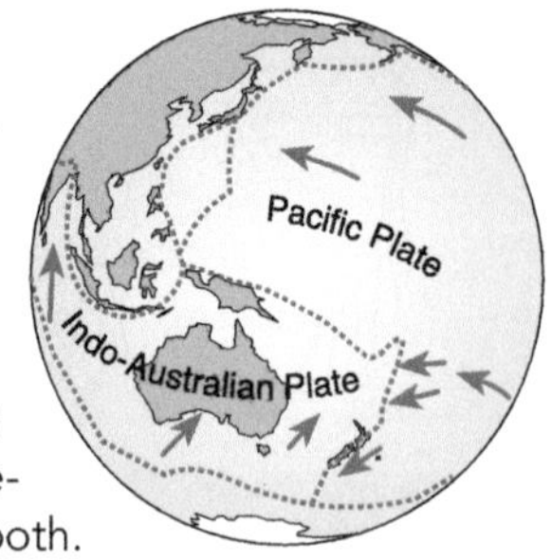
- Not only is the lithosphere cracked into about 20 tectonic plates but the plates are *moving relative to each other* at a rate of a few centimetres per year.
- The plates are moved by *huge convection currents in the mantle.* Locally heated lower mantle rock expands becoming less dense, so it rises due to its **buoyancy**. When it reaches the upper mantle it flows sideways transferring heat to the lithosphere. As it cools it contracts becoming more dense, so it sinks to the lower mantle.
- In the upper mantle these currents flow in different directions, pushing the overriding plates apart, together or past each other.

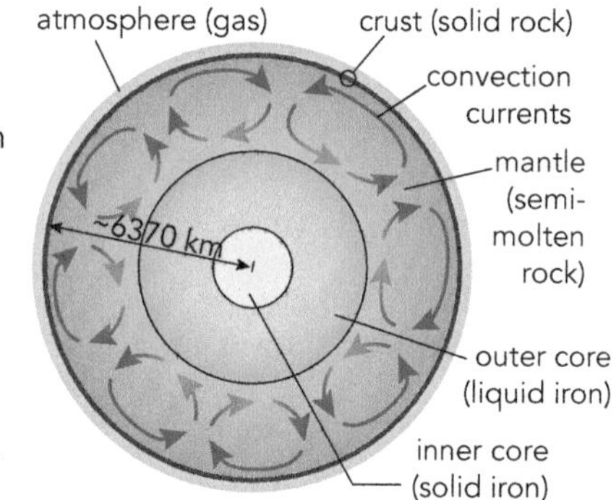

- Plates are pushed apart along mid-ocean ridges. This allows **magma** (melted rock) to flow up from the mantle, forming new oceanic crust on either side of the ridge. This geological process is called **sea-floor spreading**.
- Elsewhere, oceanic crust is **subducted** (pushed down) into the mantle and melted. This occurs in deep sea trenches alongside continents.

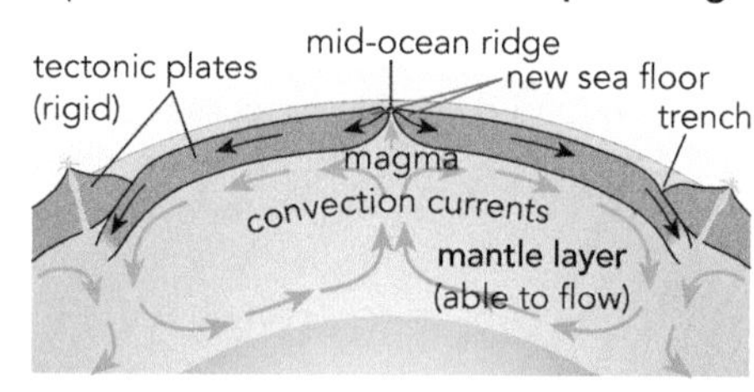

- Sea-floor spreading and subduction are related: *one creates new crust, the other destroys it.*
- The diagram shows the major directions that plates move in. Double-headed arrows show where sea-floor is spreading, and blue lines, subduction.

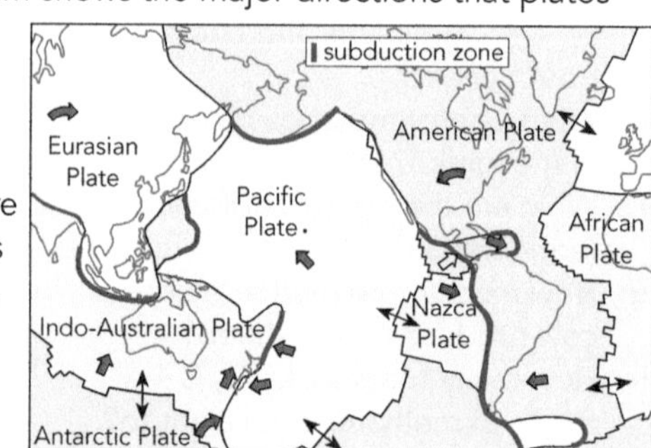

- Volcanic features, quakes and mountain building mostly occur *near plate boundaries.* What happens depends on the directions plates are moving relative to each other.

ISBN: 9780170262316

Convergent and Divergent Boundaries

- At a **divergent boundary**, such as a mid-ocean ridge or rift valley, *plates are being rafted apart*, resulting in new crust being formed in between as magma rises.
- At a **convergent boundary**, such as the Hikurangi Trough running parallel to the east coast of the North Island, two different types crust are *colliding*.
- As denser oceanic crust is forced under continental crust, the upper crust is folded to form the North Island mountain ranges.
- As downgoing and over-riding slabs *jolt past each other, earthquakes occur*.
- The Pacific slab melts as it enters the mantle, and the rising magma has resulted in the chain of volcanoes from Mount Ruapehu to White Island.

Transform Boundaries

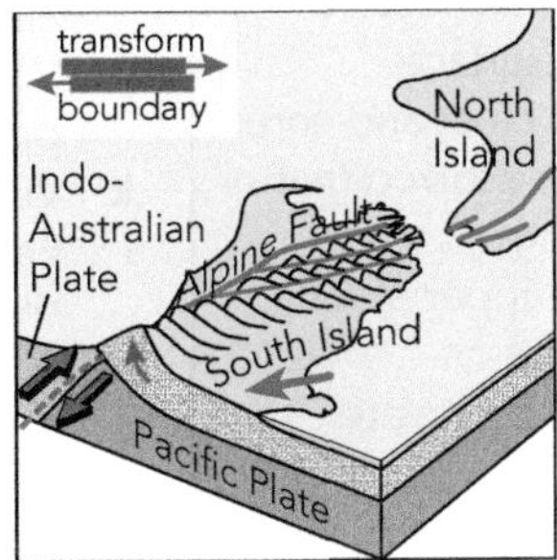

- Along a **transform boundary**, such as the 600 km Alpine Fault running from Milford Sound to Marlborough, *two plates slip sideways* (laterally) past each other rather than colliding.
- The land on the eastern side of the South Island is moving southward relative to the West Coast at a rate of about 30 metres per 1000 years. This occurs in a series of major earthquakes.
- As slippage occurs along the Alpine Fault, the crust on the eastern side is also uplifted. This mountain-building process, happening at ~1.6 metres per 1000 years, has raised the Southern Alps over millions of years.

Volcanic Eruptions

- A **volcano** is a vent in the crust, out of which gases escape, magma flows as **lava**, and vaporised magma is ejected, cooling to form ash.
- Pressure forces semi-molten rock from the top of the mantle to the surface, where lower pressure lets it melt fully.

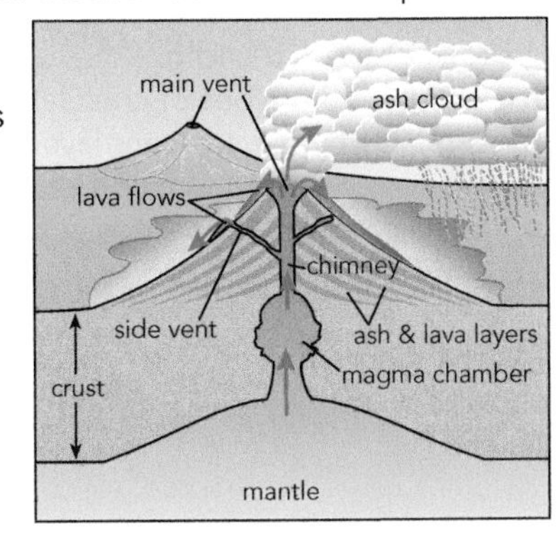

- As rock expands when it melts, the magma is less dense than the surrounding rock, so it rise in cracks, gathering in a magma chamber.
- Pressure builds as more magma arrives from below and gas pockets expand in the chamber. Eventually magma is forced up the chimney and out a vent.
- Volcanoes erupt in different ways depending on the **viscosity**, gas content and pressure of the magma.
- Some volcanoes have *low viscosity* basalt magma, which comes from melted mantle rock. They ooze streams of red-hot lava, building gentle slopes.
- Explosive volcanoes (eg Mt Tarawera) have *high viscosity, gassy magma* from subducted crust. If the slow-flowing magma solidifies blocking the chimney, pressure builds up due to heating and gas expansion.
- Eventually the blockage is blown out in a violent eruption. Gassy lava froths out, huge amounts of ash and gases (H_2O, CO_2 and SO_2) escape into the air. The viscous lava and ash *build steeply sloping volcanoes*.
- When a magma chamber empties, the volcano may collapse to form a flooded **caldera**, like Lake Taupo.
- Although most volcanic activity occurs near plate boundaries (eg the Pacific Rim of Fire), some occurs away from the edges, particularly as crust moves over mantle hot spots, eg Auckland's hotspot volcanoes.

Earthquakes and Fault Lines

- **Earthquakes** occur due to the sudden movement of adjacent blocks of the crust relative to each other.
- Along convergent or transform boundaries, the brittle *crust is subject to huge strain forces*, compressing or stretching rock, giving it elastic potential energy. The rocks cannot glide past each other because of friction.
- Eventually, when the *strain force exceeds the strength of the rock*, a sudden fracture occurs in the crust, causing a massive jolt as one block moves relative to another. Blocks move horizontally and/or vertically.
- The point where the fracture started is the focus of the quake; the point on the surface above is the **epicentre**; and the break between the two blocks is called a **fault**.
- When stressed rock fractures, a huge amount of elastic energy is transformed into kinetic energy of the moving blocks. The vibrations associated with the jolt generate body and surface **seismic waves** (see page 47) that distribute the energy widely.

External Geological Process

- The exposed rocks of mountain ranges and volcanoes are affected by physical, chemical and biological forces. These **weathering** processes slowly *transform solid rock into small particles and dissolved chemicals*.
- Rock expands and fractures as overlying material is removed. Water seeps into fractures and cracks the rock as it freezes. Plant roots penetrate cracks, wedging rocks apart. Lichens degrade rocky surfaces to obtain nutrients. Acidic rainwater dissolves minerals in rocks.
- The process by which the products of weathering are removed, transported and then deposited in basins where sediments accumulate is called **erosion**.
- Erosion agents include: gravity causing rockfalls and landslides; winds blowing fine particles; streams and rivers tumbling stones and carrying dissolved materials; and waves wearing away rock.
- The rock cycle links internal and external geological processes (see page 153).

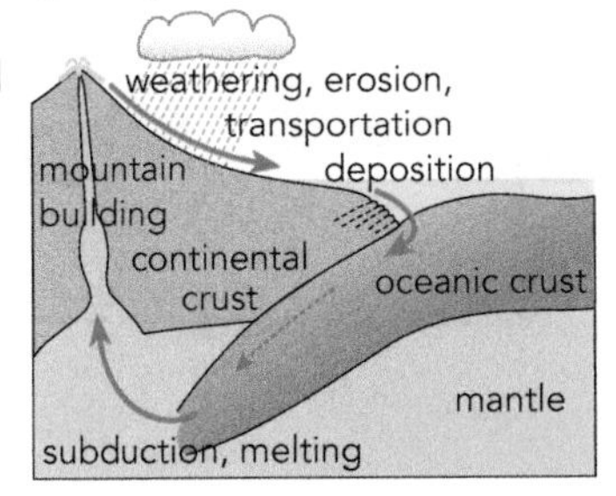

Revision Activities

1 **Match up the descriptions with the terms.**

a crust	A where plates slide sideways past each other
b mantle	B huge slabs that together form the lithosphere
c continental crust	C magma that has reached Earth's surface
d oceanic crust	D sudden jolt caused by rapid relocation of blocks
e lithosphere	E forming new oceanic crust at a mid-ocean ridge
f tectonic plates	F break between blocks caused by rock shearing
g buoyancy	G where tectonic plates are being pushed apart
h magma	H tendency of an object to float, rise or sink
i sea-floor spreading	I thin crust underlying oceans, made of basalt
j subduction	J depression caused by a volcano that has collapsed into its empty magma chamber
k divergent boundary	K thickness or stickiness of a liquid
l convergent boundary	L thin outer layer of Earth, made of light brittle rock
m transform boundary	M surface location directly above initial facture
n volcanic eruption	N molten rock found inside Earth's surface
o lava	O thick semi-molten layer between crust and core
p viscosity	P breaking rock into particles and dissolved matter
q caldera	Q where tectonic plates are colliding
r earthquake	R shell consisting of crust fused to upper mantle
s epicentre	S waves produced by an earthquake
t fault	T oceanic crust is forced under another plate
u seismic waves	U expulsion of volcanic material from a vent
v weathering	V thick crust that forms continental masses

2 **Explain the differences between the terms in each of the following items:**

- **a** internal and external geological processes
- **b** continental and oceanic crust
- **c** sea-floor spreading and subduction
- **d** divergent, convergent and transform boundaries
- **e** a volcanic crater and a caldera
- **f** weathering and erosion.

White Island

3 **Name the geological feature shown in each photo and identify the internal or external process causing it.**

4 **Decide whether the following statements are true or false. Rewrite the false ones to make them correct.**

- **a** Crustal rock is brittle but mantle rock is more pliable.
- **b** Nuclear reactions provide the energy to move plates.
- **c** Oceanic crust is thicker and denser than continental crust.
- **d** The plates form the lithosphere.
- **e** Convection currents in the core move the plates.
- **f** Subduction destroys what sea-floor spreading creates.
- **g** Plate edges are geologically stable.
- **h** Chains of volcanoes occur parallel to a convergent boundary.
- **i** Along a transform fault uplift may also occur.
- **j** Volcanic eruptions release the pressure in the magma chamber.
- **k** Viscosity doesn't affect eruptions.
- **l** Earthquakes release rock stress by shearing and shifting blocks.
- **m** A fault line is where the fault emerges on the surface.
- **n** Weathering has the opposite effect to mountain building.

5 **The wedge diagram, which is not drawn to scale, shows the layers of Earth.**

Label structures a to i.

- **j** State the main element found in layer **a** and describe its influence (see page 32).
- **k** Compare the nature of layers **b** and **c** and suggest why they differ.
- **l** Describe the heat transfer process occurring in layer **d** and state the heat source.
- **m** Describe the key difference between the state of the rock in layers **d** and **g**.
- **n** Compare the types of crust labelled **e** and **f**.
- **o** Explain why the adjacent layers identified by label **g** are considered to function as a single unit.
- **p** Explain why the two kinds of tectonic plate labelled **h** and **i** are so different.

6 **The diagram shows how mantle and crustal rock moves.**

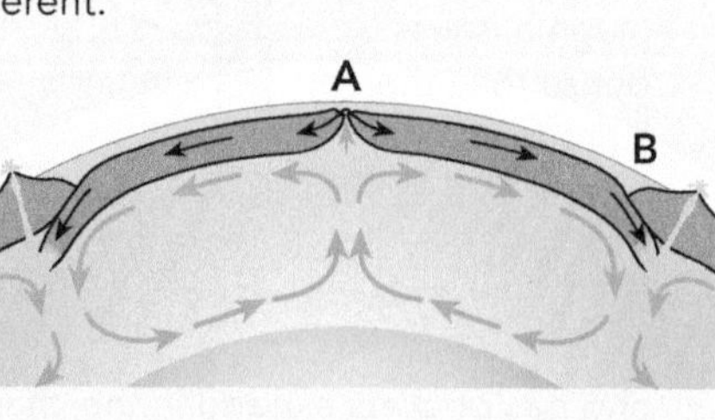

- **a** Describe the differences between the nature of mantle and crustal rock.
- **b** Explain what causes mantle convection currents.
- **c** Describe how the currents affect the plates above.
- **d** Compare and contrast the processes occurring at locations **A** and **B**.
- **e** Explain why oceanic crust is recycled but not continental.

ISBN: 9780170262316

7 The diagram shows different kinds of plate boundaries.

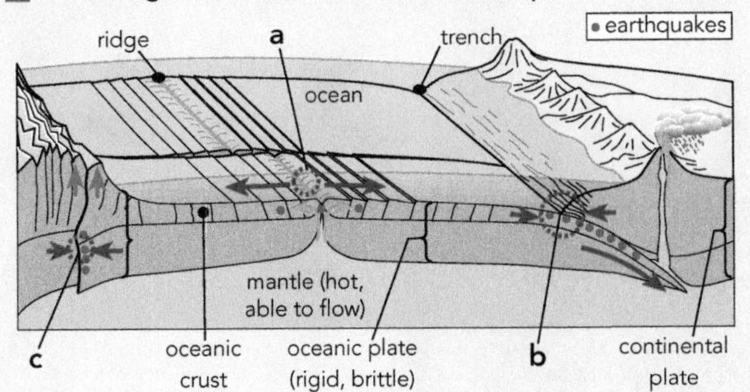

For each of the plate boundaries labelled a, b and c:

- i identify the type of boundary
- ii identify the types of crust on either side
- iii name the main geological process involved
- iv describe associated geological events
- v explain the underlying cause of the main process.

8 The diagram shows the earthquake zones of the world.

- a Where are quakes mostly located?
- b What two processes cause the quakes? Explain why.

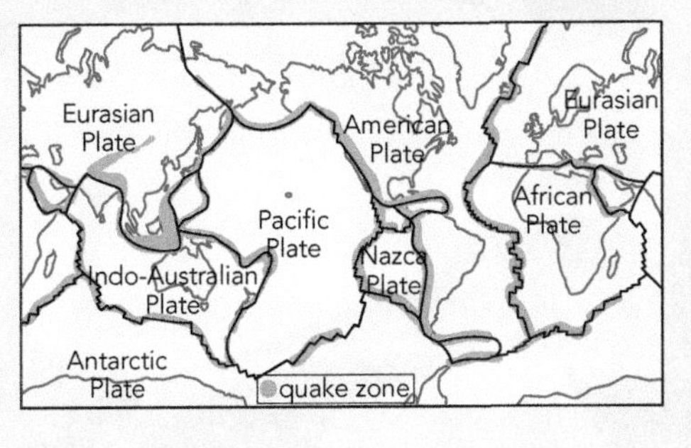

9 The diagram shows the location of the main branch of the Alpine Fault.

- a Which plate is Hokitika on?
- b What about Christchurch?
- c What kind of boundary is it?
- d What type of crust is found on either side of the fault?
- e What two movements occur along this fault?
- f When do they occur?
- g How have the Southern Alps been formed?

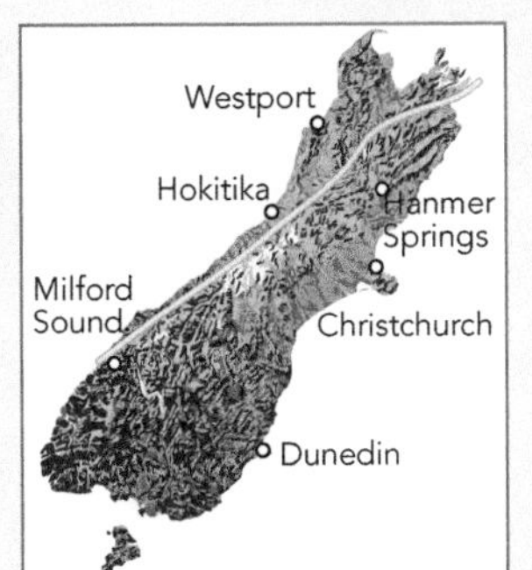

10 Read the article opposite, then answer the questions below.

- a How does water constantly reshape the landscape?
- b Why are flood plains created by gently flowing rivers?
- c How does a fast-flowing river create a V-shaped valley?
- d How does the gradient affect the flow of water in a river?
- e What is a ria (sound) and what geological event produces one?
- f How is it that the ice in a glacier can flow even though it is a solid?
- g How does a slow-flowing glacier create a U-shaped valley?
- h What are the features of a fiord and how is one formed?
- i Why is Milford Sound misnamed?

11 Use the diagram to explain how Auckland's hotspot volcanoes (eg Rangitoto, Mt Eden and One Tree Hill) were formed.

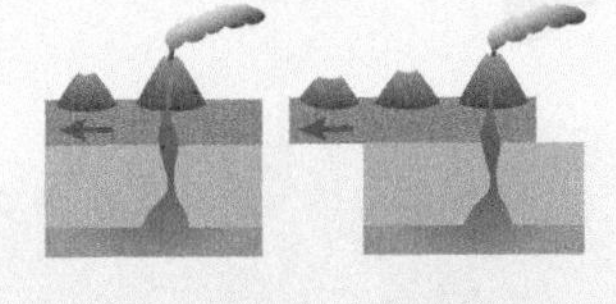

12 New Zealand is the visible part of a partially submerged continent called Zealandia, which sits astride two plates.

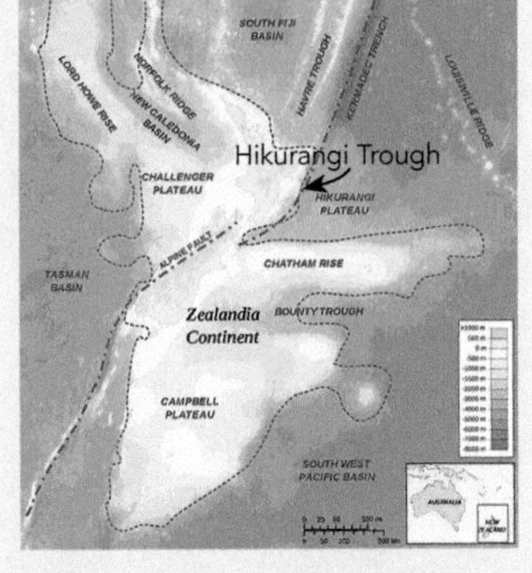

- a What are the plates?
- b What boundary type is the Hikurangi Trough and what happens along it?
- c What North Island geological features have resulted?
- d What boundary type is the Alpine Fault and what occurs?
- e What South Island geological features have resulted?

13 Study the rock cycle diagram and describe the internal and/or external geological processes that form:

- a plutonic rock
- b volcanic rock
- c sedimentary rock
- d metamorphic rock.

Metamorphic means 'changed'.
Plutonic means 'underground'.

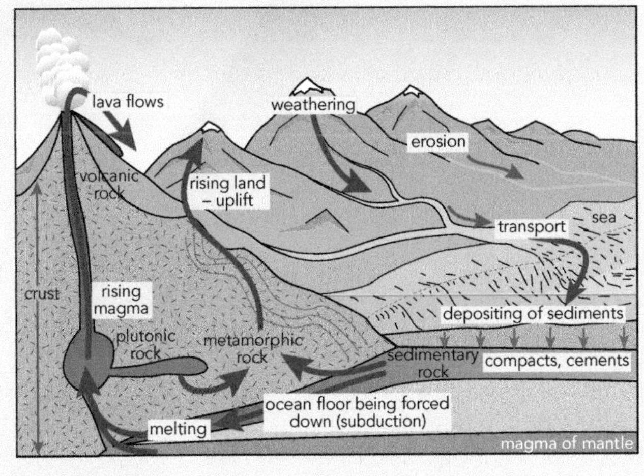

Valleys, Sounds and Fiords

Moving water, whether in a liquid or solid state, is a powerful force that continuously reshapes the landscape.

Rivers with gentle gradients, such as the Manawatu River as it crosses the plains, are wide and build up large flood plains with all the sediment they deposit.

Rivers with steep gradients, such as the water flowing through the Manawatu Gorge, are V-shaped in cross-section. The fast-flowing water wears away rock in a central channel to give steep walls.

When a network of V-shaped river valleys is drowned by rising seawater during the period between ice ages, it forms a ria or sound, such as the Queen Charlotte Sound in Marlborough.

Glaciers are huge, wide bodies of ice that move slowly downhill because of force of gravity. As they push their way forward, the ice grinds away the rock on each side, creating glacial valleys that are U-shaped in cross-section. The flat bottoms of these wide valleys are clearly visible when the ice retreats, as can be seen at the head of lakes such as Wakatipu and Tekapo.

When a glacial valley drowns due to rising sea levels, a wide but steep-sided fiord is formed. Milford Sound in Fiordland was originally a glacial valley but now its mountainous sides rise steeply and spectacularly out of the sea.

ISBN: 9780170262316

THE Nature
of SCIENCE

Practical Investigations

Learning Outcomes - On completing this unit you should be able to:
- explain what is distinctive about a scientific investigation
- describe the differences between experimental and observational methods
- formulate an aim, question or hypothesis to be investigated
- plan a controlled experiment, carry it out, then collect, process and analyse data
- interpret the results, draw valid conclusions and evaluate the investigation.

Scientific Investigations

- For NCEA level 1 practical investigation standards, the emphasis is on using investigational skills to determine the relationship between variables, to conduct a fair test, or to identify patterns in the field.
- A scientific investigation must use appropriate methods to ensure results are *objective* (obtained without bias) and *reliable* (repeatable by others), and the conclusion is *valid* (logically derived from the results).

Scientific Methods

- Different methods are used in different sciences to obtain **scientific data**.
- For the physics, chemistry and biology investigation standards you are likely to use an **experimental method** in which one variable is changed while the effect on another variable is measured. For example:
  - *how the length of a spring changes as the force applied to it is increased*
  - *how the rate of a chemical reaction changes as the concentration of a reactant is increased*
  - *how the action of an enzyme changes as the temperature of the solution increases.*
- Other variables that could have a significant effect on results need to be **controlled** (kept constant) during the experiment, so that a valid conclusion is reached.
- For physics, you will also be asked determine the **linear equation** that connects the two variables.
- For biology, you may be given the option of using an **observational method** rather than an experimental one, involving gathering data in the field using sampling methods. Sufficient samples must be collected either randomly or in a structured way to ensure that data is *unbiased* and *representative*, otherwise valid conclusions cannot be drawn.
- Either way, you will be collecting **primary data** (your own data) using an appropriate scientific method.
- Practical work will usually be done in a small group, with the analysis and interpretation done individually.

A Practical Investigation

- The investigation will either be a standard NCEA practical, or one adapted specially for your school.
- The practical investigation standards require you to 'carry out a practical ... investigation, *with direction*'.
- 'With direction' means an instruction sheet that:
  - states the purpose of the investigation
  - outlines a method or technique you can use
  - lists the equipment and materials available.
- The instructions will be of a *general nature*, so you have to decide how you will carry out the investigation after you are familiar with the method or technique.
- You may be given some training in using a technique (eg reducing parallax errors when taking readings) or procedure (eg measuring gas volumes, using a quadrat).

The Purpose of the Investigation

- You will need to be clear *why* you are being asked to do the investigation. The **purpose** may need to be formulated as a **task** to be achieved, eg:
  Find the mathematical relationship between the length of a spring and the mass hanging from it.
  or as a **question** to be answered, eg:
  What effect does changing the pH of a solution have on the action of the enzyme catalase?
  or as an **hypothesis** to be tested, eg:
  Increasing the temperature of hydrochloric acid will increase its reaction rate with magnesium ribbon.
- An **hypothesis** usually makes a statement about the relationship between two variables, in the form:
  Changing variable A in a regular way will cause variable B to change in a certain way.
- The hypothesis may be a **prediction** based on scientific concepts (eg the collision theory of reactions), or an educated guess based on observations.

The Context of the Investigation

- As the **context** of the investigation will relate to a standard you have studied (eg Chemical Reactions), you should be familiar with the scientific ideas involved (eg reaction rates, particle collision theory).

The Stages of an Investigation

- You will need to work through the following stages:
  1. Planning how to carry out the investigation
  2. Developing and specifying the method
  3. Carrying out the practical work
  4. Collecting, processing and analysing the data
  5. Interpreting the results and drawing a conclusion
  6. Evaluating your investigation
  7. Writing up a formal report of the investigation.

ISBN: 9780170262316

Planning Your Investigation

- A write-on template should be provided to help you plan the investigation. The table below is a generic format that you can copy and use.

Purpose: (task, question or hypothesis)		
What are the key variables?	(two variables)	
Which will you manipulate?	(independent variable)	
How will you change it?	(technique)	
What values will you use?	(range and number)	
Which will you measure?	(dependent variable)	
How will you measure it?	(method and units)	
What other variables could affect the results. How?	(controlled variables)	
How will you control each?	(fix the level)	
How will you ensure you get reliable results?	(repetitions, etc)	

- The chemistry example is used to illustrate some key questions you should consider whatever the standard.
  - **a** What are the **key variables** in your investigation?
    Temperature of the acid and the reaction rate
  - **b** Which is the **independent variable** you will alter?
    The temperature of the hydrochloric acid
    What range of values will you use and why?
    20°, 30°C, 40°C, 50°C, 60°C; as that is sufficient
    How will you obtain those values?
    By adding hot water to a beaker of cold water until the required temperature is reached
  - **c** Which is the **dependent variable** you will measure?
    The time for the magnesium strip to disappear
    How will you measure it and in what units?
    Using a watch and measuring in seconds
  - **d** What other variables will need to be **controlled** to Ensure results are valid, and how will you do that?
    - *length of magnesium strips (cut all to 10 mm)*
    - *thickness of strips (all strips from same ribbon)*
    - *volume of acid used (use 10 mL acid each time)*
    - *strength of acid (use same strength dilute acid)*
  - **e** How could measurement accuracy be improved?
    Place the thermometer inside the test-tube
  - **f** How could you produce more reliable data?
    Repeat three times at each temperature, then find the average result for that temperature
  - **g** What equipment will you need?
    250 ml beaker, 3 boiling tubes, thermometer, 10 mL measuring cylinder, dilute hydrochloric acid, magnesium ribbon, ruler, scissors, watch

Developing and Specifying Your Method

- Before specifying your method, try out the equipment, technique or procedure that you will be using, as that will help you to spot and resolve possible problems.
- Imagine working through all of the steps involved and jot down notes on what your group will need to do.
- When you are ready to describe how you will carry out the investigation, write down the steps involved. Your **method** needs to be detailed enough so that a student from another group would do an identical experiment if they followed your instructions. The chemistry example at the top of the next column illustrates the level of detail needed.

Method:

1. Cut 15 pieces of ribbon, each 10 mm in length.
2. Measure and pour 10 mL of dilute hydrochloric acid (ie excess acid) into each of three clean test-tubes.
3. Half fill a 250 mL beaker with cold tap water and slowly add some hot water until the temperature is just above 20°C.
4. Pour off excess water so the beaker is not too full.
5. Place the three test-tubes of acid in the water and leave for 3 minutes. Measure and record the temperature of the acid in the test-tubes.
6. Drop a magnesium strip into each test-tube and record the time it takes for each to disappear completely.
7. Wash out the test-tubes and repeat steps 2 to 6 for these temperatures: 30°, 40°C, 50°C, 60°C.

- Before carrying out the experiment, you should submit your method to your teacher, who will check that there are no safety or health issues.

Preparing a Data Table

- Before you actually do the practical part of the investigation, you need to draw up a **table** in which to record the data. The table below shows a suitable format for recording the data from the chemistry example.

Temperature of acid (°C)	Time for the bubbling to cease (s)			
	Trial 1	Trial 2	Trial 3	Mean
19	136	187	152	158
etc.				

- The values for the independent variable are entered in the left-hand column. Each column should have a heading stating what the variable is and its units.

Completing the Practical Work

- When you carry out the practical work, follow the method step-by-step, and record the data as you go.
- You can modify your method to improve accuracy or reliability, but make a note of any changes and the reasons why. For example: *We switched to recording the time it took for the bubbling to cease as it was hard to spot when all of the ribbon had gone.*
- It is important to consider how you could improve the accuracy of your measurements. If you spot a way of doing so, redo the experiment. For example: *We then used a stopwatch to increase the accuracy of our measurements and to reduce maths errors.*

Processing and Analysing Your Data

- If you get an 'extreme' measurement that is clearly inconsistent with the others, you could ignore it or repeat that part. Extreme results are called **outliers**.
- When several measurements have been taken, you must calculate the **mean** value by adding measurements and dividing by the number of repeats. For example: $(136 + 187 + 152) \div 3 = 475 \div 3 = 158.33333$.
- When you average a set of measurements the answer needs to be rounded off to the accuracy of the original measurements, eg: *158.33333 is rounded off to 158.*

ISBN: 9780170262316

- The table below gives the **raw** and **processed data** from the chemistry example.

Temperature of acid (°C)	Time for the bubbling to cease (s)			
	Trial 1	Trial 2	Trial 3	Mean
19	136	187	152	158
32	143	133	150	142
41	105	158	127	130
50	97	113	99	103
58	84	75	110	90

- **Numerical data** is usually displayed on a **graph** so that any trend can be readily identified. If the independent variable has values arranged in **classes**, a bar graph is used. If both variables have **discrete values** such as in the table above, then a line graph is used.
- Graphs need to be drawn according to conventions.

Graphing Conventions

1 Rule two axes on suitable graph paper.
2 Identify the variable to go on each axis (the independent goes on the horizontal **x** axis).
3 For each axis, label the variable and unit.
4 Decide on a suitable range for each variable.
5 Mark up a linear (even) scale along each axis.
6 Accurately plot the data using small crosses.
7 Draw a best-fit straight line or curve through the majority of the points.
8 Add an appropriate title to the graph.

- The data from the table above has been graphed below.

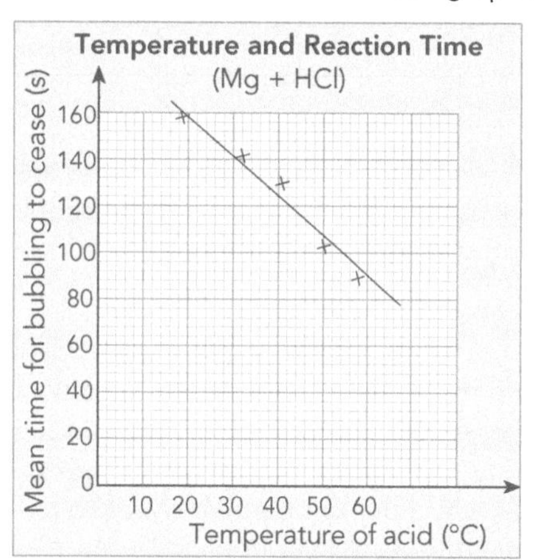

- In physics investigations you will need to find the **mathematical relationship** between the two variables. Work out the **gradient** m (= rise/run) of the line of best-fit and the value of c where it intercepts the vertical (*y*) axis. The equation for the relationship is:
Dependent variable = m x *independent variable* + c

Interpreting Your Data

- Having collected, processed and graphed your data, you now need to interpret it and draw a valid conclusion that relates to the purpose of the investigation.
- Any trend, pattern or relationship between the variables will be shown by the line of best-fit. Check whether the dependent variable increases, decreases or changes in a regular way as the independent variable is altered. For example: *The mean time for bubbling to cease decreased regularly as the acid temperature increased.*
- Next you must interpret what the trend means, for example: *The bubbling ceases when the magnesium ribbon has disappeared. Magnesium is a reactant and so the faster the strip disappears, the faster the reaction must be going. So the graph implies that the reaction rate increased as the acid temperature was increased.*

Drawing a Conclusion

- Your conclusion must link your findings to the purpose of the investigation. If the example had:
  a an *aim* or *question*, the conclusion could be: *Increasing the temperature increased the reaction rate between magnesium and hydrochloric acid as evidenced by the increasingly rapid disappearance of the magnesium ribbon at higher temperatures.*
  b an *hypothesis*, the conclusion could be: *The increasingly rapid disappearance of the ribbon at higher temperatures supported the hypothesis that increasing the temperature of hydrochloric acid increases its reaction rate with magnesium.*
- In the conclusion you should also link your findings to relevant scientific concepts, for example: *Increasing the temperature would make the acid particles move faster and therefore more would collide with magnesium atoms and have sufficient energy to react.*

Evaluating Your Investigation

- This is the final thinking stage of the investigation.
- Consider the limitations of the experiment, for example: *The results do not prove that all reactions go faster at higher temperatures, nor that Mg will react faster at acid temperatures above the range tested.*
- Review the problems and implications, for example: *It was hard to judge when strips finally disappeared, which made timing inaccurate. Strips floating on the surface seemed to disappear more slowly than immersed ones. The reaction released heat, but this would cause a similar temperature rise each time.*
- Justify changes you made to your method, for example: *Switching to recording when bubbling ceased was a more clear-cut indication of the end of the reaction.*
- Identify possible sources of error, for example: *Water temperatures varied a bit in the three test-tubes.*
- Consider unexpected results, eg: *Some Mg remained.*
- *Suggest improvements, for example: Use a water bath with a temperature control to get acid temperatures closer to the 20°, 30°, 40°, 50° and 60° values.*
- Finally, in the light of the above, evaluate:
  a the validity of your method and conclusion, for example: *The method clearly showed the time for the reaction to reach completion.*
  b the reliability of your results, for example: *The results were reasonably reliable given their consistency.*

Reporting Your Investigation

- Present your findings in a **report** using these headings:
  1 **Purpose** (aim, question or hypothesis)
  2 **Method** (steps, equipment and materials)
  3 **Results** (data table, graph, equation, interpretation)
  4 **Conclusion** (relate findings to aim and science ideas)
  5 **Discussion** (limitations, problems, justified changes, further improvements, reliability, validity, etc.)

ISBN: 9780170262316

Research Projects

Learning Outcomes - On completing this unit you should be able to:
- respond to NCEA expectations regarding conducting source research
- access, record and reference information from a variety of sources
- process information and the science relating to an issue, application or event
- interpret information and evaluate issues from a scientific perspective
- *communicate your findings effectively in a selected format.*

NCEA Investigations

- Many NCEA Science standards and a few of the Biology, Chemistry and Physics standards require you to **investigate**. In this context, 'to investigate' means to carry out an inquiry, and there are two kinds.
- The first involves carrying out a practical investigation using either an experimental or observational method (see page 156). This approach is sometimes called **primary research**.
- The second is a project that involves researching existing information and scientific knowledge. This process is called **secondary** or **source research**.
- Some standards require you to carry out primary research, most give you the option of doing either or both, and a few require secondary research only.
- The previous unit focused on primary research; this one provides guidelines for secondary research.
- Some standards require you to *investigate* a particular scientific **phenomenon** (eg electricity and magnetism, wave behaviour, heat, combustion, metallic properties, chemical reactions, life processes, human-microbe interactions), in which case, you need to have *studied the relevant science topic before investigating*.

Investigating Implications and Issues

- If a standard requires you to investigate *the implications of the phenomenon for everyday life,* there is an expectation that the implications will relate to an issue affecting 'individuals, groups of people, society in general, the environment or natural phenomena'.
- An **issue** is a topic on which people hold different viewpoints or opinions, and you will need to consider the 'different views, positions, perspectives, arguments, explanations, or opinions' involved.
- Typically, the requirement for:
  - *Achievement* involves an investigation that collects information about the issue, relates scientific evidence to the issue, and describes opposing views
  - *Merit* involves an in-depth investigation that collects sufficient relevant information and applies scientific knowledge to the issue in order to evaluate views
  - *Excellence* involves a comprehensive investigation that explicitly relates scientific knowledge and evidence to the issues, leading to a critical evaluation of views. This may involve elaborating, justifying, relating, evaluating, comparing and analysing.

Investigating Applications and Events

- Some standards focus on investigating how and why chemical/physical principles, relationships, laws or theories relate to a specific **technological application**.
- For chemistry, an 'application' refers to any use of chemistry to meet the needs of society (eg cosmetics, detergents, beverages). For physics, an 'application' must operate in a way that involves physics principles.
- For those standards, assessment is based on how well you understand the physics or chemistry involved and how it is made use of in the application.
- Other standards require you to investigate an event (eg an ecological, astronomical or geological event). The focus is on the nature of the event, the scientific principles that account for the event or its impact.
- When investigating an application or event, there is no 'issue' with conflicting views that you evaluate.

Data, Scientific Knowledge and Information

- The word '**data**' refers to measurements made when applying experimental or observational methods.
- **Primary data** are measurements that you make; **secondary data** comes from other sources.
- There are two kinds of **scientific knowledge** that you will need to consider in your research project:
  1. general scientific knowledge relating to the topic, including processes, laws, relationships, concepts, facts and theories
  2. specific research findings relating to the topic that have resulted from the work of scientists.
- Not all of the **information** that relates to an issue will come from scientists. Some of the information you collect is likely to involve observations, analysis, comments and opinions of non-scientists.

Conducting the Research

- Although you may be allowed to work together in a group collecting information, you will be required to process and present your research individually.

Selecting a Topic

- Your teacher may set the topic, provide a list for you to choose from, or allow you to freely choose a topic. If the latter, make sure it's a topic that requires significant scientific explanations.

ISBN: 9780170262316

- The topic we will use as an example is *Coal Exports*, as it's clearly contentious and there are well-articulated opposing views. Also, it could be the focus of a Physics, Chemistry, Biology or Earth Science topic.

Sourcing Information

- You will be expected to access information from several different **sources**, such as: reference texts; science internet sites; other websites; YouTube; resource sheets; interviews with scientists and spokespeople; newspaper and magazine articles.
- If you are going to use written texts or the internet as sources of information, you will need to identify the **key words** that you will look up in a book index or type into a search engine.
- Think about individual words or phrases that relate to your topic and the relevant science. Make your key words as specific and precise as you can (eg combustion, greenhouse gases). The most productive key words are not always immediately obvious but should emerge as your research proceeds.
- Scan web pages to see if they are relevant and comprehensible before reading in detail. Bookmark the most useful ones so you can revisit and reference them.
- The Wikipedia website is often a first port of call when it comes to learning more about a topic or the science that relates to the topic. However, some Wikipedia pages have warnings indicating that statements are unsupported or do not present a neutral point of view.

Processing Information

- Whatever sources you access, you will need to work through these steps in order to process information:
  1. *Select* material that is directly relevant to the topic and/or the science involved.
  2. *Record* relevant information (eg photocopy/ print useful articles and highlight, make summary notes).
  3. *Reference* sources of information as you use them.
  4. *Classify* the information you collect as either about the topic or the science relating to it.
  5. *Identify* information according to the point of view it supports, if you are investigating an issue.
- Develop a **concept map** (see page 3) that identifies the key points you have discovered in your source research and the relationships between them, particularly the connections between the science involved and the topic itself.

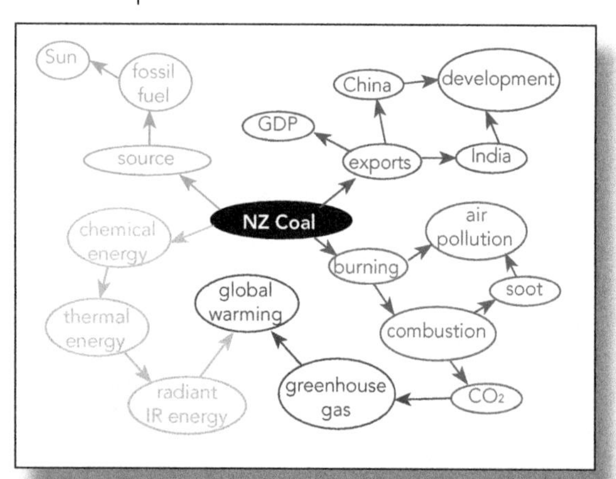

- Use your concept map to construct a set of headings and sub-headings. For an issue topic, you could use the following set of headings: The Issue; Who or What it Affects; First Point of View; Second Point of View; The Science Involved; Research Findings.
- Next, write brief **research notes** under each of your headings, summarising what you have found out using your own words. Don't copy text unless you intend to use it as a quote, otherwise it's plagiarism!
- Keep all the information you collect with your research notes, as you may be asked to hand it in when you communicate your findings.

Referencing Your Sources

- You must reference the sources you used for your research in a way that enables others to locate them.
- Standard referencing formats are shown below.

Source	Format
book	<Name of author(s)>, <*Book title*>, <Name of publisher>, <date of publication>
article	<Name of author(s)>, <*Article title*>, <Name of newspaper or magazine>, <date of publication>
web page	< title of web page>, <full url>, viewed on <date>
video clip	<full title>, YouTube, viewed on <date>

Interpreting and Evaluating

- This will be the most challenging part of your investigation. You have sourced relevant information and summarised it in your own words in a structured format, but to achieve Merit or Excellence, you must explain how 'the science' connects with 'the topic'.
- Whether you are investigating an issue, application or event you need to think about how or why scientific principles, relationships, laws and theories apply, focusing on the links between scientific ideas as well.
- If you are investigating an issue, you will also need to evaluate the claims made by the opposing viewpoints in light of the relevant scientific knowledge and research findings. If you decide to support a viewpoint, you will need to articulate the reasons why.

Communicating Findings

- Depending on the standard and your teacher, you may have to communicate your findings in a particular way, or you may have options such as:
  - writing a formal report
  - designing an informational poster
  - developing a digital presentation
  - preparing an oral report.
- To ensure that your work is *authentic*, you may be asked to bring your source material and research notes to class, then prepare your report, poster, presentation or speech within a set period of time (usually two to three hours). At the end of that period of time, you hand in your work along with your notes and the information you collected.
- Some brief guidelines for using each of the above communication modes are included on the next page. Most likely, your teacher will provide you with more specific guidance and even a template for you to use to plan how you will communicate your findings.

ISBN: 9780170262316

Written Report

- The **report** should be between two and four pages long, handwritten or typed in the form of paragraphs with supporting diagrams, graphs, maps, photos, etc.
- Headings and sub-headings will help you to structure your report. For example, for the *Coal Exports* issue, the heading structure could be:
  - Should NZ Leave its Coal in the Ground? (title)
  - Introduction (describing the nature of the issue)
  - Implications of Selling Coal (how the issue relates to the global environment and NZ's GDP)
  - Opposing Views (outlining the environmental and economic viewpoints)
  - Scientific Explanations (eg combustion, greenhouse gases, radiation trapping, global warming)
  - Evaluating Viewpoints (a critical look at the claims of both points of view from a scientific perspective)
  - Conclusion (your personal view after considering the issues and the scientific evidence)
  - References (sources you used and/or quoted).
- Make sure you proofread your report and correct any spelling and grammatical errors before you hand it in.
- Check that any text you have copied is in quote marks and the author or source is listed in the references.

Informational Poster

- You can use a large **poster** format to communicate your research. A poster is primarily a *visual means of communication*, and definitely shouldn't be the pages of your written report glued onto a large sheet of paper.
- The text sections of a poster could be under similar headings to those used for a written report, but they will need to be much briefer and to the point.
- Whoever looks at your poster should be able to absorb the key messages easily, and read all of the text in under five minutes if they choose to.

Title: briefly states the issue, readable from 3 m
Your name

Introduction or summary
The single 'big idea' you want people to remember *(they may start reading here)*

The issue
- A brief description of what you were researching
- Why this issue is important
- . . .

Evidence and information from your research
- Brief; sentences easy to read and to understand.
- Use photographs and graphs where possible.
- Present the key viewpoints on this issue and the evidence each 'side' uses.
- Your interpretation of your research findings, including implications of different viewpoints.
- Use colour sparingly (so it does not distract from your ideas).
- Could someone read it all under 5 minutes?

Atmospheric CO_2 (ppm)
Year

Acknowledgements
List all sources of information and assistance

Your conclusions or recommendations
What would you like people to do after reading your poster? *(they may skip details and finish reading here)*

- An effective poster has the right balance of text and graphics, with well-placed elements and readable text.
- Here are some tips for effective poster-making:
  - a wide format is easier to work with than a tall one
  - identify the key elements (eg text blocks, diagrams, charts, photos, graphs, etc.) that will go on your poster, hopefully no more than eight
  - sketch draft arrangements of elements on A4 paper
  - work in several columns, perhaps a wide and a narrow column
  - consider the order in which people might look at or read elements and place the more important elements in eye-catching positions
  - be consistent in the font style, size and colour you use for the title, headings, subheadings, body text, etc., whether you are working on screen or paper
  - make sure the body text is readable at a distance, most likely you will need to prune your text.

Digital Presentation

- If you choose to present your findings in a digital presentation, you will need to decide whether it will be a stand-alone **presentation** that you hand in or a classroom presentation that you will give.
- If it is the latter, then the words on the screen should be the briefest of notes, which you use as 'talking points'. Basically they are memory-joggers you expand on.
- The image shows a sample screen for a classroom presentation on *Coal Exports*.

The Greenhouse Effect
a solar radiation penetrates the atmosphere reaching the surface
b light radiation heats up the surface
c the surface of the planet re-radiates the energy as infrared radiation
d greenhouse gases absorb the outgoing infrared radiation
e greenhouse gases re-radiate some of the infrared radiation back to the surface, making the planet warmer

- Here are some tips for effective presentations:
  - use a font set at 16-18 point for the main text
  - make sure that headings and body text are done in consistent styles, in terms of font, size and colour
  - bullet points work well as talking points
  - restrict yourself to four or five points per screen
  - include diagrams and images but ensure they are big enough to be easily seen
  - prepare notes to refer to as you expand on your talking points
  - include an audio commentary if you are submitting rather than presenting.
- When presenting, don't just read what's on the screen. Your audience is capable of doing that themselves.

Oral Report

- An **oral report** is a brief talk of up to five minutes in length, which doesn't rely on a digital presentation.
- You will need to prepare a series of small speaking cards you can hold comfortably in your hand. Each card is a series of talking points, not a script to read out loud.
- You can structure your talk using similar headings to those that you would use for a written report.
- Most successful speeches follow the pattern of having an engaging opening, followed by the body of your argument, and finishing off with a clear and hopefully memorable conclusion.
- You can also use visuals aids to illustrate your talk.

Answers

Unit 2: Speed and Acceleration

1 a E b G c H d L e O f B g R h A i J j C k I l F m D n Q o R p N q K r M

2 a a speedometer measures the instantaneous speed of an object/ an accelerometer measures the instantaneous acceleration of an object b acceleration occurs when an object's speed is increasing/ deceleration occurs when an object's speed is decreasing c the instantaneous speed of an object is its speed at one point in time/ average speed is the mean speed of an object over a journey d negative acceleration is when an object is slowing down/ positive acceleration is when an object is speeding up

3

distance	d	metres	m
time	t	seconds	s
speed	v	metres per second	ms^{-1}
acceleration	a	metres per second squared	ms^{-2}

4 10 kmh^{-1} = 2.8 ms^{-1}; 50 kmh^{-1} = 14.0 ms^{-1}; 80 kmh^{-1} = 22.4 ms^{-1}; 100 kmh^{-1} = 28.0 ms^{-1}; 3.6 kmh^{-1} = 1 ms^{-1}; 18.0 kmh^{-1} = 5 ms^{-1}; 36.0 kmh^{-1} = 10 ms^{-1}

5 a true b true c true d false - means it is slowing down e true f false - measures instantaneous speed g true h true i true j false - gives its acceleration k true

6

Event	**Distance**	**Fastest Time**	**Speed** (ms^{-1})
swimming	4000 m	2610 s	1.5
cycling	120 km	9605 s	12.5
running	32 km	9913 s	3.2

a so that the speeds can be compared b kmhr^{-1} c cycling - running - swimming d swimming is so slow because water is a dense medium to move through/ running and cycling are faster as air is a less dense medium to pass through/ cycling is fastest because the gearing multiplies the distance travelled e 1.5 km swum, 44 km biked and 10 km run f overall distance = 54.5 km = 54 500 m, Alistair Brownlee's winning time of 1 h 46 min 5 s = 6365 s, therefore average speed = 54 500 ÷ 6365 = 8.6 ms^{-1} to 1 dp

7 a distance-time graph b constant speed c sprinter is accelerating d Δd = 88 m and Δt = 8 s, so slope = $\Delta d \div \Delta t$ = 88 m ÷ 8 s = 11 ms^{-1} e $v = d \div t$ = 88 m ÷ 8 so v = 11 ms^{-1} f $v = d \div t$ = 100 m ÷ 10 s = 10 ms^{-1} g because the average speed for the whole journey includes the first two seconds in which the athlete is accelerating to top speed, whilst during the last 8 seconds he is running at top speed all of the time

8 b car A accelerates at a faster rate than car B/ both cars accelerates at constant rates c for car A: Δv = 185 kmh^{-1} = 51.8 ms^{-1} and Δt = 10 s, so slope = $\Delta v \div \Delta t$ = 51.8 ms^{-1} ÷ 10 s = 5.2 ms^{-2}; for car B: Δv = 135 kmh^{-1} = 37.8 ms^{-1} and Δt = 10 s, so slope = $\Delta v \div \Delta t$ = 37.8 ms^{-1} ÷ 10 s = 3.8 ms^{-2} d the acceleration of car A is 5.2 ms^{-2} whilst the acceleration of car B is 3.8 ms^{-2} e because the athlete's body has to begin responding/ because of his reaction time

9 a fastest average speed v = 100 m ÷ 9.58 s = 10.44 ms^{-1} to 2 dp b because there is a period at the start of the race when the athlete's speed is less than maximum speed so average speed must be less than maximum speed c about 40 m d thrust (or the reaction to thrust) e the athlete's body compressing air in front, which pushes back against the athlete f it increases g remains constant

10 a speed-time graph b acceleration c acceleration in first five seconds, constant speed for two seconds, then deceleration for last three seconds d acceleration = $\Delta v \div \Delta t$ = (7.5 ms^{-1} – 0 ms^{-1}) ÷ 5 s = 7.5 ms^{-1} ÷ 5 s = 1.5 ms^{-2} e zero f acceleration = $\Delta v \div \Delta t$ = (0 ms^{-1} – 7.5 ms^{-1}) ÷ 3 s = –7.5 ms^{-1} ÷ 3 s = –2.5 ms^{-2} g distance travelled equals the area under graph line = area of left triangle + area of middle rectangle + area of right triangle = (0.5 x 5 x 75) + (2 x 75) + (0.5 x 3 x 75) = 187.5 +150 +112.5 = 450 m

11 a Task: find average speed in ms^{-1} to 1 dp Quantities: t = 12 min and d = 3250 m Formula: $v = d \div t$ Modify units: t = 12 x 60 s = 720 s Substitute: v = 3250 m ÷ 720 s Calculate: 3250 ÷ 720 = 4.5138888 Round off: 4.5138888 becomes 4.5 Answer: v = 4.5 ms^{-1}
b Task: find acceleration in ms^{-2} to 2 dp; Quantities: t = 11 s, v_i = 62 ms^{-1}, v_f = 140 ms^{-1}; Formula: $a = \Delta v \div \Delta t$; Modify units: OK; Substitute: a = (140 ms^{-1} – 62 ms^{-1}) ÷ 11 s; Calculate: (140 – 62) ÷ 11 = 7.090909; Round off: 7.090909 becomes 7.09; Answer: a = 7.09 ms^{-2}
c Task: find acceleration in ms^{-2} to 2 dp; Quantities: t = 25 s, v_i = 100 kmh^{-1}, v_f = 0 kmh^{-1}; Formula: a = $\Delta v \div \Delta t$; Modify units: 100 kmh^{-1} = 28 ms^{-1} and 0 kmh^{-1} = 0 ms^{-1}; Substitute: a = (0 ms^{-1} – 28 ms^{-1}) ÷ 25 s; Calculate: –28 ÷ 25 = –1.12; Round off: –1.12 stays as –1.12; Answer: a = –1.12 ms^{-2}

Unit 3: Force, Mass and Acceleration

1 a U b C c G d J e M f B g I h N i E j R k T l D m H n K o A p L q Q r O s F t S u P

2 a with a contact force, the object causing the force must touch another object to make it move/ with a non-contact force, the object causing the force does not need to touch another object to make it move b the magnitude of a force is the size or strength of the force/ the direction of a force is the direction in which it will move an object c mass is the amount of matter in an object/ weight is the force of gravity acting on an object d balanced forces do not produce any change in the motion of an object/ unbalanced forces will cause a free object to accelerate, decelerate or change direction e pressure is the force exerted per unit area/ force is the overall push applied to the object

3 a If the net force on an object increases, then its acceleration will increase. b If the total mass of an object increases, then its acceleration will decrease. c If the mass of an object increases, then the weight force acting on it will also increase. d If the area on which a force acts decreases, then the pressure will increase.

4 a net force = 5 N downward b net force = 21 N upward c net force = 2 N to left d net force = 28 N to left

5 a true b true c false - is sometimes required to keep an object in motion d true e false - transfer kinetic energy to objects f false - example of a non-contact force g true h true i true j false - to accelerate, decelerate or change its direction k true l false - newton m true n true

6 a thrust is F_1 b weight is F_2 c friction is F_3 d support is F_4 e friction opposes the thrust force/ friction must be the smaller force f net force in the vertical plane is zero g none h net force in the horizontal plane is 7 N toward the right i object will accelerate toward the right j balanced in the vertical plane and unbalanced in the horizontal plane k balanced forces cause no change in an object's motion l unbalanced forces will cause a free object to accelerate, decelerate or change direction

ISBN: 9780170262316

7 a because he weight is spread over a wide area b $F_g = mg$ = 65 x 10 = 650 N c $P = F/A$ = 650/0.2 = 3250 P = 3.25 kP

8 a slope is a straight line b as the net force increases, the acceleration of the trolley increases c acceleration doubles too d acceleration is directly proportional to net force e curve slopes downward rapidly then levels off f about 0.8 ms^{-2} g as the mass of the trolley is increased, the acceleration of the trolley decreases h acceleration decreases from 5 ms^{-2} to 2.5 ms^{-2}/ the acceleration is halved i acceleration is inversely proportional to mass

9 a gravitational potential energy b unbalanced force is weight force c kinetic energy d because the 'elastic force' of the stretched latex rope slows her down/ also air friction will slow her down a small amount e the 'elastic force' in the stretched latex rope f the more the latex rope is stretched, the larger the force in the rope g if an object is stationary then the forces on it must be balanced h gravitational potential energy to kinetic energy to elastic potential energy to kinetic energy to gravitational potential energy, etc

10 a weight force: $F_g = mg$ = 80 kg x 10 ms^{-2} = 800 N b support force = 800 N acting upward from the road c yes, because the bike is travelling at a constant speed on a level road d thrust force = 60 N e unbalanced forces, because the net force is not zero f acceleration: $a = F/m$ = 160 N ÷ 80 kg = 2 ms^{-2} g net force: $F = ma$ = 80 kg x 1.5 ms^{-2} = 120 N backwards h weight = 800 N downward/ support = 800 N upward/ both thrust and friction forces are zero

11 a air friction or drag slows down the jet skier/ friction between the underside of the ski and the water slows down the skier/ friction helps the skier maintain his hand and feet grip b air friction or drag will slow down sprinters/ friction between soles of shoes and track will provide grip as they run c air friction or drag will help by slowing down parachutist

Unit 4: Energy, Work and Power

1 a I b N c M d Q e F f E g A h L i S j K k R l B m P n C o H p D q G r J s O

2 a energy is defined as having the capacity to do work/ work is done when a force moves an object b kinetic energy is the type of energy possessed by a moving object/ gravitational energy is the type of energy an object gains when it is lifted c work is the transfer of energy by the application of a force/ power is the rate at which work is done d an independent variable causes a change in the dependent variable

3 a the cyclist does no work because he is applying no force b the tug does no work because the ship does not move c the linesman does work because he is applying a force to lift his body and he moves up the pole

4 a $W = Fd$ = 80 x 2000 = 160 000 J = 160 kJ b $W = Fd$ = 960 x 9 = 8640 J = 8.64 kJ c $W = Fd$ = 800 x 2400 = 1 920 000 J = 1920 kJ d $F = W/d$ = 240/12 = 20 N

5 a true b false - you do no work c true d false - depends on the size of the force and the distance the object is moved e true f true g true h false - does not depend on how long you take i true j true k true

6 a If the force used is increased, then the work done will increase. b If the distance an object is moved is increased, then the work done will increase. c If the mass of an object is increased, then its kinetic energy will increase. d If the speed of an object is increased, then its kinetic energy will increase rapidly. e If the height of an object is increased, then its potential energy will increase. f If the time taken to do work is increased, then power will decrease.

7 a yes, because a force is used and the bag moves upward b $W = Fd = mgh$ = 5 x 10 x 1 = 50 J c gravitational potential energy d 50 J, as gain in energy equals work done e because a force is being used and the trolley has moved f $W = Fd$ = 12 x 15 = 180 J g kinetic energy h 180 J, as gain in energy equals work done

8 a straight line sloping upward b as the height the box is lifted increases its potential energy increases c potential energy gain is proportional to height lifted d curves downward then levels off e as time taken to do work increases, the power output decreases f power is inversely proportional to time taken to do work g rises rapidly h as speed increases, the object's kinetic energy increases rapidly i kinetic energy is proportional to the square of the object's speed

9 a the sum of the reaction and braking distances b the time it takes you to spot the danger and apply the brakes c reaction time and speed d 56 m e a force must be applied in the opposite direction to the car's motion to slow it down f transformed into heat energy of the brakes and tyres g power of the brakes/ tyre tread/ road surface/ speed h because you will have four times as much kinetic energy to get rid of i 15 m x 4 = 60 m

10 a (own graph) b speed is the independent variable and braking distance is the dependent variable c slopes upward rapidly d braking distance increases rapidly e braking distance is proportional to the square of the speed f (own graph) g because speed is the independent variable - the one that was altered in the investigation h slopes upward rapidly i as speed increases the car's kinetic energy increases rapidly j kinetic energy is proportional to the square of the speed k very similar-shaped curves l as speed increases, both kinetic energy and breaking distance rise rapidly/ both kinetic energy and breaking distance are proportional to the square of the object's speed

Unit 5: Static Electricity and Charging

1 a E b F c P d J e O f I g N h C i H j Q k G l L m S n A o K p U q B r R s M t D u T

2 a electron b nucleus c proton d neutron e four negative charges f three positive charges g surplus of electrons h the electron labelled b because it is in the outer shell i one negative/ –1 j it would have attracted an electron from another atom k ion

3 a protons are found in the nucleus and are positively charged; neutrons are found in the nucleus and are uncharged; electrons move rapidly around the nucleus and are negatively charged b an atom is electrically neutral overall; an ion is a charged atom having lost or gained an electron c a conductor has outer shell electrons that are free to roam from atom to atom; an insulator has no free electrons roaming from atom to atom d uniform charge distribution occurs when the charge is evenly spread across the surface of an object; non-uniform charge distribution occurs when the charge is concentrated at a certain location on the surface of the object e friction charging occurs when

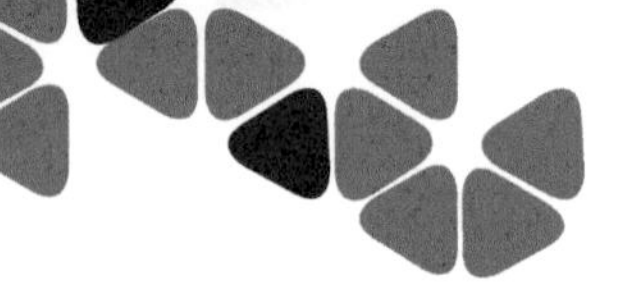

one material is rubbed by a different kind; conduction charging occurs when an insulator or an insulated conductor is touched by a charged object; induction charging occurs when a charged insulator is brought close to a neutral object

4 a the suspended rod swings toward the other rod
b the suspended rod swings away from the other rod
c the suspended rod swings toward the other rod
d the suspended rod swings away from the other rod

5 a false – either repel or attract each other b true c false – uncharged particles d true e true f false – negatively charged g true h true i false – insulator j true k false – repel each other l true m true n true

6 a friction charging b insulator c electrons are transferred from the atoms of one object to the atoms of the other d (own example, eg charging your shirt when you pull off your jersey) e conduction/contact charging f so that the charge does not escape off the conductor g electrons move from the tip of the rod and flow over the surface of the sphere h induction charging i no j free electrons are attracted to the side of the sphere nearest to the rod

7 Unlike charged objects attract each other but like charged object repel each other.

8 a because her hair is standing on end b it must be a conductor otherwise the girl's hair would not have become charged c because they are all being forced apart d a spark will discharge her e her hair would not have stood on end as moisture in the air would cause the charge to leak off her

9 a the glass rod would become positively charged and the polyester cloth negatively charged b the cling film would become negatively charged and the fur positively charged c no charging would occur because they are too close in the series d cling film because it will remove electrons off all the other materials e hair because it will lose electrons to all the other materials

10 a the pen became charged by friction charging and when it was brought close to the pieces of paper it induced the opposite charge on the closest edge of the paper, which was then attracted to the paper b your skin became charged by friction charging, which was then discharged by a spark that occurred between your finger and the metal tap, which is a conductor c your body became charged by friction charging through rubbing against the seat, the charge is then discharged by a spark between your finger and the metal frame of the car

a because it generates an electrical charge b static charge c uniformly distributed positive charge d from off the earthed metal comb at the bottom of the diagram e friction charging and conduction charging f because some of the charge will leak off the person g friction droplets in the humid air, as water is a very good conductor h they would get a nasty shock as the large charge on their skin suddenly discharges through a spark

11 a positive charge b because it is an insulator and charge will therefore not escape onto the skin of the hand with which you are holding the rod c because it is a conductor and any charge that is transferred to it will escape d onto your skin then to the ground e hold it with an insulating material such as a rubber glove f conductors have free outer electrons that roam from atom to atom; the outer electrons of insulators can be transferred to other atoms but they can't roam freely from atom to atom

12 a both will gain a negative charge b because that is the only charge that can be transferred from the charged rod c

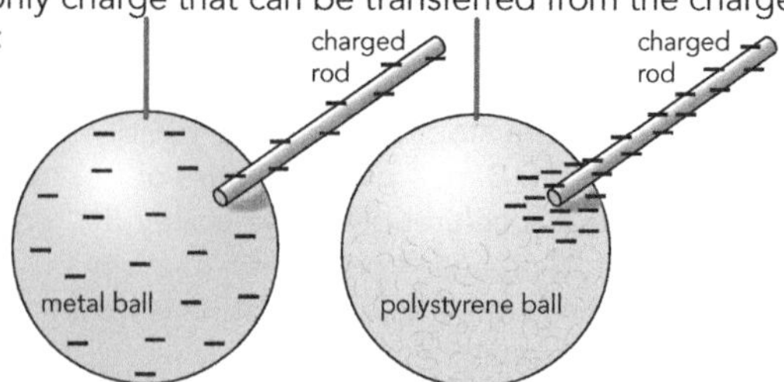

d the metal sphere because the excess electrons spread out evenly over the surface of the sphere to minimise repulsions e the plastic sphere because the excess electrons transferred to one side will stay there as electrons can't freely roam from atom to atom across the surface of an insulator

13 a otherwise the excess charge would not be able to travel to the wire and then down to the ground b the massive charge from lightning striking the unearthed rod would damage the house and possibly kill the inhabitants

Unit 6: Direct Current Electricity

1 a H b L c F d O e T f Q g A h D i P j B k N l E m I n G o J p C q S r R s K t M

2 a a conductor allows a current to pass through it/ an insulator does not allow a current to pass through it b a solar cell converts light energy into electrical energy/ a chemical cell converts chemical energy into electrical energy c a series circuit has all components in a single loop/ a parallel circuit has components on different pathways d an ammeter measures the size of the current flowing/ a voltmeter measures the amount of electrical energy gained or lost by the current e a cell transforms chemical energy into electrical energy/ a battery can consist of a single cell or several cells

3 a three metal prongs are conductors/ plastic body of plug acts as an insulator b two metal prongs are conductors/ plastic body of plug and plastic coating on wires act as insulators c copper wires are conductors/ plastic coating on wires acts as an insulator d silvery metal dimples on base of the light bulb are conductors/ glass bulb acts as an insulator e terminals act as conductors/ plastic coating of the battery acts as an insulator f the terminals are the conductors/ the casing of the batteries acts as an insulator g the prongs are the conductors/ the plastic handle and the base are the insulators

4 a true b true c false - electrons are the particles d false - an electrical conductor will have many free electrons e true f true g true h false - from the negative terminal of a battery around to the positive i true j false - in parallel k true

5 a cell, lamp, wire, switch b cell c wire d lamp e yes - because the lamps are glowing

6 a power pack b ammeter c switch d lamp e cell or battery f voltmeter g resistor h diode i light-emitting diode or LED

7 a to vary the resistance in the circuit/ to vary the current in the circuit/ to dim the lamp b series c because the lamp is glowing d see below e to provide electrical energy/ produce electricity f no g same current in each

ISBN: 9780170262316

h see below i turn current on or off j cells in series/ bulbs in series too k other bulb goes out/ circuit broken/ no current anywhere l see below m in parallel n in series o other bulb would still glow p see below

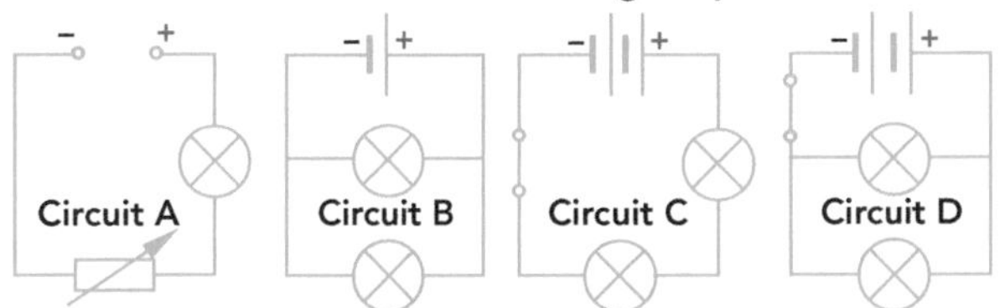

8 a because it is a high energy electrical current b no free electrons c tops of clouds are positive/ bottoms are negative d convection currents in clouds e because the bottoms of clouds are negatively charged and the ground is positively charged f turns air molecules into ions g the supply of free electrons when the air is ionised h because leaders from the ground usually leap up from tall objects i as they will be positive ions they will move upward toward the bottoms of the clouds

9 a series b to limit the current flowing in the circuit c go out d three e in series f in parallel g both go out h top lamp glows and bottom one goes out i power pack j series k current decreases l gets brighter m in series n in parallel o positive terminal p yes

Unit 7: Current and Voltage in Circuits

1 a D b H c K d M e B f R g A h P i C j O k F l J m Q n T o L p E q I r G s S t N

2 a charge is a property that is either positive or negative/ current is the flow of charge b an analogue meter has a scale to read off/ a digital meter provides the reading in digits c current is a flow of electrons/ voltage is the energy gained or lost by electrons in the current d voltage supply is the electrical energy gained by electrons as they pass through the power supply/ voltage loss in the electrical energy lost by electrons as they pass through a component e a component is any part of an electrical circuit/ a terminal is a connection point on a component

3 a charge/ electrons/ conductors/ I / ampere/ amp/ ammeter/ force field/ power b electrical/ electrons/ current/ power/ drop/ gain/ V / volt/ voltmeter

4 a ammeter b analogue c amperes or amps d because it has two different ranges e two outside terminals f left and centre terminals

5 a true b true c false - a digital meter d true e true f false - smallest share of the current g false - red terminal of the power supply h true i true j true

6 a ammeter/ in amperes or amps b 0-5 A c interval is equal to 0.1 A d to the nearest 0.1 A e meter A = 0.3 A / meter B = 4.7 A / meter C = 1.1 A f so that the reading is correct

7 a in series next to the component b in parallel around the component c to the positive terminal of the power supply d so that the meter is not damaged by an excessive current or voltage

8 a in series b positive/ red terminal c needle has moved forward on the scale d about 0.5 A e in parallel f positive/ red terminal g about 5.5 V h 6 V i less than/ meter reading inaccurate or voltage lost elsewhere in circuit, eg across the ammeter j As a current of about half an amp passes through the lamp, a voltage drop of five and a half volts occurs. k light and heat energy

9 a a device that transforms another type of energy into electrical energy b a chemical cell transforms chemical energy into electrical energy/ a solar cell converts light energy into electrical energy c 4.2 J d because other power sources would become exhausted after a time e last a long time/ free energy source f can't work in the dark/ large surface area needed for a reasonable voltage gain g substance that conducts electrons, but not as well as metals h when light waves are absorbed, the two layers develop different charges i a current is produced and the electrons gain electrical energy

10 a in series b 2 A in both meters c in parallel d 6 A e the current in a series circuit is the same at all points/ the current in a parallel circuit is shared between the branches f 3 V g 6 V

11 a both lamps and cells in series b 1.5 V + 1.5 V = 3 V c 1.5 V d lamps in parallel/ cells in series e 3 V, as voltage drop across parallel components is identical to the voltage gain

Unit 8: Resistance, Power and Energy

1 a D b L c N d B e R f T g I h K i A j Q k G l U m O n S o F p C q H r J s H t E u M

2 a an insulator does not allow electrons to travel through it/ a resistor opposes the flow of electrons but does allow some electrons to flow through b a fixed resistor has a set resistance/ the resistance of a rheostat can be varied c resistance is a component's opposition to the current/ power is the rate at which a component transforms energy d total energy is the energy used by a component over a period of time/ power is the amount of energy used per second by a component

3

Quantity	Symbol	Unit	Unit Symbol
Current	I	ampere	A
Voltage	V	volt	V
Resistance	R	ohm	Ω
Power	P	watt	W
Energy	E	joule	J

4 a $R = V \div I$ b $P = VI$ c $I = V \div R$ d $V = P \div I$

5 a false - some energy is lost b false - less current will flow c true d true e true f true g true h true i false - proportional to current and voltage j true

6 a current increases b yes c see below

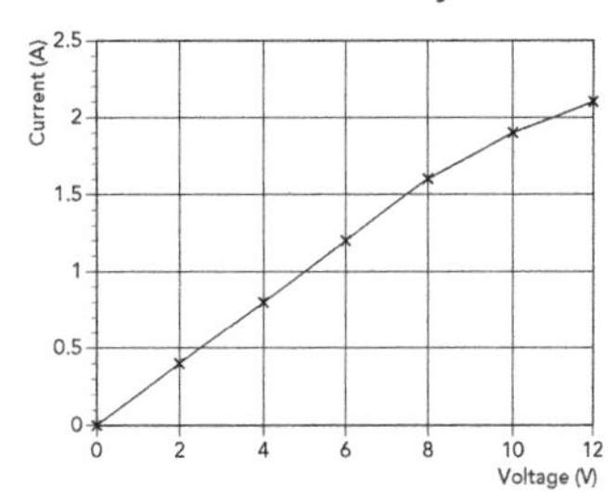

d a straight sloping line initially, but then it curves e because the resistance in the circuit is increasing f as the bulb heats up, its resistance increases

7 a $R = V \div I$ b R = 10.2 ÷ 2.5 = 4.08 = 4.1 Ω c $P = VI$ d P = 10.2 x 2.5 = 25.5 W e $E = Pt$ f E = 25.5 x 90 = 2295.0 J g 10% of 2295 J = 0.1 x 2295 = 229.5 J h 90% of 2 295 J = 0.9 x 2295 = 2065.5 J

8 toaster 960 W / television 48 W / shaver 3 W / radio 1.5 W/ heater 2400 W fan/ 480 W / lamp 72 W / torch 28.8 W

9 a electricity that comes out of a three-pin power point b kilowatt-hour c 3 600 000 J d 20 kilowatt-hours e 20 x 3 600 000 = 72 000 000 J f 20 x 12¢ = $2.40 g advantage - save on power bill/ disadvantage - hot

ISBN: 9780170262316

water cannot be reheated in daytime

10 toaster 60 Ω / television 1200 Ω / shaver 12 Ω / radio 6 Ω / heater 24 Ω / fan 120 Ω / lamp 80 Ω / torch 5 Ω

11 a 1.5 V b $R = V \div I = 1.5 \div 0.5 = 3\ \Omega$ c total resistance $= 3\ \Omega + 3\ \Omega = 6\ \Omega$ d $I = V \div R = 3 \div 3 = 1$ A e total current = 1 A + 1 A = 2 A

Unit 9: Magnetism and Electromagnetism

1 a F b J c M d E e C f O g Q h S i A j P k G l R m B n L o I p D q N r K s H

2 a northwards b because it is attracted toward a south magnetic pole c a south magnetic pole d a north magnetic pole

3 a the two fields diverge and the magnets repel each other b the two fields converge and the magnets attract each other c the two fields converge and the magnets attract each other d the two fields diverge and the magnets repel each other

4

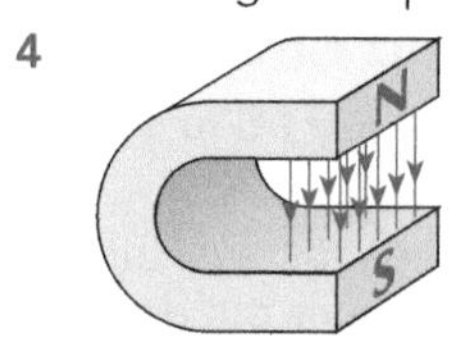

5 a false - attract other magnets and iron objects b true c true d true e false - closely spaced lines f false - like poles repel and unlike poles attract g false - always diverge h true i true j true k false - current direction l true m true

6

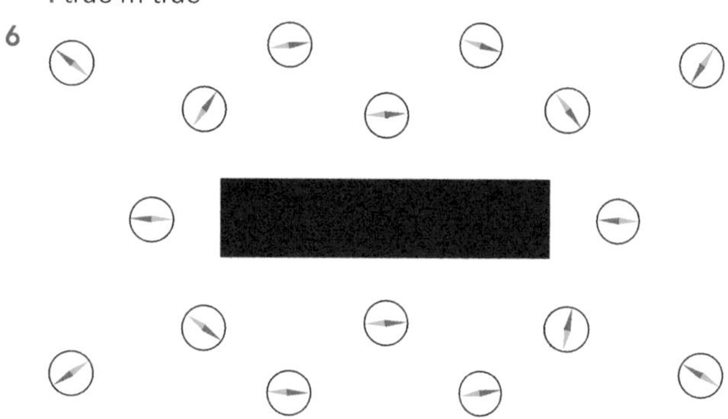

7 a anticlockwise b $B = kId$ c $B = 2 \times 10^{-7} \times 8/0.12 = 0.00001$ T
d increase the size of the current/ loop the wire

8 a

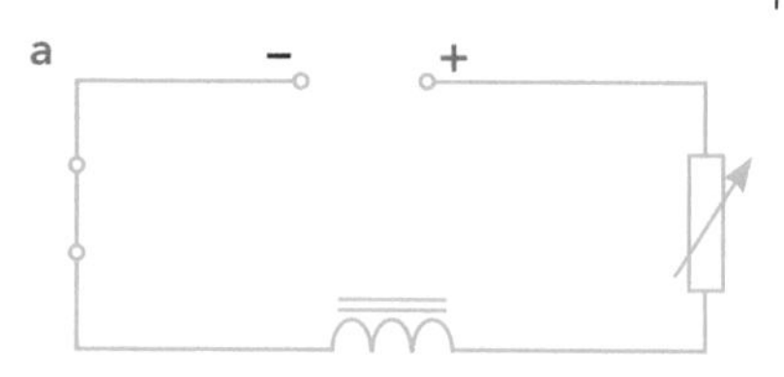

b in series c to increase the strength of the magnetic field d to increase the strength of the magnetic field e by increasing the current f reducing the resistance would increase the current and the magnetic field strength/ increasing the resistance would decrease the current and the magnetic field strength g fall off

9 a right end b left end c 2 and 4 point right, 1 and 3 point left d much stronger inside coil than outside e reverse the direction of the current f opposite direction/ clockwise

10 a because the globe appears to be defying gravity b mains electricity c electrical energy into magnetic energy into kinetic energy d South Pole e because it needs to find a position where the forces of gravity and magnetic repulsion are balanced f a push force and a turning force

11 a attract b the fields around the two wires converge resulting in attraction c repel d the fields around the two wires diverge resulting in repulsion

12 a northward b south magnetic pole c southward d north magnetic pole e giants electrical currents in Earth's outer core of molten iron f it repels the solar wind coming from the Sun thus protecting living things from harmful ultraviolet radiation g geographic poles are found along the axis of Earth's spin but magnetic poles are found where Earth's force field is strongest

Unit 10: Thermal Energy and Heat Flow

1 a D b F c I d K e M f P g O h A i G j B k L l C m Q n J o N p E q H

2 a atoms are extremely small particles that all matter is made of/ molecules are groups of atoms bonded together b thermal energy is the total kinetic energy of all the particles in a substance/ heat is the amount of thermal energy transferred c temperature is proportional to the average kinetic energy of particles/ thermal energy is the total kinetic energy of all the particles in a substance d melting point is the temperature at which a particular solid melts into a liquid/ freezing point is the temperature at which a particular liquid freezes into a solid e boiling point is the temperature at which a particular liquid boils into a gas/ condensing point is the temperature at which a particular gas condenses back into a liquid f heat capacity is about how much heat flow is required to change the temperature of a substance by 1 °C/ latent heat is about how much heat flow is needed to change the state of a substance

3 The molten lava is in a liquid state. It will cool to a certain temperature and then solidify as heat flows from it to the cooler surrounding lava that has already solidified.

4 a temperature fall, freezing and contraction
b temperature rise and expansion c temperature rise and melting d temperature rise and expansion
e temperature rise then change of state

5 a false - from hotter to colder objects b true c true d true e true f true g false - at the same temperature h true i false - heat up slowly j false - will stay the same

6

Feature	Solid	Liquid	Gas
Spacing	close	close	far apart
Fixed	fixed	partly free	free
KE	low	moderate	high
Vibrating	yes	yes	rapidly
Rotating	no	yes	rapidly
Moving	no	slower	faster

7 a boiling b condensing c melting d freezing
e sublimating f evaporating (in the key, energy required should have a red box and energy released a blue box)

8 a melting b stayed constant c yes d used to separate the solid particles so they are free to move about in a liquid state e liquid f rise in temperature of the liquid water g stayed constant h boiling i used to separate the liquid particles so that they are free to fly about in the gas state j a rise in temperature alternates with a change of state when the temperature remains constant

9 a Initially as the test tube cooled the temperature fell because the liquid naphthalene particles transferred thermal energy to the cooler surroundings. The

ISBN: 9780170262316

temperature then remained constant at 80 °C while the liquid naphthalene solidified by transferring more thermal energy. After this the temperature of the solid naphthalene fell as more thermal energy was transferred to the surroundings.

10 a solid and gas states b because it turns directly into a gas, there is no liquid state c It will keep the ice-cream colder for longer without messy puddles d 0.735 kJ/kg/°C e 571 kJ/kg f subliming because 571 kJ/kg will be transferred from the ice-cream as compared to only 0.735 kJ/kg/°C when the solid dry ice warms g the frozen goods are warmer that the subliming dry ice so heat flows to the dry ice thus cooling the ice-cream h because the temperature difference will be much greater and therefore the heat flow will be larger making the dry ice sublime quicker

11 a $\boldsymbol{Q = mc\Delta t}$ = 0.5 x 2.9 x 40 = 58 kJ b $\boldsymbol{Q = mL}$ = 0.5 x 210 = 105 kJ c 268 kJ, much of the heat would have gone into heating the container and surroundings

12 a alcohol liquid; ammonia gas; lead solid; nitrogen gas; oxygen gas; water liquid b $\boldsymbol{Q = mL_f}$ = 5 x 23 = 115 kJ c $\boldsymbol{Q = mL_v}$ = 5 x 871 = 4355 kJ d $\boldsymbol{m = Q/L_v}$ = 3420 ÷ 855 = 4 kg e much more heat is needed to boil a liquid substance than is needed to melt the same mass of the solid substance; this is because much more thermal energy is needed for liquid particles to begin to fly about freely in the gas state than is required for solid particles to begin moving about in the liquid state f nitrogen gas, liquid oxygen

Unit 11: Heat Transfer and Insulation

1 a F b D c L d K e M f N g A h I i U j C k R l E m T n P o H p Q q B r S s O t J

2 a thermal energy is the total kinetic energy of the particles making up a substance/ temperature is related to the average kinetic energy of the particles making up the substance b heat is the amount of thermal energy transferred/ thermal energy is the total kinetic energy of the particles making up a substance c a thermal insulator is a poor conductor of heat/ a thermal conductor is a good conductor of heat d conduction is the transfer of thermal energy by kinetic energy being passed on from atom to atom/ convection is the transfer of thermal energy by the bulk flow of heated gas or liquid from a hotter to a colder location/ radiation is the transfer of thermal energy by infrared waves e infrared radiation consists of invisible electromagnetic waves/ light radiation consists of visible electromagnetic waves f power is the rate at which heat transfer occurs/ heat transfer is the total amount of thermal energy transferred

3 a conduction (from element to water) and convection (from bottom to top of water) b radiation (from element to air) c radiation (from microwave generator to food) d convection (from above radiator to top of the room) e conduction (from hot gas to saucepan bottom to water in pan) f conduction (from base of iron to material) g radiation (from sun to hand)

4 a true b true c true d true e false - only occur in liquids and gases f true g false - cannot be seen h true i false - good absorbers and emitters j false - reduces the rate at which thermal equilibrium occurs k true l true

5 a B b C c A d radiation will be the main mode of heat transfer from the stove element to the pie, conduction will occur particularly along the metal racks, and a convection current may occur when the oven is heating up

6 a $\boldsymbol{P = Q/t}$ = 209 000 J ÷ 90 seconds = 2322 W b $\boldsymbol{\Delta t = Q/mc}$ = 209 kJ ÷ (1 kg x 4.18 kJ/kg/°C) = 50 °C water temperature = 22 + 50 = 72 °C

7 a (own graph) b volume of water, temperature of water, size of opening, same thermometer, time intervals c surface colour of the cans d the black can cooled rapidly, the silver can cooled at a slower rate and the white can slowest of all e black can f a black can is better at radiating heat than a white or silver can g all three will reach the same temperature as the air in the room/ they will reach thermal equilibrium with the air in the room

8 a by radiation b to allow solar radiation to be transmitted through it c the black surface would radiate some of the heat to the air thus not heating the water as much d they cause the water molecules to vibrate more energetically e to absorb infrared radiation that has been transmitted through the water f to reduce heat loss to the ground g most solar radiation would be reflected into the air and the water would not heat very much h wrap the bag in a thermal insulator such as a sleeping bag

9 a evaporation b absorbed c the heat is absorbed from your body thus cooling it d the sweat would evaporate more slowly and less heat would be absorbed from your body e on a windy day because evaporation occurs more rapidly when air with water vapour above your skin is replaced with drier air

10 a condensation b releases thermal energy c it warms up d because the glass is not cold enough to enable thermal heat transfer to occur

11 a 100 °C while the water is boiling b condensation c latent heat as the phase change occurs d run cool water over your hand for 20 minutes

Unit 12: Properties and Types of Waves

1 a M b Q c E d S e I f P g A h G i O j L k C l D m P n R o J p K q B r H s V t U u F v N

2 a a mechanical wave requires a physical medium which it temporarily deforms or displaces as it passes/ an electromagnetic wave consists of interacting, vibrating electrical and magnetic fields that propagate b as a transverse wave passes, the disturbance is perpendicular to the direction of propagation/ as a longitudinal wave passes, the disturbance is parallel to the direction of propagation c a plane wave front is a straight wave front that is perpendicular to the direction of propagation/ a circular wave front is one that radiates out in an ever-increasing circle from the wave source d a surface wave propagates across the surface of a substance/ a body wave propagates through a substance radiating out in all directions e a crest is the highest point of a water wave and a trough is the lowest point f the amplitude of a wave is the distance between a crest (or trough) and the resting position/ the wavelength is the distance between two successive crests (or troughs) g the period of a wave is the time that one wave cycle takes/ the frequency of a wave is the number of wave crests that pass per second h a compression is a region where the medium is compressed as a longitudinal wave passes/ a rarefaction is a region where the medium expands as a longitudinal wave passes

ISBN: 9780170262316

3 a mechanical, transverse body wave b mechanical, longitudinal surface wave c electromagnetic, transverse wave d mechanical transverse wave e mechanical, transverse, surface wave

4 a false - no waves b false - some waves c true d false - circular wave fronts e true f true g true h true i false - will not j true k true

5

Water	Light	Sound
yes	no	yes
mechanical	electromagnetic	mechanical
transverse	transverse	longitudinal
both	not applicable	circular
surface	not applicable	both
force	brightness	loudness

6 a lower frequency source generates a longer wavelength wave; higher frequency source generates a shorter wavelength wave b higher frequency source generates a higher pitched sound; lower frequency source generates a lower pitch sound

7 a 1.08 m b the energy transferred would be quartered as energy transferred is proportional to the square of the amplitude ($\frac{1}{2} \times \frac{1}{2} = \frac{1}{4}$) c none d $f = 1/T = 1 \div 4 = 0.25$ Hz e $v = \lambda f = 2.5 \times 0.25 = 0.62$ ms^{-1} f $v = 0.62 \times 2 \div 3 = 0.41$ ms^{-1} g $\lambda = v/f = 0.41 \div 0.25 = 1.64$ m

8 a four times as much energy would be transferred b the larger the amplitude of the wave, the louder the sound c $T = 1/f = 1 \div 264 = 0.004$ s d $\lambda = v/f = 340 \div 264 = 1.29$ m e the higher the frequency, the higher the pitch f more than four times as fast ($1497 \div 340 = 4.4$) g because the particles are much closer together enabling the compression waves to travel faster h $\lambda = v/f = 1497 \div 264 = 5.67$ m i no, as pitch depends only on frequency

9 a waves associated with earthquakes b generated by sudden fractures in the rock beneath Earth's surface c elastic energy of the stretched or squashed rock d P- and S-waves are body waves; L-waves are surface waves e P-waves are the fastest, S-waves are next fastest, and L-waves are the slowest f S- and L-waves are transverse; P-waves are longitudinal g first a forward and back movement , followed by an up and down movement, then finally a sideways movement h because they transfer so much energy with their larger amplitudes and longer duration

10 a $f = v/\lambda = 300000 \div 750 = 400$ THz b 800 THz c $\lambda = v/f = 300000 \div 800 = 375$ nm d the wavelength and frequency are directly proportional e for red light $\lambda = v/f = 200000 \div 400 = 500$ nm; for violet light $\lambda = v/f = 200000 \div 800 = 250$ nm

11 a wind b kinetic energy c kinetic and gravitational d kinetic and gravitational as the beach particles are moved about

Unit 13: Reflection and Refraction of Light

1 a H b J c D d L e F f L g M h R i A j P k H l V m N n T o C p J q Q r E s O t S u U v B

2 a an opaque medium will not transmit any light waves/ a transparent medium will allow light waves to be transmitted through it/ a reflective medium will 'bounce' light waves off its surface b a light wave is an electromagnetic disturbance that propagates/ a light ray is an arrow that shows the direction in which a wave propagates c an image is a view of an object at a location where it is not/ an object is the source of the incident light waves d reflection occurs when light waves are sent back into the incident medium/ refraction occurs when light waves change their direction and speed as they enter a different medium e the angle of incidence is the angle between the incident ray and the normal/ the angle of reflection is the angle between the reflected ray and the normal/ the angle of refraction is the angle between the refracted ray and the normal f the (mass) density of a substance is its mass per unit volume/ the optical density of a substance is a property that affects the speed at which light is transmitted through a transparent substance g refraction occurs when a light wave abruptly changes direction and speed as it enters a different medium/ dispersion occurs when white light is separated into the visible spectrum because frequency affects how much a wave is refracted

3 The transparent parts of the wings are transmitting light waves coming from behind the butterfly. The black parts of the wings are absorbing the light waves reaching them. The red parts of the wings are reflecting the red light that is part of sunlight.

4 a diffraction b reflection c transmission d reflection e absorption (dark area of lenses) and transmission (light areas)

5 a true b false - can occur c true d true e true f false - not identical in every way g true h false - more slowly i true j true k false - toward l true m true

6 a if the surface was rough the light waves would not be reflected in parallel and the image would be distorted b they all lie in the same plane c they are always equal d because a light ray does not actually travel from the image to the mirror e behind the mirror; same size as the object; same distance behind the mirror as the object is in front; same way up as the object; left and right sides are swapped

7 a changes them into plane wave fronts b it bends toward the boundary c if the optical density decreases then the angle of refraction will be greater than the angle of incidence or converse d increased e wave speed increases in a less optically dense medium f increased g no

8 a and b

angle of incidence
angle of reflection
incident ray
reflected ray
glass (more dense)
air (less dense)
refracted ray
normal
angle of refraction

c the angle of reflection is equal to the angle of incidence; the angle of refraction is greater than the angle of incidence (and the angle of reflection) d (see diagram above) e the reflected light wave will travel at the same speed as the incident wave because it is moving in the same medium; the refracted wave will travel faster than the incident and reflected waves as it is moving in a less dense medium f the angle of incidence is less than the critical angle for air because although the refracted ray is bent away from the normal it is not yet parallel with the boundary

9 a because it enters and leaves the prism straight on to the boundary and not at an angle b the angle of incidence c because it has been bent that far away from

ISBN: 9780170262316

the normal **d** 90° **e** because the angle of incidence is equal to the critical angle for air **f** the angle of refraction would be reduced and the light wave would travel through the glass **g** light would no longer be refracted into the glass, all light waves would be reflected back into the air **h** total internal reflection

10 **a** when waves have to go around a small obstacle or through a narrow gap **b** yes **c** (own drawing) **d** transmission of sounds through the wall, reflection of sound waves that enter through the doorway off walls **e** little diffraction would occur **f** little diffraction would occur **g** the waves would be strongly diffracted

11

Wave	Incident	Reflected	Refracted
Speed	300 000	300 000	200 000 kms^{-1}
Direction	fixed	'mirror' angle	closer to normal
Colour	red	red	red
Frequency	500 THz	500 THz	500 THz
Wavelength	600 nm	600 nm	400 nm

12 **a** slowest in diamond; fastest in a vacuum **b** toward the normal when entering a more optically dense medium; away from the normal when entering a less optically dense medium **c** it would bend toward the normal **d** it would bend away from the normal **e** it would bend away from the normal at the diamond-air boundary then toward the normal at the air-water boundary **f** diamond-vacuum boundary

13 **a** the violet ray because it has a higher frequency (670 THz) than red light (480 THz) and a higher frequency wave will refract more than a lower frequency wave **b** the higher the frequency, the greater the angle of refraction (or the converse) **c** dispersion **d** the frequency of the wave and the difference in optical density between the two media **e** the white light would be separated into the spectrum of visible colours

Unit 15: Atoms and Elements

1 **a** G **b** N **c** D **d** J **e** C **f** L **g** R **h** F **i** P **j** B **k** M **l** K **m** E **n** A **o** O **p** I **q** Q **r** H **s** S

2 **a** a nucleus is the dense central area of an atom containing protons and neutrons/ an electron cloud is the space around the nucleus occupied by the electrons **b** a proton is a positively charged particle found in the nucleus/ a neutron is an uncharged particle found in the nucleus/ an electron is a negatively charged particle that moves around the nucleus **c** the atomic number is the number of protons in the nucleus/ the neutron number is the number of neutrons in the nucleus/ the mass number is the number of protons and neutrons in the nucleus **d** an atom is an extremely small particle/ an element is a substance made of one type of atom only **e** a period is a row/ a group is a column

3

Particle	Mass	Charge	Location
proton	heavy	positive	nucleus
neutron	heavy	neutral	nucleus
electron	light	negative	electron cloud

4 **a** 5 protons **b** 6 neutrons **c** 5 electrons **d** atomic number = 5 **e** neutron number = 6 **f** mass number = 5 + 6 = 11 **g** yes, equal numbers of protons and electrons **h** boron **i** period 2 **j** group 13

5 **a** true **b** true **c** false - protons have a positive charge and electrons a negative charge **d** true **e** false - like charges repel and unlike charges attract **f** false - equal numbers of protons and electrons **g** false - the further it will be from the nucleus **h** true **i** true **j** true **k** true **l** true

6 **a** 9 electrons **b** 9 protons **c** atomic number = 9 **d** fluorine **e** mass number = 9 + 10 = 19 **f** nonmetal **g** group 17 **h** chlorine

7 The left diagram shows electrons moving in orbits like planets around a sun, which is incorrect. Electrons with a particular level energy are likely to be found in a particular region as shown in the right diagram.

8 **a** F **b** Na **c** Fe **d** Br **e** Ca **f** Pb **g** Hg **h** Ag **i** I **j** Zn **k** Cu **l** Ni **m** Cl **n** K **o** Ar **p** Al **q** C **r** N **s** Ne **t** Li **u** He **v** Mg **w** Zn **x** Si **y** P **z** O

9 **a** iron **b** calcium **c** potassium **d** lead **e** zinc **f** hydrogen **g** beryllium **h** argon **i** silver **j** silicon **k** potassium **l** sulfur **m** nitrogen **n** neon **o** helium **p** carbon **q** bromine **r** lithium **s** oxygen **t** fluorine **u** sodium **v** magnesium **w** mercury **x** copper **y** aluminium **z** gold

10 **a** hydrogen **b** helium **c** carbon **d** nitrogen **e** neon **f** sodium **g** magnesium **h** aluminium **i** sulfur **j** chlorine **k** calcium **l** iron **m** copper **n** zinc **o** lead

11 **a** 1 **b** 6 **c** 12 **d** 7 **e** 13 **f** 20 **g** 16 **h** 9 **i** 3 **j** 11 **k** 17 **l** 30

12 **a** 26 **b** 82 **c** 16 **d** 8 **e** 1 **f** 12 **g** 6 **h** 10 **i** 80 **j** 2 **k** 35 **l** 7

13 **a** because the elements are arranged in periods **b** 74 elements **c** in order of their atomic numbers from left to right across the periods **d** a row of elements **e** Li, Be, B, C, N, O, F and Ne **f** a column of elements **g** Be, Mg, Ca, Sr, Ba and Ra **h** He, Ne, Ar, Kr, Xe and Rn **i** metals on the left and in the middle **j** nonmetals on the right **k** group 18 **l** group 1 **m** group 17

14 **a** $\mathbf{Z} = 3$, $\mathbf{N} = 4$ and $\mathbf{A} = 4 + 3 = 7$
b $\mathbf{Z} = 6$, $\mathbf{N} = 6$ and $\mathbf{A} = 6 + 6 = 12$
c $\mathbf{Z} = 11$, $\mathbf{N} = 12$ and $\mathbf{A} = 11 + 12 = 23$

15 **a** neutral **b** not neutral **c** neutral

16 **a** 2, 1 **b** 2, 3 **c** 2, 8, 1

17 **a** lithium, group 1 **b** carbon, group 14 **c** sodium, group 1

18

Element	Atomic no.	Neutron no.	Mass No.
lithium	3	4	7
carbon	6	6	12
fluorine	9	10	19
aluminium	13	14	27
chlorine	17	18	35

19 **a** argon **b** 2, 8, 8 **c** group 18

Unit 16: Ions and Compounds

1 **a** R **b** C **c** D **d** I **e** M **f** A **g** N **h** E **i** L **j** H **k** Q **l** P **m** F **n** G **o** B **p** K **q** T **r** O **s** J **t** S

2 **a** atoms are the basic building blocks of matter/ ions are charged atom (or molecules) **b** a negative ion has more electrons than protons/ a positive ion has more

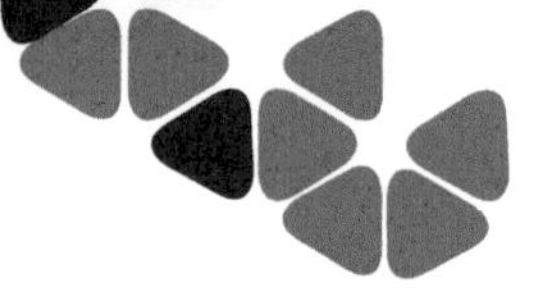

protons than electrons **c** an element is a substance made of one kind of atom only/ a compound is a substance made of different kinds of atoms that are bonded together **d** a lattice is a regular array of atoms/ a molecule is a group of atoms involved in sharing electrons **e** an ionic compound is a substance in which different kinds of atoms are held together by ionic bonds/ a molecular compound is a group of non-identical atoms sharing electrons **f** an ionic bond occurs when oppositely charge ions are attracted together/ a covalent bond occurs when two atoms are held together by the attraction they have for shared electrons

3

Element	Symbol	No. of electrons	Elect config.	Valence no.	Group no.
Hydrogen	H	1	1	1	1
Helium	He	2	2	2	18
Lithium	Li	3	2, 1	1	1
Beryllium	Be	4	2, 2	2	2
Boron	B	5	2, 3	3	13
Carbon	C	6	2, 4	4	14
Nitrogen	N	7	2, 5	5	15
Oxygen	O	8	2, 6	6	16
Fluorine	F	9	2, 7	7	17
Neon	Ne	10	2, 8	8	18
Sodium	Na	11	2, 8, 1	1	1
Magnesium	Mg	12	2, 8, 2	2	2
Aluminium	Al	13	2, 8, 3	3	13
Silicon	Si	14	2, 8, 4	4	14
Phosphorus	P	15	2, 8, 5	5	15
Sulfur	S	16	2, 8, 6	6	16
Chlorine	Cl	17	2, 8, 7	7	17
Argon	Ar	18	2, 8, 8	8	18
Potassium	K	19	2, 8, 8, 1	1	1
Calcium	Ca	20	2, 8, 8, 2	2	2

a He, Ne and Ar **b i** Li, Na and K **ii** Be, Mg and Ca **iii** B and Al **iv** O and S **v** F and Cl **c i** lose electrons **ii** share electrons **iii** gain electrons **iv** none of the options

4 **a** false - are bonded **b** true **c** false - stable **d** true **e** true **f** true **g** false - different kinds of atoms **h** true **i** true **j** false - opposite charge **k** true **l** true **m** false - large numbers

5 **a** sulfur 2, 8, 6; magnesium 2, 8, 2 **b** sulfur 6; magnesium 2 **c** gain two valence electrons **d** lose two valence electrons **e** sulfur atom **f** two valence electrons are transferred from each magnesium atom to a sulfur atom **g** sulfide ion 2, 8, 8; magnesium ion 2, 8

6 **a** He, Ne and Ar **b** they keep their electrons **c** they tend to lose their valence electrons **d** two valence electrons **e** they share valence electrons **f** two electrons **g** they can do both **h** nonmetal atoms take, share, or just keep valence electrons depending on how full their shells are **i** O F S Cl

7 **a** F^- **b** Ca^{2+} **c** O^{2-} **d** Mg^{2+} **d** Al^{3+} **e** K^+

8 **a** $CaCl_2$ has 1 calcium ion to every 2 chloride ions
b $Mg(OH)_2$ has 1 magnesium ion to every 2 hydroxide ions
c $Mg(NO_3)_2$ has 1 magnesium ion to every 2 nitrate ions
d K_2O has 2 potassium ions to every 1 oxide ion
e $CaCO_3$ has 1 calcium ion to every 1 carbonate ion
f NH_4OH has 1 ammonium ion to every 1 hydroxide ion
g Na_2SO_4 has 2 sodium ions to every 1 sulfate ion
h $Al(OH)_3$ has 1 aluminium ion to every 3 hydroxide ions
i $ZnSO_4$ has 1 zinc ion to every 1 sulfate ion

9 **a** metal **b** nonmetal **c** the chloride ion has one more; the sodium ion has one less **d** an electron is transferred from each sodium atom to a chlorine atom **e** chlorine atom **f** the are in a regular array with each sodium ion surrounded by 6 chloride atoms and vice versa **g** because the negatively charged ions would all repel each other **h** the electrostatic attraction between oppositely charged ions/ ionic bonds

10 **a** H^+ **b** Li^+ **c** Be^{2+} **d** O^{2-} **e** F^- **f** Na^+ **g** Mg^{2+} **h** Al^{3+} **i** S^{2-} **j** Cl^- **k** K^+ **l** Ca^{2+}

11 **a** hydroxide **b** nitrate **c** carbonate **d** bicarbonate **e** sulfate **f** ammonium

12 **a** 1 Na^+ : 1 Cl^- => NaCl
b 1 K^+ : 1 Cl^- => KCl
c 1 Ca^{2+} : 1 O^{2-} => CaO
d 1 Mg^{2+} : 1 O^{2-} => MgO
e 1 Mg^{2+} : 2 Cl^- => $MgCl_2$
f 1 Cu^{2+} : 2 OH^- => $Cu(OH)_2$
g 1 Mg^{2+} : 1 CO_3^{2-} => $MgCO_3$
h 1 Ca^{2+} : 2 OH^- => $Ca(OH)_2$
i 1 NH_4^+ : 1 Cl^- => NH_4Cl
j 1 Mg^{2+} : 2 NO_3^- => $Mg(NO_3)_2$
k 1 Al^{3+} : 3 OH^- => $Al(OH)_3$
l 2 Al^{3+} : 3 CO_3^{2-} => $Al_2(CO_3)_3$
m 3 Li^+ : 1 N^{3-} => Li_3N

Unit 17: Properties of Metals

1 **a** C **b** F **c** H **d** N **e** L **f** P **g** S **h** G **i** R **j** I **k** A **l** O **m** C **n** E **o** K **p** Q **q** D **r** M **s** B

2 **a** metals are strong, shiny, silvery solids that are good conductors and can be readily worked into new shapes/ nonmetals do not have these properties **b** a physical property is a feature that does not involve a chemical reaction/ a chemical property relates to chemical reactions that the substance undergoes **c** ductile means able to be stretched/ malleable means able to be hammered or squashed into a new shape **d** oxidation occurs when a substance reacts with oxygen gas/ combustion occurs when a substance is rapidly oxidised when heated **e** a pure metal is made of one kind of metallic element/ a metal alloy consists of a metal with a smaller amount of another element (or elements) mixed into it

3 **a** Cu and Au **b** hardest Fe, softest Pb **c** Hg, K and Na **d** K, Li and Na **e** Au **f** best Ag, worst Hg **g** best Ag, worst Hg **h** strongest Fe, weakest Pb **i** Al, Au and Mg

4 **a** true **b** false - most metals **c** false - do not include the reactivity of the metal **d** true **e** true **f** true **g** false - a mixture **h** false - become positive ions **i** true **j** false - some metals **k** true **l** false - an unreactive metal tightly holds on to its free electrons

5 **a** most metals are dense because the atoms are tightly packed **b** the free electrons allow metals to conduct electricity and heat well **c** metals are mostly solid because of the strength of the metallic bonding **d** metals mostly have high MPs because a lot of heat is required to free the atoms from the metallic bonding **e** metals are ductile because metallic bonding holds the metal atoms together as the metal is stretched **f** metals are malleable because the layers of atoms slide over each other **g** metals are shiny because the valence electrons reflect light well **h** metals are fairly hard because of the strength of the metallic bonding holding the atoms together

ISBN: 9780170262316

6 a fishing sinker - lead, as it is heavy b bridge - iron (steel) for strength and cheapness c aircraft wing - aluminium for lightness (magnesium is too reactive) d electrical wire - either copper or aluminium (gold and silver are too expensive) e thermometer - mercury, as it is the only liquid metal at room temperature f jewellery - silver and gold because they are resistant to corrosion and are expensive g ladder - aluminium, as it is light and strong h car body - iron (steel) for strength i saucepan - copper, as it is a very good heat conductor j scissors - iron (steel) for strength

7 a Al, Cu and Pb b Ca > Mg > Zn >Fe > Al, Cu and Pb c aluminium d because it has a thin, resistant oxide coating that prevents other reactions e hydrogen gas, insert a flame and the gas pops f calcium chloride $CaCl_2$ g iron chloride $FeCl_2$ h magnesium chloride $MgCl_2$ i zinc chloride $ZnCl_2$
j Calcium + sulfuric acid → calcium sulfate + hydrogen gas; $Ca + H_2SO_4 \rightarrow CaSO_4 + H_2$
k Iron + hydrochloric acid → iron chloride + hydrogen gas; $Fe + 2HCl \rightarrow FeCl_2 + H_2$
l Magnesium + hydrochloric acid → magnesium chloride + hydrogen gas; $Mg + 2HCl \rightarrow MgCl_2 + H_2$
m Zinc + sulfuric acid → zinc sulfate + hydrogen gas; $Zn + H_2SO_4 \rightarrow ZnSO_4 + H_2$

8 a alloys b a mixture of several elements where the main element is a metal; because they have more useful properties c by mixing in other elements when the metal is in a molten state during smelting d hardness, strength, corrosion resistance, melting point, etc e bronze is an alloy of copper and tin, which gave humans a reasonably strong metal for weapons and tools f brass is an alloy made from copper and zinc; it was used to make important machine parts on steam engines during the Industrial Revolution g steel is much stronger and is more resistant to rusting and corrosion h used for cutlery, knives and saucepans i 'designer alloys' are special alloys developed for very specific purposes

9 a metals lose their free electrons in reactions b positive ions c a metal oxide d Metal + oxygen → metal oxide e hydrogen gas and a metal hydroxide f Metal + water → metal hydroxide + hydrogen gas g hydrogen gas and a salt h Metal + acid → salt + hydrogen gas

10 a because they are either explosively or violently reactive b as you go from right to left across the table the metal atoms tend to lose valence electrons much more readily/ as you go from left to right across the table the atoms hold on to their valence electrons much more tightly c because heat makes most reactions occur more readily and oxygen gas has much more oxygen than air has (20%) d because heat makes most reactions occur more readily e aluminium, because of its protective oxide coating

Unit 18: The Extraction and Uses of Metals

1 a G b K c M d B e N f H g L h S i P j E k J l Q m S n A o R p C q O r I s F

2 a an element is a pure substance made of one kind of atom only/ a compound is a pure substance in which different kinds of atoms are bonded b a neutral atom has equal numbers of electrons and protons/ an ion is a charge atom, with more or less electrons than protons c an ore is a rock or mineral mined from Earth's crust/ a native element is an element that occurs in an uncombined state in nature d electrolysis involves the reduction of metal ions to metal atoms using electricity and heat/ smelting involves the reduction of metal ions to metal atoms using carbon and heat e tarnishing involves a metal reacting with chemical(s) in the air to form an impervious coating that protects the metal beneath/ rusting involves iron reacting with oxygen and moisture in the air to form an orange-red oxide coating that flakes off exposing the iron beneath to further corrosion f a pure metal has no other element alloyed with it/ an alloy is a metal that has had other elements mixed into it when it was in a molten state, in order to give it specific properties g galvanising involves coating steel in a thin layer of zinc, which acts as a protective coating, and also provides sacrificial protection if the metal is scratched/ sacrificial protection involves attaching a block of a more reactive metal to steel; the block then supplies the steel with electrons to prevent it from rusting

3

Reactivity	Elements	Extraction
Highly reactive	K, Na, Li, Ca, Mg, Al	electrolysis
Moderately reactive	Zn, Fe, Pb	smelting
Low reactivity	Cu, Ag, Au	digging

a ions of reactive metals must be forced to accept electrons, as the resulting neutral atoms have higher energy levels and are therefore less stable, which means that the more reactive the metal, the more energy is need to reduce its ions

4

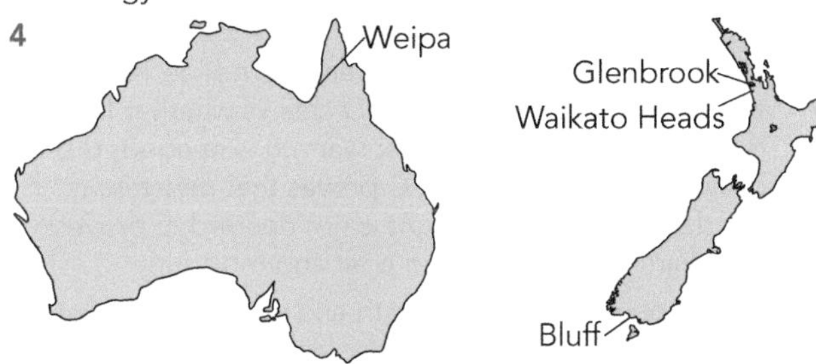

5 a false – few metal elements b true c false - more reactive d true e true f true g false – protects h true i true j false - may require k false – more reactive l true

6 a because the ores are on or near the surface of the ground b the ores are dug up by huge excavators c quartz sand d magnets attract the magnetite, separating it from the nonmagnetic quartz sand e because bauxite is red and alumina is white f using chemical reactions

7 a ironsand is being mined at the mouth of the Waikato River b magnetite is being reduced at high temperature by carbon monoxide in the rotary kiln c sponge iron is being melted and purified into pig iron in the melter d molten pig iron is being transported to the steel-making furnace

8 a electrolysis b because a great deal of energy is required to force the aluminium ions to accept electrons/ to reduce aluminium ions to neutral atoms c aluminium oxide, Al_2O_3 d Al^{3+} e it dissolves the alumina to give a very hot solution of aluminium ions and oxide ions; 970 °C f the negative terminal, because the aluminium ions are positively charged g each ion is 'forced' to accept three electrons to become a neutral atom/ they are reduced to aluminium atoms h $Al^{3+} + 3e^- \rightarrow Al$ i reduction j because the relatively denser molten aluminium will sit in the bottom of the furnace k to capture the CO_2 gas (and other harmful gases such

as fluorides that need to be treated or released safely)

9 a iron oxide, Fe_3O_4 b by using large magnets that separate the magnetic iron ore from the nonmagnetic quartz sand c to generate carbon monoxide and heat d Fe^{2+} and Fe^{3+} e around 950 °C f the combustion of carbon and carbon monoxide g electrons h smelting i because it is cheaper than using electrolysis/ because iron is a less reactive metal, so it can be reduced without using an electrical current j $Fe^{2+} + 2e^- \rightarrow Fe$ and $Fe^{3+} + 3e^- \rightarrow Fe$ k Carbon monoxide + oxygen → carbon dioxide l oxidation m to get rid of impurities n 'sponge iron' is iron in a powdery, spongy form; 'pig iron' was so named because the solidified ingots looked like piglets attached to a sow because of the shape of the mould

10 a rock formed when lava cools either above or below ground b because the rising heated geothermal water would have created hot pools, mud pools and possibly geysers c because dissolved gold crystallised out in rock fractures when earthquakes caused sudden pressure drops d concentrating the gold and silver particles then extracting them from the surrounding material e 1000 tonnes; 33 tonnes f turn the gold and silver atoms into ions g as it is highly toxic it could poison river or ground water if it leaches out of the ponds or tailings h to attract the gold and silver ions so that they can be separated from the other chemicals i electrolysis gives electrons back to the gold and silver ions while the reaction with hydrogen cyanide removed those electrons j to separate the gold from the silver

11 a A b B to E c B checks whether air is needed; C checks whether moisture is needed; D checks whether light is needed; E checks whether warmth is needed; d B proves that air is needed; C proves that moisture is needed; D proves that light is not needed; E proves that warmth is not needed e air and moisture

12 a steel because of its strength and weight b aluminium because of its light weight, ductility and corrosion resistance c aluminium because of its light weight, malleability and corrosion resistance d copper because of its excellent thermal conductivity e wrought iron because of its strength and cheapness f lead because of its weight and cheapness g gold because it is valuable, shiny, malleable and corrosion-resistant h silver because of its excellent electrical conductivity i stainless steel because of its hardness and corrosion resistance j sodium because the gas phase will emit a bright yellow light when ionised by an electrical current k magnesium because it readily burns with a bright white flame l lithium because the metal can act as the positive electrode of the battery

13 a enamelling b galvanising c sacrificial protection d plastic coating e oiling f painting g anodising h alloying (stainless steel) i painting j galvanising k painting l copper plating m alloying (argentium silver) n it will form its own resistant coating (patina) after a while o alloying (magnesium with some aluminium) p alloying (stainless steel) q anodising

Unit 19: Properties and Uses of Nonmetals

1 a K b O c R d G e I f E g N h B i H j M k C l J m P n A o D p Q q L r F

2 a metal elements are shiny, solid elements that are good conductors of heat and electricity with high melting and boiling points/ most nonmetal elements have low melting and boiling points; in the solid state they are dull coloured, brittle and poor conductors of heat and electricity b an element is a pure substance made of one kind of atom only/ a compound is a pure substance in which several different kinds of atoms are chemically combined c ionic bonding is the attraction between oppositely charged ions/ covalent bonding is the attraction that the nuclei of atoms in a molecule have for shared electrons d a crystalline solid consists of a regular array of atoms, ions or molecules/ in an amorphous solid the particles are not arranged regularly so there is no crystalline structure e an ionic compound is a substance in which charged atoms or molecules are held together by the attraction between oppositely charged ions/ a molecular compound is a substance in which the atoms are held together by the attraction that the nuclei have for shared electrons f incomplete combustion refers to burning a carbon compound in a limited supply of oxygen, resulting in the formation of carbon monoxide/ complete combustion refers to burning a carbon compound in an abundant supply of oxygen, resulting in the formation of carbon dioxide g an acidic solution results from dissolving a nonmetal oxide in water; it will turn blue litmus red/ an alkaline solution results from dissolving a metal oxide in water; it will turn red litmus blue

3

Property	Metal elements	Nonmetal elements
Density	heavier	lighter
Appearance	shiny	dull when solid
Conduction	very good	mostly insulators
MP and BP	high	low (often gases)
Strength	strong and flexible	weak and brittle
Allotropes	some have	quite a few

4 a true b true c false - most d true e false - except for hydrogen f true g false - more stable h true i true j true k true l true m false - four electrons n true o true

5 a Br, as it is a liquid b 4 c He, Ne and Ar d C e C as diamond f a greater variety of colours as nearly all metals are silvery-grey g C, S and I h P i solid graphite sublimes into a gas rather than melts into a liquid j boiling point as the gas state is much less dense than the liquid or solid states k diamond, because it is 10 on the Moh's scale, which is the hardest level l C as graphite m very poor thermal conductors/ nearly all are thermal insulators n because C as diamond is a very good thermal conductor

6 a A, because it is burning with a clear blue flame b Sulfur + oxygen → sulfur dioxide; $S_8 + 8O_2 \rightarrow 8SO_2$ c C, because it occurs only in the intense heat of jet and car engines d Nitrogen + oxygen → nitric oxide; $N_2 + O_2 \rightarrow 2NO$ e A, because the carbon is burning with a yellow flame f because it would be burning with a blue flame if it was complete combustion g Carbon + oxygen → carbon monoxide; $2C + O_2 \rightarrow 2CO$ h Carbon + oxygen → carbon dioxide; $C + O_2 \rightarrow CO_2$

7 a as nonmetal atoms tend to share electrons they mostly exist as molecules with low BPs, which means they are gases at room temperature b as nonmetal atoms do not have free electrons they do not reflect light well and are therefore not shiny c as nonmetal atoms tend to share electrons they mostly exist as molecular compounds that melt at low temperatures d as nonmetal atoms tend to share electrons they mostly exist as molecular compounds that boil at low temperatures e nonmetals are mostly light because

ISBN: 9780170262316

most of them are gases owing to their atoms sharing electrons to form discrete molecules **f** nonmetals are mostly soft because they lack the free electrons that result in the metallic bonding that gives metals their strength **g** nonmetals are very poor electrical conductors because they lack the free electrons that enable metals to conduct electricity **h** nonmetals are mostly very poor thermal conductors because they lack the free electrons that enable metals to conduct heat

8 **a** to prevent the spread of infectious water-borne diseases **b** because both are very reactive elements that destroy micro-organisms when in solution **c** it kills bacteria and deactivates viruses **d** by the electrolysis of water or air **e** because it kills micro-organisms effectively and produces no pollutants or objectionable odours **f** as liquid chlorine for public baths and as safer chlorine-releasing compounds for private pools **g** hypochlorous acid **h** chlorine kills micro-organisms effectively but may result in an objectionable smell or an unpleasant taste

9 **a** orthorhombic crystals are like two steep pyramids attached at their bases but with their tops cut off; monoclinic crystals are long, thin, skewed cuboid shapes **b** in orthorhombic crystals the ring molecules are all stacked the same way up; in monoclinic crystals the ring molecules are stacked in opposing directions **c** allotropes **d** melting and boiling points, density, hardness and colour **e** because the atoms in the molecules can be bonded in different ways/ because the molecules can be packed in different ways in the solid state

10 **a** A, because blue litmus turns red in an acidic solution **b** nonmetal oxide, because nonmetal oxides form acidic solutions **c** B, because red litmus turns blue in an alkaline solution **d** metal oxide, because metal oxides form alkaline solutions

11 **a** sulfur trioxide gas reacts with liquid water to give liquid sulfuric acid; Sulfur trioxide + water ⟶ sulfuric acid; $SO_3 + H_2O \rightarrow H_2SO_4$
b nitrogen gas reacts with hydrogen gas to give ammonia gas; Nitrogen + hydrogen ⟶ ammonia; $N_2 + 3H_2 \rightarrow 2NH_3$
c ozone gas changes into oxygen gas; Ozone ⟶ oxygen; $2O_3 \rightarrow 3O_2$

Unit 20: Reaction Rates and Particle Collisions

1 **a** K **b** M **c** N **d** B **e** O **f** J **g** A **h** G **i** P **j** F **k** R **l** H **m** C **n** L **o** D **p** E **q** Q **r** I

2 **a** reactant particles are the original substances used up in a reaction/ product particles are the new substances formed in a reaction **b** a word equation describes a chemical reaction using the names of the chemicals/ a symbol equation describes a reaction using the formulas of the chemicals **c** an exothermic reaction releases heat to its surroundings, although it may need some heat to start with/ an endothermic reaction will continue to require heat from its surroundings **d** a dilute solution has few particles of the substance per unit volume/ a concentrated solution has many particles per unit volume **e** a catalyst is a chemical that speeds up a reaction without being consumed/ an enzyme is a special type of catalyst made by a living organism

3 **a** Hydrogen gas + chlorine gas ⟶ hydrogen chloride gas **b** Copper oxide solid + hydrogen gas ⟶ copper metal + steam **c** Hydrogen gas + oxygen gas ⟶ steam **d** Magnesium solid + hydrochloric acid ⟶ hydrogen gas + magnesium ions + chloride ions **e** Sodium solid + chlorine gas ⟶ sodium chloride solid

4 **a** $H_2 + Cl_2 \rightarrow 2HCl$ **b** $2H_2 + O_2 \rightarrow 2H_2O$ **c** $C + O_2 \rightarrow CO_2$ **d** $2CO + O_2 \rightarrow 2CO_2$ **e** $N_2 + O_2 \rightarrow 2NO$ **f** $2SO_2 + O_2 \rightarrow 2SO_3$ **g** $SO_3 + H_2O \rightarrow H_2SO_4$

5 **a** true **b** true **c** true **d** false - only some collisions are effective **e** false - products are the new substances **f** true **g** false - is produced **h** true **i** true **j** false - has less particles **k** true **l** false - natural catalysts

6 **a i** iodine gas I_2 and hydrogen gas H_2 **ii** hydrogen iodide gas HI **iii** hydrogen gas + iodine gas ⟶ hydrogen iodide gas **iv** $H_2 + I_2 \rightarrow 2HI$
b i nitrogen gas N_2 and oxygen gas O_2 **ii** nitric oxide gas NO **iii** nitrogen gas + oxygen gas ⟶ nitric oxide gas **iv** $N_2 + O_2 \rightarrow 2NO$
c i ozone gas O_3 **ii** oxygen gas O_2 **iii** ozone gas ⟶ oxygen gas **iv** $2O_3 \rightarrow 3O_2$
d i nitric oxide gas NO and ozone gas O_3 **ii** nitrogen dioxide gas NO_2 and oxygen gas O_2 **iii** nitric oxide gas + ozone gas ⟶ nitrogen dioxide gas + oxygen gas **iv** $NO + O_3 \rightarrow NO_2 + O_2$

7 **a** hydrogen peroxide liquid is the reactant, steam and oxygen gas are the products **b** Hydrogen peroxide ⟶ steam + oxygen gas **c** $2H_2O_2 \rightarrow 2H_2O + O_2$ **d** to speed up the reaction/ to act as a catalyst/ to reduce the amount of activation energy required **e** catalysts **f** $2H_2O_2 \xrightarrow{MnO_2} 2H_2O + O_2$

8 **a** the rate will increase as having more zinc surface particles exposed will increase the collision frequency **b** the rate will increase because having more acid particles per unit volume will increase the collision frequency **c** the rate will decrease because both the frequency and force of collisions will decrease **d** the rate will increase because both the frequency and force of collisions will increase **e** the rate will decrease because with fewer acid particles per unit volume the frequency of collisions will decrease

9 **a** it will increase the reaction rate as more collisions will occur because more reactant particles are present per unit volume **b** it will increase the reaction rate as faster moving particles will result in more collisions and more energetic particles will result in more effective collisions **c** it will increase the reaction rate as more surface particles will be exposed to collisions by other reactants **d** it will increase the reaction rate because the catalyst will reduce the amount of activation energy reactant particles require for effective collisions

10 **a** the rate at which reactants are being consumed in grams **b** the high rate at which reactants are consumed is maintained during the first 60 s, then it slows until finally no more reactants are consumed after 90 s **c** initially there are plenty of colliding reactant particles to be consumed, but after 60 s most Mg particles have been used up, so fewer and fewer collisions occur until the reaction ceases after 90 s

11 **a** about 37°C **b** catalysts that are made by living organisms **c** several thousand **d** you would starve to death/ you would die as the reactions that sustain life could not occur **e** amylases break down starches into sugars; proteases break down proteins into amino acids; lipases break down fats into fatty acids **f** so that food molecules are small enough to pass through the gut wall into the blood system **g** each enzyme catalyses

only one particular kind of reaction **h** enzymes are specific; they work best at body temperature; different enzymes work best at different pHs **i** no

12 **a** to capture a pure sample of the gas produced/ to allow the less dense gas to be collected as it forces the water out of the cylinder **b** deliver the gas into a test-tube of clear limewater, which will go milky if the gas is CO_2

13 **a** reactants are solid calcium carbonate and dissolved hydrochloric acid; products are carbon dioxide gas and dissolved calcium and chloride ions **b** calcium carbonate solid + dissolved hydrochloric acid ⟶ carbon dioxide + water + calcium ions + chloride ions; $CaCO_3 + 2HCl \rightarrow CO_2 + H_2O + Ca^{2+} + 2Cl^-$

c

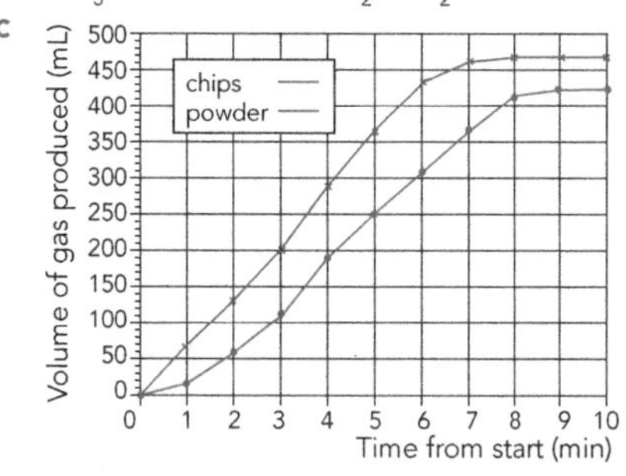

d the chip and acid curve increases steadily then levels off; the powder and acid curve increases rapidly then levels off earlier **e** acid and powder reaction **f** because the powder had more exposed surface particles for acid particles to collide with **g** because all of the calcium carbonate had been consumed in the reaction **h** because it may have been difficult to weigh exactly two grams of chips given their size

Unit 21: Types of Chemical Reactions

1 **a** G **b** K **c** N **d** D **e** H **f** M **g** I **h** A **i** O **j** C **k** E **l** L **m** P **n** L **o** B **p** F

2 **a** an element is a pure substance made of one type of atom only/ a compound is a pure substance in which different kinds of atoms are chemically combined (bonded) **b** an ionic compound is a substance in which different kinds of atoms are held together by ionic bonds/ a covalent compound is a substance in which atoms are held together by covalent bonds **c** thermal decomposition involves breaking up a compound into simpler compounds or elements using heat/ catalytic decomposition involves breaking up a compound into simpler compounds or elements using a catalyst **d** spectator ions are dissolved ions that do not participate in the reaction/ participant ions are involved in the reaction **e** soluble ionic compounds separate into ions when mixed into water/ insoluble ionic compounds do not separate into ions when mixed into water

3 **a** a decomposition reaction in which a compound breaks down into simpler compounds or elements **b** a displacement reaction in which one element replaces another which is part of a compound **c** a combination reaction in which two elements combine to form a single compound **d** a precipitation reaction in which dissolved ions from two different solutions meet and form an insoluble compound

4 **a** Na is a solid **b** H_2O is a liquid **c** OH^- is dissolved in water **d** H_2 is a gas **e** the sodium ion has one positive charge overall **f** the hydroxide ion has one negative charge overall **g** in the ratio of species there are two sodium atoms involved **h** two hydrogen atoms are part of the water molecule **i** that there are multiple reactants and products **j** that the reactants on the left of the arrow are changed into the products on the right **k** yes, because there are equal numbers of each type of atom on both sides of the equation **l** because the charges on ions are shown **m** yes, because the total charge on both sides of the equation is the same, zero

5 **a** false - reactant and product chemicals **b** true **c** true **d** true **e** false - shared **f** true **g** true **h** true **i** true **j** true **k** true

6 **a** carbon dioxide, CO_2 **b** hydrogen, H_2 **c** oxygen, O_2 **d** steam, H_2O

7 **a** Magnesium + oxygen ⟶ magnesium oxide; $Mg(s) + O_2(g) \rightarrow MgO(s)$
b Copper + oxygen ⟶ copper oxide; $Cu(s) + O_2(g) \rightarrow CuO(s)$
c Hydrogen + oxygen ⟶ steam; $2H_2(g) + O_2(g) \rightarrow 2H_2O(g)$
d Sodium + chlorine ⟶ sodium chloride; $Na(s) + Cl_2(g) \rightarrow NaCl(s)$

8 **a** Hydrogen peroxide ⟶ oxygen + water; $2H_2O_2(s) \xrightarrow{MnO_2} O_2(g) + 2H_2O(l)$
b Magnesium hydroxide ⟶ magnesium oxide + steam; $Mg(OH)_2(s) \rightarrow MgO(s) + H_2O(g)$
c Copper carbonate ⟶ copper oxide + carbon dioxide; $Cu(CO_3)(s) \rightarrow CuO(s) + CO_2(g)$

9 **a** $Cu(s) + 2Ag^+(aq) \rightarrow 2Ag(s) + Cu^{2+}(aq)$
b $Zn(s) + Cu^{2+}(aq) \rightarrow Cu(s) + Zn^{2+}(aq)$
c $Mg(s) + Fe^{2+}(aq) \rightarrow Fe(s) + Mg^{2+}(aq)$

10 **a** the ions are more strongly attracted to the water molecules than they are to each other **b** the ions are less strongly attracted to the water molecules than they are to each other

11 **a** $Cu^{2+}(aq) + 2OH^-(aq) \rightarrow Cu(OH)_2(s)$
b $Ca^{2+}(aq) + 2OH^-(aq) \rightarrow Ca(OH)_2(s)$
c $Ca^{2+}(aq) + SO_4^{2-}(aq) \rightarrow CaSO_4(s)$

12 when nonmagnetic sulfur powder and magnetic iron filings are reacted, a solid nonmagnetic compound called iron sulfide is formed

13 **a** the heat makes the iron atoms move faster pushing them further apart, thus making the metal expand **b** so that when the rails expand on hot days they don't buckle because the gap is larger than the expansion **c** because heating reduces the tension they are under rather than causing expansion **d** a reaction in which a more reactive metal replaces the ions of a less reactive metal **e** because aluminium is a very reactive metal, releasing much heat when it gives away electrons **f** the oxide ions in the compound iron oxide are displaced by the newly formed aluminium ions **g** any other metal that is less reactive than aluminium - Zn, Pb, Cu, Ag **h** the extreme heat makes the products molten and they then separate because of their different densities (the lighter molten aluminium oxide floats to the top of the heavier molten iron) **i** as pure iron is not very strong, carbon is added to form the strong alloy steel

14 **a** limewater/ calcium hydroxide solution **b** it will go milky because insoluble calcium carbonate is formed which is insoluble **c** K_2CO_3, Na_2CO_3 and Li_2CO_3 **d** as they are carbonates of very reactive metals they are very stable compounds and will not decompose even when strongly heated

15 **a** yes, copper hydroxide, $Cu(OH)_2$ **b** yes, barium sulfate, $BaSO_4$ **c** yes, magnesium carbonate, $Mg(CO_3)$ **d** yes,

ISBN: 9780170262316

lead iodide, PbI_2 **e** yes, calcium carbonate, $Ca(CO_3)$ **f** no **g** no **h** yes, lead chloride, $PbCl_2$ **i** no

16 **a** catalytic decomposition **b** Water → hydrogen + oxygen, $2H_2O(l) \rightarrow 2H_2(g) + O_2(g)$ **c** the 'pop' test could be used to identify hydrogen gas and the 'lighting a glowing splint' test to identify oxygen gas **d** because two molecules of hydrogen gas are produced for every molecule of oxygen produced **e** to enable an electrical current to flow through the solution

Unit 22: Acids, Bases and Neutralisation

1 **a** K **b** C **c** F **d** N **e** G **f** O **g** Q **h** J **i** S **j** B **k** T **l** L **m** H **n** D **o** M **p** A **q** E **r** R **s** P **t** I

2 **a** an acid is a chemical that releases hydrogen ions when dissolved in water/ a base is a chemical that reduces the number of hydrogen ions in a solution **b** an alkali is a soluble base/ a base is any chemical that reduces the number of hydrogen ions in a solution **c** a strong acid ionises completely/ a weak acid ionises partially **d** a strong alkali ionises completely/ a weak alkali ionises partially **e** an acidic solution has more H^+ ions than OH^- ions/ an alkaline solution has more OH^- ions than H^+ ions **f** a hydrogen ion is a positively charged monatomic ion/ a hydroxide ion is a negatively charged polyatomic ion **g** a concentrated solution has many particles per unit volume of water/ a dilute solution has few particles per unit volume of water **h** a salt is an ionic compound without any H^+ or OH^- ions/ an ionic compound may be made of H^+ or OH^- ions

3

Property	Acidic solution	Alkaline solution
Taste	bitter	sour
Feel	sticky	soapy, slippery
Strong solution	corrosive	caustic
Ions	$H^+(aq)$	$OH^-(aq)$
Litmus	blue goes red	red goes blue
pH	less than 7	more than 7
Electricity	good conductor	good conductor
Base reaction	forms salt + H_2O	N/A
Acid reaction	N/A	forms salt + H_2O
Metal reaction	salt + H_2	not covered
CO_3 reactions	salt + H_2O + CO_2	not covered

4 most acids are already liquids before they dissolve in water; an exception is citric acid, which is a solid/ most bases are solids; an exception is ammonia, which is a gas

5 **a** true **b** true **c** true **d** true **e** false - it separate into ions **f** true **g** false - chemical opposites **h** false - soluble bases **i** true **j** true **k** true **l** true **m** true **n** false - equal **o** true

6 **a i** hydrochloric acid **ii** HCl **iii** release **b i** nitric acid **ii** HNO_3 **iii** release **c i** ammonia **ii** NH_3 **iii** remove **d i** sulfuric acid **ii** H_2SO_4 **iii** release **e i** CH_3COOH **ii** acetic acid **iii** release

7 **a** as dilute HCl has far fewer H^+ ions it just irritates the skin rather than acts corrosively **b** as concentrated HCl has many more H^+ ions it is corrosive

8 **a** alkaline with a pH of 10 **b** relatively strong **c** acidic solution **d** when the universal indicator paper has just turned green **e** it would turn acidic/ have a pH of less than 7

9 **a** no **b** yes **c** yes **d** yes **e** no **f** no **g** yes **h** no

10

Acid +	Salt?	Water?	Gas?
Metal	yes	no	H_2
Metal oxide	yes	yes	none
Metal hydroxide	yes	yes	none
Metal carbonate	yes	yes	CO_2
Metal bicarbonate	yes	yes	CO_2

11 **a** colourless **b** pH = 7 **c** to prevent more being added than is needed to neutralise the acidic solution **d** $Na^+(aq)$ and $Cl^-(aq)$ **e** soluble salt **f** by evaporating the water **g** $H^+(aq) + Cl^-(aq) + Na^+(aq) + OH^-(aq) \rightarrow NaCl(s) + H_2O(l)$

12 **a** enzymes **b** acidic conditions **c** between 1.5 and 3.5 **d** the lining secretes a layer of mucus that prevents the H^+ ions from reaching the cells of the stomach wall **e** the acidic solution reacts the lining of your mouth **f** when some of the contents of the stomach re-enters the oesophagus **g** medications that contain a base, such as a metal hydroxide, carbonate or bicarbonate **h** by reducing the acidity of the stomach contents **i** no, because it does not result in a neutral solution; it is, however, an example of an acid-base reaction

13 **a** hydrogen **b** use the 'pop' test (the gas pops when ignited) **c** because all the H^+ ions in the dilute hydrochloric acid solution will have been used up in the reaction **d** zinc ions and chloride ions **e** $Zn(s) + 2H^+(aq) \rightarrow Zn^{2+}(aq) + H_2(g)$ **f** soluble salt **g** evaporate the remaining liquid **h** $Zn(s) + 2H^+(aq) + 2Cl^-(aq) \rightarrow ZnCl_2(s) + H_2(g)$

14 **a** carbon dioxide **b** use the limewater test (goes milky when CO_2 is bubbled through it) **c** because all the H^+ ions in the dilute hydrochloric acid solution will have been used up in the reaction **d** calcium ions and chloride ions **e** $CO_3^{2-}(aq) + 2H^+(aq) \rightarrow H_2O(l) + CO_2(g)$ **f** soluble salt **g** evaporate the remaining liquid **h** $Ca^{2+}(aq) + CO_3^{2-}(aq) + 2H^+(aq) + 2Cl^-(aq) \rightarrow CaCl_2(s) + H_2O(l) + CO_2(g)$

15 **a** in a strong alkaline solution the alkali will have completely ionised, so there will be sufficient OH^- ions to neutralise all of the H^+ ions **b** in a strong acidic solution all of the acid molecules will have ionised, so there will be sufficient H^+ ions to neutralise all of the OH^- ions **c** in a weak acidic solution only some of the acid molecules will have ionised, so there will be insufficient H^+ ions to neutralise all the OH^- ions

Unit 23: Properties of Carbon Compounds

1 **a** D **b** J **c** F **d** L **e** O **f** A **g** H **h** M **i** E **j** N **k** B **l** R **m** C **n** Q **o** K **p** P **q** G **r** I

2 **a** an alkane is a hydrocarbon with single bonds only/ an alkene is a hydrocarbon with a double bond between two carbon atoms/ an alcohol is a carbon compound that consists of a hydrocarbon chain bonded to an O atom with an attached H atom **b** a saturated hydrocarbon is unable to form bonds with any more atoms because all of its bonds are single bonds/ an unsaturated hydrocarbon is able to form bonds with more atoms because it has a double or triple bond **c** a chemical formula specifies the number or ratio of different types of atoms found in a compound using symbols/ a structural formula is a diagram showing which atoms each atom is bonded to and how many bonds are involved by using lines to represent bonds **d** complete combustion of carbon compounds occurs in abundant oxygen and results in the formation of carbon dioxide and water only/ incomplete combustion of carbon compounds occurs in limited oxygen and results in the formation of soot, carbon monoxide and water **e** monomers are identical small molecules that

ISBN: 9780170262316

can be joined up into long-chain molecules/ polymers are the long-chain molecules that are formed when monomers are linked up

3 **a** butane, C_4H_{10} **b** methanol, CH_3OH **c** ethene, C_2H_4 **d** methane, CH_4 **e** pentane, C_5H_{12} **f** ethane, C_2H_6 **g** ethanol, C_2H_5OH

4 ball-and-stick models show the number of bonds between atoms, eg each molecule has a double bond between two of the carbon atoms; the models also show the orientation of the bonds and the underlying shape of the molecules, eg the ethene molecule is planar (flat) while the propene molecule is more three-dimensional

5 **a** true **b** true **c** true **d** false - saturated hydrocarbons **e** false - ball and stick **f** true **g** true **h** true **i** true **j** false - are not hydrocarbons **k** true **l** true

6 **a** C_9H_{20} **b** $C_{10}H_{22}$ **c** C_4H_8 **d** C_5H_{10} **e** C_3H_7OH **f** C_4H_9OH

7 **a** Ethane + oxygen ⟶ carbon dioxide + water; $2C_2H_6 + 7O_2 \rightarrow 4CO_2 + 6H_2O$
b Propane + oxygen ⟶ carbon dioxide + water; $C_3H_8 + 5O_2 \rightarrow 3CO_2 + 4H_2O$
c Ethene + oxygen ⟶ carbon dioxide + water; $C_2H_4 + 3O_2 \rightarrow 2CO_2 + 2H_2O$
d Propene + oxygen ⟶ carbon dioxide + water; $2C_3H_6 + 9O_2 \rightarrow 6CO_2 + 6H_2O$
e Methanol + oxygen ⟶ carbon dioxide + water; $2CH_3OH + 3O_2 \rightarrow 2CO_2 + 4H_2O$
f Ethanol + oxygen ⟶ carbon dioxide + water; $C_2H_5OH + 3O_2 \rightarrow 2CO_2 + 3H_2O$

8 **a** Methane + oxygen ⟶ carbon monoxide + water; $2CH_4 + 3O_2 \rightarrow 2CO + 4H_2O$
b Methane + oxygen ⟶ carbon + water; $CH_4 + O_2 \rightarrow C + 2H_2O$
c Ethene + oxygen ⟶ carbon monoxide + water; $C_2H_4 + 2O_2 \rightarrow 2CO + 2H_2O$
d Ethene + oxygen ⟶ carbon + water; $C_2H_4 + O_2 \rightarrow 2C + 2H_2O$
e Methanol + oxygen ⟶ carbon monoxide + water; $CH_3OH + O_2 \rightarrow CO + 2H_2O$

9 **a** the open air-hole allows more oxygen to be mixed with the fuel as air is 'sucked' into the burner **b** the closed air-hole only allows oxygen in the air around the top of the pipe to be mixed with the fuel **c** because it is more visible and not as hot

10 **a** MP = –50 °C, BP = 150 °C **b** the MP would be about 20 °C; the BP would be about 270 °C **c i** those alkanes with between 1 and 4 carbon atoms are gases **ii** those alkanes with between 5 and 14 carbon atoms are all liquids **iii** none of those alkanes would be solid at room temperature **d** the boiling points steadily increase as the number of carbon atoms in the alkane increases; this is because heavier molecules require more heat energy to break free of the attractive forces holding the molecules together in the liquid state (and also because there are more attractive forces between larger molecules than smaller ones) **e** in general, melting points increase as the number of carbon atoms in the alkane increases **f** the melting points for the first two alkanes appear to hardly increase at all, even falling slightly; after the alkane with three carbons, the melting point of every second alkane does not increase as much as the others do

11 **a** ethanol, because it is burning with a nearly invisible blue flame **b** butane, because if is burning with a yellow flame **c** ethanol, because it is undergoing complete combustion

12 **a** ethanol with various additives **b** as a cleaning fluid, as a grease solvent, and as fuel or spirit burners **c** to discourage people from drinking the highly concentrated alcohol **d** because it was thought that the presence of a toxic alcohol would stop people from drinking meths **e** it can cause blindness, organ damage and death **f** an aniline dye; so that people wouldn't accidentally drink it **g** because people who were desperate to drink any alcohol were getting poisoned **h** the extremely bitter taste would make it very difficult to drink the meths

13 (the diagram in part **a** could be drawn as on page 89)

a
```
CH3 H
 |  |
 C = C
 |  |
 H  H
```

b
```
 CH3 H  CH3 H  CH3 H  CH3 H
  |  |   |  |   |  |   |  |
- C - C - C - C - C - C - C - C -
  |  |   |  |   |  |   |  |
  H  H   H  H   H  H   H  H
```

c each of the double bonds is replaced by two single bonds, one of which links two adjacent monomers **d** heat, pressure and a suitable catalyst

14 **a** mix a small amount of each with an equal volume of water; pentane would form a layer of top of the water as pentane and water are immiscible but ethanol would dissolve in the water **b** if the two samples were combusted in the air, ethanol would burn with a clear blue flame but pentane would burn with a yellow-orange flame because not all of the carbons would get oxidise (the product soot causes the yellow flame)

15 **a** all have carbon and hydrogen atoms; the H-C bonds are all single bonds; the atoms are spaced as far apart as possible **b** the left molecule has a double bond between two carbon atoms; the middle molecule has an oxygen atom with a hydrogen atom attached to it; the right molecule, has single bonds only between the carbon and hydrogen atoms it is made of **c** the left molecule is an alkene; the middle molecule is an alcohol; the right molecule is an alkane **d** the left molecule is ethene, C_2H_4; the middle molecule is ethanol, C_2H_5OH; the right molecule is ethane, C_2H_6 **e** ethene (2C, 4H), ethane (2C, 6H), ethanol (2C, 6H, 1O)

Unit 24: Production and Uses of Carbon Compounds

1 **a** B **b** H **c** Q **d** D **e** S **f** G **g** K **h** F **i** N **j** T **k** P **l** I **m** C **n** O **o** M **p** A **q** J **r** E **s** R **t** L

2 **a** hydrocarbons are made of carbon and hydrogen atoms only/ alcohols are carbon compounds in which an alkane chain is bonded to an oxygen atom with an attached hydrogen atom **b** diesel is a liquid fuel used by trucks, ships and some cars/ petrol is a liquid fuel used by motorbikes and most cars/ LPG is a liquid fuel used by some vehicles **c** boiling occurs when a liquid is heated sufficiently to turn it into a gas/ condensing occurs when a gas is cooled sufficiently to turn it into a liquid **d** the kerosene fraction has hydrocarbons with between 12 and 16 carbons and it is used to make

ISBN: 9780170262316

jet fuel/ the gasoline fraction has hydrocarbons with between 4 and 10 carbons and it is used to make petrol **e** complete combustion results in a fuel being fully oxidised, with only carbon dioxide and water as the products/ incomplete combustion results in a fuel being only partially oxidised, with carbon monoxide and soot being produced as well as carbon dioxide and water **f** cracking involves splitting up a large alkane molecule into a smaller alkane molecule and an alkene molecule/ polymerisation involves joining up numerous small identical alkene molecules into a very long chain molecule

3

Fuel	State	Source	kJ/g	Used in:
paraffin wax	solid	crude oil	41.4	candles
butane	liquid	crude oil	49.5	lighters
petrol	liquid	crude oil	47.0	cars
kerosene	liquid	crude oil	45.4	jet planes
diesel	liquid	crude oil	45.0	trucks, ships
meths	liquid	natural gas	29.7	camping stoves
natural gas	gas	natural gas	54.0	gas stoves
LPG	gas	natural gas	49.6	barbecues, cars

4 **a** because butane lights easily and can be readily liquefied **b** because LPG combusts well and large amounts of the fuel can be stored in a liquid form in a tank **c** because natural gas combusts well and it can be supplied by pipe directly to the house **d** because paraffin wax is a solid fuel and burns readily (although combustion is incomplete) **e** because meths is already a liquid and it combusts well **f** because kerosene has suitably sized hydrocarbons for combustion in jet engines **g** because petrol has suitably sized hydrocarbons for efficient combustion in car engines

5 **a** true **b** true **c** false - more heat **d** true **e** true **f** true **g** false - fall **h** false - fewer waste products **i** true **j** true **k** false - occurs in the absence of oxygen **l** true

6 **a** refinery gas, butane, petrol, kerosene, jet fuel, diesel, paraffin **b** commercial natural gas, LPG (liquefied petroleum gas)

7 **a** $CH_4 + 2O_2 \rightarrow CO_2 + 2H_2O$;
$2CH_3OH + 3O_2 \rightarrow 2CO_2 + 4H_2O$
b $2C_2H_6 + 7O_2 \rightarrow 4CO_2 + 6H_2O$;
$C_2H_4 + 3O_2 \rightarrow 2CO_2 + 2H_2O$;
$C_2H_5OH + 3O_2 \rightarrow 2CO_2 + 3H_2O$
c $C_3H_8 + 4O_2 \rightarrow C + 2CO_2 + 4H_2O$
d $C_4H_{10} + 6O_2 \rightarrow CO + 3CO_2 + 5H_2O$

8 **a** the temperature of the water to start with; the volume of water; the height of the container above the burner **b** the heat gained by the water can be calculated using the formula for heat ($Q = mc\Delta T$, where m is the mass of the water, ΔT is the rise in temperature, and c is the specific heat capacity of water); the energy content of the fuel can then be estimated by dividing the heat gained by the water by the mass of fuel consumed **c** no, because a lot of the heat released from the fuel would have heated the air and the container **d** yes, provided the conditions mentioned in **a** above were the same in each case

9 **a** fractional distillation **b** to record the temperature of the vapour so that you know when a particular liquid is being turned into a gas **c** to cool the vapour in the inner tube so that it will liquefy **d** the water will heat up to 78 °C and will then stay at this temperature until all the ethanol has boiled off, at the same time the ethanol vapour will be liquefying in the condenser, the water will then heat up to 100 °C and stay at this temperature while the water boils, the water vapour will condense in the condenser as it cools to below 100 °C **e** collect one sample in a test-tube when the thermometer was reading about 78 °C and another when the thermometer was reading 100 °C **f** you should be able to ignite the alcohol and the water will turn blue cobalt chloride paper pink

10 **a** so that all hydrocarbons with less than 25 carbons are vaporised **b** they will liquefy in the lower, hotter trays because they have higher boiling points **c** they will liquefy when they reach the higher cooler trays because they have lower boiling points **d** by passing through trays of different temperatures, which results in fractions of similar-sized hydrocarbons that liquefied together because their boiling points were within a certain temperature range **e** to produce petrol **f** to produce diesel and industrial oil **g** it is refined into jet fuel

11 **a** because its boiling point is 0°C **b** because it is under pressure **c** butane gas (vapour) **d** the button opens a valve **e** the spark created by the metal wheel striking a flint **f** you keep pressing on the button, which holds the valve open and some of the pressurised butane vaporises to give a continuous stream of gas **g** because it doesn't get mixed with much oxygen **h** because the flame is yellow **i** carbon monoxide, carbon dioxide and water **j** butane + oxygen $\rightarrow$ carbon monoxide + carbon dioxide + water
k $C_4H_{10} + 6O_2 \rightarrow CO + 3CO_2 + 5H_2O$ (equations with more carbon monoxide molecules and less carbon dioxide molecules are also possible)

12 **a** (own graph) **b** carbon dioxide **c** otherwise the natural gas would not burn **d** methane and ethane **e** propane and butane **f** butane **g** because they are liquids under normal atmospheric pressure **h** because they are under pressure in the reservoirs deep underground

13 **a** hexane $\rightarrow$ butane + ethene; $C_6H_{14} \rightarrow C_4H_{10} + C_2H_4$
b pentane $\rightarrow$ ethane + propene; $C_5H_{12} \rightarrow C_2H_6 + C_3H_6$

14 **a** in the first stage methane gas reacts with steam to give carbon monoxide and hydrogen gases
$CH_4(g) + H_2O(g) \rightarrow CO(g) + 3H_2(g)$
b in the seconds stage hydrogen gas reacts with carbon monoxide gas to give methanol gas
$2H_2(g) + CO(g) \rightarrow CH_3OH(l)$
c because pure methanol needs to be separated from by-products formed in the reaction and unreacted reactants

15 **a** fermentation requires warmth and the enzymes produced by yeast/ the chemical process requires high temperatures and a catalyst **b** glucose is the raw material for fermentation/ ethene and water are the materials for the chemical process **c** ethanol and carbon dioxide are the products of fermentation/ ethanol is the sole product of the chemical process
d Glucose $\rightarrow$ ethanol + carbon dioxide;
$C_6H_{12}O_6(aq) \rightarrow 2C_2H_5OH(l) + 2CO_2(g)$
Ethene + water $\rightarrow$ ethanol;
$C_2H_4(g) + H_2O(g) \rightarrow C_2H_5OH(l)$

Unit 26: Traits, Genes, Chromosomes and DNA

1 **a** H **b** J **c** C **d** U **e** E **f** L **g** S **h** I **i** N **j** T **k** B **l** Q **m** D **n** G **o** K **p** P **q** R **r** F **s** M **t** A **u** W **v** V **w** O

2 **a** acquired variation results from the actions of the organism or from the effects of its environment/

ISBN: 9780170262316

genetic variation is caused by different alleles being passed on from one generation to the next **b** traits are features of organisms/ genes are inherited instructions that determine traits/ alleles are alternative forms of a gene **c** genotype refers to the two alleles an organism possesses for a particular trait/ phenotype refers to how the genotype is expressed **d** a homozygous phenotype has two identical alleles/ a heterozygous phenotype has two different alleles **e** a chromosome is a very long DNA molecule that is wrapped around special proteins/ a DNA molecule consists of two-linked strands that are twisted into a double helix **f** base-pairs are the paired bases that connect the two strands of a DNA molecule/ base-triplets are groups of three consecutive bases found along the sense strand of a DNA molecule **g** a base sequence is the sequence of bases (or base-triplets) found along the length of a gene/ an amino acid sequence is the sequence of amino acids forming a protein **h** protein synthesis occurs when a cell uses the base sequence of a gene to construct the corresponding protein/ DNA replication occurs when the cell 'unzips' a DNA molecule and builds new sense and anti-sense strands on the exposed bases, which results is duplicated DNA molecules and therefore chromosomes

3 **a** sitting in the sun is a functional trait that enables the tuatara to increase its body temperature, allowing it to be more active/ its grey skin colouration allows it to blend in with its environment/ the spikes are defensive structures **b** the moth's colouration and patterning allows it to blend in with its environment **c** the snake's venom is a functional trait that enables it to paralyse prey/ its sharp teeth allow it to puncture the skin of its prey in order to inject venom **d** the monkey's sharp teeth enable it to engage in aggressive behaviour/ they also enable it to grasp food **e** the stick insect's shape and colour enable it to blend in with the twigs of the plants that it lives on

4 **a** true **b** true **c** false - all features **d** true **e** false - two copies **f** true **g** true **h** false - different alleles **i** false - only when they are coiled up tightly **j** true **k** true **l** true **m** true **n** true

5 **a** amino acids **b** in two linked chains **c** a protein **d** because proteins are folded chains of amino acids **e** the order of amino acids in the two chains **f** the sequence of bases (base-triplets) in the gene that codes for insulin **g** the INS gene, which is found along one of the chromosomes **h** different base-triplets code for particular amino acids, so the sequence of particular triplets in the INS gene codes for the specific sequence of amino acids in the insulin molecule **i** a space-filling model **j** carbon (black spheres), hydrogen (white spheres), oxygen (red spheres) nitrogen (blue spheres), and sulfur (yellow spheres) **k** the space-filling (bottom) diagram, because it shows the location and bonding of all of the atoms in the molecule **l** the sub-unit diagram, because it shows the underlying structure of the two amino acid chains and how they are linked together

6 **a** the gene codes for the protein **b** the protein determines the trait **c** two **d** alleles **e** defective insulin molecules would be made, which may not keep blood sugar levels within a safe range **f** one of the bases in a triplet may have been replaced by another base, which resulted in a different amino acid being specified

7 **a** straight hair **b** curly hair **c** two 'curly hair' alleles **d** two 'straight hair' alleles **e** a 'curly hair' allele and a 'straight hair' allele **f** it could be curly or straight or even wavy (but see Unit 26)

8 **a** in the cell nucleus **b** thread-like and invisible **c** they coil up very tightly to form visible rods **d** they will have been duplicated **e** eight **f** two **g** four

9 **a** acquired variation, because it resulted from the effects of the environment, ie being forced to grow within a cubic container **b** because their genes have not be altered, so when the seeds are allowed to grow and develop normally they will have the normal rounded watermelon shape

10 **a** because it enables individuals to detect potentially toxic chemicals in foods **b** to prevent herbivores from eating them **c** because we have a variety of taste receptors **d** the trait is the ability to taste PTC; the protein involved is called 'Taste receptor 2 member 38' or TAS2R38 for short; the gene involved is the TAS2R38 gene **e** the 'taster' allele and the 'non-taster' allele **f** two **g** an individual who is homozygous for the 'taster' allele finds PTC intensely bitter; an individual who is homozygous for the 'non-taster' allele usually can't taste PTC at all; an individual who is heterozygous for the alleles finds PTC moderately bitter **h** because it is an industrial chemical not found in nature **i** supertasters might detect and avoid a newly introduced toxic plant, which would help those members of the population to survive

11 **a** 15 **b** ACT-TGG-TAC-AGT-GGG **c** five **d** five **e** Thr-Trp-Tyr-Ser-Gly **f** TGA-ACC-ATG-TCA-CCC **g** because it acts as a template on which a new coding strand is built when the DNA is replicated **h** the two strands of the DNA molecule 'unzip' and new strands are assembled using the existing strands as templates; this results in duplicated chromosomes

12 **a** 46 **b** 23 **c** one member of each pair came from the mother and the other member from the father **d** they are different sizes and shapes/ one is long and the other is much shorter **e** the X and Y chromosomes/ the sex chromosomes **f** there would be two long chromosomes of the same size and shape **g** at the same location on the other chromosome of pair 11

Unit 27: Reproduction, Growth and Variation

1 **a** K **b** D **c** T **d** E **e** G **f** R **g** A **h** M **i** C **j** C **k** N **l** B **m** P **n** L **o** Q **p** F **q** H **r** O **s** I **t** S

2 **a** sexual reproduction involves gametes from two parents combining to form genetically different offspring/ asexual reproduction involves a single parent producing genetically identical offspring **b** gametes are reproductive (sex) cells) with half the normal number of chromosomes/ body cells include all non-reproductive cells in the body, they all have the same full set of chromosomes **c** sperm are small, mobile male gametes/ eggs are large, immobile female gametes **d** mitosis is a form of cell division that produces two genetically identical body cells/ meiosis is a form of cell division which produces genetically different gametes, either eggs or sperm **e** maternal chromosomes are those inherited from the female parent/ paternal chromosomes are those inherited from the male parent **f** the phenotype refers to the physical appearance or state of a trait/ the genotype refers to the two alleles the organism possesses for the trait

 ISBN: 9780170262316

3 a meiosis in the testicles produce sperm with only 23 chromosomes b meiosis in the ovaries produce eggs with only 23 chromosomes c a sperm reaches an egg d fertilisation of the egg by a sperm results in a zygote with 46 chromosomes e the zygote undergoes cell division by mitosis f cells begin to specialise in the embryo g meiosis h otherwise the number of chromosomes would double every generation/ to keep the chromosome number constant from generation to generation i meiosis shuffles maternal and paternal chromosomes and fertilisation combines set of chromosomes from two different organisms j genetically different offspring k fertilisation l mitosis m cell specialisation n trillion of cells and over 300 different types of cells

4 a true b true c false - all human body cells d true e true f true g false - phenotypes differ h false - do not i true j false - separates k true l false - the sperm

5 **Sexual reproduction advantages**: genetically diverse offspring have variation that may help them to survive adverse conditions
Sexual reproduction disadvantages: requires two parents; sperm must reach the egg
Asexual reproduction advantages: only one parent required; enables a population to rapidly increase in numbers in favourable conditions
Asexual reproduction disadvantages: genetically identical offspring lack the variation that may help them to survive adverse conditions

6 a one b two c four chromosomes d mitosis e because only two new cells are produced and they have the same number of chromosomes as the original cell f genetically identical g i chromosomes are being duplicated just before division starts ii nucleus disappears, duplicates remain attached at the centromere, spindle starts to form iii chromosomes become attached to the spindle at the centromere iv spindle pulls duplicated chromosomes apart v cell pinches into two new cells and a nucleus forms in each with a full set of chromosomes

7 a two b four c two chromosomes d meiosis, because the number of chromosomes has been halved e genetically different f i chromosomes are duplicated ii after the nucleus disappears the spindle begins to pulls homologous chromosomes apart iii original cell pinches into two daughter cells iv spindle begins to pull duplicated chromosomes apart v a nucleus with half the normal number of chromosomes forms in each of the four cells

8

Feature	Mitosis	Meiosis
Purpose of division	growth	reproduction
Location of division	anywhere in body	testicle/ ovary
Original cell type	body cell	body cell
No. of divisions	one	two
No. of cells formed	two	four
Type of new cells	body cells	gametes
Chromosome no.	unchanged	halved
Chromosome sets	identical	different
Homologous pairs	remain together	separate
Allele pairs	remain together	separate
Subsequent event	cell specialisation	fertilisation

9 a crossing-over, involving swapping sections of homologous genes, results in new combinations of alleles for different alleles lying along the same chromosome b meiosis randomly separates homologous chromosomes thus creating different chromosome collections in the resulting gametes c fertilisation randomly combines gametes to produce offspring, each with a different collection of chromosomes and therefore of alleles for the several genes that control most traits

10 a eye colour and shape, face shape, lip thickness, etc. b dark eyes, narrow eye shape, thin lips c 46 d because he started life from an egg and sperm that were different to the egg and sperm that the two girls started from e a single sperm fertilised a single egg, which then developed into a ball of cells that split into two, with the two smaller balls of cells developing into identical twin girls f two g because he inherited a distinctly different collection of alleles from those inherited by the girls; because he inherited a Y and an X chromosome and they inherited two X chromosomes h because they inherited identical sets of chromosomes i if they experience different environments (eg sun exposures) or engage in different habits (eg dieting or not)

11 a 46 b no, because there may be other species with 46 chromosomes (eg introduced European hares in NZ) c there is an unpaired chromosome at the end of the second and fourth rows d male, because it has a Y and an X chromosome e females have two alleles for a larger number of genes along their sex chromosomes and males have two alleles for a smaller number of genes along their sex chromosomes (the others genes will have one allele only)

12 a the plant has both male and female sex organs b ovary c anther d pollen e transferred by animal pollinators f inside the ovary when a pollen tube has reached an egg g meiosis → pollination → fertilisation → mitosis → germination → growth

Unit 28: Patterns of Inheritance and Adaptive Features

1 a D b H c K d N e O f B g M h E i P j F k L l Q m G n J o R p C q A r I

2 a genes are inherited instructions that are involved in determining traits/ alleles are alternative forms of a gene that result in different forms of a trait b genotype refers to the pairs of alleles inherited for a trait/ phenotype refers to the actual form the trait takes c a homozygous genotype consists of two identical alleles/ a heterozygous genotype consists of two different alleles d gametes are sex cells that have only one of each type of chromosome/ zygotes result when gametes fuse, they have two of each type of chromosome e a dominant allele is expressed if only one is present/ a recessive allele is only expressed if two are present f pedigree charts are used to show the inheritance of a plant or animal trait over several generations/ family trees do the same thing for human families g phenotype probabilities refers to the chances of an individual having different phenotypes/ phenotype ratios refer to the expected or actual numbers of offspring showing different phenotypes

3 a white and coloured b two c the W allele because only one copy is required for it to be expressed/ the W allele because it prevents any other colours from being expressed d the w allele because two are required for it to be expressed e WW, Ww and ww f WW is the homozygous dominant genotype, ww

is the homozygous recessive genotype, Ww is the heterozygous genotype **g** the homozygous dominant genotype is white, the homozygous recessive genotype is coloured, and the heterozygous genotype is white **h** ww **i** WW or Ww **j** mate it with a coloured cat and if all of the offspring are whit, then it is likely to be pure breeding **k** mate it with a coloured cat and if any of the offspring are coloured, then it must have a recessive allele for expressing coat colour

4 **a** false - two copies of each gene **b** true **c** true **d** true **e** false - not affected in any way **f** true **g** true **h** false - three expressing the dominant allele **i** false - homozygous recessive individual **j** true

5 **a** white coat colour, deafness and no iris pigmentation **b** that it affects whether other genes are expressed or not **c** because it results in no pigments being formed/ because it can cause deafness as well **d** in either an egg or a sperm **e** either because it does not affect whether an individual survives and reproduces, or because the phenotype provides some advantage, or because breeders ensure that the white masking allele persists

6 **a** white coat and coloured coat **b** the male would be either WW or Ww, as W is a dominant allele, the female cat would be ww as the allele for colour is recessive **c** five **d** a coloured coat allele, as the mother has only recessive alleles to pass on in her eggs, while the father could pass on either a white coat allele or a coloured coat allele

7 **a** homozygous recessive **b** both heterozygotes **c** because they both carry an allele for cystic fibrosis, but it is not expressed as it is recessive **d** because the allele for cystic fibrosis is not expressed when you only have one of them **e** one chance in four **f** one chance in four as well **g** one chance in two/ two chances in four **h** one chances out of four **i** (personal opinion)

8 **a** because it is inherited and increase the chances that offspring will survive and reproduce **b** because it enables the lactose in milk to be digested and used by the body, thus increasing the baby's chances of survival **c** because offspring are usually weaned at an early stage **d** probably through a genetic mutation that prevented the lactase gene from being switched off as would normally occur **e** because it enabled humans to take advantage of an abundant food source (cow's milk) after infancy, thus increasing their chances of surviving and reproducing

9 **a** to transport oxygen around the body **b** so that they will deform readily when passing through narrow capillaries and then regain their shape **c** it results in a form of haemoglobin that crystallises readily, deforming the shape of red blood cells into a rigid sickle shape **d** the sickle cells do not last as long as normal red blood cells; this results in a shortage of red blood cells and insufficient oxygen being delivered to body cells **e** one chance in four **f** one chance in four still **g** because heterozygous individuals produce some abnormal haemoglobin, but not sufficient to give cells a sickle shape **h** because the presence of a single allele reduces the effect of a malarial infection

10 **a** the introduction of predatory mammals, such as such as cats, rats, stoats and weasels **b** when only small populations survive, genetic diversity is reduced **c** because it might result in inbreeding/ because there might be little variation amongst the breeding pairs **d** by ensuring that genetically different individuals are mated **e** there may be no natural immunity amongst the limited number of organisms **f** because some members of the species may have traits that enable them to survive and reproduce in changing environmental conditions

11 **a** all are tall **b** three smooth seeds for every wrinkled seed **c** all produce green seeds **d** equal numbers of purple- and white-flowered plants **e** all will have pinched pods **f** it isn't possible to tell as the parent plants could be homozygous or heterozygous **g** three plants with axial flowers for every plant with terminal flowers

Unit 29: Bacteria, Fungi and Viruses

1 **a** G **b** J **c** P **d** D **e** L **f** U **g** A **h** I **i** R **j** T **k** C **l** M **m** B **n** Q **o** E **p** H **q** N **r** S **s** F **t** O **u** K

2 **a** producers make their own food using raw materials from their environments/ consumers obtain their food by eating other organisms/ decomposers obtain their food by eating dead organisms or foodstuffs **b** parasites live in or on other organisms, feeding off them/ saprophytes feed on dead organisms **c** bacterial cells are smaller and lack a nucleus/ fungal cells are larger and have nuclei **d** secretion occurs when a cell releases useful chemicals (eg digestive enzymes) into its environment/ absorption occurs when a cell takes in useful substances from its environment **e** aerobic conditions occur when oxygen is available; it results in lots of energy and carbon dioxide gas/ anaerobic conditions occur when oxygen is unavailable; it produces little energy and waste products such as ethanol or methane gas **f** asexual reproduction occurs when a single parent produces genetically identical offspring/ sexual reproduction occurs when two parents combine their genes to produce genetically different offspring **g** binary fission is the method of asexual reproduction used by bacteria, in which the original cell divides into two genetically identical daughter cells/ budding is the method of asexual reproduction used by yeasts, in which the original cell buds off a smaller genetically identical cell **h** inoculation occurs when spores or active micro-organisms are transferred onto sterile agar/ incubation occurs when the micro-organisms are grown in the laboratory at an optimum temperature

3 **a** Petri plate/ plastic plate **b** nutrient agar **c** fungal colony, because it is furry **d** bacterial colony because it is shiny **e** when they develop into colonies **f** bacterial colonies are usually shiny in appearance because their cells have a slime capsule/ fungal colonies are usually furry or fuzzy in appearance because of the growth of the tiny feeding filaments called hyphae **g** all the nutrients they need for growth and reproduction **h** because viruses cannot 'grow' on agar, they can only reproduce inside living cells **i** usually at a temperature of around 25 °C, with the plates upside down to prevent the colonies being flooded by condensation **j** the lid should not have been removed after incubation

4 **a** the shiny, zigzag-shaped colony is bacterial; the fuzzy, circular colony is the penicillin mould **b** because the object (eg cotton bud) used to collect microbes would have wiped in a zigzag shape across the agar **c** the penicillin mould has prevented the growth of the bacterial colony around it

5 **a** true **b** true **c** true **d** false - much smaller **e** false -

ISBN: 9780170262316

externally **f** true **g** true **h** false - less **i** true **j** false - by the air **k** true **l** false - often **m** true **n** true

6 **a** bacterium **b** fungus/ mould **c** virus **d** cell wall - supports the cell/ gives the cell a shape **e** cell membrane – controls the entry and exit of chemicals **f** flagellum – propels the cell forward **g** cytoplasm – provided a medium in which metabolism can occur **h** chromosome – bears the genes of the bacteria **i** sporangium (spore case or capsule) – protects the spores as they mature, then releases the spores **j** spores – enable the mould to spread widely as the microscopic cells disperse readily in air currents **k** hyphae - enable the fungus to grow through the tissue of its host and feed off it **l** protein coat (capsid) – encloses the genes, allows the virus to attach to a host cell, enables the virus to enter the host cell **m** DNA or RNA molecule – bears the genes of the virus

7 **a** diplococcus **b** coccus **c** streptococcus **d** staphylococcus **e** diplobacillus **f** bacillus **g** streptobacillus **h** flagellated bacillus **i** spirillum **j** vibrio

8 **a** binary fission,as the cell is about to divide into two across the middle **b** two genetically identical 'daughter' cells

9

Feature	Bacteria	Fungi	Viruses
Living	yes	yes	no
Made of cells	yes	yes	no
Cell nucleus	none	yes	–
Life processes	all	all	one only
Feeding role	wide variety	parasite, saprophyte mutualist	–
Digestion	extra-cellular	extra-cellular	–
Excretion	diffusion	diffusion	–
Anaerobic metabolism	produces methane	produces alcohol	–
Asexual reproduction method	binary fission	spores, budding	viral replication
Dormant spores	yes	yes	no
Source of genetic variation	mutation, gene transfer	mutation, sexual reproduction	mutation

10 **a** binary fission **b** asexual reproduction **c i** chromosomes are duplicated and one copy is pulled to each end of the cell **ii** the cell pinches in half **iii** two separate bacteria are formed **d** by DNA replication **e** 262 144 bacteria **f** in favourable conditions it enables bacteria to multiply very rapidly, and as all of the offspring are genetically identical, they will be well suited to the conditions

11 **a** at least three, because there are three fuzzy/furry colonies **b** at least seven, because there are seven separate shiny colonies

12 **a** a fungi (yeast) and bacteria (lactic acid bacteria) **b** the best rate of growth and reproduction possible **c** to enable the microbes to carry out anaerobic metabolism under optimum conditions **d** fermenting/fermentation **e** lactic acid bacteria are able to grow across a temperature range of 4 to 41 °C, and optimum growth occur around 33 °C/ yeast are able to grow across a temperature range of 8 to 35 °C, and optimum growth occur around 28 °C **f** lactic acid bacteria **g** because they produce lactic acid which causes acidic conditions **h** because salt is sometimes added to sourdough

13 **a** extra-cellular digestion **b** to break large, insoluble food molecules down into small, soluble ones **c** chemical bonds at specific parts of the large molecules making up the food source are broken, resulting in smaller soluble molecules **d** they diffuse through the cell membrane (from a region of higher concentration outside the cell to a region of lower concentration inside) **e** no

14 **a** spore production **b** asexual reproduction **c i** spores are produced by mitosis inside a protective sporangium (spore case or capsule) **ii** when the spores are mature the sporangia split **iii** the spores are dispersed by air currents **iv** spores germinate on suitable food sources **d** an identical set of chromosomes/genes to the original mould **e** because huge numbers are produced and the very light, tiny spores can be carried great distances **f** in favourable conditions it enables the fungi to multiply rapidly, and as all of the offspring are genetically identical, they will be well suited to the conditions **g** because the sexually produced spores may have genetic variations that are useful in different environments and their resistant coats mean that they can survive for long periods of time until new opportunities arise

15 **a** viral replication **b** asexual as no mixing of genes from two sources is involved **c i** virus attaches itself to the surface of a suitable host cell **ii** the virus enters the host cell **iii** the virus takes over the host cell, using it to produce thousands of identical copies **iv** the host cell releases the new viruses into the environment **d** an identical copy of the genes of the original virus **e** it enables the virus to rapidly create huge numbers of genetically identical copies, which can then enter other cells of the host or infect another organism

Unit 30: Interactions between Humans and Micro-organisms

1 **a** E **b** I **c** P **d** B **e** L **f** Q **g** T **h** K **i** U **j** C **k** R **l** A **m** O **n** S **o** H **p** D **q** J **r** G **s** M **t** F **u** N

2 **a** aerobic conditions occur when oxygen is available; it results in lots of energy and carbon dioxide gas/ anaerobic conditions occur when oxygen is unavailable; it produces little energy and waste products such as ethanol or lactic acid **b** pasteurisation is the brief heat treatment of food to kill all pathogenic bacteria/ fermentation is an anaerobic process carried out by microbes **c** pathogens are disease-causing micro-organisms/ saprophytes are micro-organisms that feed on dead organisms and organic wastes **d** sterilising involves killing all microbes using heat, radiation or chemicals/ inoculating involves adding active microbes or spores during food production **e** disinfectants are powerful chemicals used to kill microbes on surfaces/ antiseptics are milder chemicals applied to wounds to kill bacteria without killing cells **f** antigens are specific marker chemicals on pathogens/ antibodies are proteins produced by lymphocytes that recognise antigens on particular pathogens and attach themselves **g** natural immunity arises from a previous infection because the required antibodies are already present in the blood/ artificial immunity occurs when a person is injected with a dead or weakened strain of a pathogen, which ensures the required antibodies are produced to fight a real infection by that pathogen

3 **a** freezing - the cold temperature prevents microbes from multiplying/ sealing - prevents airborne microbes

ISBN: 9780170262316

from reaching the food **b** acidifying - the low pH kills or inhibits the growth of microbes **c** sterilising - boiling the jam kills active microbes and spores **d** salting – the high concentration of salt stops microbe growth **e** dehydrating – the absence of water/ moisture prevents microbes from multiplying **f** sealing - the absence of oxygen prevents aerobic microbes from multiplying **g** bottling after heat sterilising - heating kills microbes and sealing prevents entry of airborne microbes **h** chilling - the low temperature inhibits the growth and reproduction of microbes

4 **a** true **b** true **c** false - produces ethanol and lactic acid **d** false - heating milk at 72 °C for 15 seconds **e** true **f** true **g** false - antiseptics **h** true **i** true **j** false - artificial immunity **k** false - will not **l** true **m** true **n** true

5

Feature	Wine-making	Yoghurt-making
Raw materials	grapes	milk
Sugars involved	glucose, fructose	lactose
Sterilisation used	vats sterilised	milk pasteurised
Type of microbes	fungi	bacteria
Microbe name	yeast	Lactobacillus
Process name	fermentation	fermentation
Type of respiration	anaerobic	anaerobic
Incubation temp.	varies	about 45 °C
Products	ethanol and CO_2	lactic acid
Preserving product	SO_2 added	low pH

6 **a** about 42 °C **b** it's about five degrees warmer than body temperature (37 °C) **c** above about 47 °C and below about 30 °C **d** at 20 °C growth would be very slow or have ceased/ at 50 °C growth would have ceased (Campylobacter are killed by temperatures above 55 °C) **e** chilling would slow growth and reproduction; freezing would prevent any growth and reproduction **f** cooking would kill the bacteria (they don't form heat-resistant spores) **g** to ensure that internal temperatures are hot enough to kill all the Campylobacter bacteria

7 **a** because the mould will damage those materials **b** moisture, warmth and a suitable food source **c** ensure rooms are well ventilated/ dehumidified/ kept warm

8 **a** because the milk powder is completely dehydrated and is sealed to keep out moisture from the air **b** because the powder is dry **c** no need to because microbes won't grow on the powdered milk, unless it is a very humid climate **d** fresh milk has been pasteurised to kill pathogenic bacteria but other bacteria or air-borne microbes will grow if the milk is not refrigerated

9 **a** because they feed on dead matter and organic wastes **b** warm, moist and well aerated **c** aerobic respiration would occur producing smelly waste products and cause only partial decomposition **d** they recycle chemical elements back into the environment for plants to absorb

10 **a** to ensure that the bacteria carry out aerobic respiration, breaking down wastes more rapidly **b** moisture, suitable temperature and pH **c** nutrition, respiration and excretion **d** carbon dioxide, water and other nutrients (eg nitrate ions) **e** it is usually discharged into a river, lake, estuary or sea **f** dissolved nitrate ions could over-enrich the body of water

11 **a** Acquired Immune Deficiency Syndrome, Human Immunodeficiency Virus **b** a virus **c** it has been transferred from one species (monkey) to another (humans) **d** through sexual intercourse, blood transfusions, shared syringes, birth and breast-feeding **e** by using condoms, testing blood, setting up needle exchanges for drug addicts, providing medicinal drugs for infected mothers and their babies, infected mothers avoiding breast-feeding **f** they are infected with HIV/ they have specific antibodies in their blood (the test is 99% accurate) **g** because the viruses usually become dormant once they have entered lymphocytes **h** infections that occur when the immune system is compromised/ not functioning properly **i** because the virus mutates rapidly, which means that artificial immunity gained from vaccination would not be effective for long **j** because it has spread to a very large number of people all over the world and many people have died from it

12 **a** in the atmosphere as part of nitrogen molecules **b** from nitrate ions in the soil **c** they release nitrate ions when they digest the proteins found in dead organisms and organic wastes **d** because they are an essential part of the proteins that all living things are made out of and are needed as enzymes to enable metabolism to occur

13 **a** they absorb antigens that separate from the surface of pathogens and use them to create antibodies that recognise antigens still attached to pathogens **b** they recognise the shape of specific antigens **c** it prevents the virus from entering a host cell **d** they 'spot' the attached antibodies then engulf and digest the viruses

14 **a** from a mutation caused by a DNA copying error, radiation or chemicals **b** the offspring of the resistant bacterium survive antibiotic treatments and steadily increase in numbers while vulnerable bacteria in the population are eliminated over time **b** develop new antibiotics

Unit 31: Ecological Features and Processes

1 **a** E **b** J **c** N **d** Q **e** A **f** S **g** H **h** U **i** L **j** B **k** M **l** F **m** O **n** C **o** P **p** K **q** T **r** D **s** R **t** G **u** I **v** V

2 **a** a key species either forms the physical environment of most other species in a community or has a major influence on most other species present/ an indicator species indicates the quality of the environment in which it lives **b** predators eat other animals, killing them in the process/ parasites feed on other organisms without killing them **c** saprophytes consume organic matter (dead organisms, plant matter and wastes) by extra-cellular digestion/ detritivores consume organic matter by eating it then digesting it internally **d** a food chain is a sequence of species, each being eaten by the next/ a food web consists of all of the interconnected food chains in a community **e** in a mutualistic relationship both species benefit/ in a commensal relationship one species benefits but the other is not harmed **f** energy flow is the one-way movement of energy from the Sun, through a community, then into the environment/ nutrient cycling is the cyclical movement of nutrients from the environment, through a community, then back into the environment where they can be reused **g** distribution refers to how species are spread/ density refers to how abundant species are/ diversity refers to how many different species there are in an ecosystem

3 **a** grazing (caterpillar eats leaf) **b** commensalism (plants perch in the tree fork to get more light) **c** predation (plant digests insect) **d** detritivore (worm eats detritus) **e** competition (anemones compete for attachment space) **f** mutualism (plant feeds butterfly, butterfly pollinates plant) **g** saprophyte (fungi digest stick insect)

ISBN: 9780170262316

h parasitism (mosquito feeds on human blood)

4 a true b true c false - lower d true e false - different ways f true g true h false - light energy i true j true k false - plants l true m true

5 a sea lettuce, seagrass, phytoplankton b zooplankton, cockles (but they also feed on the zooplankton) c mud snails, crabs, marine worms d fish and birds e detritus is organic matter derived from plants, dead animals and their wastes f phytoplankton are floating micro-organisms that are producers/ zooplankton are floating micro-organisms that are consumers g a flood could cover the mud in silt enriching the mud with nutrients for the producers to absorb; a landslide could cover up part of the mudflat eliminating the habitat of some species h heavy metal pollution from factories could poison sensitive species; farm effluent or sewage works discharge could overload the environment with nutrients causing algal blooms; urban development could increase the amount of fresh water flowing onto the mud from streams, making it difficult for species that require high salinity levels

6 a rural stream b rural stream c urban stream because stoneflies, which are an indicator species for good quality water (clear, fast flowing and well oxygenated) are absent d a rural stream might be affected by cowshed effluent or fertiliser run-off; an urban stream might be affected by a high volume of water entering as rainwater is not absorbed by the sealed roads, etc.

7 a (own judgement) b field - flooding, landslide drought; mudflat - silting, change in the tidal flow; stream - flood, drought, landslide c field - application of fertilisers, species introduction, pest control, trampling; mudflat - pollution, application of fertiliser to nearby farmland, urban development close by; stream pollution, application of fertiliser to nearby farmland, urban development close by, introduced fish species

8 a light energy from the Sun is captured by producers and transformed into chemical energy, which is passed along food chains, where it is converted into heat energy by consumers and decomposers, which warms the environment b nutrients in the environment are absorbed by producers, becoming part of the biomolecules that get passed along food chains; eventually those molecules are digested by decomposers, releasing the inorganic nutrients back into the environment where they can be reused by producers

9 a fertilising fields would increase the productivity of the producers (crop or grasses) b the lake water would be enriched by the nutrients, which could result in the lake being able to support a large biological community, but silting may also make the lake shallower and shallower till it could no longer support an aquatic community c the heavy rainfall could leach nutrients from the soil, which would limit the growth of grasses d removing the trees and shrubs would mean that large amounts of nutrients would not be returned to the soil, thus limiting the productivity of the new ecosystem e planting legumes would eventually increase the number of nitrate ions in the soil, allowing a greater variety of plants to thrive f the effluent would overload the water with nutrients, resulting in the excessive growth of algae

10 a an ecosystem that involves a body of water, such as a stream, river, lake, estuary, etc. b the water turns an intense colour (green, brown or red) c because it is often in short supply in aquatic ecosystems d when excessive amounts of phosphate ions enter an aquatic ecosystem e many of the nutrients are transferred to the bottom when the algae die, and the amount of dissolved oxygen in the water decreases when the decomposers digest the dead algae f because the oxygen level in the water is below what they can tolerate g the algal bloom removes so many other nutrients that other plants species don't grow, which eliminates some food chains; also the low oxygen content of the water may kill off many invertebrate species h some algal blooms produce toxins that are absorbed by shellfish; the shellfish tolerate the toxins in their bodies but the toxins poison humans when they eat the shellfish i farm effluent, fertiliser run-off, treated (and untreated) sewage

11 a the drought would reduce water levels, and the ecosystem would not be able to support so many plant and algal species, which would impact on the community as a whole through food chains that no longer exist b silting would decrease the stream depth and water clarity, and affect water flow; this could eliminate the habitats of some species and/or allow new species to colonise the stream c increased farm effluent would overload the stream with nutrients which could result in an algal bloom that changes the community dramatically d didymo increases rapidly covering suitable rock surfaces with a thick mat, eliminating the habitats of some species and the food sources of other species

12 a a quadrat is a square frame that defines a standard area; the quadrat is placed on the ground and the number of individuals of different animal species are counted (or the percentage cover of different plant species is estimated) b plants and fixed or slowing-moving animals c fast-moving animals d a transect is a line that is placed across an ecosystem, often along an environmental gradient (the transect line is usually marked at regular intervals to indicate where quadrats should be placed) e exposure to the air when the tide retreats f moisture content of mud, temperature of surroundings, available oxygen, possibly pH g i the vertical height (width) of the kites indicates density ii the horizontal length of the kites shows how the species is distributed iii the numbers of different species found in the quadrats shows the diversity

13 a very few mud snails would be found in areas polluted with heavy metal ions b because they are very sensitive to the presence of heavy metal ions in their bodies and die off, so their absence is a good indicator that pollution may be occurring, although it is not the only reason why the mud snails could be absent

Unit 32: The Life Cycle of Flowering Plants

1 a W b G c K d Q e D f N g A h L i I j C k S l O m F n M o E p B q U r H s R t J u V v P w T

2 a sexual reproduction involves combining the genes of two parent organisms to produce genetically different offspring/ asexual reproduction involves a single parent organism producing genetically identical offspring b cross-fertilisation occurs when a flower is fertilised by pollen from another plant of the same (or a closely related species) resulting in genetically different offspring/ self-fertilisation occurs when a flower is fertilised by pollen from the same plant, resulting in

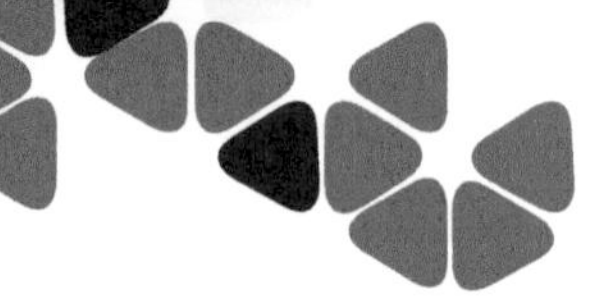

offspring with no significant genetic diversity **c** meiosis is the type of cell division that produces genetically different gametes/ fertilisation is the process in which two gametes (sperm and egg) fuse to form a genetically unique zygote **d** seeds are the new plants that are in a state of dormancy/ fruits are the ripened ovaries that enclose the seeds **e** the radicle is the part of the plant embryo that forms the first root, which eventually develops into the root system/ the plumule is the part of the plant embryo that forms the first shoot, which eventually develops into the shoot system **f** dormancy occurs when seeds are neither metabolising nor growing/ germination occurs when the dormant seeds starts to metabolise and grow

3 **Sexual reproduction advantage**: genetically diverse offspring have variation that may help them to survive adverse conditions
Sexual reproduction disadvantage: requires two parents; pollen must be transferred to flowers on other plants
Asexual reproduction advantage: only one parent required; enables a plant population to rapidly spread and increase in numbers in favourable conditions
Asexual reproduction disadvantage: genetically identical offspring lack the variation that may help them to survive adverse conditions

4 **a** tuber **b** bulb **c** cutting **d** stolon **e** runner

5 **a** false - the same **b** true **c** true **d** false - asexual reproduction results in uniformity and sexual reproduction in diversity **e** true **f** false - more pollen **g** true **h** true **i** true **j** false - ovaries become fruits and ovules, seeds **k** true **l** true

6 **a** reproduction **b** to produce offspring **c** to produce gametes and ensure they meet/ to produce seeds **d** petals - to visually attract pollinators **e** filament - to hold up the anther **f** anther - to produce and release pollen grains containing male gametes **g** stigma - to receive pollen grains **h** style - to hold up the stigma and to allow pollen tubes to grow down to the ovule **i** ovary - to bear the ovules and to develop into a fruit after fertilisation occurs **j** ovule - to bear the egg cell and to develop into the seed after fertilisation occurs

7 **a** the manuka flowers are white to visually attract bees; they provide nectar and pollen to feed the bees; the male and female organs are located so that bees readily transfer pollen from the anthers of one flower to the stigma of another **b** the bright colours of the flax flowers visually attract the tui; the shape of the flowers enable the tui to drink the nectar; the male and female organs are located so that the tui readily transfers pollen on its forehead from the anthers of one flower to the stigma of another **c** the anthers of the male corn flowers on the right produce lots of pollen and dangle out of the flower so that pollen can be released into the wind; the long silky stigmas of the female flower on the right provide a lot of surface area for wind-blown pollen to land on

8 **a** seed in soil: germination → growth → flowering → meiosis → pollination → fertilisation → seed development → ripening → dispersal **b** **germination** - embryo plant develops into a seedling; **growth** - plant develops a shoot and root system; **flowering** - floral buds develop into flowers; **meiosis** - body cells divide to become gametes; **pollination** - pollen is transferred from anthers to stigmas by pollinating agents; **fertilisation** - gametes fuse to become the first cell of a new plant; **seed development** - zygote undergoes multiple cell divisions then develops organs; **ripening** - ovary ripens into a fruit and ovules mature into seeds; **dispersal** - seeds are dispersed away from the parent plant by dispersal agents **c** pollination and dispersal **d** both involve moving plant structures (pollen or seeds) and rely on external agents, such as animals or the wind: pollination enables genes from two different plants to be combined, resulting in genetic variation; seed dispersal enables offspring to colonise new areas and it reduces the competition for resources **e** flowering - photoperiod and temperatures; germination - temperature, moisture levels

9

Feature	**Insect**	**Bird**	**Wind**
flower size	varies	large	small
petals	present	present	absent
petal colour	mostly white	coloured	–
scent	scented	little scent	no scent
nectar	often	lots	none
anthers	inside	project out	dangle
pollen	sticky	sticky	smooth
pollen amount	less	less	abundant
pollen size	larger	larger	small, light
stigma	compact	compact	feathery
stigma surface	sticky	sticky	–

10 **a** **i** germination **ii** growth **iii** flowering **iv** meiosis **v** pollination **vi** fertilisation **vii** seed development **viii** dispersal **b** by birds **c** because they have large colourful petals and are shaped to accommodate the long beak of birds like tui

11 **a** reproduction **b** to develop into a new plant; to aid in the dispersal of the species **c** plumule - develops into the first shoot **d** radicle - develops into the first root **e** seed coat - protects the embryo inside **f** cotyledons provide food for the germinating seed **g** plumule, radicle and cotyledons **h** the seed coat forms a tough protective surface that makes it difficult for consumers or decomposers to damage the embryo inside **i** the cotyledons contain a food store, which supplies nutrients, such as sugars, to the germinating embryo

12 **a** the kernels are seeds **b** the male flowers are found in the tassels at the top of the stalk; the female flowers are the ears found at the base of leaves lower down on the plant **c** the male flowers are much smaller and have dangling sex organs; the female flowers (ears) are miniature cobs and have long silky stigmas **d** to reduce the chances of cross-pollination occurring **e** wind **f** dangling anthers, masses of light pollen, long trailing stigmas to catch pollen **g** because the pollen tubes have to grow a long distance **h** because the seeds are easily digested **i** farmers and gardeners have become the dispersal agent for the corn seeds when they plant them

13 **a** dandelion/wind - seeds have parachutes that enable theme to be carried away in a breeze **b** gorse/self-dispersal - dried out seed pods suddenly split open, throwing out the seeds **c** chestnuts/animals - food-rich seeds are collected by squirrels that transport them to other locations where some germinate rather than getting eaten **d** burdock/animals - hook-shaped seeds catch on the fur of mammals **e** sycamore/wind - wings enable seeds to travel further in the wind **f** mangrove/ water - seeds float in the tide **g** strawberry/animals - succulent fruits are eaten but tough seeds pass through gut unscathed **h** water lily/wind - light seeds are shaken

ISBN: 9780170262316

out of the pods in a breeze **i** plants need to disperse their seeds so that they can spread into new locations and to reduce competition among offspring or between offspring and the parent plant for resources

14 **a** 7 °C **b** 5, 6 and 7 °C **c** 7 °C because germination reaches 100% most rapidly **d** because slightly higher temperatures could result in even faster germination

Unit 33: Plant Processes that Maintain Life

1 **a** H **b** K **c** E **d** Q **e** L **f** C **g** G **h** N **i** O **j** A **k** J **l** M **m** B **n** F **o** T **p** D **q** V **r** S **s** I **t** U **u** P **v** R

2 **a** cells are the smallest living objects/ tissues are masses of cells that carry out a particular function/ organs are complex structures constructed out of tissues, which carry out a particular function **b** xylem vessels are hollow, thick-walled tubes running the length of the plant, which transport water and minerals/ phloem tubes are columns of living cells that transport dissolved sugars around the plant **c** primary growth occurs at root tips, and shoot tips and buds; it elongates the plant and forms new organs/ secondary growth occurs within the stem and root of woody plants and produces new xylem and phloem, thus increasing the girth of the plant **d** photosynthesis is the process that uses sunlight, carbon dioxide and water to make energy-rich glucose and oxygen gas/ respiration is the process that releases the energy stored in glucose for metabolism **e** palisade mesophyll consists of a layer of tightly packed, cylindrically shaped cells that are found near the top surface of a leaf and which contain many chloroplasts/ spongy mesophyll consists of a layer of loosely packed, irregularly shaped cells that are found near the bottom surface of a leaf and which contain fewer chloroplasts **f** gas exchange is a life process that involves the exchange of O_2 and CO_2 gases between an organism and its environment/ excretion is a life process that involves disposing of cell waste products into the organism's environment

3 **a** leaves are green or yellow-green because they contain the pigment chlorophyll, which absorbs sunlight **b** leaves are thin so that light can penetrate all layers and so that CO_2 can rapidly diffuse from the stomata to all photosynthetic cells **c** leaves are flat rather than curved or rolled up so that they intercept the maximum amount of light **d** leaves have a large amount of surface area so that they intercept the maximum amount of sunlight **e** leaves have an extensive network of veins to supply all photosynthetic cells with water and to strengthen the leaf **f** leaves have lengthy stalks, which spread and hold up the leaves to maximise the interception of light

4 **a i** xylem cell **ii** to transport water **iii** it forms part of a hollow tube consisting of stacked dead xylem cells **b i** palisade cells **ii** to photosynthesise **iii** they have many chloroplasts/ they are cylindrical in shape to maximise light absorption as light passes down the cells **c i** root hair cell **ii** absorb water and minerals from the soil **iii** the long thin extension increases the surface area for absorbing water and minerals **d i** epidermal cells **ii** protect delicate tissue **iii** it is a flat layer covering other cells beneath and produces the waxy water-tight cuticle that reduces water loss **e i** guard cells **ii** control the entry and exit of gases **iii** they can inflate with water to open the stomata or deflate to close it

5 **a** true (except for movement) **b** true **c** false - made of tissues, which are made of cells **d** true **e** false - sugar is transported upward and downward **f** false - never cease dividing until the plant dies **g** true **h** true **i** true **j** false - food and oxygen **k** false - into chemical energy **l** true **m** true **n** true

6 **a** internal transport **b** to transport water and minerals from the roots to stems, leaves, flowers and fruits, and to transport sugars made in the leaves to all other parts of the plant **c** the blue tissue is the phloem tissue; the red tissue is the xylem tissue **d** dissolved sugars are carried in the red tissue; water and dissolved mineral ions are carried in the blue tissue **e** they move upward in the red tissue, and both upward and downward in the blue tissue **f** in the roots in cross-section, the xylem tissue is star-shaped with the phloem tissue found between the arms; in the stems in cross-section, the xylem tissue is found on the inside, and phloem tissue on the outside of multiple vascular bundles that form a circle; in the leaves in cross-section, the xylem tissue sits on top of the phloem tissue in the veins that spread throughout the leaf

7

Feature	Primary Growth	Secondary Growth
Type of plant	all types	trees and shrubs
Location	tips and buds	inside stem
Meristem name	apical meristem	cambium
Height of plant	increases	no effect
Girth of plant	no effect	increases
Organs formed	leaves, flowers	none
Vascular supply	not increased	increased
Main function	reach light, water	support & transport

8 about 50 years old, as each growth ring represents one year of growth

9 **a** to capture the energy required to carry out life processes and to make the biomolecules needed to construct cells **b i** the root system absorbs water and minerals **ii** the shoot system transports water up the plant **iii** the leaves capture light energy and manufacture biomolecules **c** the root system absorbs the water and minerals needed for photosynthesis and anchors the stem; the stem transports the water and minerals to the leaves and holds them up to the light; the leaves use the water and minerals, carbon dioxide from the air and sunlight to manufacture energy-rich biomolecules

10 **a i cuticle ii** to make the leaf water-tight/ to transmit light **iii** made of wax/ it's transparent **b i upper epidermis ii** to protect delicate cells inside leaf/ to transmit light **iii** covers those cells/ transparent cell contents **c i palisade mesophyll ii** to carry out photosynthesis **iii** the tightly packed, upright cylindrical cells allow more light to be absorbed by the abundant chloroplasts **d i spongy mesophyll ii** to allow carbon dioxide to rapidly diffuse to all photosynthetic cells **iii** the loosely packed cells allow gases to circulate freely **e i lower epidermis ii** to protect delicate cells inside leaf **iii** covers those cells **f i guard cells ii** to control entry and exit of gases **iii** cells can inflate with water to open the stomata and deflate to close them **g i vein ii** to help support the leaf **iii** cells have thick cell walls that provide rigidity **h i phloem tube ii** to transport dissolved sugars to other parts of plant **iii** consists of stacked cells that actively transfer the solution **ii xylem vessel ii** to transport water and dissolved mineral to all leaf cells **iii** hollow tube that water flows through **j i chloroplast ii** to carry out photosynthesis **iii** it contains the pigment chlorophyll

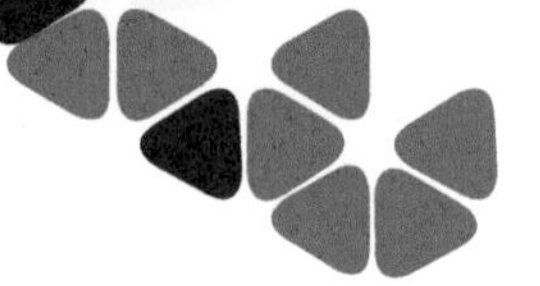

which absorbs light energy
k the transparent cuticle (a) and upper epidermis (b) allow light to penetrate through to the palisade mesophyll cells (c), which are packed with chloroplasts (j) that carry out photosynthesis
l the guard cells (f) open the stomata to allow in as much CO_2 gas as possible and the widely spaced spongy mesophyll cells (d) allow CO_2 to rapidly diffuse to all photosynthetic cells
m the vein (g) helps strengthen the leaf so that it is flat and able to intercept more light, the xylem vessel (h) supplies the water needed for photosynthesis, and the phloem tube (i) transports dissolved sugars to feed the other parts of the plant, which are then able to continue supporting and supplying the leaf
n the epidermal layers (b and e) produce the waxy cuticle (a), which makes most of the leaf water-tight, and the guard cells (f) are able to close to prevent the leaf from wilting in hot, dry, windy conditions

11 a the leaf inside the plastic bag would not go blue-black at all as no starch would have been made; the other leaf would only go blue-black in the green area that was exposed to light but the green parts of the leaf under the strip would not go blue-black as no starch would have been made b that light, green chlorophyll and CO_2 are needed for photosynthesis to occur c water, warmth

12 a CO_2 and O_2 b the orange line shows the CO_2 level; the purple line, the O_2 level c a pattern that repeats itself on a daily basis d the CO_2 level falls during the day and rises during the night; the O_2 level rises during the day and falls during the night e the CO_2 level falls during the day because the plants are using up CO_2 as they photosynthesise in the light f the O_2 level rises during the day because the plants are producing O_2 as they photosynthesise g during the day CO_2 enters and O_2 exits the plants; during the night O_2 enters and CO_2 exits the plants h during both the daytime and night-time O_2 enters and CO_2 exits the microbes i because otherwise the CO_2 levels in the water would be more closely related to the concentration of CO_2 in the air j the O_2 level would keep falling and the CO_2 level would keep rising as the microbes respire, eventually the microbes may die when O_2 levels fall to low

13 a the bubbles are caused by the oxygen gas produced by the water weed carrying out photosynthesis, so the rate of bubble production indicates the rate of photosynthesis b i **light intensity** - by altering the distance between the light source and the plant, and counting the rate of bubble production after the plant has adapted to the new conditions ii move the light closer to the plant iii the rate should increase c i **CO_2 level** - by altering the amount of sodium bicarbonate added to the water, and counting the rate of bubble production after the plant has adapted to the new conditions ii add more sodium bicarbonate than before iii the rate should increase d i **water temperature** - by altering the temperature of the water in the water bath, and counting the rate of bubble production after the plant has adapted to the new conditions ii add warmer water iii the rate should increase

14 a a stem b because it is circular in cross-section and there are separate vascular bundles arranged in a circle c A is the cambium; B is the xylem tissue; C is the phloem tissue; D is a vascular bundle d where the cambium is located/ between the existing xylem and phloem tissues e a leaf f because it is flat in cross-section with a mid-rib g E is xylem tissue; F is phloem tissue; G is spongy mesophyll; H is palisade mesophyll; E is a vein h structure D, the stem vascular bundle connects with structure I, the leaf vein

Unit 34: Food and the Digestive System

1 a E b H c M d O e F f S g A h W i J j W k G l D m K n Q o V p l q C r U s N t T u R v P w L

2 a consumers must eat other organisms to obtain the organic nutrients and energy they need to make the molecules required for life processes/ producers are able to manufacture the molecules they need using light energy and inorganic chemicals from their environment b ingestion involves getting food into the gut/ digestion involves breaking down food into small soluble molecules that can be absorbed c absorption involves the movement of the products of digestion through the villi and into the blood system/ assimilation refers to the products of digestion being absorbed into cells and becoming parts of cell structures or being used as a source of energy d herbivores eat only plant matter/ carnivores eat only animal matter/ omnivores eat both plant and animal matter e physical digestion is the crushing and grinding of food into small particles by the teeth/ chemical digestion is the breaking up of large food molecules into small soluble ones by enzymes f acidic conditions occur when the pH is less than 7/ neutral conditions occur when the pH is close to 7/ alkaline conditions occur when the pH is greater than 7

3 a i **amylase** ii salivary glands iii mouth iv 6.5-7.5 v maltose b i **maltase** ii wall of the small intestine iii small intestine iv 7.0-8.5 v glucose c i **lipase** ii pancreas iii small intestine iv 7.0-8.5 v fatty acids and glycerol d i **pepsin** ii stomach wall iii stomach iv 1.5-4 v peptides e i **trypsin** ii pancreas iii small intestine iv 7.0-8.5 v amino acids

4 a true b true c true d false - soluble molecules e true f false - harder g true h true i true j false - cellulose k false - and temperature l true m true n false - active transport o true

5

Feature	Dog	Rabbit	Chimp
Feeding role	carnivore	herbivore	omnivore
Food	meat	grass	fruit & meat
Canine shape	sharp, pointy	none	pointy, long
Canine role	stab animal	–	tear food
Incisor shape	chisel shape	chisel shape	chisel shape
Incisor role	scrape meat	tear leaves	cut food
Molar shape	scissor blade	flat-topped	flat-topped
Molar role	cut meat	grind food	grind grass

6 a stomach b small intestine c caecum d large intestine e right gut f because it has an extensive caecum g the caecum contains bacteria that breaks down cellulose, providing the herbivore with sugars; the lengthy large intestine ensures that most of the water in the gut is reabsorbed h left gut i because it has a very small caecum and a short large intestine j it only has enzymes for digesting the proteins and fats found in animal matter; its short gut means that food remains in the gut for a short period of time reducing putrefaction k left gut

7 a **mouth** - it forms a moist enclosed space where food is physically and chemically digested b **tongue** - its flexibility allows it to position food for chewing

ISBN: 9780170262316

and swallowing c **salivary glands** - produces saliva to moisten food and enzymes to digest starch d **oesophagus** - its muscular wall squeezes food down to the stomach e **stomach** - its muscular wall churns food; it secretes the enzyme pepsin which begins the digestion of proteins; it secretes hydrochloric acid, which optimises the pH for pepsin f **small intestine** - its length allows plenty of time for digestion to occur; its walls secrete enzymes to complete the digestion of food g **liver** - it makes bile which enables fats to be digested; it acts as a chemical factory processing the products of digestion h **gall bladder** - it acts as a container to store bile until it is needed i **pancreas** - it is a gland that produces and secretes a range of digestive enzymes into the small intestine j **large intestine** - its length enables water to be reabsorbed and waste food to be solidified k **rectum** - it acts as a container to store faeces l **anus** - it is an opening that allows faeces to be expelled m **blood vessels** - they enable the products of digestion to be transported in the blood to the liver n salivary glands - saliva moistens food so that it can be swallowed and commences the chemical digestion of starch o stomach wall - gastric juice acidifies the contents of the stomach and supplies digestive enzymes that continue the digestion of all three food types p liver - bile emulsifies fats and helps to make the contents of the small intestine slightly alkaline, so that digestive enzymes that prefer that pH can work efficiently q pancreas - pancreatic juice supplies a range of digestive enzymes that help complete the digestion of all three food types; it also contains bicarbonate ions that help to make the contents of the gut more alkaline r appendix - thought to have no function (but may have a role in the immune system or in sheltering mutualistic bacteria)

8 a bicarbonate ions in bile help to neutralise the acidic liquid released by the stomach into the small intestine, making it slightly alkaline b bile emulsifies fat blobs, turning them into tiny oil droplets c the pancreatic juice contains a lipase that digests lipids into separate fatty acid and glycerol molecules d by turning the blobs of fat into tiny oil droplets, bile greatly increases the number of fat molecules exposed to the lipase, which enables more efficient digestion to occur; by helping to make the contents of the small intestine slightly alkaline, bile, along with pancreatic juice, provides the optimum pH for the lipase to work, which enables more efficient digestion to occur

9 a villus (singular of villi) b microvilli c blood vessel/ capillaries d lymph vessel/ lacteal e absorption of products of digestion f they greatly increase the surface area for absorption of products of digestion g to receive simple sugars and amino acids; the network of blood vessels within villi maximise absorption h to receive fatty acids and glycerol; the lacteals penetrate all villi to maximise absorption

10 a all types: carbohydrates, lipids, proteins, minerals and vitamins b because those large molecules cannot be absorbed into the blood c because it contains a sugar called lactose d each molecule consists of a galactose unit bonded to a glucose unit e it breaks it down into separate galactose and glucose molecules f the villi actively transport them into the blood system g a milkfat molecule consists of three fatty acid units bonded to a single glycerol unit h it breaks in down into separate fatty acid and glycerol molecules i the villi actively transport them into the lymph system j a casein molecule consists of a folded-up chain of a large number of different amino acids k it cuts it up into shorter chains of amino acids, called peptides l it breaks down the peptides into individual amino acids m the villi actively transport them into the blood system

11 a enamel b because they are living objects c so that damage can be detected d the cement e (see below)

Feature	Incisors	Canines	Molars
size	smaller	smaller	larger
roots	single	single	multiple
shape	chisel shaped	pointy	flat-topped
sharpness	very sharp	not very sharp	blunt
role	scrape & cut	grasp & tear	crush & grind
action of teeth	top & bottom teeth meet to cut off food	top & bottom teeth stab and hold food	top & bottom teeth grind together

12 a the rate rises to a peak as the temperature increases to 40 °C and then falls to zero as the temperature increases further to 100 °C b about 40 °C c from about 0 °C to about 80 °C d because the foods we ingest vary hugely in temperature

13 a incisor teeth b canine teeth c molar teeth d lip e tongue f salivary glands g gum h physical digestion i it breaks them down into smaller particles j chemical digestion k it breaks them down into glucose and maltose l effective chewing grinds food into numerous small lumps, which increases the number of food molecules that are exposed to digestive enzymes, thus making chemical digestion more efficient

Unit 35: Respiratory, Circulatory and Excretory Systems

1 a F b K c S d O e E f L g A h Q i B j H k R l N m V n J o C p U q D r P s I t T u G v M

2 a respiration is the release of energy from glucose molecules, which occurs in all cells/ gas exchange is the exchange of respiratory gases between an organism and its environment which occurs in the lungs b inhalation is the intake of air into the lungs (breathing in)/ exhalation is the expelling of air from the lungs (breathing out) c arteries are thick-walled blood vessels (tubes) that carry blood away from the heart/ capillaries are very fine blood vessels that go past cells/ veins are thin-walled blood vessels (tubes) that carry blood to the heart/ d diffusion is the natural movement of chemicals from an area of high concentration to an area of low concentration/ active transport is the movement of chemicals in or out of a cell from a low to a high concentration area, which requires energy e secretion involves cells or a gland releasing useful chemicals into an organ or the environment/ excretion involves the release of cellular waste products into the organism's environment f oxygenated blood is bright red in colour because its haemoglobin molecules are carrying a lot of oxygen/ deoxygenated blood is dark red in colour because its haemoglobin molecules are carrying little oxygen g aerobic conditions occur when oxygen is plentiful; it releases much energy from glucose (produces many ATPs)/ anaerobic conditions occur when oxygen is absent, it releases little energy from glucose (produces only a few ATPs)

3 nutrients are needed to supply chemicals and energy for growth, reproduction, movement and keeping warm, as well as all other life processes

ISBN: 9780170262316

4 a **trachea** - it is a stiffened tube that allows air to flow freely in an out of the body b **lungs** - they are made of spongy material that enables air to come in close contact with the blood c **ribs** - they form a strong bony cage that protects the lungs d **diaphragm** - it is a muscular layer that expands the volume of the chest when it contracts, the resulting pressure drop causes air to flow into the lungs e **rib muscles** - when they contract this expands the volume of the chest, the resulting pressure drop causes air to flow into the lungs f **bronchus** - this tube allows the air to move in and out of one of the two lungs g **bronchioles** - these increasingly fine tubes enable air to move in and out of multiple clusters of microscopic alveoli h **alveoli** - these microscopic, balloon-like structures greatly increase the surface area for gas exchange i to supply the blood with oxygen and to remove carbon dioxide from the blood j the inner lining of the alveoli k otherwise oxygen would not dissolve and could not pass through cell membranes, similarly dissolved carbon dioxide would not be able to reach the gas exchange surface to escape into the air l it reduces the water lost by evaporation from the gas exchange surface/ it conserves water m the inflatable balloon-like microscopic alveoli greatly increase the area of the gas exchange surface, allowing a much greater number of respiratory molecules to enter or leave the organism n they diffuse through the lining of the alveoli, moving from the side with the higher concentration to the side with the lower concentration

5 a true b true c false - lungs d true e true f true g false - except for the pulmonary artery h true i true j true k true l false - aerobic metabolism is 15 times more efficient than anaerobic metabolism m false - amino acids n true

6 a inhalation b oxygen c diffusion d oxygenated blood e deoxygenated f diffusion g carbon dioxide h exhalation i because they have been oxygenated

7 a (own graph) b both heart rate and blood pressure stay high for several minutes before falling back to resting levels c they both respond in a similar way d during the sprint a significant amount of lactic acid would have been formed in muscles, which would have required oxygen to dispose of it, so heart rate and blood pressure remain high for a while to supply the muscles with lots of oxygen

8 a when the subject inhales, air is drawn through the left test-tube and when the subject exhales, air is blown out through the right test-tube b the limewater in the left test-tube will remain clear as there is very little carbon dioxide in air; the limewater in the right test-tube will go cloudy because there is a significant amount of carbon dioxide in exhaled air

9 a i glucose, oxygen, amino acids ii carbon dioxide, ammonia b i carbon dioxide, glucose ii oxygen c i oxygen ii glucose, amino acids, carbon dioxide d i oxygen, glucose, amino acids, ammonia ii carbon dioxide, urea e i urea, glucose, oxygen ii carbon dioxide f to supply nutrients and oxygen to all cells and to remove their waste products g heart - pumps blood; arteries - carry blood to organs; capillaries - carry blood to individual cells; veins - return blood to the heart h the 'blue' blood represents deoxygenated blood and the red blood, oxygenated blood

10 a heart; because it consists of two separate pumps b all organs of the body, other than the lungs; because the capillaries are receiving oxygenated blood c lungs; because the capillaries are receiving deoxygenated blood d veins; because they are carrying blood to the heart e arteries; because they are carrying blood away from the heart

11 a the function of the circuit going to the lungs is to oxygenate the blood; the function of the circuit going to all other body organs is to supply glucose and oxygen to the cells of those organs, and to remove their wastes b if the blood pressure is too low, the blood moves sluggishly in the veins/ circulation is poor c by contracting strongly and forcing the blood out into the arteries d the pressure would be very low if the blood had to pass through the lungs then directly to all other organs e there is only a single pressure drop as the blood passes through each of the two circuits (except for blood going from the gut to the liver)

12 a aerobic respiration b because the contraction of muscles generates a lot of heat energy c cells and tissues will not function well and heat exhaustion may set in d radiation and sweating e the water in sweat evaporates when it absorbs heat from the body, thus cooling the body f it becomes dehydrated, which can lead to collapsing g by reducing the amount of water excreted in urine h because sweat is salty i muscle cramps, dizziness and lack of co-ordination may occur j by drinking lots of water and by drinking liquids containing salts/electrolytes

13 a to package energy in small amounts that cells can use b because all life processes require energy in an accessible form c glucose molecules d oxygen e ATP f because it transports small amounts of energy to all locations within cells g carbon dioxide and water h Glucose + oxygen → carbon dioxide + water + energy i $C_6H_{12}O_6 + 6O_2 \rightarrow 6CO_2 + 6H_2O + 30$ ATPs

14 a aerobic metabolism produces carbon dioxide, water and lots of ATPs; anaerobic metabolism produces lactic acid and few ATPs b because all of the energy originally in glucose molecules is released in aerobic conditions but only a fraction is in anaerobic conditions c because it can supply some energy for muscle movement when insufficient oxygen is reaching those cells d because oxygen is needed to deal with toxic lactic acid (it is converted back to pyruvic acid)

15 a **artery** - it is a tube that takes blood with urea, oxygen and glucose to the kidneys for filtration b **kidney** - it extracts waste products from the blood and helps to keep the water content of the blood constant c **filtering units** - these are special structures that separate wastes from useful chemicals d **ureter** - it is a tube that transports wastes and water to the bladder e **bladder** - it is a hollow organ that enlarges as it fills with urine and empties when its muscular wall contracts f **urethra** - it is a tube that allows urine to drain out of the body (when the sphincter muscle is relaxed) g **vein** - it is a tube that takes filtered blood back to the main vein h to remove cellular waste products from the blood i urea, oxygen and glucose, also salt j water with small dissolved molecules enters the filtering units and wanted chemicals are actively transported back into the blood k as the kidneys would not be working properly there would be a build-up of toxic chemicals in the blood causing a range of symptoms, and ultimately damaging cells and tissues, leading to death l to keep the water content of the blood within healthy limits m otherwise cells would be damaged if they absorbed

ISBN: 9780170262316

or lost too much water

16 The digestive system breaks down food into small absorbable nutrients that diffuse into the blood. The respiratory system dissolves oxygen molecules on its moist gas exchange surface and they diffuse into the blood, where they become attached to haemoglobin molecules inside red blood cells. The circulatory system transports the nutrients dissolved in the plasma and the oxygen in the red blood cells to all organs of the body. When the blood capillaries pass individual cells, oxygen and nutrients diffuse into the cells, and carbon dioxide and cellular wastes diffuse into the blood plasma. The circulatory system carries the carbon dioxide to the lungs where it evaporates from the gas exchange surface and is exhaled into the air. The other wastes are transported to the liver to be processed, then taken to the kidney to be excreted in urine.

Unit 36: Effects of Astronomical Cycles on Planet Earth

1 a G b J c C d M e P f A g T h S i R j B k F l O m Q n H o D p K q I r L s U t E u N

2 a the apparent motion of an object is an illusion caused by the movement of the observer/ actual motion occurs when an object really is moving b a planet is a large heavenly body that orbits a star/ a star is a huge heavenly body made of gases undergoing nuclear reactions/ a moon is a large heavenly body that orbits a planet c summer solstice occurs on the longest day of the year, when the Sun 'crosses' the sky on its southernmost pathway / winter solstice occurs on the shortest day of the year, when the Sun 'crosses' the sky on its northernmost pathway d a lunar day lasts for 29.5 Earth days/ an Earth day last for 24 hours e with a waxing moon the illuminated area is increasing/ with a waning moon the illuminated area is decreasing

3 a 'sunset': caused by Earth's spin leaving the Sun out of sight b tides: caused by Earth rotating 'beneath' the Moon c iceberg or ice shelf: the curvature of Earth means that polar regions receive very little solar radiation and hence are cold enough to form large masses of frozen water d autumn: the seasonal change is caused by the tilted spin of Earth as it orbits the Sun

4 a watts per square metre b (own graph) c the amount of radiation falls off slowly to start with then much more steeply as the latitude increases d at higher latitudes less radiation is received per square metre because the Sun 'travels' lower in the sky

5 a false - Earth spins eastward b true c false - is visible during the day and night d true e true f false - less g true h false - there is no friction to slow Earth i true j true k false - tilted axis l true m true n false - longer o true p true

6 a apparent b actual c apparent d apparent e actual f apparent

7 a equator b a line around Earth equally distant from both poles c North and South Poles d located where the polar axis emerges through the surface e along a line running between the North and South Poles f eastward g 24 hours h because there are no significant forces to slow it down i because the planet spins in the vacuum of space

8 a visible and infrared radiation b heats it up c because the surface radiates heat back into space in the night d ultraviolet radiation can cause skin cancers e the surface of the tropics faces the Sun more directly as it passes overhead; the curvature of Earth means that the surface of regions outside of the tropics are not directly facing the Sun and therefore do not receive as much solar radiation f latitude, time of the day, and day of the year

9 a Sun b a moon orbits a planet and a planet orbits a star c elliptical d when the planet is closer to the Sun it receives more solar radiation and vice versa e gravitational attraction f 365 days; it is the length of one Earth year g 30000 metres per second, because everything around us travels at the same speed and we have no fixed reference point h 29.5 Earth days; it is approximately the length of a month i as Earth hurtles sideways past the Sun, gravitational attraction pulls it into a curved path around it j as the Moon hurtles sideways past the Earth, gravitational attraction pulls it into a curved path around it

10 a Earth collided with a smaller planet b dramatically different seasons occur outside of the tropics c no seasons would occur because each latitude would receive a constant amount of radiation throughout the year on a daily basis

11 a summer b more of the northern hemisphere is exposed to solar radiation and the exposure is for a longer time c summer d more of the southern hemisphere is exposed to solar radiation and the exposure is for a longer time e less of the southern hemisphere is exposed to solar radiation and the exposure is for a shorter time f because the amount of solar radiation the tropics receive over the course of a year does not vary greatly g a wet season when the Sun is more directly overhead

12 a the gravitational pull of the Moon and of the Sun b because the Sun is much further away from Earth than the Moon c the gravitational pull of the Moon when it is overhead d the gravitational pull of the Moon is weakest when it is close to the horizon e a centrifugal effect that 'pushes' the sea outwards when the Moon is on the opposite side of Earth f their gravitational pulls are in such directions that they combine to create a greater pull g because one occurs when the Moon is close to the Sun at new moon and another when it is on the opposite side to the Sun at full moon h at first quarter and last quarter the sideways pull of the Sun to some extent flattens out the bulge caused by the Moon's pull i as Earth rotates eastward tides rise and fall at a particular location because the pull of the moon changes depending on whether the Moon is above, near the horizon, or behind Earth

13 a in midsummer/ at summer solstice b Antarctica is exposed to solar radiation for 24 hours a day in midsummer because the southern hemisphere is tilted toward the Sun at that time of the year

14 a convection b tropical areas receive more solar radiation and the heated air rises; subtropical areas receive less radiation so cooler air from there flows toward the equator to replace the heated air c subtropical air pressure is higher than the tropical air pressure d the Coriolis effect/ the eastward rotation of the planet

15 a because the illuminated part is exposed to solar radiation and the dark part isn't b the orbit of the Moon around Earth

16 a a full moon rises in the east at sunset because it is

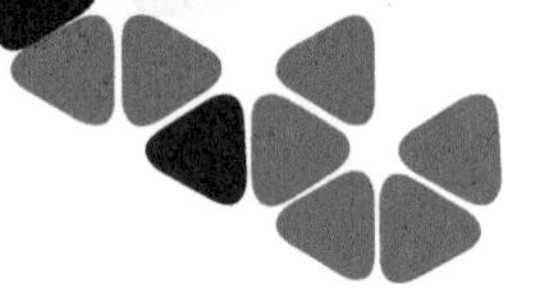

on the opposite side of the planet to the Sun; a new moon rises in the east at sunrise because it is located close to the Sun in terms of its angle **b** in the southern hemisphere a first quarter moon is illuminated on the left side and a last quarter moon on the right side

Unit 37: Recycling Carbon Atoms

1 **a** Q **b** H **c** N **d** D **e** J **f** B **g** K **h** I **i** F **j** O **k** E **l** R **m** M **n** A **o** T **p** U **q** G **r** C **s** L **t** P **u** S

2 **a** producers make food using inorganic chemicals and sunlight/ consumers eat other organisms/ decomposers feed on dead bodies and organic matter **b** photosynthesis uses light energy to convert carbon dioxide and water into energy-rich organic molecules/ respiration releases energy from energy-rich organic molecules, producing water and carbon dioxide as waste products **c** aerobic metabolism uses oxygen to fully release the energy stored in organic molecules, resulting in the waste products carbon dioxide and water/ anaerobic metabolism partially releases the energy stored in organic molecules, resulting in waste products such as methane gas **d** combustion (burning) is the rapid release of energy from a substance in a chemical reaction that requires oxygen/ respiration is the controlled release of energy from a substance in a process that requires oxygen and is catalysed by enzymes **e** visible radiation includes all the light waves that our eyes are able to detect/ infrared radiation is longer-wave radiation that heats objects but which our eyes cannot detect

3 **a** biological **b** chemical **c** physical or geological **d** physical or chemical **e** physical **f** chemical **g** biological **h** physical **i** chemical

4 **a** deforestation adds carbon **b** decomposition adds carbon **c** planting trees removes carbon **d** dissolving removes carbon **e** combustion adds carbon

5 **a** true **b** false - producers **c** false - billion tonnes **d** true **e** false - removes **f** true **g** false - opposite **h** true **i** true **j** false - incomplete **k** true **l** true **m** false - warm **n** true

6 **a** photosynthesis by plants **b** plant respiration **c** consumption **d** death and loss of plant matter **e** deforestation **f** consumer respiration **g** death and wastes **h** aerobic decomposition **i** anaerobic decomposition **j** oxidation of methane **k** sedimentation **l** the orange boxes represent trophic levels and the pink boxes, reservoirs **m** the blue numbers represent the amount of carbon transferred and the green numbers the amount of carbon stored in gigatonnes **n** 120 gigatonnes are being added and 120 gigatonnes are being removed **o** that it is in a steady state with the amount of carbon in the atmosphere staying relatively constant **p** the combustion of fossils fuels is greatly increasing the amount of atmospheric carbon dioxide

7 **a** organic matter falling to the swamp floor does not fully decompose so it builds up in thick layers that are known as sediments **b** the lack of oxygen in the boggy ground **c** pressure and heat **d** millions of years **e** it removes carbon from the atmosphere and stores it for long periods of time

8 **a** photosynthesis by phytoplankton **b** respiration by phytoplankton **c** consumption by zooplankton **d** death and loss of plant parts **e** anaerobic decomposition **f** consumer respiration **g** death and wastes **h** aerobic decomposition **i** sedimentation **j** thermohaline circulation **k** photosynthesis, respiration and decomposition **l** sedimentation, thermohaline circulation **m** carbon dioxide gas (the product of aerobic respiration) dissolves in deep ocean water/ methane gas (the product of aerobic respiration) becomes frozen on the sea floor **n** because it is at the bottom of the ocean and the phytoplankton are near the surface **o** thermohaline circulation over decades or centuries **p** fossil fuel combustion increases the amount of CO_2 in the atmosphere, 40% of which dissolves in the ocean

9 **a** it sinks to the bottom of the sea forming thick layers **b** the absence of oxygen in the sea floor sediments prevents complete decomposition **c** heat, pressure and impervious layers to trap the oil and gas **d** millions of years **e** it removes a significant amount of carbon from the environments that producers exist in **f** it returns a huge amount of carbon to the atmosphere **g** it occurs over very short time scales, much faster than the additional carbon in the atmosphere can be absorbed by producers

10 **a** they have complex stomachs with four compartments, one of which is called a rumen **b** because plant cells all have cell walls made of hard-to-digest cellulose **c** it is an enzyme that digests cellulose into simple sugars **d** the bacteria are provided with a habitat and food in the rumen, and the ruminant receives extra nutrition from the cellulose digested by the bacteria **e** anaerobic metabolism, because there is very little oxygen **f** fatty acids and methane gas **g** about 11 times more **h** 22 tonnes **i** because we have a huge number of cows and sheep **j** by breeding more efficient feeders, adding urea to their diet, feeding them tannin-rich plants and using antibiotics

11 **a** air-sea CO_2 exchange **b** chemical weathering of silicate rock **c** reversible chemical reactions **d** shell formation and death **e** rock formation **f** subduction **g** volcanic eruption **h** uplift **i** chemical weathering of carbonate rock **j** air-sea CO_2 exchange **k** chemical weathering occurs over years; reversible chemical reactions **l** it increases acidity/lowers the pH of the oceans as it releases excess hydrogen ions; the increased acidity damages coral reefs, weakens shells and affects marine food chains **m** shell formation and death occur over the lifespan of organisms in days, months or years **n** rock formation, uplift and subduction over thousands or millions of years; volcanic eruptions over days or week **o** air-sea exchange, and chemical weathering when carbon dioxide dissolves in rainwater **p** air-sea exchange and volcanic eruptions

12 **a** solar radiation penetrates the atmosphere reaching the surface **b** light radiation heats up the surface **c** the surface of the planet re-radiates the energy as infrared radiation **d** greenhouse gases absorb the outgoing infrared radiation **e** greenhouse gases re-radiate some of the infrared radiation back to the surface, heating it more than would have occurred if there were no greenhouse gases **f** the amount of solar radiation, how much radiation is reflected by the planet, and the greenhouse effect **g** carbon dioxide and methane **h** the major part of solar radiation is visible light; earthly radiation is mostly infrared radiation **i** if it did not occur then the planet's surface would be completely frozen and there would be no life on Earth **j** fossil fuel combustion and deforestation **k** livestock farming (cows and sheep burp methane produced by bacteria digesting plants in their stomachs)

ISBN: 9780170262316

Unit 38: Internal and External Geological Processes

1 a L b O c V d I e R f B g H h N i E j T k G l Q m A n U o C p K q J r D s M t F u S v P

2 a internal geological processes are those driven by convection currents in the mantle/ external processes are those that wear down or reshape surface features b continental crust is thicker and less dense, forming the mass of continents/ oceanic crust is thinner and denser, forming the sea floor c sea-floor spreading is the production of new oceanic crust at mid-ocean ridges/ subduction is the destruction of oceanic crust as it is forced down into the mantle d at a divergent boundary, plates are moving apart/ at a convergent boundary they are colliding/ at a transform boundary they are slipping sideways past each other e a crater is the mouth of a volcano/ a caldera is a depression formed when a volcano collapses into an empty magma chamber f weathering is the breaking down of rock into smaller pieces and dissolved chemicals/ erosion is the removal, transportation and deposition of rock fragments and dissolved chemicals

3 a lava flow, caused by the pressure of magma welling up from the chamber b weathering, caused by the waves wearing away the rock surfaces c a fault line, caused by a fracture in the crust d ash cloud, caused by the sudden release of pressure in the volcano e hot pool, caused by heat rising from subducted rock melting beneath the crust f rock slip/ landslide caused by weathering making a slope unstable g mountain building, caused by colliding plates crumpling up layers of rock

4 a true b true c false - thinner d true e false - mantle f true g false - active h true i true j true k false - does affect l true m true n true

5 a core b inner core c outer core d mantle e continental crust f oceanic crust g lithosphere h oceanic plate i continental plate j iron, it acts as a dynamo creating Earth's magnetic field k the outer core is liquid iron because of the temperature, but the inner core is solid because of the extreme pressure l heat is transferred by convection currents, in which heated lower mantle rock rises as it expands and cooling upper mantle rock sinks as it contracts; the heat source is primarily radioactivity m hotter upper mantle rock is semi-molten and therefore able to flow slowly/ cooler lithosphere rock is solid and brittle n continental crust is thicker and less dense than oceanic crust o because the bottom of the crust is fused to the top layer of the mantle p oceanic plate is thin because is very dense and formed on the sea-floor/ continental plate is thick because it is less dense and gets uplifted when plates collide

6 a mantle rock is hot enough to be semi-molten and therefore able to flow/ crustal rock is cool enough to be solid, and therefore brittle and unable to flow b heated lower mantle rock rises as it expands and cooling upper mantle rock sinks as it contracts c the currents move the plates above in the same direction d at the divergent boundary A sea-floor spreading is creating oceanic crust/ at the convergent boundary B subduction is destroying oceanic crust e when plates collide. the dense oceanic crust is forced down into the mantle to be melted and the rock recycled but less dense continental crust does not get subducted, and is therefore not recycled in the same way as oceanic rust

7 a i divergent ii both are oceanic iii sea-floor spreading iv earthquakes, volcanic eruptions v mantle convection currents carrying the plates apart b i convergent ii left - oceanic crust; right - continental crust iii subduction iv earthquakes, volcanoes, mountain building v mantle convection currents moving the plates together, pushing the less dense oceanic plate under c i convergent ii both are continental iii mountain building iv earthquakes v mantle convection currents pushing the plates together but neither will subduct

8 a along or close to plate boundaries b subduction causes earthquakes when a slab of the down-going plate slips suddenly past the overriding plate; sea-floor spreading causes earthquakes as blocks of crust shear either parallel or perpendicular to the mid-ocean ridge

9 a Indo-Australian Plate b Pacific Plate c transform fault d continental crust on both sides e the eastern block shifts southwards and upward relative to the western block f during earthquakes f by ongoing uplift over millions of years

10 a by wearing it down b because the river is wide and shallow often flooding the surrounding areas c it cuts a channel deep into the underlying rock d the steeper the gradient, the deeper the cut e a drowned network of V-shaped rivers; the ending of an ice age causing the sea level to rise f because it is not fused to the rock beneath, the force of gravity pulls it in a 'downhill' direction g glaciers are wide and grind down the side walls of valleys rather than cutting into the floor h steep-walled mountains rising out of the sea; by rising sea levels drowning a glacial valley i because it is actually a fiord, not a sound (ria)

11 There is an underlying hot-spot at the top of the mantle and magma forces its way up through the crust, forming a volcano. Over geological time the crust moves relative to the location of the hot spot because of plate tectonics, forming a string of volcanoes.

12 a Indo-Australian and Pacific Plates b convergent boundary along which oceanic crust from the Pacific Plate is being forced under the continental crust of the North Island which sits on the Indo-Australian Plate c a chain of volcanoes running from Ruapehu to White Island; volcanic activity around Rotorua; a spine of mountain ranges running up the North Island from Wellington to the East Cape d transform boundary along which the continental crust on the eastern side moves southward and upward relative to the continental crust on the westward side and e Southern Alps and earthquakes along the fault

13 a plutonic rock is rising magma that cools slowly within the crust b volcanic rock is lava or ash that has been ejected from a volcano due to rising pressure in the magma chamber below, caused by the melting of subducted oceanic crust beneath c sedimentary rock develops when sediments weathered from rock and transported to the sea floor by rivers become compacted by the weight of sediments above d metamorphic rock occurs when other types of rock are heated under pressure deep in the crust

ISBN: 9780170262316

Glossary

A

absorption (biology) - occurs when molecules or ions enter a cell or organism by passing through a membrane

absorption (physics) - occurs when thermal radiation heats a substance; occurs when light waves excite the surface atoms of a substance and are not re-emitted

acceleration - the change in an object's speed (or direction) caused by unbalanced forces; rate at which an object's speed is changing

acceleration due to gravity - approximately 10 ms^{-2} on Earth

accelerometer - an instrument used to measure instantaneous acceleration

acid - a substance that forms an acidic solution when it dissolves in water/ a substance that releases hydrogen ions when it dissolves in water

acid rain - rain that has a pH of less than 7 owing to dissolved nonmetal oxide gases such as sulfur dioxide

acidic solution - a solution with a pH of less than 7/ a solution that turns blue litmus paper red

acidify - make a solution more acidic/ reduce the pH of a solution

acquired variation - differences between organisms caused by the environment or by actions

action force - the force that you apply to an object

activation energy - the amount of energy required for a reaction to start happening/ the amount of energy required to break reactant bonds in order to form products

active transport - moving selected molecules across a cell membrane from an area of low concentration to an area of high concentration, which requires energy

actual motion - this occurs when an object is moving, rather than the observer (cf apparent motion)

adaptations - see *adaptive features*

adaptive features - inherited features that aid individuals to survive or reproduce in their environment

aerobic respiration - cellular respiration that occurs in the presence of oxygen; it releases a lot of energy and produces carbon dioxide gas and water

agar - jelly-like substance containing nutrients, which is used to culture bacteria and fungi

alcohols - carbon compounds that consist of a hydrocarbon chain bonded to an O atom with an attached H atom

alkali - a base that is soluble in water

alkaline solution - a solution with a pH greater than 7/ a solution that turns red litmus paper blue

alkanes - hydrocarbons with single bonds only

alkenes - hydrocarbons with a double bond between two of the carbon atoms

alleles - alternative forms of a specific gene that have significantly different base sequences

allotropes - alternative forms of an element

alloy - a mixture of several elements, the main one being a metal

alloying - mixing other elements into a molten metal

Alpine Fault - the large fault running most of the length of the South Island

alumina - the geological name for aluminium oxide

alveoli - the microscopic, balloon-like structures within the lungs; the inner lining is the gas exchange surface

amino acids - the small organic molecules that proteins are assembled out of; there are 20 different kinds

ammeter - an instrument that measures the size of an electrical current

ammonia - a toxic chemical produced when amino acids are used for cellular respiration

amorphous solid - a solid that does not have a crystalline structure

ampere or **amp** - the scientific unit for electrical current; one ampere is one coulomb of charge passing each second

amplitude - distance between the resting position and a crest or trough

anaerobic metabolism - metabolism that occurs in the absence of oxygen; little energy is released from glucose and waste products such as lactic acid, ethanol or methane gas are formed

analogue meter - a meter that has a scale and a marker

angle of incidence - the angle between the incident ray and the normal

angle of reflection - the angle between the reflected ray and the normal

angle of refraction - the angle between the incident ray and the normal

animals - large, complex multicellular organisms that acquire energy and nutrients by consuming other organisms

anodising - making the outer oxide layer of aluminium thicker and more protective

Antarctic Circle - the southernmost region of the planet that experiences 24 hours of light in midsummer and complete darkness in midwinter

anther - part of the male sex organ of a flower, which produces male sex cells (gametes) called pollen

antibiotic resistance - occurs when a strain of bacteria is no longer affected by a particular antibiotic

antibiotics - medicines, often derived from natural products, that are used to kill or inhibit bacteria

antibodies - chemicals that recognise and help destroy pathogens

antigens - marker chemicals on the surface of pathogens

antiseptics - chemicals applied to a wound to prevent infection; they do not kill human cells

anus - opening through which solidified waste food is expelled from the body

apparent motion - an illusion that an object is in motion, caused by the motion of the observer (cf actual motion)

application - any device or process that operates on the basis of scientific principles

Arctic Circle - the northernmost region of the planet that experiences 24 hours of light in midsummer and complete darkness in midwinter

artery - thick-walled blood vessels (tubes) that transport blood from the heart to other organs

ISBN: 9780170262316

artificial immunity - immunity to a disease that is gained by being vaccinated with dead or weakened pathogens

asexual reproduction - the production of genetically identical offspring by a single parent

ash - formed when vaporised magma ejected from a volcano solidifies as it cools

assimilation - occurs when the products of digestion are absorbed into cells and become part of structures or are used as energy sources

astronomer - a scientist who studies astronomy

astronomy - the study of the origin, nature and motion of heavenly bodies

atmosphere - the layer of gases surrounding a planet

atomic number - the number of protons in the nucleus of an atom

atoms - the basic building blocks of matter/ extremely small particles that all matter is made of

ATP (adenosinetriphosphate) - the chemical that acts as the energy-carrier within cells

aurora australis - amazing lights seen at night at higher southern latitudes

average acceleration - the mean acceleration of an object over part of a journey

average speed - the mean speed of an object over part of a journey

axis - an imaginary line running from pole to pole, around which a heavenly body rotates

B

bacteria - very small micro-organisms that do not have a cell nucleus

bacterium - an individual bacterial organism

balanced forces - occurs when two forces on an object are equal in size and opposite in direction

balanced symbol equation - for each type of atom involved in the reaction, the number of reactant and product atoms in the symbol equation are the same

basalt - dense, dark grey or black rock that comes from melted mantle rock

base (chemistry) - the chemical opposite of an acid/ a chemical that can neutralise an acidic solution / a chemical that removes hydrogen ions from a solution

base-pairing rule - C pairs with G and A with T

base sequence - the sequence of bases (or base triplets) along a DNA molecule, which codes for a single gene

base triplets - groups of three consecutive bases found along the sense strand of a DNA molecule

bases (DNA) - the four chemicals that are used to encode genes

basic oxide - a metal oxide that reacts with an acid to give a salt

battery - either a single chemical cell or a series of chemical cells that can generate electricity

bauxite - a reddish ore containing alumina (aluminium oxide)

behavioural adaptation - any inherited animal behaviour or plant response, which assists the species to survive and reproduce in its environment

bicarbonate compound - a compound containing bicarbonate ions

bicarbonate ion - the ion HCO_3^-

bile - the liquid produced by the liver and stored in the gall bladder, which emulsifies fats and helps to reduce the acidity of the contents of the small intestine

binary fission - cell division process by which a single bacterium divides into two bacteria that are genetically identical to the original bacterium

biodiversity - the number and range of species present in an ecosystem

biogeochemical cycle - a nutrient cycle involving biological, chemical and geological processes

biological community - all of the interacting species within a given location

biological factor - any feature or behaviour of other species that affects an individual or population

biological principle - a general rule that applies to all organisms

biological process - a series of enzyme-controlled steps that result in a particular outcome

biological pump - the transfer of carbon dioxide from surface waters to the ocean floor, which results from photosynthesis, the sinking of organic matter and decomposition

biologist - a scientist who studies living organisms and their relationship with their environment

biology - the scientific study of organisms

biomass - the total mass (weight) of living matter in a population, trophic level or community

biomolecules - complex organic chemicals made by organisms

biosphere - that part of the planet that is inhabited by life

biotic factor - see *biological factor*

bladder - a hollow organ that enlarges as it fills with urine and empties when its muscular wall contracts

blood - the fluid circulating in arteries and veins, which transports dissolved substances and red and white blood cells

blood pressure - the pressure that forces blood through blood vessels, which is caused by heart muscle contracting

blood system - the circulatory system of mammals

body (somatic) cells - all cells in an organism other than reproductive cells/ cells that have the normal number of chromosomes

body wave - a wave that travels through a substance

boiling point - the temperature at which a heated liquid boils into a gas

bond, chemical - the force of attraction that holds atoms together in a molecule or lattice

bronchi - the major airways branching off the trachea into the lungs

bronchioles - the increasingly narrow, branching airways that eventually connect with the alveoli

budding - occurs when a yeast cell reproduces asexually by forming a small genetically identical cell that detaches itself

buoyancy - the tendency of matter to rise, float or sink in a liquid, depending on its relative density

C

caecum - an organ branching off near the junction of the small and large intestines, containing bacteria that digest cellulose particularly in herbivores

caldera - a depression created when a volcano collapses because its magma chamber is empty

cambium - a continuous cylinder of dividing cells in the stems and roots of woody plants, which produces new vascular tissue

capillaries - very fine blood vessels that run past individual cells

carbohydrates - a class of food molecules, including sugars and starch, that either provide energy, store energy, or form part of the structure of cells

carbon cycle - a cycle in which the element carbon is cycled through a variety of physical, chemical, biological and geological processes

carbon compounds - compounds in which carbon atoms are covalently bonded to other kinds of atoms, particularly hydrogen

carbonate compound - a compound containing carbonate ions

carbonate ion - the ion CO_3^{2-}

carbonate rock - rock made out of carbonate compounds

carnivores - organisms that obtain the energy and nutrients needed for life by killing and eating other animals

carpels - the female sex organs of flowering plants

catalyst - a chemical that speeds up the reaction rate without being used up/ a chemical that reduces the amount of activation energy required to start a reaction

catalytic decomposition - breaking down a compound into simpler compounds or elements using a catalyst

categorical data - data that can be arranged into discrete categories

caustic solution - an hydroxide that chemically burns or destroys living tissue

cell division - the process by which a single cell divides into two genetically identical 'daughter' cells or into four genetically different gametes

cell (electricity) - see *chemical cell*

cell membrane - the living layer that encloses a cell and controls the entry and exit of chemicals; it consists of a double layer of lipids

cell nucleus - a complex internal cell structure that contains the chromosomes

cell specialisation (differentiation) - occurs when different types of cells are formed by specific sets of genes being expressed

cell wall - structure found outside of the cell membrane in plants, fungi and bacteria, which provides some rigidity and determines the shape of the cell

cells (biology) - the basic building blocks of organisms; the smallest biological object considered to be living

cellular respiration - a process that releases the energy stored in glucose molecules in small amounts that cells are able to use

cellulose - the main component of plant cell walls, which is difficult for herbivores to digest

centromere - part where duplicated chromosomes are connected, which then becomes attached to the spindle

change of state - changing from a solid, liquid or gas state to a different state

charge - a property of subatomic particles that is either positive or negative; see also *electrical charge*

charge distribution - how electrical charge is spread across the surface of a conductor or an insulator

charge separation - occurs when electrons are transferred from one object to another or from one location on an object to another location

chemical - a substance studied by chemists

chemical cell - an object that transforms chemical energy into electricity

chemical digestion - the splitting of large organic molecules into small soluble ones by digestive enzymes

chemical energy - see *chemical potential energy*

chemical formula - a shorthand way of showing the ratio of ions present in an ionic compound or the actual number of atoms present in a discrete molecule

chemical plant - a factory that carries out commercial chemical reactions

chemical potential energy - potential energy stored in the bonds that hold atoms together

chemical process - a series of chemical reactions that result in a particular product

chemical property - a property that relates to how a substance reacts

chemical reaction - occurs when new chemicals are formed from existing ones/ occurs when electrons are transferred between atoms or when atoms start sharing electrons

chemical weathering - the breakdown of exposed rock through chemical reactions

chemist - a scientist who studies the properties of chemicals and how they react

chemistry - the scientific study of chemicals and their reactions

chlorophyll - a green pigment found in chloroplasts, which is able to absorb light energy and transform it into chemical energy

chloroplasts - complex structures found inside some plant cells, which carry out the process of photosynthesis

chromosome - a complex cell structure that consists of a double-stranded DNA molecule wrapped around special proteins

circuit - a complete conducting pathway

circuit diagram - a drawing of an electrical circuit

circular wave front - a wave front that radiates out in an ever-increasing circle

circulation - the life process that involves transporting substances to and from all parts of an organism

circulatory system - the organ system that circulates substances around the body of the animal or plant

ISBN: 9780170262316

clones - offspring that are genetically identical to a single parent organism

coal - black or dark brown sedimentary rock formed out of the remains of ancient swamp vegetation that has been compressed over millions of years

coding strand - the DNA strand that codes for genes

collisions - these occur when moving particles such as atoms, ions or molecules encounter each other

colon - see *large intestine*

combination reaction - reaction in which two substances combine chemically to form a new substance

combustion - the process of burning a fuel in the presence of oxygen/ the rapid oxidation of a substance

commensalism - an ecological relationship that benefits one species but does not harm the other

community, biological - all of the plant, animal and microbe populations living in a particular location

competition, resource - occurs when two species living in the same location require the same limited resource

complete combustion - the combustion of a substance in an abundant supply of oxygen, resulting in the substance being fully oxidised

complete dominance - occurs when the phenotype expression of one allele completely masks the expression of another allele

component, chemical - a chemical species that is part of a solid, liquid or gas mixture

component, electrical - a part of an electrical circuit

compound - a pure substance in which different kinds of atoms are bonded together/ a pure substance in which different kinds of atoms are chemically combined

compressed natural gas or **CNG** - natural gas that is transported in a very compressed form in order to reduce its bulk

compression - region where the particles of a medium are squashed together

concentration - the number of molecules or ions per unit volume

concept, scientific - an idea that explains a natural phenomenon

condensation point - the temperature at which a cooling gas liquefies

condensing - occurs when a gas turns into a liquid as it cools

conduction - the transfer of heat or electrical energy through a substance

conduction charging - occurs when a neutral object is charged by contact with a charged object

conductivity - a substance's ability to conduct heat or electrical energy

conductor, electrical - a substance or object that allows a current to flow through it freely

conductor, thermal - a substance or object that allows heat energy to move through it freely

conservation of energy - energy cannot be created or destroyed, it can only be transformed from one type to another

constant acceleration - acceleration that is unvarying

consumers - animals that eat other species (or organic matter made by them) in order to obtain the energy and organic nutrients needed for life

contact charging - see *conduction charging*

contact force - occurs when the object applying the force touches the other object

continental crust - the thicker, less dense crust that forms the mass of continents

continuous data - data that can have any value

continuous variation - inherited variation that can take on a range of values, as opposed to discrete variation

control - see *controlled variable*

controlled variable - a variable that is kept constant during an experiment

convection - the transfer of thermal energy (heat) from a hotter to a colder location through the bulk flow of a heated gas, liquid or semi-molten substance

convection current - the roughly circular bulk flow of a gas, liquid or semi-molten substance, caused by a density difference arising from heating at one location

conventional current direction - the direction that positively charged particles would move around a circuit if they were free to move

convergent boundary - a boundary at which two tectonic plates are colliding

co-ordination - the life process that involves integrating the activities of organ systems of complex organisms through hormones or nerve impulses

core (geology) - the central part of Earth, which is mostly made of iron

Coriolis effect - the apparent deflection of trade winds in a westward direction, caused by Earth's rotation

corrosion - slow destruction of a metal through chemical reactions

corrosive chemical - a chemical that attacks and breaks down substances

cotyledons - structures found in the seeds of flowering plants that supply food to the germinating embryo and sometimes function as the first leaves

coulomb - the scientific unit for charge

covalent bonding - occurs when atoms are sharing electrons in order to have more stable electron arrangements

covalent compound - a compound in which atoms are bonded together because of the attraction that the atomic nuclei have for shared electrons

cracking - splitting a larger alkane into a smaller alkane and an alkene

crater - a depression created by a volcano

critical angle - the angle of incidence at which refracted waves travel across the surface of the new medium

crossing over - occurs when sections of homologous chromosomes are swapped during meiosis

cross-pollination - the transfer of pollen from the male to the female sex organs of the same flower or plant

crude oil - a mixture of liquid hydrocarbons found inside Earth's crust

crust - the solid outer layer of rock forming Earth's surface

crystal - a solid substance with a regular geometrical shape
crystalline solid - a solid made out of crystals/ a solid whose atoms are arranged in a lattice
crystallise - the formation of crystals from molten rock or from a solution
culturing - growing micro-organisms on nutrient agar in a warm environment
current, electrical - a flow of charged particles, usually electrons; a flow of electrical charge
current, thermal - the bulk flow of a liquid or a gas from a hotter to a colder location, due to differences in temperature and density
current electricity - a flow of charged particles through a conductor
cytoplasm - the jelly-like interior of a cell in which metabolism occurs

D

data - measurements or observations made in a systematic way
deceleration - occurs when the speed of an object is decreasing, see also *negative acceleration*
decomposers - organisms, such as fungi and bacteria, that obtain energy and nutrients by feeding on dead organisms, organic wastes or exposed foodstuffs
decomposition (biology) - the biological breakdown of the bodies of dead organisms by bacteria and fungi
decomposition reaction - breaking down a compound into simpler compounds or elements
deep sea trench - trench running along the edge of a continent where subduction occurs
deforestation - the clearing of all trees of an area of land, often by burning
degree Celsius - an everyday scientific unit in which temperature is measured
denature - the unfolding of an enzyme so that it no longer works
density (matter) - the mass of a specified volume of matter
density, species - the number of individuals of a species in a defined area or volume
dependent variable - a variable that alters because of a change in another
deposit, mineral - a layer of rock or mineral laid down over a long period of time
deposition - the laying down of sediment on a surface such as the sea floor
detritivores - animals that feed on detritus by consuming lumps and digesting it internally
detritus - organic matter, such as plant material, dead animals and wastes, often found on the bottom surface of an ecosystem
development - a life process that involves the specialisation of cells into tissues and the formation of organs made of different tissues
diaphragm - a muscular sheet that separates the chest and stomach cavities; when it contracts the volume of the chest increases, causing air to flow into the lungs
diffusion - the movement of chemicals from where they are more concentrated to where they are less concentrated; occurs because particles move randomly
digestion - the breakdown of large food molecules into smaller soluble ones by digestive enzymes; or the cutting, crushing and grinding of food into smaller particles
digestive enzymes - catalysts produced by organisms that break down large food molecules into smaller soluble ones
digestive system - the organ system that processes the food taken into the organism, breaking down lumps of food into small soluble molecules that are absorbed into the blood
digital meter - a meter that provides the reading in digits on a small screen
diode - component that allows electricity to flow in one direction only
direct current or **DC** - a current that flows in one direction only
discharge - see *electrical discharge*
discrete data - data that can only have particular values
discrete variation - inherited variation that exists either as one option or another, as opposed to continuous variation
disease - an illness with particular symptoms that affect plants, animals or humans, see also *pathogen*
disinfectants - powerful chemicals that kill micro-organisms but not resistant spores
dispersion - separation of different-frequency waves, separating white light into the visible spectrum
displacement reaction - a reaction in which atoms of a free element replace ions of a less reactive metal
dissolving - occurs when the particles of a substance separate when mixed into a liquid, or when a gas mixes into a liquid
distance - describes how far an object has travelled
distance-time graph - a graph on which the distance gone is plotted against the time elapsed
divergent boundary - a boundary at which two tectonic plates are being pushed apart
diversity (ecology) - see *biodiversity*
DNA or **deoxyribonucleic acid** - the complex molecule that is the main component of a chromosome, and which encodes genes
DNA replication - occurs when a cell separates the two strands of a DNA molecule and builds new sense and anti-sense strands on the exposed bases, which results in duplicated DNA molecules
dominant allele - an allele that will always be expressed whether one or two are possessed by the organism
dormancy - a state in which neither growth nor metabolism is occurring
double bond - occurs when two atoms share two pairs of electrons in order to have more stable states
double helix - the shape that two linked DNA strands form together
drag force - an opposing force created when an object passes through a body of water or air; a form of friction
ductile - able to be stretched out

ISBN: 9780170262316

E

earthing - using a conducting wire to connect an object to the ground

earthquake - a sudden jolt caused by the rapid movement of blocks of crust relative to each other

ecological relationships - relationships between species living in the same environment

ecologist - a scientist who studies ecology

ecology - the study of the interactions of species with their environment and with other species

ecosystem - a biological community and its environment interacting as a system

effluent - liquid waste from an industrial plant or run-off from a farm, containing urine and excrement

egestion - the expulsion of undigested waste food from the digestive system

eggs - female reproductive cells (sex cells or gametes), which are immobile and larger than sperm; they have half the normal number of chromosomes

elastic potential energy - the energy that is stored in stretched or squashed objects

electrical charge - a property of subatomic particles that can be either positive or negative

electrical circuit - see *circuit*

electrical conductivity - the capacity of a substance to conduct electricity

electrical conductor - a substance or object that allows a current to flow through it

electrical current - see *current electricity*

electrical discharge - loss of an object's charge by electron transfer

electrical energy - the energy possessed by charged particles moving in a current

electrical force field - provides the force that drives a current around a circuit; area in which electrical forces are experienced

electrical insulator - a substance or object that does not allow a current to flow through it

electrically neutral - neither positively nor negatively charged overall

electricity - a form of energy associated with charged particles

electrode - a conductor through which a current enters or leaves a solution

electrolysis - reducing metal ions to atoms using electricity and heat/ the decomposition of a compound by passing a current through a molten solution

electromagnet - current-carrying wire coiled (solenoid) around a soft-iron bar

electromagnetic effect - charged particles moving through a conductor create a magnetic force field

electromagnetic radiation - waves emitted by atoms that travel at the speed of light and don't need a medium

electromagnetic waves - self-perpetuating waves emitted by atoms, which consist of interacting electrical and magnetic fields

electron cloud - the space around the nucleus that is occupied by an atom's electrons

electron configuration - an annotation describing how electrons are distributed in electron shells

electron shell - a space around the nucleus that electrons with equal energy occupy

electrons - extremely small, negatively charged particles that rapidly vibrate around the nucleus/ negatively charged particles that flow in an electrical current

electrostatic force - an attractive or repulsive force occurring in an electrical force field/ between charged objects

element - a substance made of identical atoms/ a pure substance made of one kind of atom only

elliptical orbit - the oval-shaped orbit of a planet or moon

embryo - an early stage in the development of a new organism when organs first become recognisable

emission - occurs when excited atoms release an electromagnetic wave

emulsify - converting blobs of fat into tiny oil droplets

endothermic reaction - a reaction that continues to absorb heat from its surroundings

energy (biology) - required for organisms to live, grow and reproduce

energy (physics) - defined as the capacity to do work

energy efficiency - the percentage of input energy that becomes useful energy

energy flow - the one-directional movement of energy from the sun, through a biological community then into its environment

energy input - the energy supplied in an energy transformation

energy output - the energy produced in an energy transformation

energy transfer - occurs when the type of energy that is passed on is the same type

energy transformation - occurs when the type of energy that is passed on is a new type

environment - an organism's surroundings, both living and non-living/ the medium in which a community exists and the physical and chemical conditions

environmental factor - any aspect of the surroundings of an organism or a community that affects its functioning

environmental gradient - this occurs when the level of an environmental factor, such as tidal exposure, changes gradually across a location

environmental impact - the consequences of a natural event or human action on the physical environment, a species or a community

environmental issue - an issue related to how a human action impacts on the natural environment

enzymes - biological catalysts, speeding up the reactions that occur inside cells

epicentre - the point on Earth's surface directly above the initial fracture in an earthquake

epidemic - an outbreak of a pathogenic disease that spreads rapidly in a population

equator - an imaginary circle around Earth found at an equal distance from both poles

erosion - the removal, transportation and deposition of rock fragments and dissolved chemicals

ethanol - an alcohol produced when yeast carry out fermentation, an anaerobic process
evaporation - occurs when a liquid or dissolved substance changes into a gas at temperatures below the boiling point
event - an observable occurrence
evolution - the modification of a species over many generations as it adapts to a changing environment
excretion - the life process that involves disposing of the waste products of cells into the organism's environment
excretory system - the organ system that extracts cellular wastes from the blood and excretes them
exothermic reactions - reactions that release heat to the surroundings; most need some heat to start
exotic species - see *introduced species*
expansion - an increase in the size of an object due to heating
experiment - a scientific procedure used to make a discovery, test an hypothesis or confirm a scientific fact
experimental method - a scientific procedure that involves manipulating one variable and observing or measuring the effect on another
extra-cellular digestion - digesting food sources outside of cells by secreting digestive enzymes

F

fact, scientific - an observation or measurement that can be confirmed by other scientists
faeces - solidified waste food expelled through the anus
fault line - where the fault emerges through the ground surface
family tree - another name for a pedigree chart
fatty acids - organic acids that lipids are constructed out of
fault - a break between two blocks of the crust caused by the shearing of rocks, along which more earthquakes may occur
fermentation - an anaerobic process that results in waste products such as methane gas (bacteria), ethanol (yeast) or lactic acid (muscles)
fertilisation - occurs when the nucleus of a sperm enters an egg to form the zygote
field direction - see *magnetic field direction*
field strength - see *magnetic field strength*
fieldwork - observational or experimental work carried out in the outside world rather than the laboratory
fiord - a drowned glacial valley
flowers - reproductive organs of flowering plants
foetus - name given to a human organism after about eight weeks of development; the embryo develops into a foetus
food - complex molecules that provide energy and nutrients for organisms when digested
food chain - sequence of species, each eaten by the next
food poisoning - disease or illness caused by eating food contaminated by pathogenic microbes
food web - the linked-up food chains of a community
force - something required to change the motion or shape of an object; a push or a pull
formula (chemistry) - see *chemical formula*
formula (physics) - a mathematical relationship between different quantities
fossil fuel - fuels such as coal, oil and natural gas, which are formed out of the undecomposed remains of ancient organisms; they are rich in hydrocarbons
fraction - a mixture of similar-sized hydrocarbons produced in a fractional distillation tower
fractional distillation - separating a complex mixture of compounds into fractions of similar-sized molecules on the basis of differences in boiling/condensation points
fracture - a break in the crust that occurs when rock shears under pressure
free electrons - mobile outer shell electrons that move from atom to atom in a metal
frequency - the number of wave crests passing per second
friction - a force caused by two surfaces rubbing against each other, which converts kinetic energy into heat
friction charging - occurs when an insulator is rubbed by a different insulating material, which transfers electrons
fruit - the ripened ovary of a flower, enclosing the seeds
fuel - a substance that is combusted to generate heat
function (biology) - the task that a biological structure carries out/ the purpose of a biological process
fungi - a group of immobile organisms that feed on dead or living organisms, and exposed food sources

G

g - the acceleration that the force of gravity will cause
gall bladder - the organ that stores bile from the liver
galvanising - coating steel in a thin protective layer of zinc, by either hot-dipping or electrolysis
gametes - reproductive cells, either eggs or sperm, which are used to combine genes from two organisms
gas exchange - the life process that involves the exchange of respiratory gases between an organism and its environment
gastric juice - the acidic fluid secreted by the lining of the stomach, which contains digestive enzymes
gene - an inherited instruction involved in determining the nature of a trait; a section of a DNA molecule that codes for a protein
genetic disease - an inherited disease caused by the expression of a defective allele
genetic variation - occurs when different alleles for a gene (or different combinations of alleles for several genes) exist within a population
geneticist - a scientist whose studies genetics
genetics - the study of how genes are inherited and expressed
genome - the total collection of genes (rather than alleles) possessed by an organism
genotype - the two alleles an organism possesses for a specific trait
genus - a group of closely related species
geographic poles - found along the axis of Earth's spin at the surface
geological event - event that transforms rock, usually over a long period of time

ISBN: 9780170262316

geological processes - internal and external forces that reshape the physical structure of the planet

geologist - a scientist who studies geology

geology - the study of Earth's structure and rocks, and the processes that shape them

geomagnetic field - the magnetic field that surrounds the planet, created by giant electrical currents in the outer core of molten iron

germination - occurs when a seed, spore or pollen grain begins to metabolise and grow

gigatonne - a billion tonnes

glacier - a large body of ice that deforms and flows downhill because of force of gravity

gland - an organ that produces and secretes useful chemicals

global warming - the gradual increase in the average global surface temperature due to increasing amounts of greenhouse gases in the atmosphere trapping more heat energy

global winds - the major flows of surface air over the globe

glucose - the simple sugar molecule that is the building block of complex carbohydrates such as starch, glycogen and cellulose; the main source of energy used by cells carrying out respiration

glycerol - the organic molecule that is used to link the three fatty acids in a lipid molecule

glycogen - a multi-branched complex carbohydrate made of glucose units; it serves as an energy store in animals

glycolysis - the process by which glucose molecules are split into pyruvic acid molecules by enzymes

gradient - the rate of change of a graph line

gram - the basic scientific unit for mass

gravitational potential energy - the energy stored in an object held above the ground/ the type of energy an object gains when lifted

gravity - the force of attraction that exists between all objects that have mass

grazing - eating plant tissue, such as leaves and grass

greenhouse effect - the process by which Earth's surface is warmed by gases that absorb outgoing infrared radiation and re-radiate some of it back to the surface

greenhouse gases - gases in the atmosphere that absorb infrared radiation

group (chemistry) - a column of the Periodic Table; elements in a group undergo similar chemical reactions

growth - the life process that involves cell division and enlargement, and also cell specialisation into tissues and organs in multicellular organisms

growth ring - an annual ring in a tree trunk

guard cells - pairs of cells found in the epidermis of a leaf, which control the entry and exit of gases

gut - see *digestive system*

H

habitat - the type of environment in which members of a species are typically found

health impact - the effect of an enterprise or chemical process on the health of workers or on members of the local community

haemoglobin - a chemical found in red blood cells, whose function is to transport oxygen molecules from the lungs to body cells

heart - a hollow, muscular organ that pumps blood around the body as it rhythmically contracts

heat - the amount of thermal energy transferred

heat capacity - the ratio of heat flow to or from a substance to its change in temperature

heat conductor - see *thermal conductor*

heat content - the amount of thermal energy that will be released when a fuel undergoes complete combustion

heat flow - the transfer of thermal energy from a hotter to a colder region

heat of combustion - the amount of thermal energy released when a standard mass of a fuel undergoes complete combustion

heat transfer - the transfer of thermal energy from one location to another

heating curve - a graph of temperature versus the time a substance is heated for

heavenly body - a large mass located in space

hemisphere - half of the planet, either above or below the equator

herbivores - animals that get the energy and organic nutrients they need by feeding on plant matter

hertz - the basic unit of frequency

heterozygous genotype - occurs when the two alleles an organism possesses for a trait are different

homologous chromosomes - chromosomes that bear the same genes in the same order/ chromosomes that are the same size and have the same patterning

homozygous dominant - occurs when an organism has two dominant alleles for a trait

homozygous genotype - occurs when the two alleles an organism possesses for a trait are identical

homozygous recessive - occurs when an organism has two recessive alleles for a trait

hormones - chemical messengers that are transported around the organism

hydrocarbons - carbon compounds consisting of covalently bonded carbon and hydrogen atoms only

hydrogen carbonate ion - see *bicarbonate ion*

hyphae - the network of fine feeding filaments of a fungus that invade the host organism or food source

hypothesis - a statement about natural phenomena that can be tested experimentally

I

igneous rock - rock formed when magma or lava solidifies

immiscible liquids - liquids that will not mix at all but form two separate layers

immunity - the ability of the human body to prevent the occurrence of a pathogenic disease

incident ray - incoming ray from a light source or object

incomplete combustion - the combustion of a substance in a limited supply of oxygen, resulting in partial oxidation

incubation - growing micro-organisms at a temperature that will result in optimal growth

independent variable - the variable that is altered or set by the experimenter/ a variable that causes a change in a dependent variable

indicator - a liquid or paper that changes colour as the pH of a solution changes

indicator species - a species whose presence or absence provides information about the quality of the environment

induction charging - a temporary effect that occurs when a charged insulator is brought close to a neutral object

inert gases - the gases found in group 18 of the Periodic Table which have full outer electron shells and are therefore very unreactive

inertia - the resistance of an object to any change in its motion

infection - the invasion of tissues by pathogenic microbes/ the process by which pathogenic microbes are transferred from one organism to another

infrared radiation - see *infrared waves*

infrared waves - invisible electromagnetic radiation that heats an object when absorbed by its atoms

ingestion - the intake of food into the digestive system, through the mouth

inherited variation - differences between organisms that are caused by genetic differences

inner core - the solid, extremely hot centre of Earth, which is mostly composed of iron

inoculation - the introduction of spores or active micro-organisms, in order to culture them

inorganic carbon - carbon that is not part of organic molecules

inorganic chemicals - chemicals that lack carbon and hydrogen atoms; chemicals formed by geological processes

inorganic nutrients - inorganic chemicals required by organisms; usually they are ions

insoluble compound - a compound that does not dissolve when mixed into a liquid

instantaneous acceleration - the acceleration of an object at one point in time

instantaneous speed - the speed of an object at one point in time

insulated conductor - a conductor that is electrically insulated from its surroundings

insulator, electrical - a substance or object that does not allow a current to flow through it/ a substance that does not have free outer electrons

insulator, thermal - a substance or object that does not allow heat to pass through it

insulin - a hormone produced by the pancreas, which is involved in controlling blood sugar levels

internal transport - see *circulation*

introduced species – a plant or animal species that has been introduced either deliberately or accidentally in a location where it does not exist naturally

inversely proportional - the doubling of one variable, say, causes the other to be halved

invertebrates - animals that lack a backbone

investigation - an enquiry that involves either primary or secondary research

ionic bonding - the attraction between oppositely charged ions that holds the ions together in a lattice

ionic compound - a substance in which charged atoms or charged molecules are held together by ionic bonds

ionic equation - a symbol equation in which the charges on ions are shown

ionisation - occurs when a compound separates into ions, either when dissolved in water or when in a molten state

ions - charged atoms or molecules

iron sand - black sand, which is rich in the ore magnetite

issue - a topic about which people hold different opinions or viewpoints

J

joule (J) - the scientific unit for energy; 1 joule is defined as the work done when a 1 newton force moves an object a distance of 1 metre

K

karyotype - an image of all of the chromosomes of an organism arranged in pairs of decreasing size

key species - a species that forms the physical environment of most other species in a community or has a major influence on most of the species

key word - a word or phrase used when searching for information in either a book or on the internet

kidneys - the excretory organs of vertebrates, which extract waste products from the blood

kilogram - a scientific unit for mass; equal to one thousand grams

kilojoule - a scientific unit for energy; equal to one thousand joules

kilonewton - a scientific unit for force; equal to one thousand newtons

kilopascal - a scientific unit for pressure; equal to one thousand pascals

kinetic energy - the energy possessed by moving objects, includes vibrational, rotational and free motion

L

lacteal - part of the lymph system that absorbs fatty acids and glycerol from the villi lining the small intestine

lactic acid - an organic acid that is the waste product of anaerobic metabolism, particularly in muscles

large intestine or **colon** - the last part of the digestive system, which reabsorbs water, solidifying the contents of the gut

latent heat of fusion - heat required to melt or freeze one kilogram of a substance

latent heat of vaporisation - heat required to boil or condense one kilogram of a substance

lateral inversion - the reversal of an object as seen in a mirror; left and right sides of an object are swapped in the image

ISBN: 9780170262316

latitude - the angular distance above or below the equator

lattice - a regular array of large numbers of atoms or ions

lava - the molten rock (magma) that flows out of a volcano

law, scientific - a statement based on extensive observations and/or experiments that describes some aspect of the natural world but does not explain it

law of energy conservation - energy cannot be created or destroyed, only transformed from one form to another; the total amount of energy is the same after as before an energy transformation

law of heat flow - thermal energy flows spontaneously from a hotter to a colder region

law of magnetism - unlike magnetic poles attract each but like magnetic poles repel each other

leaching - the removal of soluble nutrients from the soil

leaf primordium - mass of immature cells in a bud that will eventually develop into a leaf

legume - plant species that has special root nodules containing nitrogen-fixing bacteria

life cycle - series of stages that an organism passes through in its lifetime

life processes - the processes that all organisms must carry out in order to live, grow and reproduce

light - electromagnetic waves that are visible to the human eye; a form of energy

light emitting diode or **LED** - a component that lights up if the current is flowing in the right direction

light energy - see *light*

light radiation - see *light*

light ray - an arrow used to show the direction of wave propagation

limestone - sedimentary rock formed from the shells of marine creatures accumulated in layers on the sea floor over millions of years

limewater - a saturated solution of calcium hydroxide that turns a milky colour when it reacts with carbon dioxide gas as it forms insoluble calcium carbonate

limiting factor - an environmental factor that limits growth or reproduction of a population or species

linear equation - an equation that is associated with a straight line on a graph

liquefied petroleum gas or **LPG** - a mixture of propane and butanes gases that have been liquefied under pressure

lipids - energy-rich type of food (fats and oils); lipid molecules consists of three fatty acids attached to a glycerol unit

lithosphere - the upper shell of Earth formed out of crust fused to the upper mantle

litmus - a chemical that turns red in acid and blue in alkali

liver - a large digestive organ, which produces bile and processes the products of digestion

longitudinal wave - a wave in which the direction of the disturbance is parallel to the propagation direction

lunar eclipse - occurs when the Moon moves into Earth's shadow

lunar month - the period of time from full moon to full moon

lungs - the main organs of the respiratory system; their function is to transfer respiratory gases between the air and the organism

lustre - the nature of the shine on the surface of a solid

lymph system - system that absorbs digested fats from the gut and drains into the blood system; it also drains fluid that has escaped from the blood back into the blood system

lymphocytes - white blood cells that produce antibodies that mark or destroy pathogens

M

magma - hot molten rock inside the crust

magma chamber - a place beneath a volcanic vent where magma accumulates

magnet - object that attracts things made of iron; object that has a magnetic force field surrounding it

magnetic field direction - the direction in which a compass needle will point at a particular location

magnetic field strength - strength of the magnetic field at a particular location, measure in teslas

magnetic force - a pull or a push on an object caused by a magnetic force field

magnetic force field - region in which a magnetic force is experienced by an iron object or another magnet

magnetic poles - two locations on a magnet where the magnetic field strength is strongest

magnetism - a property of an object that affects a nearby compass needle

magnetite - a mineral rich in iron oxide, found in black ironsand

magnitude - the size or strength of a quantity

malleable - able to be hammered or squashed into a new shape

mammals - animals that have hair or fur, give birth to live young, and feed them on milk from mammary glands

mammary glands - the milk-producing organs of mammals

mantle - the deep layer of hot semi-molten rock beneath the crust, which is able to flow in convection currents

mass - the amount of matter in an object; a large body

mass number - the number of protons and neutrons in the nucleus of an atom

maternal chromosomes - chromosomes inherited from the female parent

matter - a substance that has mass and occupies space

mean - the average of a set of measurements

mechanical wave - a wave that requires a medium to propagate and which temporarily deforms or displaces the medium as it passes

mechanics - the study of how forces affect the motion of objects

medium (ecology) - the air, water or soil in which a community exists

medium (microbiology) - a nutrient-rich substance used to culture micro-organisms such as fungi and bacteria

medium (physics) - a substance needed for some types of waves to propagate; a transparent substance through which a transmitted wave passes

ISBN: 9780170262316

meiosis - a type of cell division that produces gametes, which are cells with half the normal number of chromosomes; each gamete has a different set of chromosomes

melting point - the temperature at which a particular solid melts into a liquid

membrane - see *cell membrane*

meristem - a zone of a plant that continues to undergo cell division throughout the life of the plant

mesophyll - the internal tissues of a leaf, which carry out photosynthesis

metabolism - the complex set of chemical reactions necessary for life, which occur inside all cells

metal bicarbonate - a compound made of metal ions bonded to bicarbonate ions

metal carbonate - a compound made of metal ions bonded to carbonate ions

metal compound - a chemical in which positive metal ions are bonded to other negative ions

metal hydrogen carbonate - see *metal bicarbonate*

metal hydroxide - a compound made of metal ions bonded to hydroxide ions

metal oxide - a compound made of metal ions bonded to oxide ions

metallic bonding - the attraction between positive metal atoms and the free electrons moving around them

metalloid - elements that have both metal and nonmetal properties

metals - strong, shiny solids that are good conductors of heat and electricity

metamorphic rock - rock that has been transformed under high temperature and pressure

methane ice - a substance found on the sea floor, which is a mixture of frozen methane and ice

microbes - another name for micro-organisms

micro-organisms - very small organisms that are usually only visible individually under the microscope

microvilli - extremely small extensions of villi in the gut, which greatly increase the surface area for absorption

mid-ocean ridge - a divergent plate boundary where oceanic crust is formed through sea-floor spreading

mineral - a naturally occurring solid inorganic substance with a definite composition and structure

mineral extraction - the removal of a mineral resource from Earth's crust

mineral nutrients - inorganic chemicals (usually ions) needed by organisms, obtained from their physical environment or by eating other organisms

mining - the process of extracting useful minerals from Earth's crust

mitochondria - sausage-shaped cell structures that carry out aerobic respiration

mitosis - a type of cell division that results in two cells, each with an identical set of chromosomes, and therefore alleles

molecular compound - a compound consisting of molecules rather than ions

molecule - a discrete group of atoms that are bonded together because pairs of atoms share electrons

molten - in a liquid state due to being heated to a high temperature

monatomic ion - an ion consisting of a single atom only

monerans - unicellular micro-organisms that lack a cell nucleus

monohybrid cross - the pattern of inheritance associated with a single gene

monohybrid inheritance - the inheritance of a single gene (involving a pair of alleles) considered in isolation from other genes

monomer - small identical molecules that can be chemically combined to form long chain molecules called polymers

moon - a large heavenly body that orbits a planet

motion - occurs when an object's location is changing

mountain building - the crumpling up of the crust when tectonic plates slowly collide over millions of years

multicellular organisms - organisms made of a large number of cells, usually of different types

mutations - random changes in the base sequence of a gene, caused by radiation, chemicals or copying errors when DNA or RNA is replicated

mutualism - an ecological relationship that benefits both species

N

native element - an element found in a chemically uncombined state in Earth's crust

native species - a plant or animal species that occurs naturally in a particular location

natural gas - a mixture of gases found in Earth's crust, formed from the undecomposed remains of ancient plants and animals that have been compressed and heated over millions of years

natural immunity - immunity to a particular pathogen because of the presence of specific antibodies owing to a previous infection by the same pathogen

natural selection - a process in which nature 'selects' phenotypes that are best suited to survive and reproduce in the environment, which results in the alleles that cause those phenotypes increasing in frequency in the population

neap tide - a very low high tide that occurs twice in a lunar month

negative acceleration - occurs when an object is slowing down

negative ion - an atom or molecule that has more electrons than protons overall

net force - the overall effect of combining the forces acting on an object

neutral, chemically - a solution that is neither acidic nor alkaline; a solution with a pH of 7

neutral, electrically - particles that are uncharged or have no overall charge

neutral atoms - atoms with the same numbers of electrons as protons

neutralisation - a reaction in which a chemically neutral solution is formed

neutralise - to make neither acidic nor alkaline

ISBN: 9780170262316

neutron - an uncharged particle in the nucleus of an atom

neutron number - the number of neutrons in the nucleus of an atom

newton - the scientific unit for force

Newton's 1st law - an object's motion remains unchanged if the net force on it is zero

Newton's 2nd law - a net force will cause an object to accelerate in the direction of that force

Newton's 3rd law - for every action force there is an equal but opposite reaction force

nitrogen-fixing bacteria - bacteria found in the root nodules of legume plants, which are able to convert nitrogen gas from the atmosphere into nitrate ions

non-contact force - the object applying the force does not need to touch another object in order to act on it

nonmetal - an element that does not have metallic properties

normal - an imaginary line perpendicular to where a wave strikes a boundary between two media

north pole, magnetic - the pole of a magnet that a compass needle points toward

nuclear force - the force that holds protons and neutrons together in the nucleus of an atom

nuclear reactions - reactions in which the nucleus of atoms are changed, which either requires or releases a very large amount of energy

nucleus, atomic - the central area of an atom where protons and neutrons are found/ the positively charged central region of an atom

nucleus, cell - see *cell nucleus*

numerical data - data that is in the form of numbers

nutrient cycle - the recycling of the atoms of a chemical element by an ecosystem

nutrients - inorganic chemicals (usually ions) or organic chemicals required by organisms in order for them to live, grow and reproduce

nutrition - the life process by which organisms obtain the energy and nutrients required for living, growing and reproducing

O

object (optics) - the source of the incident light waves

objective data - measurements or observations that have been made without bias and can be repeated by others

observational method - a procedure that involves collecting data or making observations without any variables being altered, usually sampling is involved

ocean acidification - the increasing acidity of the ocean, due to increasing amounts of carbon dioxide being absorbed from the atmosphere

oceanic crust - thinner, denser crust formed on the sea-floor

oesophagus - part of the digestive system that moves food from the mouth to the stomach

ohm - the scientific unit of resistance; one ohm is the resistance of an object that limits the current to one ampere when a voltage drop of one volt occurs

Ohm's law - states that the voltage across a component is proportional to the current flowing through it, provided the temperature of the component remains constant

oil - see *petroleum*

omnivores - animals that eat both plant and animal matter

opaque - does not allow light waves to be transmitted

optical density - property of a transparent medium that affects how fast light waves are transmitted

optimum growth - best rate of growth and reproduction

optimum level - environmental conditions that result in optimum growth

orbit - the pathway of a planet around the Sun, or a moon around a planet

orbital plane - the flat two-dimensional surface in which one object orbits another

ore - a mineral or a combination of minerals found in a deposit in the crust

organ - part of a multicellular organism, made of different tissues, which performs a specific function

organ system - a group of organs that work together to perform a specific task or function

organelle - microscopic structures within cells

organic acids - organic chemicals that have acidic properties

organic carbon - carbon in compounds made by organisms or in compounds derived from them

organic chemicals - chemicals produced by organisms, which have a 'backbone' of carbon atoms

organic matter - matter derived from living plants and animals, their waste products, and dead organisms

organic molecules - complex carbon-based molecules made by organisms

organic nutrients - molecules produced by organisms, which are made out carbon, hydrogen, oxygen and nitrogen atoms

organisms - individual living things/ objects that carry out life processes

outer core - the very hot region of molten iron that surrounds the inner core of Earth

outlier - a data value that is well outside the range of other measurements, often due to an error

ovary - female sex organ of a plant or animal, which produces female gametes called eggs

ovule - part of the female sex organ of flowering plants, which contains an egg and develops into a seed after fertilisation

oxidation - a reaction in which oxygen atoms are combined with other atoms, ions or a compound/ a reaction in which atoms, ions or a compound lose electrons

oxide - a compound in which oxygen atoms are bonded to other kinds of atoms

P

Pacific Rim of Fire - the edges of the Pacific Ocean, where numerous volcanoes and earthquakes occur

ISBN: 9780170262316

palisade mesophyll - layer of closely packed, upright cylindrical cells found in a leaf, which carry out photosynthesis

pancreas - a digestive organ that produces and secretes a range of enzymes into the small intestine; it also produces insulin, which controls blood sugar levels

pancreatic juice - the liquid secreted by the pancreas into the small intestine, containing a range of digestive enzymes and also bicarbonate ions

pandemic - a pathogenic disease that has spread over a very large area of the world and infected many people

parallel circuit - a circuit in which components are connected in several branches; circuit in which the current can travel through several different pathways

parallel components - components that are connected on different pathways

parasitism - an ecological relationship in which one species lives in or on a larger species, feeding off it but without killing it

participant ions - ions that are actually involved in a reaction

particles - the atoms or molecules that all matter is made of

pascal - the scientific unit for pressure; equal to a force of one newton acting on an area of one square metre

pasteurising - heating food at a specific temperature for a brief period of time in order to kill all pathogenic bacteria

paternal chromosomes - chromosomes inherited from the male parent

pathogen - a disease-causing micro-organism

pedigree chart - a chart used to specify the phenotypes of several generations and identify genotypes

peptides - short chains of amino acids

period (astronomy) - the length of time it takes a heavenly body to complete one orbit or rotation

period (chemistry) - a row of elements across the Periodic Table

period (physics) - time required for a single wave cycle

Periodic Table - a chart with the elements arranged in periods and groups

petroleum - a thick liquid formed underground from the undecomposed remains of ancient marine organisms buried millions of years ago then transformed under heat and pressure

pH - the degree of acidity or alkalinity of a solution

pH scale - the scale used to indicate acidity or alkalinity of a solution

phagocytes - white blood cells that engulf and digest pathogens

phase change - process of changing from one state of matter to another

phases of the Moon - the appearance of the Moon during different stages of a lunar month

phenomenon - an aspect of the natural world that can be observed and/or measured

phenotype - the physical appearance or state of a trait, which results from the expression of the genotype

phenotype ratio - the expected ratio of the different phenotypes that are possible/ the actual ratio of phenotypes of offspring produced by a mating

phenotype variation - the existence of a variety of phenotypes within a population

phloem tube - column of living cells that transport dissolved sugars around a plant

photoperiod - the period of time during which a plant receives light from the sun

photosynthesis - the process occurring in the green parts of producers, which transforms carbon dioxide and water into energy-rich glucose using sunlight energy

physical digestion - cutting, crushing and grinding food into small particles

physical factor - any non-living factor that can affect an organism or a population

physical process - a process in which only physical changes are involved

physical property - any property of a substance that does not involve a chemical reaction

physicist - a scientist who studies physics

physics - the scientific study of motion and energy

physiological adaptation - any inherited process carried out by individuals, which assists the species to survive and reproduce in its environment

phytoplankton - floating microscopic producers

pigment - a biomolecule with a characteristic colour

pitch - the highness of lowness of a sound

plane boundary - a flat boundary between two different media

plane wave front - a straight wave that is perpendicular to the direction of propagation

planet - a large heavenly body that orbits a star such as the Sun

plant variety - the result of generations of breeding involving the selection and mating together of plants with desirable features

plants - large, complex multicellular organisms that are able to make their own food by photosynthesis

plasma - the liquid part of the blood, which transports many dissolved substances

plastic - substances consisting of very large molecules; they are light, corrosion-resistant, strong, and have low melting points, which enables them to be readily moulded

plate boundary - the junction between neighbouring tectonic plates of Earth's crust

plumule - the part of a plant embryo that develops into the first root

plutonic rock - rock formed when magma solidifies inside the crust

polar axis - the axis around which a heavenly body spins

pole, geographic - point on the surface of Earth that the polar axis passes through

pole, magnetic - the location of either the north or the south magnetic pole on the surface of Earth

pollen - microscopic grains produced by flowers, which contain the male sex cells (sperm)

pollen tube - a tube that grows down to the egg when a pollen grain germinates on the stigma

ISBN: 9780170262316

pollination - the process of transferring pollen from the male sex organ to the female sex organ of plants

pollutant - a solid, liquid or gaseous chemical released into the environment, which has an adverse effect on a species or a community

polyatomic ion - a molecule with an overall electrical charge

polymer - a long-chain molecule formed when many small identical molecules called monomers are bonded together

polymerisation - the process of chemically combining monomers to form polymers

population - a group of organisms belonging to the same species and living in the same location

positive acceleration - occurs when an object is speeding up

positive ion - an atom or molecule that has less electrons than protons overall

potential difference - the difference in electrical energy across a component in an electrical circuit

potential energy - energy that is stored in some way

power, electrical - the rate at which energy is supplied or used; the rate at which work is done

power, thermal - the rate at which heat transfer occurs/ the rate at which thermal energy is emitted or absorbed by a substance

power rating - the wattage or power of a component

practical investigation - an enquiry that involves carrying out either an experiment or fieldwork

precipitate - an insoluble compound formed when two ionic solutions are mixed

precipitation reaction - a reaction in which ions from two different solutions combine to form an insoluble compound

precipitation (weather) - rainfall

predation - an ecological relationship in which one species kills and eats another species

pressure - the force applied per unit of surface area of the object

primary data - scientific data collected by oneself

primary growth - the growth that occurs at root tips, and at shoot tips and buds, which elongates the plant and forms new organs

principle, scientific - a statement that expresses a well-established scientific relationship or law

concise verbal or mathematical statement of a relation that expresses a fundamental principle of science, like Newton's law of universal gravitation.

prism - a block of glass with the same cross-section from top to bottom

probability - the chances of an event occurring, usually expressed as a fraction or decimal

producers - organisms that are able to make food using inorganic chemicals and energy from their environments

productivity - the rate at which producers create new biomass

products - the new substances formed in a chemical reaction

propagation - the movement of waves away from the wave source

proportional to - means that the doubling of one variable, say, will cause the other to double as well

proportional to the square of - means that one variable increases in proportion to the square of the other

protein synthesis - the process by which a cell uses the base sequence of a particular gene to construct a specific protein; it occurs in cell structures called ribosomes

proteins - a food class needed for growth, repair and health; proteins molecules consist of lengthy chains of amino acids folded into special shapes; the molecules that genes code for

protists - microscopic unicellular organisms that do have a nucleus

proton - a particle with a positive charge found in the nucleus of an atom

Punnett square - a diagram used to predict the expected ratio of phenotypes amongst offspring

pure-breeding - organisms that are homozygous for a dominant allele

Q

quantity - a physical property of an object that is measurable

quarantine - the isolation of a person or animal with a serious infectious disease

R

radiant heat - the transfer of heat by infrared waves

radiation - the transfer of energy by electromagnetic waves

radicle - the part of a plant embryo that develops into the first shoot

radioactive decay - occurs when atoms break down into smaller atoms, releasing radioactivity

rarefaction - a region where the medium expands as the wave passes

rate - how fast a process or event is occurring

reactants - the substances that are used up in a reaction

reaction, chemical - occurs when existing chemicals are changed into new ones/ occurs when electrons are transferred between atoms or when they start sharing electrons

reaction force - the equal but opposite direction force that arises automatically when a force is applied to an object

reaction progress graph - a graph showing the progress of a chemical reaction in terms of either reactants being consumed or products being formed

reaction rate - the speed at which reactants are changed into products

reactivity - a measure of how reactive a chemical is with different reactants, such as oxygen, water and acids

reactivity series - a list of different metals arranged according to the strength of their reactions with oxygen, water and acids

ISBN: 9780170262316

recessive allele - an allele whose expression is masked by the expression of a dominant allele/ an allele which is only (fully) expressed if the organism possesses two copies

red blood cells - very small, flexible biconcave cells, whose function it is to transport oxygen around the body

reduction - a reaction in which oxygen is removed from a compound/ a reaction in which atoms, ions or a compound gain electrons

reference - a source of information

refinery - an industrial plant where physical and chemical processes are carried out in order to produce useful chemicals in a purified form

reflected ray - a ray showing how a light wave has 'bounced' off the surface of a new medium; a ray showing how a light wave has been re-emitted by the surface atoms of a medium back into the original medium

reflection - occurs when waves 'bounce' off a surface

refracted ray - a ray showing how a light wave has been abruptly bent as it crossed the boundary into a new medium

refraction - the abrupt bending of light waves toward or away from the normal as they enter a different medium

relationship - how one variable affects another

relative density - the mass of an object compared to the mass of a similar volume of water

reliability - the extent to which data are accurate, as indicated by the consistency of results when an experiment or series of observations are repeated

replication - the production of two identical DNA molecules; the production of multiple copies of a virus

reproduction - the life process that involves the production of new organisms, either sexually or asexually

reproductive cells - see *sex cells* or *gametes*

research - obtaining either primary or secondary data/ carrying out an enquiry into an issue or topic

reservoir - part of the environment in which organic or inorganic nutrients are stored

resistance - the opposition to the flow of electrons through a substance

resistor - an object or substance that opposes the flow of current; a component that is designed to limit current flow

resource competition - an ecological relationship in which two species living in the same location compete for the same limited resource

respiration - the release of energy from food molecules that occurs within all cells all of the time

respiratory enzymes - enzymes that control (catalyse) each step of the process of respiration

respiratory gases - oxygen and carbon dioxide

respiratory system - the organ system that draws oxygen into the organism and transfers it to the blood, and removes carbon dioxide from the blood and forces it out into the air

rheostat - an electrical component whose resistance can be varied using a slider

ria (geology) - a drowned network of V-shaped rivers, often called a sound

ribosomes - tiny cell structures in the cytoplasm that assemble proteins

rift valley - a large, wide valley formed when continental crust is being separated

right-hand grip rule - gives the direction of the magnetic force field surrounding a current-carrying wire

RNA or **ribonucleic acid** - a complex molecule similar to DNA, which is also able to encode genes

rock - the solid material that forms the bulk of Earth's crust

rock cycle - the transformation of rock into different forms

rock strata - layers of rock containing different minerals and perhaps fossils

room temperature - a standard temperature that is taken to be 20°C

rounding off - reducing the number of decimal places to a set number

runners - a stem that grows across the ground, taking root in various places, eventually forming separate plants

rusting - the reaction of iron with oxygen and moisture to form an orange-red oxide layer that flakes away exposing the metal underneath to further corrosion

S

sacrificial protection - involves attaching a block of a more reactive metal to steel, which then supplies the steel with electrons to prevent it from rusting

salt - the crystals left after an acid has been neutralised by a base (or an acid has reacted with a metal) and the liquid has evaporated

sample, biological - a representative subset of a population or a biological community

sampling - the process of obtaining and measuring randomly selected but representative samples

saprophytes - fungal and bacterial species that obtain energy and nutrients by feeding on dead organisms and their wastes, or foodstuffs

saturated hydrocarbon - a hydrocarbon whose carbon atoms cannot bond with any more atoms because all bonds are single bonds

science - the systematic investigation of natural phenomena using scientific methods; the body of knowledge that results

scientific data - data that has resulted from applying a scientific method

scientific method - a way of collecting data that ensures the results are objective, reliable and valid

sea-floor spreading - the formation of new oceanic crust along a mid-ocean ridge where two oceanic plates are being moved apart

seam - a geological layer or stratum

secondary data - data that has been collected by other individuals/ data obtained from other sources

ISBN: 9780170262316

secondary growth - the growth that occurs within the stem and root of a woody plant, producing new (secondary) xylem and phloem tissues, thus increasing the girth of the plant

secondary research - researching existing information and scientific knowledge and findings

secretion - the release of useful chemicals from cells, glands or organs

sediment - rock fragments, sand, mud and the remains of organisms that settle onto the sea floor in layers

sedimentary rocks - the rocks formed when sediments on the sea floor harden under the weight of material above

seeds - new plants in a state of dormancy, consisting of an embryo inside a protective seed coat

seismic activity - earthquakes and tsunami

seismic waves - waves generated by an earthquake

self-pollination - the fertilisation of a flower by pollen from the same flower or plant

sensitivity - a life process that involves the organism sensing changes in its external or internal environment

series circuit - a circuit in which all the components are connected on the same pathway

series components - the components are connected on the same pathway

sewage - waste matter and excrement transported through sewers in a liquid form

sex cells (gametes) - cells that have half the normal number of chromosomes; egg and sperm

sex chromosomes - the chromosomes that determine the sex of an individual, either an X or a Y chromosome

sex-determination - how the sex of offspring is determined in a species

sexual reproduction - the production of varying offspring by gametes from two parents combining/ the combining of genes from two organisms

silicate rocks - rocks that are rich in silicon and oxygen; most of Earth's crust is made of silicate rocks

single bond - occurs when two atoms share a pair of electrons, one of which came from each atom

slag - impurities floating on the surface after a metal has been smelted

small intestine - the lengthy part of the gut between the stomach and the large intestine, where chemical digestion is completed and absorption occurs

smelting - the extraction of a metal from its ore by heating the ore with a source of carbon

solar cell - an object that transforms light energy into electricity

solar eclipse - occurs when the Moon moves between the Sun and Earth

solar energy - see *solar radiation*

solar radiation - the radiant energy generated by nuclear reactions in the Sun, which includes visible light, infrared and ultraviolet radiation

Solar System - the Sun and its planets along with their moons

solenoid - a long coil of wire carrying an electrical current

solidify - to change from a molten to a solid state

solubility - the amount of a substance that will dissolve in a specified volume of a particular liquid

solubility rules - rules specifying whether metal compounds are soluble or insoluble

soluble - able to dissolve in a liquid

soluble compound - a compound that dissolves when mixed into a liquid/ an ionic compound that separates into ions when mixed into a liquid

solute - the solid that dissolves into a liquid

solution - the result when a substance dissolves as it is mixed into a liquid

solvent - the liquid that a solid dissolves into

sound - pressure waves radiating out from a vibrating object through a medium

sound energy - energy that is transferred in waves by the compression and expansion of air

sound wave - see *sound*

source research - research that involves finding information from a range of sources

south celestial pole - the point in outer space directly above the south geographic pole

south pole, magnetic - the pole of a magnet that a compass needle will point toward

space, outer - the region of space beyond Earth's atmosphere

species, biological - a group of similar organisms capable of interbreeding to produce fertile offspring

species, chemical - the reactant and product chemicals of a reaction

specific heat capacity - the amount of heat needed to change the temperature of 1 kg of a substance by 1°C

spectator ions - ions that are present in a solution but which are not actually involved in the reaction

spectrum - the distribution of light frequencies

speed - describes how fast an object is travelling

speed of light - speed at which electromagnetic waves travel

speed-time graph - a graph on which the current speed is plotted against the time elapsed

sperm - male sex cells (gametes), which are mobile and much smaller than eggs; they have half the normal number of chromosomes

spindle - cell structure that separates homologous chromosomes or duplicated chromosomes

spongy mesophyll - region of loosely packed, irregularly shaped cells in a leaf, which carry out photosynthesis

sporangium - the part of a fungus that produces spores; also called a spore case or spore capsule

spores - very small reproductive cells produced in large numbers, which disperse the micro-organism

spring tide - a very high 'high tide' that occurs twice in a lunar month

stable atoms - atoms that have full outer occupied electron shells or electron arrangements with minimal energy

stamens - the parts of the male sex organs of flowers that produce pollen grains by meiosis

star - a gigantic glowing ball of gas which is undergoing nuclear reactions

ISBN: 9780170262316

starch - a complex carbohydrate consisting of molecules that are long chains of bonded glucose sugar units

state of matter - the form in which a substance normally exists, either a solid, liquid or gas

static electricity - occurs when there is a build-up of charge on the surface of an object

sterilisation - killing of all microbes using heat, chemicals or radiation

stigma - the part of the female sex organ of a flower where pollen grains land and germinate

stolons - stems that curve downward until they touch the soil, where they form roots

stomach - a hollow, muscular organ that secretes digestive enzymes, and acidifies and churns food

stomata - microscopic openings in the surface of leaves, which allow gases to enter and exit

strata - layers of rock, particularly sedimentary layers

strong acid - an acid that ionises completely when dissolved in water

strong alkali - an alkali that ionises completely when dissolved in water

structural adaptation - any inherited physical feature or part of individuals, which assists the species to survive and reproduce in its environment

structural formula - diagram showing the number of covalent bonds between pairs of atoms in a compound

structure, biological - part of a cell, organ or organism that has a distinct role (function)

sub-atomic particles - the particles that atoms are made of; protons, neutrons and electrons

subduction - a geological process that occurs when oceanic crust is forced under another plate

sugars - sweet-tasting, simple carbohydrates such as glucose, sucrose, fructose and lactose

support - a life process which involves providing an organism's tissues with strength and/or rigidity

support force - the force that a surface applies to an object resting on it

surface wave - a wave that travels across the surface of a substance

symbol equation - a shorthand way of summarising a chemical reaction using the formulas of the chemicals involved

T

tarnishing - a reaction that occurs when metals are exposed to the air and which may form a protective layer preventing further damage

technology - the results of applying scientific principles in the design of useful objects, substances or techniques

tectonic plates - huge, slowly moving plates that Earth's crust is broken up into

temperature - how hot or cold an object is

temperature (physics) - a quantity that is proportional to the average kinetic energy of the particles making up a substance

terminal - a connection point on an electrical component

terrestrial - on land

tesla - unit of magnetic field strength

test cross - a mating of an individual expressing a dominant allele with a homozygous recessive to discover whether the individual has a recessive allele

testicles - sex organs of a male mammal, which produce sperm

theory, scientific - a well-established explanation of some aspect of the natural world, based on extensive experiments and/or observations

thermal conduction - heat transfer by the passing of kinetic energy from atom to atom

thermal conductivity - the ability of a substance to conduct heat

thermal conductor - a substance or object that conducts heat well

thermal decomposition - breaking down a compound into simpler compounds or elements using heat

thermal energy - the total kinetic energy of all of the particles making up a substance

thermal equilibrium - occurs when a substance or object is at the same temperature as its surroundings

thermal insulation - a barrier to unwanted heat transfer

thermal insulator - an object that is a poor conductor of heat energy

thermal pathway - a mode of heat transfer that is available for transferring heat from a hotter to a colder location

thermal radiation - transfer of heat through infrared waves

thermohaline circulation - ocean currents that bring dissolved carbon dioxide up from the ocean depths

thrust force - the force that acts on an object making it move

time - describes how long a journey or event has taken

tissue - a distinct group of similar cells

tonne - one thousand kilograms

total internal reflection - all light waves are reflected back into the incident medium/ all light waves are reflected and none is refracted

total resistance - the overall resistance of several resistors, which may be connected in series or parallel

toxic - poisonous or damaging to cells or organisms

toxins - chemicals produced by pathogens, which poison or damage cells

trachea - a hollow, stiffened tube that enables air to move between the nasal passages and the lungs

trade winds - winds caused by cooler subtropical air moving across Earth's surface toward the equator

trait - any feature of an organism

transect line - a line across a biological community, often along an environmental gradient, which is used to determine where samples are taken

transform boundary - a boundary along which tectonic plates move sideways relative to each other

transform fault - a fault along which blocks of crust move sideways relative to each other

translucent - light is able to travel through a substance but the light is diffuse (no image can be seen)

transmission - occurs when waves travel through a transparent medium such as glass

ISBN: 9780170262316

transparent - light is able to travel through a substance and a sharp image can be seen

transpiration - the loss of water from plants through evaporation from leaves

transport - see *circulation*

transverse wave - a wave in which the direction of the disturbance is perpendicular to the propagation direction

trophic level - all of the organisms found at the same step along food chains in a community

Tropic of Cancer - an imaginary line that marks the northernmost journey of the Sun at the time of the year when it is directly overhead

Tropic of Capricorn - an imaginary line that marks the southernmost journey of the Sun at the time of the year when it is directly overhead

tropics - the region on either side of the equator lying between the Tropic of Cancer and the Tropic of Capricorn

tubers - swollen part of a root that serves as a food store

U

ultraviolet radiation - invisible electromagnetic radiation that has a shorter wavelength than the visible spectrum

unbalanced forces - occurs when the net force on an object is not equal to zero

unicellular organism - micro-organisms that consist of a single cell or several cells joined together

unit - what a physical quantity is measured in

universal indicator - a chemical used to show the pH of a liquid by a change of colour

unsaturated hydrocarbon - a hydrocarbon that is capable of forming bonds with more atoms because of the presence of a double bond

uplift - the very slow, upward motion of parts of Earth's crust

urea - the less toxic chemical that the liver converts ammonia into, which is excreted by the kidneys

ureters - the tubes connecting the kidneys to the bladder

urethra - the tube through which urine is expelled from the body

urine - the fluid produced by the kidneys, which contains dissolved waste products from cells, particularly urea

useful energy - energy that is able to be utilised for an intended purpose

V

vaccination - inoculation with dead or weakened micro-organisms to provide immunity

vaccine - dead or weakened microbes that are to be injected into the body

vacuole - part of a cell that inflates the cell when it fills with water

valence electrons - the electrons in the outermost occupied electron shell of an atom that can participate in forming bonds with other atoms

valence number - the number of electrons in the outermost occupied electron shell of an atom that are able to form bonds

validity - the extent to which a conclusion is valid given the experimental design and how data was collected

vaporisation - the process of turning a liquid into a gas through either evaporation or boiling

vapour - the gas formed when a liquid (or a dissolved substance) evaporates or boils

variable - something that varies in magnitude (size); see also *quantity*

variable resistor - a resistor whose resistance can be altered

vascular bundle - a defined group of xylem vessels and phloem tubes

vascular plants - large complex multicellular plants that have a vascular system

vascular system - system of tubes that transport nutrients, water, gases, hormones, etc., around the organism

vein, blood - a thin-walled blood vessel that carries blood back to the heart

vein, leaf - a vascular bundle within a leaf

vein (geology) - a sheet-like deposit of an ore or mineral running through rock strata

velocity - speed of an object in a specified direction; see also *speed*

vent - a volcanic opening in the crust

vertebrates - animals that have a bony internal skeleton enclosing the spinal column

villi - tiny projections of the inner lining of the small intestine, which greatly increase the surface area for absorption

viral replication - the production of large numbers of genetically identical viruses by the host cell

virtual image - a view of an object other than where it is located

viruses - non-living objects that use host cells to make numerous copies, often destroying the host cell in the process

viscosity - the thickness or stickiness of a liquid

visible spectrum - the range of light frequencies/ wavelengths that are visible to the human eye

volcanic eruption - the release of volcanic gases, vaporised magma and lava from a volcano

volcanic rock - igneous rock formed when magma solidifies on or near the surface of the crust

volcano - an opening in the crust through which volcanic matter escapes

volt - the scientific unit for voltage gain or loss; one volt is defined as one joule of energy gained or lost by each coulomb of charge

voltage - the energy gained or lost as the current passes through a component

voltage difference - the electrical potential difference between two terminals or electrodes

voltage gain - electrical energy provided by a power supply such as a battery or the mains supply

voltage loss - the electrical energy transformed by a component

voltmeter - a meter used to measure voltage gain or loss

ISBN: 9780170262316

W

warm blooded - animals such as mammals and birds that maintain a constant internal temperature

waste energy - the output energy that cannot be used

waste products - unwanted chemicals formed by a process or chemical reaction

watt - the scientific unit of power; one watt is defined as one joule of work being done each second

wattage - another term for power or power rating

wave - a regular disturbance or vibration of a medium or force fields, which propagates a series of wave fronts

wave cycle - an event lasting from the passing of one crest to the arrival of the next

wavelength - the distance between two successive crests

weak acid - an acid which only partly ionises when dissolved in water

weak alkali - an alkali which only partly ionises when dissolved in water

weathering - the gradual breaking down and wearing away of exposed rock by physical, chemical or biological processes

weight force - the force of gravity acting on an object

white blood cells - types of blood cells that are involved in fighting disease

word equation - an equation that describes a reaction using the names of chemicals

work - something done when a force moves an object

X

X-rays - high energy, short wavelength electromagnetic waves that can cause genetic mutations

xylem vessels - hollow, thick-walled non-living tubes that run the length of the plant, transporting water and dissolved mineral nutrients

Y

yeast - microscopic organisms that produce ethanol and carbon dioxide by fermenting dissolved sugars in the absence of oxygen

Z

Zealandia - the partially submerged continent on which New Zealand sits

zooplankton - free-floating marine consumers, including micro-organisms and small invertebrates

zygote - the first cell of an organism after a sperm fertilises an egg; first cell of a new organism in sexual reproduction

ISBN: 9780170262316

NZQA Science Matrix

LEVEL 1			
Physical World	**Material World**	**Living World**	**Planet Earth & Beyond**
AS90940 **Science 1.1** Demonstrate understanding of aspects of mechanics 4 credits External	AS90944 **Science 1.5** Demonstrate understanding of aspects of acids and bases 4 credits External	AS90948 **Science 1.9** Demonstrate understanding of biological ideas relating to genetic variation 4 credits External	AS90952 **Science 1.13** Demonstrate understanding of the formation of surface features in New Zealand 4 credits Internal
AS90941 **Science 1.2** Investigate implications of electricity and magnetism for everyday life 4 credits Internal	AS90945 **Science 1.6** Investigate implications of the use of carbon compounds as fuels 4 credits Internal	AS90949 **Science 1.10** Investigate life processes and environmental factors that affect them 4 credits Internal	AS90953 **Science 1.14** Demonstrate understanding of carbon cycling 4 credits Internal
AS90942 **Science 1.3** Investigate implications of wave behaviour for everyday life 4 credits Internal	AS90946 **Science 1.7** Investigate the implications of the properties of metals for their use in society 4 credits Internal	AS90950 **Science 1.11** Investigate biological ideas relating to interactions between humans and micro-organisms 4 credits Internal	AS90954 **Science 1.15** Demonstrate understanding of the effects of astronomical cycles on planet Earth 4 credits Internal
AS90943 **Science 1.4** Investigate implications of heat for everyday life 4 credits Internal	AS90947 **Science 1.8** Investigate selected chemical reactions 4 credits Internal	AS90951 **Science 1.12** Investigate the biological impact of an event on a New Zealand ecosystem 4 credits Internal	AS90955 **Science 1.16** Investigate an astronomical or Earth science event. 4 credits Internal
AS90935 **Physics 1.1** Carry out a practical physics investigation that leads to a linear mathematical relationship, with direction 4 credits Internal	AS90930 **Chemistry 1.1** Carry out a practical chemistry investigation, with direction 4 credit Internal	AS90925 **Biology 1.1** Carry out a practical investigation in a biological context, with direction 4 credits Internal	
AS90936 **Physics 1.2** Demonstrate understanding of the physics of an application 2 credits Internal	AS90931 **Chemistry 1.2** Demonstrate understanding of the chemistry in a technological application 2 credits Internal	AS90926 **Biology 1.2** Report on a biological issue 3 credits Internal	
AS90937 **Physics 1.3** Demonstrate understanding of aspects of electricity and magnetism. 4 credits External	AS90932 **Chemistry 1.3** Demonstrate understanding of aspects of carbon chemistry 4 credits External	AS90927 **Biology 1.3** Demonstrate understanding of biological ideas relating to micro-organisms 4 credits External	
AS90938 **Physics 1.4** Demonstrate understanding of aspects of wave behaviour 4 credits External	AS90933 **Chemistry 1.4** Demonstrate understanding of aspects of selected elements 4 credits External	AS90928 **Biology 1.4** Demonstrate understanding of biological ideas relating to the life cycle of flowering plants. 4 credits External	
AS90939 **Physics 1.5** Demonstrate understanding of aspects of heat 4 credits External	AS90934 **Chemistry 1.5** Demonstrate understanding of aspects of chemical reactions 4 credits External	AS90929 **Biology 1.5** Demonstrate understanding of biological ideas relating to a mammal(s) as a consumer(s) 3 credits External	

There are exclusions between:

- Sci 1.2 and Phys 1.3
- Sci 1.3 and Phys 1.4
- Sci 1.4 and Phys 1.5
- Sci 1.6 and Chem 1.3
- Sci 1.7 and Chem 1.4
- Sci 1.8 and Chem 1.5
- Sci 1.11 and Bio 1.3

Index

ISBN: 9780170262316

E

F

G

H

I

ISBN: 9780170262316

ISBN: 9780170262316

ISBN: 9780170262316